Advanced
Inorganic Chemistry

for BSc (Hons), MSc and Research Scholars

Volume I

Advanced Inorganic Chemistry

for BSc (Hons), MSc and Research Scholars

Volume I

ASHUTOSH KAR

Professor, Emeritus, Lingaya's University, Faridabad, India
Formerly
Professor, School of Pharmacy, Addis Ababa University
Addis Ababa, Ethiopia

Dean, Chairman and Professor, Faculty of Pharmaceutical Sciences
Guru Jambheshwar University, Hisar, India

Professor, School of Pharmacy, Al Arab Medical University
Benghazi, Libya

Professor and Head, Department of Pharmaceutical Chemistry
Faculty of Pharmaceutical Sciences, University of Nigeria
Nsukka, Nigeria
Professor and Head, Department of Pharmaceutical Sciences
College of Pharmacy, Delhi University
New Delhi, India

CBS

CBS Publishers & Distributors Pvt Ltd

New Delhi • Bengaluru • Chennai • Kochi • Kolkata • Mumbai
Hyderabad • Nagpur • Patna • Pune • Vijayawada

Advanced Inorganic Chemistry

for BSc (Hons), MSc and Research Scholars

Volume I

ISBN: 978-93-86310-84-2

First Edition: 2017

Published by Satish Kumar Jain and produced by Varun Jain for

CBS Publishers & Distributors Pvt Ltd

4819/XI Prahlad Street, 24 Ansari Road, Daryaganj, New Delhi 110 002, India.
Ph: 23289259, 23266861, 23266867   Fax: 011-23243014   Website: www.cbspd.com
e-mail: delhi@cbspd.com; cbspubs@airtelmail.in.

Corporate Office: 204 FIE, Industrial Area, Patparganj, Delhi 110 092
Ph: 4934 4934   Fax: 4934 4935   e-mail: publishing@cbspd.com;   publicity@cbspd.com

Branches

- **Bengaluru:** Seema House 2975, 17th Cross, K.R. Road, Banasankari 2nd Stage, Bengaluru 560 070, Karnataka
  Ph: +91-80-26771678/79   Fax: +91-80-26771680   e-mail: bangalore@cbspd.com
- **Chennai:** No. 7, Subbaraya Street, Shenoy Nagar, Chennai 600 030, Tamil Nadu
  Ph: +91-44-26680620, 26681266   Fax: +91-44-42032115   e-mail: chennai@cbspd.com
- **Kochi:** Ashana House, 39/1904, AM Thomas Road, Valanjambalam, Ernakulam 682 016, Kochi, Kerala
  Ph: +91-484-4059061-65,67   Fax: +91-484-4059065   e-mail: kochi@cbspd.com
- **Kolkata:** No. 6/B, Ground Floor, Rameswar Shaw Road, Kolkata-700014 (West Bengal), India
  Ph: +91-33-2289-1126, 2289-1127, 2289-1128   e-mail: kolkata@cbspd.com
- **Mumbai:** 83-C, Dr E Moses Road, Worli, Mumbai-400018, Maharashtra
  Ph: +91-22-24902340/41   Fax: +91-22-24902342   e-mail: mumbai@cbspd.com

Representatives

- **Hyderabad** 0-9885175004   • **Nagpur** 0-9021734563   • **Patna** 0-9334159340
- **Pune**   0-9623451994   • **Vijayawada** 0-9000660880

Printed at India Binding House, Noida, UP, India

to

my beloved parents, sisters and brother,
others and my wife
who helped me build-up my life,
our sons, daughters-in-law and grandchildren
who thrilled my life with joy and cheers,
my esteemed teachers and taughts
(in India, Nigeria, Libya and Ethiopia),
the august authors (home and abroad)
whomsoever the book owes,
the people who meticulously toil, to make fruitful the soil,
for ever and now,
Almighty God's eternal blessings
I respectfully bow, bow and bow

"Those who aspire not to guess, and divine,
but to discover and know;
who propose not to devise mimic
and fabulous worlds of their own,
but to examine and dissect
the nature of this world itself,
Must go to facts themselves for everything."

F Bacon (1620)

PREFACE

Advanced Inorganic Chemistry (Volume 1), is solely intended to fulfill the dire needs of the teachers who would like to deliver the subject to the aspiring students who really require an in-depth deliberations of the various fundamental topics. It would certainly help the students to have a better understanding related to the electronic structure of atoms and elementary valence theory. The textual matter in this compendium has been specifically elaborated in an elementary fashion, detailing the main facts, and expatiating the various reactions in an excellent manner.

The back-bone and deep-seated basics decidedly hangs around the so-called *Baconian philosophy* with respect to all chemistry but critically **Inorganic Chemistry**. For the benefit of teachers and students alike, the present textbook essentially consists of a cocktail comprising: *Theory, principles, examples* along with an intermittent mention of the underlying hard facts in a lucid style. A sincere and earnest effort has been made to encourage the learning and teaching of **Inorganic Chemistry** in the *Boconian manner* perceptively.

There are in all *six* chapters that have been included in the **Advanced Inorganic Chemistry (Volume 1)**, namely:

Chapter–1: Atomic Structure: Quantum Mechanical Approach and Wave Mechanical Approach
Chapter–2: Modern Periodic Table: Electronic Configuration of Atom and Atomic Properties
Chapter–3: Radioactivity, Isotopes, Isobars, and Isotones
Chapter–4: Chemical Bonding: Lewis Theoretical Concepts
Chapter–5: Metallurgy
Chapter–6: Solvolytic Reactions (Hydrolysis)

A comprehensive deliberation of the *quantum mechanical approach* of hydrogen atom along with an exhaustive solution of Schrödinger equation, critical ensuing expressions of angular and radial wave functions, derivation of various shapes of orbitals *vis-a-vis* their inherent spherical harmonics have acquired a much deserved emphasis. A systematic elaboration of molecular symmetry as well as group theory has been dealt exhaustively.

Each chapter concludes with *three* highly important and useful supplements, *viz. problems with solutions, probable questions*, and *university questions* that would strengthen the students ability, understanding, and learning aptitude to a great extent.

It is earnestly hoped and believed that **Advanced Inorganic Chemistry (Volume 1),** may certainly help BSc (Hons), MSc, and Research Scholars to help in a big way to grasp and understand the subject in an appreciable manner.

I wish to place on record the excellent support by Shri Satish Kumar Jain, MD, CBS Publishers & Distributors Pvt. Ltd., New Delhi and his excellent editorial team for bringing out this textbook in a scheduled time-frame.

Ashutosh Kar

CONTENTS

2. Modern Periodic Table: Electronic Configuration of Atom and Atomic Properties 90–186

1. INTRODUCTION 90

2. SYSTEMATIC APPROACH TO MODERN PERIODIC TABLE 96

3. THE ELECTRON AFFINITY [EA] 137

3. RADIOACTIVITY, ISOTOPES, ISOBARS AND ISOTONES   187–239

3.1. INTRODUCTION   187

4. CHEMICAL BONDING: LEWIS THEORETICAL CONCEPTS   240–306

5. METALLURGY

6. SOLVOLYTIC REACTIONS [HYDROLYSIS]   361–375

INDEX   377–394

Chapter 1

ATOMIC STRUCTURE: QUANTUM MECHANICAL APPROACH AND WAVE MECHANICAL CONCEPTS

1. INTRODUCTION

The '**atom**' designates the smallest unit of an element that can exist. Importantly, atoms have long been known to be made up of the so-called *sub-atomic particulate matters* invariably termed as: **Protons, Neutrons**, and **Electrons**. Besides, the nucleus of an atom does contain one or more *proton*, that are **charged positively** and two or more **neutrons** (*except in hydrogen*) which bears no electric charge. Each proton and neutron has a mass number of 1, and are duly bound together in the nucleus by respective *exceedingly strong forces*.

It may, however, be made sufficiently clear that the *electrons* do move *around the nucleus in shells* (*i.e.*, **group of orbitals**) that are eventually determined by their:

- **inherent energy levels**, or
- **the wave functions**.

Interestingly, an *electron* designates a *negatively charged unit* (e^-) that critically behaves both as:

- **a particle**, and
- **a wave**.

Alternatively, the **electron**, may be regarded partly as '**mass**' and partly as '**energy**'.

NOTE	Each and every 'atom' essentially has the same number of *electrons* (e^-) and *protons* (H^+); and hence, turns out to be *absolutely neutral electrically*.

Proton: It is a fundamental particle of matter (a *nucleon*) having positive charge of **1.6×10^{-19} coulombs** and a mass of about **1.67×10^{-24} grammes** [*i.e.*, approximately equal to 1 atomic mass unit (amu)].

Neutron: It relates to the basic particle of matter. A **nucleon** bears essentially a *zero charge and mass* of **1.67×10^{-34} g** (a by-product of nuclear fission). In other words, a **neutron** is a particle obtained from an atomic pile and usually used to prepare *radioisotopes* (*viz.*, ^{131}I, ^{32}P).

Electron: It represents a basic particle of matter possessing a rest mass of **9×1.028 grammes** and **1 electrostatic unit (ESU) of charge (1.6×1.019 coulombs)**.

1.1. The Atomic Structure

In order to muster a proper understanding and clear concept of **the atomic structure** we may have to explore and examine the following aspects, namely:

- ❑ **Atom and Elementary Particles,**
- ❑ **Dalton's Atomic Theory,**
- ❑ **The Periodic Table,**
- ❑ **Bohr's Theory of Atomic Structure,**
- ❑ **Modern Model of Atomic Structure, and**
- ❑ **Electronic configuration of the Elements,**

which shall now be discussed individually in the sections that follows:

1.1.1. *Atom and Elementary Particles*

The term '**atom**' (derived from the **Greek: indivisible**) was believed vehemently to be the minute, indivisible particle of which practically all *material objects* were duly made. Interestingly, the sincere and vigorous search for the '**ultimate particulate matter**' has been both a continuous and concerted effort since the era of **Democritus** (between 460–370 BCE). Amazingly, even before the discovery of '*mesons*'* and '*hyprons*'**, the precise structure of matter was regarded to be a lot simple task. Thus, the nucleus was duly considered to comprise essentially of:

- **Protons** and
- **Neutrons,**

and ultimately to result into the formation of an '**atom**'; and, therefore, the '*electrons*' required to be incorporated meticulously in the so-called *external shells*.

Based on the above scientific evidences one could safely infer that: **protons, neutrons,** and **electrons** were mainly thought of to be the actual **elementary particles**.

Points to Ponder: Thus, theoretically all the **elements** present in the *Periodic Table* may be produced simply by:

- **Splitting '*neutrons*' into the respective '*electrons*' and '*protons*' and**
- **Combining these particles in proper ratios.**

Interestingly, during the past *four* decades, the particular discipline – '**Nuclear Physics**' has probed meticulously and progressively the '**atoms**' right from their periphery extended to their *centre*. Thus, the vigorous search for the so-called '**ultimate units of nuclear structure**' by virtue of the specially designed

---

* **Mesons:** Any of a group of *unstable elementary particles* with **strong nuclear interactions** and baryon number equal to zero. They are belived to comprise of a *quark* and its *antiquark*. Postive, negative, and neutral **mesons** do exist; when charged the magnitude of the charge is equal to that of the *electron*. **Mesons** are found in the **cosmic rays** and are duly emitted by *nuclei* under bombardment by the high-energy particles. **Muons** were originally termed **mesons**, but they are now classified as '**leptons**' rather than **mesons**.

** **Hyprons:** Elementary particles, **hadrons** having mass greater than that of the **neutron** and less than that of **deuteron**. **Hypron** is a collective name assigned to any **baryon** with *non-zero strangeness number*. The nomenclature **hypron** has been duly restricted to such particles that are *semi-stable i.e.*, which have fairly long **life spans** relative to 10^{-22}s, and which decay the photon emission or *via* **weaker-decay interactions**.

experiments comprising mostly of: "**bombarding the nuclei with high-energy particles**", has virtually provided a broad-spectrum of more than *100 species*, of which a large segment remained *unstable*.

    De Broglie, in 1924 pointed out that–

"**in a specific instance when light wave show explicitly the so-called *corpuscular character*, must not the particles also exhibit *wave character*?**"

    Based on the above spectacular revelations it may be accepted generally that in the particular situation of a '**photon**'* there prevail *two* **fundamental equations**, namely;

$$E = h\nu \quad \text{and}$$
$$E = mc^2$$

where, E = Energy,

        h = Planck's constant,

        ν = Frequency, and

        c = Speed of light

Now, combining both the aforesaid equations we have:

$$h\nu = mc^2$$

or $\lambda = c/\nu = h/mc = h/p$

where, p = Momentum of the '**Proton**'.

    Broglie suggested emphatically that a *similar equation* must govern and modulate the **wavelength of the ensuing electron wave**. At this point in time, it is quite interesting to observe that the **X-ray diffraction (XRD)**** study serves as a definitely good and reliable example pertaining to the usage of the wave characteristic feature of the so-called **electromagnetic radiation**.

1.1.2. *Dalton's Atomic Theory*

Dalton, in 1808, duly put forward his **atomic theory**, which is actually based on the following *three* **largely accepted and recognized generalizations**, such as:

    ❑ **Law of Conservation of Mass**,***

    ❑ **Law of Definite Proportions**,**** and

    ❑ **Law of Multiple Proportions**.*****

---

    *   **Photon:** A particle with *zero rest mass* comprising essentially of a **quantum of electromagnetic radiation**. Thus, the **photon** may also be regarded as a unit of energy equal to hf; where, 'h' is the Planck's constant, and 'f' is the frequency of radiation in *hertz*. **Photons** travel at the speed of *light*. They are indeed required to explain the inherent photoelectric effect and other phenomena that need light to have the *particle character*.

    ** Kar A., **Pharmaceutical Drug Analysis**, 3rd. ed., New Age International, New Delhi, 2016.

    *** **Law of Conservation of Mass:** This law relates to the fact that matter can neither be created nor destroyed. Precisely, the total mass of any system remains constant under all transformations. In light of the relativistic relationship between mass and energy, this law has been restated in the form that applies to the total quantity of mass and energy in a system, taken together as a single entity. (Also known as-: **law of the conservation of matter**).

  **** **Law of Definite Proportions:** Proportion of masses of constituent elements or components in any *definite chemical compound* is **fixed and constant**. Also known as-: **Law of constant composition**; **Proust's Law**.

 ***** **Law of Multiple Proportions:** When two elements combine in more than one proportion, the quantities of the *first*, which are combined with a given amount of the *second* stand in the **ratio of small whole numbers**.

Thus, one may summarize the critically important as well as essential segments of the **atomic theory** as stated under:

1. All elements are composed of **very small, discrete, indivisible particles** termed as 'atoms'.
2. All **atoms** of any **one specific element** are found to be *absolutely identical*. The modern structural theory tells us that electronic differences between the **atoms** of an element may usually take place, but their differences arise as a consequence of an **electronic excitation** perceptively. Thus, the **lowest energy state of an atom** is definitely more suitable for purposes of classification.
3. **The atoms of no two elements are found to be similar**.
4. **Atoms** invariably undergo *no basic change* in the course of a **chemical reaction**. Of course, there are *subtle changes* that eventually come into play in the so-called **electronic character of atoms**, although it fails to alter the **identity of an atom**.
5. Chemical entities (compounds) are duly formed as and when the **atoms of two or more different elements combine to form a molecule**.
6. In a broader perspective, **atoms combine categorically** *in simple, integral ratios*.

1.1.3. *The Periodic Table*

In true sense, the **Periodic Table** refers to the so-called:

> "**systematized arrangement of the** *basic elements* **strictly according to the** *atomic weight* **and the generally accepted electronic configuration.**"

In a broader perspective, one may take cognizance of the fact that the **periodic classification of the elements** is certainly one of the most:

*   **Spectacular achievement**, and
*   **Striking scientific advances**,

in generalizing broadly an array of '*isolated facts*'. Besides, it does make a *tremendous contribution* not only confined to the versatility and strength of the **atomic theory** at large, but also extends it to the newer aspects.

Therefore, the '**Periodic Table**' does provide such valuable and scientific informations – as stated under:

*   **conveniently learned summary of practically limitless information with respect to the chemical nature of the elements**; and
*   **serves to the students of chemistry a prime source of valuable scientific knowledge.**

Historical Breakthroughs: In the year 1869, *two* eminent scientists: **Mendeleev and Meyer** – published the epoch making '*Periodic Law*', that overwhelmingly received the wild-acclaim in the so-called scientific world and community as a whole. **Periodic Table** – It refers to a systematic, logical, and *scientific arrangement of the various known elements* as per the established **Periodic Law** (see the '**Modern Periodic Table Long Form**' of the elements). Nevertheless, the present arrangement is more or less *exactly the same as that of* **Mendeleev**; though there are certain intelligent **minor variations** perhaps on account of the addition of:

*   **altogether '***new elements***'**; and
*   **modern data**.

Following are the *four* **cardinal terminologies** that must be understood thoroughly and squarely pertaining to the **Periodic Table**, for instance:

> **Atomic Number (Z),**
> **Atomic Weight,**
> **Isotope,** and
> **Nuclide,**

which shall now be expatiated individually in the sections that follows:

☐ **Atomic Number (Z)** – It is the positive charge of the nucleus expressed as multiples of the **electronic charge (*e*)**.

☐ **Atomic Weight** – It relates to the **average weight** duly expressed in terms of the **atomic weight units** of the *natural atoms of an element* prevailing mostly as an **admixture of isotopes** present in the **same ratio as normally found in the nature.**

NOTE	1. **An atomic weight unit, used in chemistry, is precisely 1/16th the *average mass of the oxygen isotopes* taken in the same ratio as they occur in the nature.**
	2. **One *atomic weight unit* is equivalent to *1.000272 atomic mass units*.**

☐ **Isotopes** – An **isotope** refers to one of a group of **nuclides** of the **same element** (*i.e.*, same **Z**), having the **same number of protons present in the nucleus, but differing exclusively in the number of neutrons,** – *thereby resulting in different mass numbers.*

Alternatively, when *two* atoms having the same atomic number but different atomic weights – they are called **isotopes.**

☐ **Nuclides** – Any atom characterized by a **specific number of protons and a specific number of neutrons** (*e.g.*, carbon: 12; and carbon: 14). In other words, a **nuclide** relates to any one of the *more than 1000 species of atoms* and is duly characterized by the *number of protons and neutrons in the nucleus.*

1.1.4. *Bohr's Theory of Atomic Structure*

Niels Bohr (1913) – a Danish physicist, proposed a *theory of atomic structure for the interpretation of the atomic spectra*. In reality, his explicit description of the '**atom**' essentially inherited the **extranuclear electrons** revolving around the *nucleus* in **definite orbits.** Subsequently, these '**orbits**' were duly assigned the respective *quantum numbers* viz., 1, 2, 3, 4, 5, ..., *n* counting outward right from the **nucleus.**

Major Aspects of Bohr's Theory – There are *three* **major aspects of the Bohr's Theory**, namely: (a) In a specific instance, when an electron absorbs a *definite* **enhancement (quantum) of energy**, it is promoted successively to an orbit of *higher energy level* (excited state); and the moment it regains its position to the *original orbit*, – it distinctly emits the **radiation energy**. In this manner, the energy prevailing at the various levels in the atom may be related to the corresponding **frequency of radiation**, which gets duly *emitted either from or absorbed from the atom.*

Thus, the ensuing relationship may be expressed as:

$$\boxed{DE = E_2 \text{-} E_1 = h\nu}$$ (1)

where, ΔE = Difference of the energy (in ergs) existing between *two* levels;
h = Planck's constant (6.624 × 10^{-27} erg sec),
ν = Frequency.

Since the '**frequency** (*v*)' is equivalent to the *speed of light*, *c*, divided by wavelength, Eq. (1) may be written as:

$$DE = he/e$$

(2)

(b) In another situation, when the electrons do possess the so-called **lowest level of energy** possible, the 'atom' is said to be in the **ground state**.

Thus, the *energy* of an electron in an orbit may be expressed by:

$$E = \frac{-2\pi^2 Z^2 m e^4}{n^2 h^2}$$

(3)

where, Z = Atomic Number,

m = Mass of the electron $(9.1 \times 10^{-28}$ g),

e = Charge of the electron in electrostatic units $(4.8 \times 10^{-10}$ ESU),

n = Principal quantum number, and

h = Planck's constant.

Hence, it is quite convenient for one to calculate the '**radiation energy**' duly emitted as and when an electron falls from n_2 *orbit* to the respective n_1 *orbit*, as depicted in the following expression:

$$E_2 - E_1 = \frac{2p^2 Z^2 m e^4}{h^2} \left(\frac{1}{n_1^2} - \frac{1}{n_2^2} \right)$$

(4)

Importantly, when the value of n_2 is ∞, the Eq. (4) gives the energy very much needed for carrying out the *ionization phenomenon* exclusively.

Example: Thus, it is quite convenient and possible to calculate the *ionization potential* (**E**) of the **H-atom** as given under:

$$E_\alpha - E_1 = \frac{2 \times (3.14)^2 \times (1)^2 \times 9.1 \times 10^{-28}}{(6.624 \times 10^{-27})^2} \times (4.8 \times 10^{-10})^4 \left(\frac{1}{1^1} - \frac{1}{(\alpha)^2} \right)$$

$$= 2.18 \times 10^{-11} \text{erg}$$

$$= \frac{2.18 \times 10^{-11} \text{ erg}}{1.60 \times 10^{-12} \text{ erg. electron volt (ev)}^{-1}}$$

$$= 13.6 \text{ eV}$$

Comment: It is, however, important to observe that the '**quantum theory**' is virtually based on the underlying principle that:

"**the actual energy of an *atom* or *molecule* fails to alter continuously but only up to certain definite whole number unit of energy usually referred to as a *quantum*.**"

Electromagnetic Radiations – It is also referred to in literature as the *Electromagnetic Waves i.e.,* the precise nature of radiations duly emitted by the '*electrons*'.

The underlying theoretical aspects may be further expatiated by considering an '*object*'–that eventually either:

- **moves up and down**; or
- **undergoes vibration continuously**,

thereby releasing the *accomplished energy* in the particular **form of a** '*wave*'. The emanated '*wave*' is now meticulously being subjected to either:

- **transmission**, or
- **propagation**,

by the help of the so-called *vibrating object* to a preferred **distant place**. It has been duly observed that the emanated '*wave*' traverses almost at *right angles* to the respective **vibratory motion** with reference to the *vibrating object* and *positioned away from it*.

Fig: 1.1(A) depicts the typical production of a '*wave*' by the help of a **vibrating object** and definition of crests, troughs, and also the wavelength in a '*wave*'.

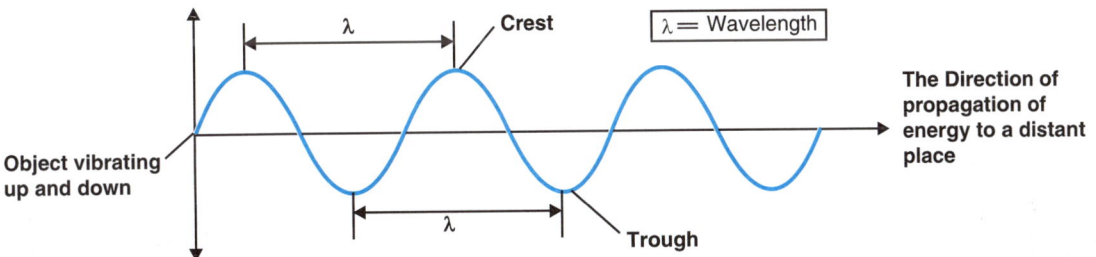

Fig. 1.1(A) : Diagramatic Representation of a Wave by a Vibrating Object and Explicit Definition of Crests, Troughs, and Wavelength (λ) in an Emanated 'Wave'.

Remarks: Importantly, the *wave* so generated by a *vibrating object* may be properly designated by a '**wave curve**'.

Crests – they represent the **tops** (or *maxima*) of the **wavy curve**.

Troughs – they represent the bottoms (or *minima*) of the **wavy curve**.

Maxwell CJ (1864) demonstrated explicitly that whenever an '*electric current*' is made to **pass** *via* **a circuit** (*viz.*, a **tungsten** *coiled coil*), – it invariably **radiates energy** in the specific form of '*waves*'.

Radiated Energy – *i.e.*, the energy radiated in such a fashion is normally known as the '**Radiated Energy**'.

Electromagnetic Waves (or *Electromagnetic Radiation*) – *i.e.*, the *waves* so generated are usually termed as the **electromagnetic waves** (or **electromagnetic radiations**).

NOTE

The '*electromagnetic waves*' virtually do not actually need either a *medium for propagation* or a *medium for transmission viz.*, light, radio waves, UV-light from sun rays etc., that ultimately the surface of the *Earth* by traversing *via* empty space (non-material medium). Hence, most of reach these '*waves*' actually travel with the *same speed of light.*

1.1.4.1. *Plausible Extension of Bohr's Theory to Systems Comprising More Than One Electron*

The spectacular success of the *simple* **Bohr's theory** applicable to **one-electron systems** there was a dire need to carry out its *extensive modification* so as to:

- **handle rather more complex species**; and
- **broaden the scope of Bohr's theory elegantly**.

Therefore, in order to circumvent the spectra of such species (*i.e.*, essentially consisting of several more lines in comparison to H-atom) – *three* additional **quantum numbers** had to be introduced intelligently.

Thus, based on the *extended* **Bohr's theory**, one may take cognizance of the fact that the *orbits* could be either:

- **eliptical in shape**; or
- **circular in dimension**.

However, the ensuing observed *ratio* $(1 + 1/n)$ affords the specific ratio of the '**minor axis**' of the **eclipse** to its respective '*major axis*'; whereas, the *second (orbital) quantum number*, could possess such values as: $0, 1, 2, \ldots, n-1$. Importantly, the critical *splitting of the* **spectral lines**, as seen in a **magnetic field*** essentially needs a *third (magnetic) quantum number*, m_1, defining explicitly the *plane*– which the **electron orbit** assumes *vis-a-vis* the **applied field**. Thus, m_1 bears such values as: $l, (l-1), \ldots 0, \ldots -(l-1), -l$. Ultimately, a *fourth (spin magnetic)* **quantum number**, m_s, is duly incorporated to expatiate the many observed *fine structure of several spectral lines***.

> **NOTE** It has been observed that the *electron* is spinning when it rotates; thus, any *electron* with *spin quantum number* 'S' equal to 1/2, but the magnetic field given by the spinning of the electron may either reinforce or *oppose the magnetic field* produced by rotation of the electron in its orbit 'm_s' possesses the value +1/2 or −1/2.

❑ **Considering a One-Electron System** – In this particular case, the observed *spectrum* may be duly related to a **single quantum number** 'n' – which explicitly means that almost *all the orbits* are intimately associated with a *given value* of n inheriting the same **energy level**.

On the contrary, for such atoms comprising **more than one electron**; nevertheless, the energies possessed by *electrons in the orbits* having the *identical* 'n' value, but *different* 'l' values are found to be **different significantly**.

Examples: These include:

➢ **Electrons in orbits** – with 'l' $= 0, 1, 2, 3$ are usually termed as s, p, d, f– **electrons** respectively; and

➢ hence, the *similar* terminology may be applicable **to the corresponding orbits**.

Thus, an **empirical** (*i.e.*, *based on experiment*) selection rule restricts the so-called '*observed transitions*' specially to those for which the value 'l' **changes by ±1**.

---

* That is, the **Zeeman Effect** [*i.e.*, the splitting of the lines in a **spectrum** when the source of the **spectrum** is duly exposed in a **magnetic field**.]

** That is, the **doublets** seen in the spectra of **alkali metals**.

Fig: 1.1 illustrates a segment of the **emission spectrum of the lithium atom,** in which case the observed **atomic spectrum of lithium** having the *transitions* commencing from the **3s, 4s** etc., *orbits* to the respective **2s orbit** do afford the *principal series* predominantly. Interestingly, the observed *transitions* emanated from the **3d, 4d** etc., **orbits** to the corresponding **2p orbit** gives rise to the *diffuse series* prominently; whereas, the *transitions* generated by the **4f, 5f** etc., **orbits** to the corresponding **3d orbit** do produce the *fundamental series*.

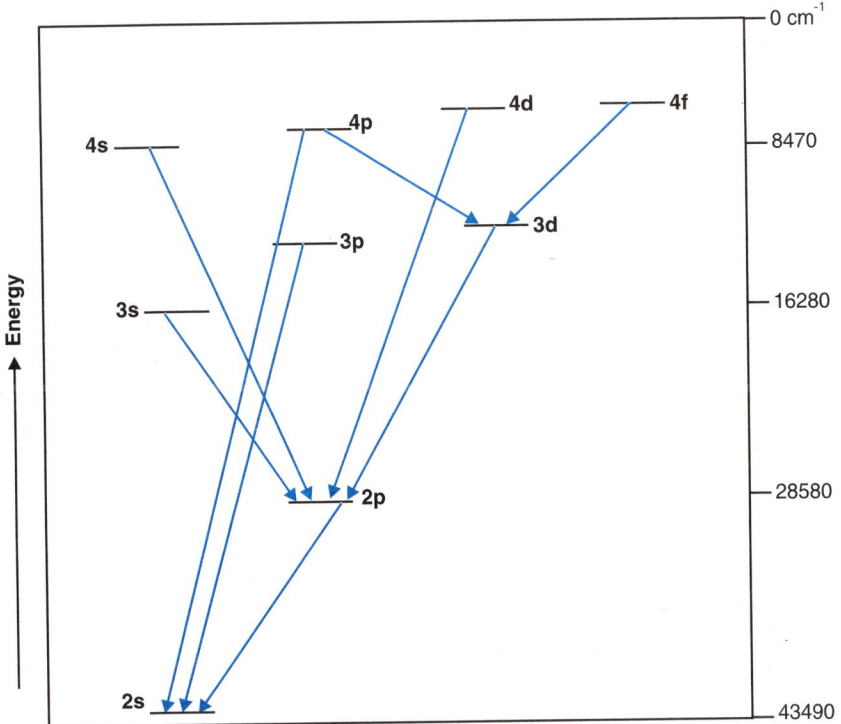

Fig. 1.1: The Emission of Lithium Atom from Energy Levels with Values of 'n' = 2, 3, or 4*.

1.1.4.2. *The Postulates of Bohr's Atomic Model*

In a broader perspective, the **Bohr's atomic model** engages the advantages of *two* essential features of the well known **Rutherford's** ** **Atomic Model,** namely:

- *Atom* bears critically a *very small positively charged nucleus* located strategically at its centre. [Besides, all the *protons* and *neutrons* are contained in the *nucleus*** .]
- *Negatively charged electrons* mostly revolve around the *nucleus* almost very much akin to the planets revolving around the Sun.

Nevertheless, the crucial application of the **Planck's Quantum Theory** to the aforesaid *revolving electrons* has substantially explained the possible explanation why the electrons:

---

* The respective 'Doublet Structure' is excluded completely.

** Rutherford E(1911)

*** That is, **maximum mass of the atom** is localized in the *nucleus* only.

"when revolving round the *nucleus* fail to lose energy level; and subsequently do no fall right inside the *nucleus* finally."

Bohr suggested *four* **different postulates** to describe explicitly his conceptualized **model of the atomic structure**, which shall now be described individually in the sections that follows:

Postulate 1: Fixed Circular Orbits – Bohr assumed thoughtfully that the '*electron*' is indeed a **material particle** which is found to be revolving very much round the **nucleus** in a manner literally means – '*concentric circular orbits*' primarily positioned at **definite distance** (*fixed distance*) from the **nucleus** and that too with a **definite velocity**.

Postulate 2: Stationary Energy Levels – It is, however, pertinent to state here that so far as an '*electron*' very much remains in a *specific* orbits, one may observe critically that:

* **it never radiates* energy**; and
* **it never gains** energy**.

Therefore, one may safely infer in a specific **orbit** the **energy** of a revolving electron remains *constant* or *static*. In other words, each of the so-called **fixed orbits** gets intimately associated with a certain definite quantum of energy level***.

Stationary Energy Levels – These are also known as: **energy levels** or **energy shells**. Importantly, such a concept of the **static energy levels** does expatiate and explain the overall **stability profile** of the *atoms*, because an '*electron*' fails to **lose energy** in a gradual manner and finally shelters inside the **nucleus**.

Principal Quantum Number – One may commonly encounter **different energy levels** being designated adequately by '*n*' that may whole number (integer) values *e.g.*, 1, 2, 3, 4, ..., ∞, commencing right from the **nucleus** itself; and thus, '*n*' has been duly termed as the **principal quantum number** (by **Bohr**).

NOTE	In usual practice, different values of '*n*' may also be duly designated by *capital alphabets viz.* K, L, M, N and the like.

The respective **principal quantum number** '*n*' and **energy level** designated by **capital letters**, – as shown under:

Principal Quantum Number (n) → 1

Energy Level (by Capital Letters) → K (First Shell) L (Second Shell) 2

M (Third Shell) 3

Inference Drawn

➢ **K shell (1st Shell):** $n = 1$;
➢ **L shell [2nd Shell]:** $n = 2$;
➢ **M shell [3rd Shell]:** $n = 3$; and so on so forth.

Besides, these *shells* are also termed as: $n = 1$ **shell**; $n = 2$ **shell**; $n = 3$ **shell** etc., respectively.

Fig: 1.2 depicts the various orbits (n = 1 through n = 5) explicitly.

---

   *   That is, emits or loses.

  **   That is, absorbs.

***   That is, with a definite **whole number of energy**.

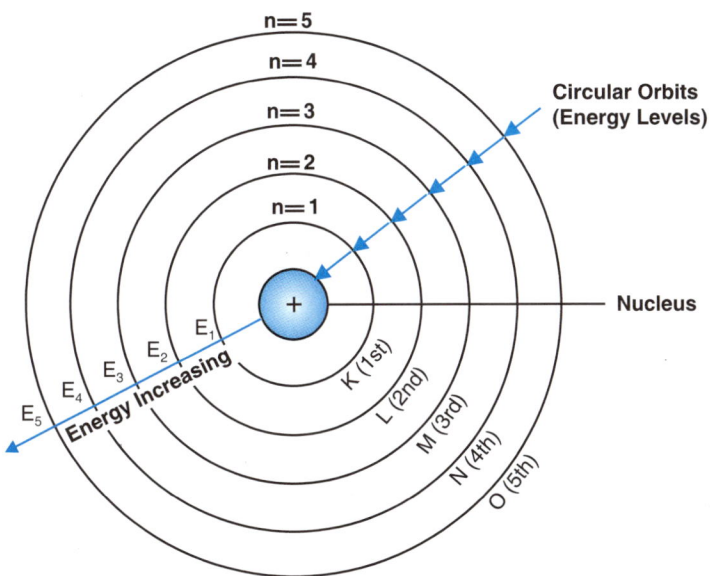

Fig. 1.2: Illustration of Various Orbits (Energy Levels) Around the Nucleus. E_1, E_2, E_3, E_4, E_5 – Energies Due to Orbits with K-Shell (n = 1), L–Shell (n = 2), M-Shell (n = 3). Energies possessed by various orbits follow the order: $E_1 < E_2 < E_3 < E_4 < E_5 <$.........

Postulate 3: Shifting of an Electron from Previous Energy Level to the Next [Ground State and Excited State of an Electron]

It is indeed a known fact that during the stay on an '*electron*' in a *specific orbit* neither:

- **radiates (*i.e.*, loses or emits)**; nor
- **absorbs (i.e., gains) energy.**

Nevertheless, in a situation when the '*electron*' gets excited right from a **lower-energy level** to a **higher-energy level**, it always *absorbs energy*. On the contrary, when the '*electron*' returns from a definite **higher-energy level** to the corresponding **lower energy level** – it predominantly **radiates (emits) energy.**

Planck's Quantum Theory – According to this theory related to radiation one may critically observe the following cardinal aspects, namely:

- **Absorption (or emission) of energy occurs almost discretely*;** and
- **Absorption (or emission) of energy comes into play in the form of *photons* (or quanta).**

Salient Features of an Atom (Electron) in Ground State – These include:

1. At **energy level '1'** *i.e.*, the lowest energy level – the '*electron*' is believed to be in the ***Ground State.***
2. The ensuing **energy level '1'** – is indeed the **most stable state of the atom**.

Salient Features of an atom (Electron) in Excited State – They include:

1. Supplement of energy to the '*electron*' located at the energy level '1' will certainly enable the '*atoms*' to move on to the respective higher levels 2, 3, 4, 5, ..., depending solely upon the quantum of energy absorbed by the '*electrons*'.

---

* That is, the absorption or emission of energy occurs specifically never as the continuous waves, but mostly as: **small bundles/packets/separate units** of the '*waves*' each of which is termed as: **Photon or Quanta.**

2. The '*electrons*' in the excited state are quite *vibrant* and *unstable*.

Postulate 4: The Underlying Principle of Quantisation of Angular Momentum of the Moving Electron–

Bohr's theory (see *Section 1.1.4*) suggested vehemently that:

"**an '*electron*' fails to move in all the orbits**".

Besides, it is capable of moving only in that orbit wherein the so-called – '**angular momentum of the electron (= *m v r*)**' moving clearly around the '***nucleus***' is found to be the *integral whole number multiple of h/2π*, for instance:

"**1h/2π, 2h/2π, 3h/2π, 4h/2π, ..., nh/2π.**"

Therefore, based on the aforesaid **Bohr's Theory**– the so called '**angular momentum of a moving electron**' may be given by the following expression:

$$mvr = n \times \frac{h}{2\pi} \qquad (i)$$

Hence, as per the **Bohr's theory** as designated by Eq. (i) the – '**momentum of a moving electron is quantitized precisely.**' From Eq. (i) we may have the following specifications:

$m =$ Mass of the electron;

$v =$ Velocity of the electron;

$h =$ Planck's constant [6.624×10^{-27} erg s];

$r =$ Radius of the orbit (in which *electron* rotates); and

$n =$ An integral* (provides the number of the orbit in which *electron* is moving).

Thus, '*n*' may have the values 1, 2, 3, 4, 5, ..., for the corresponding '**orbit**' duly numbered: 1 (K-orbit); 2 (L-orbit); 3 (M-orbit) ... from the **nucleus**. Hence, for K, L, M ... orbits $n = 1, 2, 3$, respectively.

Orbits	K	L	M	N
Principal Quantum Number (n) →	1	2	3	4

❑ **Limitations of Bohr's Theory** – There are *four* important *limitations* of **Bohr's theory** which must be understood clearly, such as:

1. **Bohr's theory** does not provide the much desired *quantitative* **explanation** for the underlying *spectra of atoms* possessing more than *one electron*.

2. **Bohr's theory** fails to attribute an acceptable calculation for the *spectral line intensities*.

3. **Bohr's theory** does not provide the actual basis for the '**periodic classification of elements**', or for the periodicity in the *characteristic features of the elements*.

4. Finally, the **Bohr's theory** cannot explain justifiably the '**line-spectrum of hydrogen**'. Thus, in a good number of instances the '*single-line*' proclaimed by **Bohr** ultimately proved to be **a cluster of closely spaced lines** (instead of the '*spectral lines*')–as determined and ascertained by more sophisticated analytical instruments *viz.*, **Electron Polarization**.

---

* That is, **principal quantum number** (by **Bohr**).

NOTE | Based on the aforesaid delebrations it may be inferred that for each orbit, one may virtually encounter several closely spaced energy levels duly present in the atom.

1.1.4.3. *Sommerfeld's Modification of Bohr's Theory: The Elliptical Orbits*

Sommerfeld modified intelligently the **Bohr's theory** by taking into active consideration the **elliptical orbitals** (and **not** the *circular orbitals* – which only served as a special case of the *elliptical orbits*). Obviously, the *nucleus* was considered to be at one of the **focii of the nucleus**. **Circular Orbit** – has the only *variable* representing the **angle of revolution (ϕ)**.

Elliptical Orbit – has both **angle of revolution (ϕ)** and the **radius vector (r)**, which may vary accordingly.

In a typical instance, having *two* **degrees of freedom**, *two* quantum restrictions are being imposed concurrently:

- *First* – on the **angular momentum (pa)**; and
- *Second* – on the **radial component of the momentum (pr)**.

❑ Interestingly, the **integer** for the observed *quantization* of '*pa*' is observed to be almost the same as in **Bohr's theory**. Thus, we may have the following expression:

$$pa = nah/2\pi = nah$$

However, the *integer* for the *quantization* of the radial component of the ensuing **momentum (n_r)** specifically involves the respective solution of the equation involving n_a, n_r, and the eccentricity of the ellipse.

❑ **Sommerfeld** vividly showed that the overall **total energy of the *electron*** under consideration depended exclusively upon $n = n_a + n_r$, which is usually termed as the '**principal quantum number**'

Fig: 1.3(a) depicts the **Sommerfeld elliptical orbitals for 'Hydrogen'**:

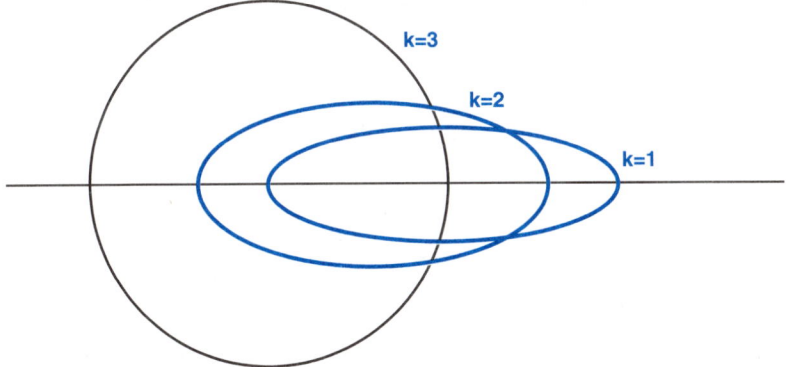

Fig. 1.3(a) : Representation of the Sommerfeld Elliptical Orbitals for Hydrogen.

Explanation: The **elliptical orbit** may be created by taking into consideration the *major axis of the ellipse* to the corresponding diameter of the **Bohr's circular orbit**; and also the ratio of the **major: minor axes as *n: na*.**

Thus, for a given *integral value* of *n*, *na* (also known as '*k*') may possess integral values 1, 2, 3, 4, …, *n*. Therefore, for the **3rd orbit**, as *n* = **3**, *na* or '*k*' may have the values 1, 2, or 3. In fact the corresponding orbits (*k* = 1, *k* = 2 and *k* = 3) are explicitly exhibited in Fig: 1.3(a).

NOTE   **A circular orbit is obtained duly when *n* = *na*.**

In a broader perspective, the tangential velocity (*V*), of the **electron** at any stage may be resolved meticulosulsy into *two* altogether divergent components, namely:

- *first*, **which is found along the *radius vector*, and is known as the 'radial velocity';** and
- *second*, **which is found to be perpendicular ($\perp$) to the respective *radius vector*, and is termed as the *transverse* or *angular velocity*.**

It has been established beyond any reasonable doubt that the aforesaid *two* **velocities** critically form the following *two* **momentums**, namely:

- **radial momentum;** and
- **angular (or azimuthal) momentum.**

Sommerfeld thoughtfully assumed that these *two* **momentums** should by all means fulfill and obey, as far as possible, the **quantum parameters squarely** *i.e.*, the said *two* **momentum** should designate the *integral* **multiple of *h*/2π.**

Hence, we may have the following *two* **expression** for each momentum separately as given under:

$$\boxed{\textbf{Radial Momentum} = n_r \cdot h/2\pi} \qquad\qquad \text{(a)}$$

$$\boxed{\textbf{Angular (or Azimuthal)} = n_\phi . h/2_\pi} \qquad\qquad \text{(b)}$$

Hence, we have the following *two* **quantum numbers**:

- ❑ **Radial Quantum Number** (integral 'n_r'); and
- ❑ **Angular (or Azimuthal Quantum Number)** (integral 'n_ϕ'), and also 'n_r' and 'n_ϕ' are found to be intimately related to the **Bohr's Quantum Number** (see *Section 1.1.4.1*) '*n*' as follows:

$$\boxed{n = n_r + n_\phi}$$

Importantly, the *two* aforesaid different quantum numbers 'n_r' and 'n_ϕ' are, infact, closely related to the '*geometry of the ellipse*' as given under:

$$\boxed{\frac{n}{n_\phi} = \frac{n_r + n_\phi}{n_\phi} = \frac{\textbf{Length of the major axis (a)}}{\textbf{Length of the major axis (b)}}}$$

Plausible Possibilities – At this material time it will be worthwhile to consider the following *four* **plausible possibilities**, such as:

1. In a situation when the **angular quantum number** integral '$n_\phi = 0$', – the prevailing length of the so-called **minor axis** of the ellipse (*i.e.*, *oval-shaped*) shall also turn out to be zero. Thus, the *electronic path* may be a **straight line** passing *via* the ***nucleus (Z)*** but this is *practically not possible*.

   Inference – From the above analogy it would be clear that n_ϕ **can never be equal to '*zero*'.**

2. In another instance, when '$n_\phi > n$' (i.e., the **angular quantum number integral** is greater than the **Bohr's quantum number**), – the ensuing length of the *minor axis* of the **ellipse** shall be greater than the actual length of the respective *major axis*.

   Inference – Hence, it strongly suggests that n_ϕ can never be greater than n.

3. In a particular case, when $n_\phi = n$, – then the so-called *major axis* almost *equalizes* to the *minor axis*; and, therefore, the '**orbit**' turns out to be *circular* in shape.

   Inference: Hence, the ultimate '**circular orbit**' thus accomplished serves as a special instance of '**elliptical orbit**'.

4. In another typical instance, when $n_\phi < n$, – the ensuing '*orbit*' would eventually remain stick to the '**elliptical**' **shape** predominantly.

Example: Having examined all the *four* **plausible possibilities** let us now consider a situation when:

- **Bohr's quantum number (n): $n = 4$;** and
- **Angular quantum number integral (n_ϕ): $n_\phi = 1, 2, 3, 4$.**

Therefore, with respect to the aforesaid *four* **values of n_ϕ** and **$n = 4$**, one may have *four* **possible orbits** – that could be appropriately designated as: $4_1, 4_2, 4_3, 4_4$ respectively.

Remarks: The above stated '*subscripts*' (*viz.,* **1, 2, 3, and 4**) do vividly designate the respective **observed values of 'n_ϕ'** – **corresponding to '$n = 4$'.**

Importantly, out of the *four* **possible orbits** (see above) the '**orbit 4_4**' happens to be *circular one*; whereas, the remaining *three* **orbits** (*viz.,* $4_1, 4_2,$ and 4_3) are found to be *elliptical* **in shape,** – as depicted under:

n	n_ϕ	n_r	Designation/Shape of Orbit
4	4	0	4_4 : A Circular Orbit
4	3	1	4_3 : An Elliptical Orbit
4	2	2	4_2 : An Elliptical Orbit
4	1	3	4_1 : An Elliptical Orbit

Fig: 1.3(b) illustrates explicitly all the *four* **possible orbits:**

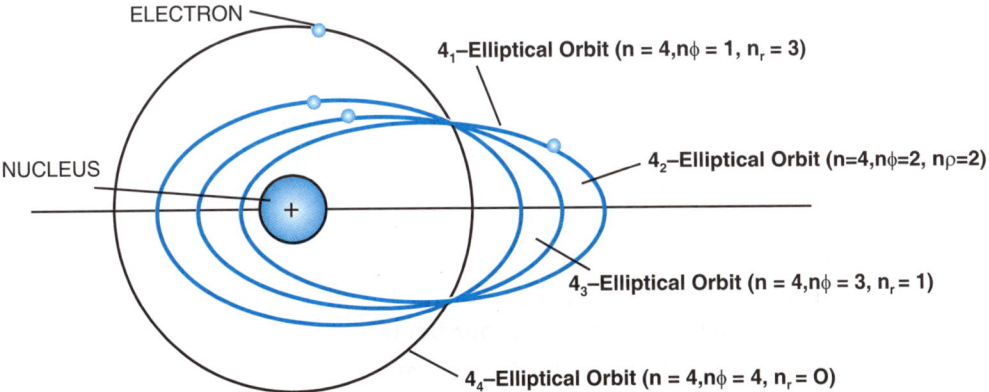

Fig. 1.3(b): Diagrammatic Representation of All the Four Possible Orbits for Bohr's Quantum Number '$n = 4$'.

1.1.5. *Modern Model of Atomic Structure*

Soonafter the spectacular promulgation and publication of the *Bohr's theory*, there was an excellent futuristic era of '**intense activity profile**' by both:

- **pure theoreticians**, and
- **experimental physicists**.

Therefore, a meticulous effort and sincere attempt was adequately made, which had the fundamental bearing upon the following *two* glaring aspects, such as:

- ➢ **mathematical principles**, and
- ➢ **significant experimental data**,

that vehemently came into view a rather –'*more definite picture of an atomic structure*'. Importantly, the so-called modern interpretation of the *atom* **is certainly a more elaborate delebration** *vis-a-vis* **the original concept of Bohr**.

Thus, there are *four* **quantum numbers** that were employed initially to expatiate as well as describe the **energy levels** or orbitals of each electrons, namely:

- **Principal Quantum Number**,
- **Azimuthal Quantum Number**,
- **Magnetic Quantum Number**, and
- **Spin Quantum Number**,

which shall now be discussed individually in the sections that follows:

- ❑ *Principal Quantum Number (n)* – It refers to an approximate measure of the *size* (*dimension*) **of the electron cloud** (π-*cloud*) i.e., the **precise order of magnitude** pertaining to the *potential energy*.

  Example: It has, for instance, the *observed values* (n) 1, 2, 3, 4, 5, 6, 7, ..., corresponding to the K, L, M, ..., Q shells of electrons.

- ❑ *Azimuthal Quantum Number (l)* – It relates to the specific shape of the **electron cloud** (π-*cloud*), – thereby determining exactly whether it bears any one of the following *geometrical configurations*, namely:

  - **spherical**,
  - **dumbbell shaped**, or
  - **complex geometrical feature**.

  However, it may invariably possess values of: 0, 1, 2, 3, ..., $(n-1)$, – corresponding with respect to the terms *s, p, d,* or *f* as employed frequently by the spectroscopists.

  Example: The particular **4d electron** will possess critically: $n = 4$ and $l = 2$.

- ❑ *Magnetic Quantum Number (m$_l$)* – It particularly refers to the orientation of the *electron cloud* (π-*cloud*) in space. It usually shows values of: **0, ± 1, ± 2, ..., ± l.** Therefore, for a *spherical cloud* there prevails only one orientation. Nevertheless, the so-called **dumbbell-shaped orbital** (O–O), for instance, may be easily oriented in altogether *three* **different directions** with respect to the *x, y,* and *z* axes of a set of **cartesian coordinates**.

- ❑ *Spin Quantum Number [S (or m$_s$)]* – It essentially provides the precise orientation of the *magnetic component* of an *electron*. As to date, there are only *two* recognized and known discrete manners by which an electron is capable of interacting with an *external magnetic field*. Obviously, very much like a '*small magnet*' it can play its role in *two* **distinct ways**, such as:

- **able to line up in the direction of the field**, or
- **prove to orient itself in the opposite direction**.

Importantly, the actual *magnetic field of an electron* was first and foremost visualized (and recorded) by virtue of the **rotation of the electron upon its axis**; and hence, for this *concrete evidence* "**an *electron* was believed solemnly to exhibit spin**". Besides, the following *two* **spin quantum numbers**, namely:

- $n = +1/2$; and
- $n = -1/2$,

were being used predominantly to describe the *two* explicitly noticeable '**spin states**'.

1.1.6. *The Electronic Configuration of the Elements*

In order to have a comprehensive understanding with regard to the **electronic configuration of the elements**, we may have to examine thoroughly the following *two* **cardinal rules**, that would ultimately go a long way to explain: "**the building up of electronic shells of elements**," – as depicted in Fig: 1.4 and Table: 1.1:

- **Pauli's Exclusion Principle**; and
- **Hund's Rule of Maximum Multiplicity**.

1.1.6.1. *Pauli's Exclusion Principle*

It relates to each electron moving round the nucleus of a neutral atom that may be characterized by *values of four quantum numbers*. Hence, the principle states that no *two* **electrons** present duly in a *neutral atom* may essentially possess the same set of *all the four quantum numbers*.

Obviously, that is analogous to the underlying principle in *"classical physics"* that clearly states that: **no two bodies can be in the same place at the same time**".

Example: Hence, *two* **electrons** located in the K shell may critically have the same *principal, azimuthal*, and *magnetic quantum numbes* [*viz.*, $n = 1$; $l = 0$; $m_1 = 0$], but altogether variant *spin-quantum members* [*viz.*, $s = \pm 1/2$ **and** $-1/2$].

1.1.6.2. *Hund's Rule of Maximum Multiplicity*

It states that as and when the orbitals do have the **same energy level**, the incumbment electrons invariably distribute themselves:

"**one to each and every orbital so as to maintain and sustain the so-called *parallel-spins***"

Example: **Oxygen (O)**–having an atomic number of eight does carry *eight* **inherent electrons**; and out of these:

- **2-electrons are present in the K-shell ($1s^2$)**; and
- **6-electrons are present in the L-shell.**

Thus, in the '**L-shell**' *two* **electrons** fill the **2s orbitals ($2s^2$)**; whereas, the remainging *four* **electrons** do fill the *2p orbitals* ($2p^4$).

Important Observation – According to the **Hund's Rule**, *three electrons* occupy predominantly: *2px*, *2py*, and *2pz* **orbitals**; and also observed to **spin in the same direction (as depicted in Fig: 1.4 shown by the direction of the arrow).**

Besides, the *fourth electron* may actually pair up with any out of *three electrons* (*viz.*, *2px* or *2py*, or *2pz*).

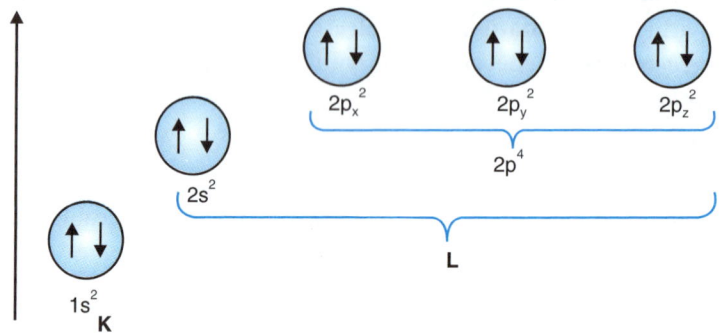

Fig. 1.4: Representation of the Electronic Configuration of an Oxygen Atom (O).

> **NOTE** **Importantly, the electronic configuration for the oxygen atom may be expressed as: $1s^2$, $2s^2$, $2px^2$, $2py^2$, $2pz^2$.**

Pauli's Exclusion Principle *Vs* Maximum Capacity of Shells

Based on the scientific evidences and logical explanations one may define comprehensively the so-called **inherent** *energy* of an electron present duly in an 'atom' only if:

"the *three* quantum numbers *viz.*, *n*, *l*, and *m*-representing its orbital and the respective spin '*S*'*–which have been specified explicitly."

> **NOTE** **Since, '*S*' has also been quantized, it may be regarded as an integral component amongst other quantum numbers'.**

Four Quantum Numbers – are, in fact, required ardently and sufficient enough to define in an elaborated manner: "the shape of the 'orbital' and hence, the inherent energy of the electron."

Multielectron System – may be adequately argued and explained based on the underlying fact that: "**each and every electron may certainly make a sincere attempt to occupy genuinely the most preferred** *stable orbital*."

Pauli (1929) introduced the most popular and widely applauded – '**exclusion principle**' (*see section 1.1.6.1*), which states vehlmently that in an '**atom**' there could be *only one electron* invariably available in any sort of a possible divergent '**energy state**'.

Points to Ponder: Because an '**orbital**' is normally defined by the help of *three* **quantum numbers**, and **two electron exclusively**; hence, it may be easily and conveniently accomodated in 'an orbital', – thereby indicating explicitly the actual capacity of '**an orbital**' to contain *two* **electrons** only.

---

*   That is, '*s*' may have a value of +1/2 only.

Based on the **Pauli's exclusion principle**, we may fix assign the inherent capacity of altogeher *divergent energy states in a shell*. As we know that a '**shell**' may essentially possess:

- **1s-orbital** • **3p-orbitals** • **5-d-orbitals** • **7f-orbitals**,

thereby suggesting that the optimum number of electrons duly present in:

➢ **s-state in a shell shall be 2;**

➢ **p-state 6;**

➢ **d-state 10; and**

➢ **f-state 14.**

Thus, we may have the actual status of some '**orbitals**' as given under:

- **σ-orbitals:** present duly in all the shells;
- **p-orbitals:** present in all the shells (except the *first* shell);
- **d-orbitals:** present in all the shells (except the *first two* shells);
- **f-orbitals:** present in all the shell having $n \geq 4$, and so on so forth.

The possible orbitals for **K, L, M, and N shells** and the individual capacity of the *shells* are as given under:

Table: 1.1: Possible orbitals for K, L, M, and N shells and individual Capacity of the shells.

S.No.	Shell	n	l	m	Designation	Exact Number of Orbitals	Max. no. of Electrons in Orbitals	Total no. of Electrons in Shell
I	K	1	0	0	1s	1	2	2
II	L	2	0	0	2s	1	2	–
			1	1, 0, –1	2p	3	6	8
III	M	3	0	0	3s	1	2	–
			1	1, 0, –1	3p	3	6	–
			2	2, 1, 0, –1, –2	3d	5	10	18
IV	N	4	0	0	4s	1	2	–
			1	1, 0, –1	4p	3	6	–
			2	2, 1, 0, –1, –2	4d	5	10	–
			3	3, 3, 1, 0, –1, –2, –3	4f	7	14	32

Explanation

1. The *first* **K-shell** has only *one orbital* (*1s*); **L-shell** has *four* (*i.e.*, **one 2s and three 2p orbitals**); **M-shell** has *nine* (*i.e.*, **1 2s, 3 3p, and 5 3d orbitals**); and **N-shell** has *sixteen* **orbitals** (*i.e.*, **1 4s, 3 4p, 5 4d, and 7 4f orbitals**).

2. Therefore, in a generalized form – the $\mathbf{n^{th}}$ **shell**, the corresponding **azimuthal quantum number** '*l*' may be 0, 1, 2, ..., $(n-1)$.

3. Thus, for each and every value '*l*' and '*m*' may have a value ranging from: 0, +1, +2, ..., +*l* thereby resulting finally in **2*l* + 1 orbitals**. Therefore, the **total number of orbitals** confined to the so-called $\mathbf{n^{th}}$ **shell** is usually given by the following expression:

$$\sum_{l=0}^{l=n-1} (2l + 1) = 1 + 3 + 5 + \cdots + (2n - 1) \, (n \text{ terms})$$

$$= \frac{n(1 + \overline{2n - 1})}{2} = n^2$$

4. Since each and every orbital may critically possess only *two* **electrons**; and hence; the *nth* **shell** may be able to accommodate comfortably $2n^2$ – electrons. In other words, the **K-shell** may possess **2 electrons**; **L-shell**: **8 electrons**; **M-shell**: **18 electrons**, and so on so forth.

5. Thus, for the **K-shell** – the specific *quantum representations* of the *two* **electrons** are: (1, 0, 0, 1/2) and (1, 0, 0, –1/2). Likewise, for the respective **L-shell** – the particular *quantum designations* are: (2, 0, 0, 1/2) and (2, 0, 0, –1/2) for the *2s* – **electrons**, (2, 1, 1, 1/2), (2, 1, 1, –1/2), (2, 1, 0, 1/2), (2, 1, 0, –1/2), (2, 1, –1, 1/2), and (2, 1, –1, –1/2) for the **6p-electrons**.

NOTE	Thus, one may safely conclude, that any electron may be regarded as an '*atom*' which may be designated appropriately by making use of the *four quantum numbers*.

2. QUANTUM MECHANICAL APPROACH

In order to have a thorough understanding of the underlying intricacies and comprehensive details with respect to the **quantum mechanical approach** we may have to look into the following *five* **allied terminologies**, namely:

- **Quantum** • **Quantum Mechanics** • **Quantum Numbers** • **Quantum State** *and*
- **Quantum Theory,**

in the sections that follows:

❑ **Quantum** – It refers to the minimum amount by which certain characteristic features, *viz.*, **energy** or **angular momentum** of a system can change perceptively. Such characteristic features perhaps do not vary continuously, but in *integral multiples of the specific relevant quantum*. Interestingly, this concept indeed forms the fundamental basis of the so-called **quantum theory**. Nevertheless, in *waves* and *fields* the **quantum** may be regarded as an **excitation**, giving a *particle-like* **interpretation** to the *wave* or *field*. Therefore, it may be inferred that the *quantum* of the **electromagnetic field** in:

- **photon**; and • **gravitation,**

designates precisely the *quantum* of the **gravitational field**.

❑ **Quantum Mechanics** – It relates to a *system of mechanics* which was duly developed from the **quantum theory**; and hence, is invariably used to expatiate the characteristic features of **atoms** and **molecules**. Now, using the *energy quantum* as a **starting point** it essentially incorporates the following *two* **cardinal aspects**, namely:

- **Heisenberg's Uncertainty Principle**; and
- **Broglie Wave Length,**

so as to establish the *wave-particle duality* upon which the **Schrödinger's Equation** is based. This form of **quantum mechanics** is usually termed as '**wave mechanics**'.

> **NOTE** An alternative but equivalent formalism, matrix mechanics, is solely based upon the *'mathematical operators'*.

❑ **Quantum Numbers** – These represent the **integral** or **half-integral numbers** which explicitly specify the *state of a system* or its *components in quantum mechanics*.

Example: An **electron (e^-)** present within an atom is specified particularly by the following *four* **quantum numbers**, such as:

1. **Principal Quantum Number (n)** – gives the main energy level and has values: 1, 2, 3, 4, ... etc. (*i.e.*, *higher the number, the farther the electron from the nucleus*). Traditionally, these *'levels'* or the *'orbits'* corresponding to them are invariably referred to as – *'shells'* and assigned alphabets: *K, L, M* etc. The *K*-shell is the one nearest to the *nucleus*.

2. **Orbital Quantum Number (l)** – that critically governs the so-called **'angular momentum'** of the *electron*. However, the most probable values of *'l'* are: $(n-1), (n-2), ..., 1, 0$. Hence, in the very *first* shell ($n = 1$) – the *electrons* may only exhibit an **angular momentum zero ($l = 0$)**. In the *second* shell ($n = 2$) – the respective observed values of *'l'* may be 1 or 0, thereby registering a definite rise to *two* subshells having slightly altogether *different energy profile*. Lastly, in the third shell ($n = 3$) – there exist predominantly *three* subshells with $l = 2, 1,$ or 0. The subshells are elegantly denoted by letters: s ($l = 0$), p ($l = 1$), d ($l = 2$), f ($l = 3$).

> **NOTE** The *orbital quantum number* is sometimes called the *azimuthal quantum number*.

3. **Magnetic Quantum Number (m)** – that particularly governs the energies of *electrons* in an **external magnetic field**. Amazingly, it may take values of: $+l, +(l-1),$ and $1, 0, -l, ... (l-1)$, $-l$. Besides, *'m'* can have values **+1.0** and $-l$ i.e., there could be *three p-orbitals* in the *p*-subshell, invariably designated as: $p_x, p_y; p_z$.

> **NOTE** Importantly, under normal circumstances these all do possess the same energy level profile.

4. **Spin Quantum Number (m_s)** – that essentially gives the *spin of the individual electrons*; and hence, may have the values: $+1/2$ or $-1/2$. According to the **Pauli's Exclusion Principle** (see *section 1.1.6*), **no *two* electrons** in an atom may have the **same set of** quantum numbers. Thus, the numbers define precisely the **quantum state of the electron**; and hence, explain correctly how the **electronic structures of atoms occur**.

❑ **Quantum State** – The condition of an ensuing physical system as described by a *'wave function'* i.e., the function may simultaneously designate:

• an *eigen-function* of one or more *quantum mechanical operator*; and
• the *eigen-values*,

are essentially the *'quantum number'* which ultimately label the state.

Example: The observed state of a **H-atom** is usually described by the *four* **quantum numbers stated** earlier *viz.*, $n, l, m,$ and n. However in the **ground state** (*i.e.*, *lower energy level*) they do have values ranging from: **1, 0, 1, and 1/2 respectively**.

❑ **Quantum Theory** – The theory that gradually grew up around **Planck's introduction** into *physics* of the:

"altogether *'newer concept'* pertaining to the discontinuity of energy".

NOTE   In fact, the prevailing *system of quantum mechanics* evolved categorically from this *theory* during *first-half of the theory* and particularly in the *first-half* of the 20th century.

2.1. Important Aspects of Quantum Mechanical Approach

Having understood the various important terminologies related to the so-called **Quantum Mechanical approach** it would be worthwhile to study exhaustively the following highly vital and equally critical aspects of the same with a view to have a rather clear concept of the present context.

2.1.1. *Microscopic Particulate Matters*

Since microscopy relates to the intensive examination of objects through the field of a sophisticated microscope (high resolution), it essentially make use of both **optical microscopy** and **electron microscopy**. Hence, it indeed becomes a lot easier to know more about the **electrons**, **protons**, **atoms**, and **molecules** exhibiting the *wave-particulate matter* duality, because they fail to offer a *positive response* to the **Newtonian mechanics*** squarely.

Besides, they specifically obey either **Quantum Mechanics** (or **Wave Mechanics**) – for which the *requisite laws* were designed and formulated in the year 1925 by a cluster of eminent researchers, namely: **Born**, **Heisenberg**, and **Jordan**.

NOTE   Based on the scientific evidence and logical explanations it has been duly ascertained that the *Schrödinger Formulation* is proved to be more widely popular and familiar to the chemists vis-a-vis *the Born-Heisenberg-Jordan Formulation.*

2.1.2. *Heisenberg Uncertainty Principle*

Based on another basic idea and concept in the domain of *quantum mechanics* one may intelligently express the **Heisenberg's uncertainty principle** by depicting an electron by taking into consideration the exact problem of determining strategically its respective *position*, '*x*', and *momentum '*mv*'.* Nevertheless, the *generalized principles* of '*optics*' do explain vividly that:

> **"it could be fairly difficult to locate the '*electron*' more precisely within ± λ *i.e.*, the ensuing wavelength of the so-called 'photon' employed to spot it accurately. "**

In such circumstances it is an absolute must to make use of the *radiation* with the **shortest possible wavelength**. However, the emerging action of the '**photon**' upon the '**electron**' ultimately aids in the virtual transfer of some of the **photon's inherent energy** to the respective **electron**, – thereby registering a resultant enhancement in its:

- **velocity**; and
- **momentum**

Furthermore, the corresponding *decrease* in the wavelength (*i.e.*, an *increase in the energy level*) of the '*photon*' vehemently increases the so-called '**margin of uncertainty**' in the **observed momentum** of the '*electron*' Δmv. Correspondingly, an increase in the wavelength of the '*photon*' causes a perceptive reduction the ensuing *degree of resolution* possible; and hence, enhances the uncertainty margin related to the exact position of the '*electron*', Δx.

---

*   **Newtonian Mechanics:** A system of mechanics based on **Newton's Laws of Motion** in which *mass* and *energy* are considered as: separate, conservative, and mechanical properties. It provides an accurate means of *determining the motions of bodies possessing ordinary velocities.*

Importantly, the overall observed product of these *uncertainties* is found to be fairly comparable in magnitude (*size*) to **Planck's constant (*h*)**; and, therefore, a rigorous treatment of the problem gives rise to:

$$\Delta mv, \Delta x \geq \frac{h}{4\pi}$$

Instead of considering the '*electron*' in a **H-atom** possessing *definite values* related to both:

- **Position**   and   •   **Momentum**,

and as a result of describing precisely the *momentum of the electron* one should critically replace the prevailing **orbits** of the **Bohr Theory** (see *section 1.1.4*) with more number of *diffuse regions* wherein the '*electron*' could be observed precisely.

> **NOTE**   It is highly desirable to consider solely the *probability* of finding the 'electron' in this region.

Points to Ponder: It is, however, pertinent to state here that the *uncertainty principle* may also be cited in terms of a few other **variables** *viz.*, **Energy (*E*), and Time (*t*)**. Thus, we may express the **product of the uncertainties ($\Delta_E \cdot \Delta_t$)** obtained from these quantities as under:

$$\Delta_E \cdot \Delta_t \geq \frac{h}{4\pi}$$

Comment: The observed transitions occurring between the *excited forms of atoms* or *nuclei of very short half-life* do provide a definite *energy change* having an appreciable **indeterminacy**; and, therefore, shows a tendency to *diffuse the lines in the suitable spectra*.

2.2. The Schrödinger Wave Equation

Preamble – The foregoing discussion amply expatiates the underlying fact that the **dynamic (moving) electrons** more or less do behave in *two* **different manners**, namely:

- **as particles**; and   •   **as waves**.

Therefore, if the *electrons* behave as a '**wave**', – there should be a *wave equation* to represent their typical behavioural pattern.

Schrödinger E (1926) put forward a **wave equation** so as to describe exhaustively the following *two* **important aspects:**

- **the wave behavioural profile of *electrons* in atoms**; and
- **inorganic molecules**.

Thus, in the **Schrödinger wave model** of an atom, the distinct energy-level patterns or the *orbits* **suggested by Bohr** have been meticulously replaced by the so-called *mathematical functions* (ψ) – that are found to be intimately related to "**the probability of locating electrons at different zones around the nucleus**".

2.2.1. Derivation of Schrödinger Equation

Let us consider the typical instance of a '***standing wave***' (*viz.*, a *vibrating string* or a *moving pendulum of a clock*) having a **wavelength** λ, – whose prevailing amplitude at any point in time along '*x*' may be duly expressed by a function $f(x)$, as given under:

$$\boxed{\frac{d^2 f(x)}{dx^2} = \frac{4\pi^2}{\lambda_2}\, f(x)}$$ (a)

Now, if an *electron* is regarded as a **wave** that moves eventually only in *one dimension*, we may have:

$$\boxed{\frac{d^2 \psi}{dx^2} = \frac{4\pi^2}{\lambda^2}\, \psi}$$ (b)

where, ψ = mathematical function.

But an *electron* may be able to move freely in all the *three* directions *viz.*, x, y, and z; and hence, it becomes:

$$\boxed{\frac{d^2 \psi}{dx^2} + \frac{d^2 \psi}{dy^2} + \frac{d^2 \psi}{dz^2} = \frac{4\pi^2}{\lambda^2}\, \psi}$$ (c)

Making use of the symbol ∇ in place of the *three* **partial differentials** x, y, and z, the above expression may be rewritten as follows:

$$\boxed{\nabla^2 \psi = \frac{4\pi^2}{\lambda^2}\, \psi}$$ (d)

The **De Broglie relationship*** states explicitly that:

$$\boxed{\lambda = \frac{h}{mv}}$$ (e)

where, h = Planck's constant,

m = mass of an *electron*, and

v = velocity of *electron*.

Putting the value of λ from Eq. (e) into Eq. (d) we have:

$$\boxed{\nabla^2 \psi = -\frac{4\pi^2 m^2 v^2}{h^2}\, \psi}$$ (f)

or we may have the following expression instead:

$$\boxed{\nabla^2 \psi + \frac{4\pi^2 m^2 v^2}{h^2}\, \psi = 0}$$ (X)

Nevertheless, the **total gross energy** of the **entire system** 'E' is comprised essentially of the **kinetic energy** 'K' plus the **potential energy** 'V'; and hence, we may have the following expression:

$$E = K + V$$

or

$$\boxed{K = E - V}$$

---

*   **See section 1.1.1 in Chapter 1.**

Since, the **kinetic energy (K) equals 1/2 mv²**, we may have the following expression:

$$1/2 mv^2 = E - V$$

or

$$V^2 = \frac{2}{m}(E - V)$$

(g)

Substituting the value of v^2 in **Eq. (X)**, we may arrive at the well-known **Schrödinger Wave Equation** – as given under:

$$\nabla^2 \psi = \frac{8\pi^2 m}{h^2}(E - V)\psi = 0$$

Conditionalities for Acceptable Solutions to the Schrödinger Wave Equation – Following are the *four* important conditionalities required for an acceptable solution to the **wave equation**, namely:

(a) Mathematical Function ψ should be continuous.

(b) Mathematical Function ψ must be a finite number.

(c) ψ should always be **single valued**.

(d) The critical probability of finding the '*electron*' spread over all the available space right from *plus* **infinity** to *minus* **infinity** should by all means be equal to '*one*'.

Thus, the probability to finding an '*electron*' at a point x, y, z is the **square of the mathematical function (ψ^2)**, and hence, we have the following expression:

$$\int_{-\alpha}^{+\alpha} \psi^2 \, dx \, dy \, dz = 1$$

2.2.2. Acceptable Solutions to the Schrödinger Wave Equation

It has been proven and ascertained that for a given type of 'atom', there exist a good number of solutions to the **Schrödinger wave equation**, that are largely acceptable. Hence, each of these *orbital* could be duly described in a unique manner by means of a *set-of-three* **quantum numbers** *viz.*, n, l, and m^*.

Subsidiary Quantum Number 'l' – It invariably elaborates precisely the **shape of the orbital** duly occupied by the *electron*. Thus, '*l*' may possess values **0, 1, 2, or 3**; and when $l = 0$; the ensuing orbital is *spherical* in shape – and is termed an '*s* orbital'. But when $l = 1$, the **orbital** thus obtained has a **dumb-bell shape** and is known as a '*p* orbital'; and when $l = 2$, the ensuing orbital attains a **double dumb-bell shape** and is called a *d* orbital. Furthermore, when $l = 3$, one may lay hands on to a more complicated *f* orbital (as illustrated in Fig: 1.5). However, the letters: **s, p, d,** and **f** – have been duly derived from the various *spectroscopic* terminologies, such as:

- **s:** *sharp*;
- **p:** *principal*;

---

* In fact, these are the *same* quantum numbers *principal*, *subsidiary*, and *magnetic* as were mostly employed in the **Bohr's Theory**.

- **d:** *diffuse*; and
- **f:** *fundamental*,

that were initially used to describe the *different lines* observed in the **atomic spectra**.

NOTE A critical examination of a detailed list of all the allowed solutions in the *wave equation* reveals that the various *orbitals* fall into *specific groups* only.

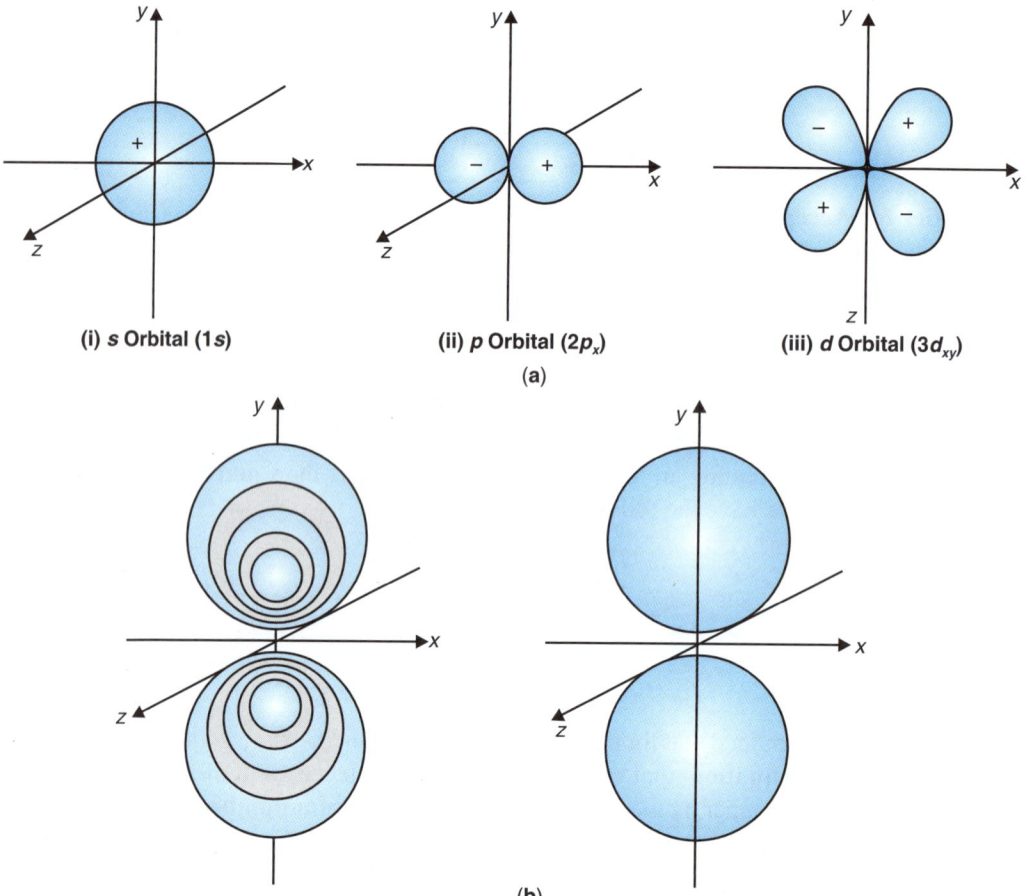

(i) *s* Orbital (1*s*)   (ii) *p* Orbital (2*p*$_x$)   (iii) *d* Orbital (3*d*$_{xy}$)

(a)

(b)

Fig. 1.5: (a) The Wave Function ψ for the *s, p,* and d Atomic Orbitals [The +ve and −ve signs refer to symmetry, no charge]; **(b)** Various Means and Ways of Designating y² for a 2*p* Orbital [As a contour Diagram or as a 90% Boundary Surface].

Remarks: The various cardinal aspects may be adequately expatiated as follows:

1. *First group of solution* clearly represent the exact value of the **wave function(ψ)**; and, therefore, the ensuing probability of determining the **electron ψ²** depends exclusively upon the **actual distance 'r' from the nucleus**; and it almost remains the same in all the **possible directions observed**. Thus, we may have the following expression:

$$\psi = f(r)$$

where, ψ = Wave function,

            f = Fundamental orbital, and

            r = Orbital distance.

❑ Obviously, it ultimately leads to **spherical orbital**; and hence, often takes place only when the *subsidiary* **Quantum Number 'l'** attains a value **zero**. Importantly, these are known as the *s* **orbitals**.

❑ But when the value of $l = 0$, the observed **Magnetic Quantum Number $m = 0$**; and hence, there remains only one such orbital for each value of n.

   2. *Second group of solution* related to the **wave equation, ψ** solely depends upon the following *two* aspects, namely:

- **Distance from the nucleus**; and
- **Direction in space** (*viz.*, *x*, *y*, or *z*)

Therefore, the **orbitals** of this type do occur only when the so-called *subsidiary* **Quantum Number** $l = 1$. These are usually known as *p* **orbitals**, that eventually give rise to *three* possible observed values of the ensuing **Magnetic Quantum Number** [*viz.*, $m = -1$, 0, and +1].

     ❑ Hence, there are *three* **orbitals** – that are critically found to possess:

- **same level of energy**;
- **identical in shape**; and
- **similar in size**,

but differ only with regard to their *direction in space*. Thus, one may express these *three* **solutions** in the **wave equation** as stated below:

$$\psi_1 = f(r) \times f(x)$$
$$\psi_2 = f(r) \times f(y)$$
$$\psi_3 = f(r) \times f(z)$$

❑ **Orbitals** that are identical in their respective *energy levels* are invariably called as 'degenerate'; and hence, the *three* observed *degenerate p orbitals* do take place for each of the values of $n = 2, 3, 4, \ldots$

   3. *Third group of solution* related to the wave equation depend exclusively upon the distance right from the **nucleus 'r'**, besides *two directions* in space.

Thus, we may have the following expression:

$$\psi_1 = f(r) \times f(x) \times f(y)$$

❑ Interestingly, this particular group does possess $l = 2$; and hence, these are termed as *d* **orbitals**. Therefore, we may have *five* **solutions** with respect to: $m = -2, -1, 0, -1, $ and -2; and incidently all of these do possess an **equal level of energy**. Hence, one may encounter the so-called **degenerated *d* orbitals** that specifically come into being *for each of the values* of $n = 3, 4, 5, \ldots$

❑ In addition, a further *set-of-solutions* does occur when *l* = **3**; and hence, these are known commonly as *f* **orbitals**. However, there are in all *seven* **values** of *m* = –3, –2, –1, 0, +1, +2, **and +3**; and, therefore, may give rise to the formation of **seven '*f*' degenerate orbitals** only when n = 4, 5, 6, ... Table: 1.2 records the different **important atomic orbitals** pertaining to various **principal quantum numbers**.

Table: 1.2: Different Atomic Orbitals for Various Princpal Quantum Numbers.			
Principal Quantum Number (*n*)	**Subsidiary Quantum Number (*l*)**	**Magnetic Quantum Number (*m*)**	**Symbol**
1	0	0	1*s* (one orbital)
2	0	0	2*s* (one orbital)
2	1	–1, 0, +1	2*p* (three orbitals)
3	0	0	3*s* (one orbital)
3	1	–1, 0, +1	3*p* (three orbitals)
3	2	–2, –1, 0, +1, +2	3*d* (five orbitals)
4	0	0	4*s* (one orbital)
4	1	–1, 0, +1	4*p* (three orbitals)
4	2	–2, –1, 0, +1, +2	4*d* (five orbitals)
4	3	–3, –2, –1, 0, +1, +2, +3	4*f* (seven orbitals)

[*Adapted from:* Lee JD: **Concise Inorganic Chemistry**, 2nd ed. Wiley, New Delhi (India), 2013]

2.3. POSTULATES OF QUANTUM MECHANICS

The various formulation(s) based on the phenomenon of '**quantum mechanics**' which have been meticulously conceptualized for the so-called:

"**wave mechanical treatment of the basic structure of an atom**"
rests exclusively upon certain vital and important *postulates* meant for a particular system moving forward in *one direction viz.*, the **α-coordinate**, as expatiated under:

A survey of literature would reveal that there are in all *six* **cardinal postulates** pertaining to the **Quantum Mechanics** which shall now be discussed briefly and separately in the sections that follows:

❑ *First* **Postulate:** It refers to the physical state of a given system at **time '*t*'**; and is being described explicitly by the wave function: ψ(*x*, *t*).

❑ *Second* **Postulate:** In this specific instance, the **wave function** ψ(*x*, *t*), and its respective *1st derivative* and *2nd derivative:*

- ∂ψ (*x*, *t*)/∂*x*, and
- ∂²ψ (*x*, *t*)/∂*x*²,

are found to be conspicuously *continuous*, *single-valued*, and *finite* for practically all values of '*x*'. Besides, the **wave function**, ψ(*x*, *t*), is rendered duly **normalized**.

Thus, we may have the following expression:

$$\int\limits_{\infty}^{+\infty} \psi^*(x, t)\, \psi(x, t)\, dx = 1 \qquad\qquad\text{(a)}$$

where, ψ^* = complex conjugate of ψ formed*.

❑ *Third* **Postulate:** It relates to a definite *physically* **observed quantity** that may be precisely represented by a *Hermitian Operator*. The *'operator A'* is said to be *Hermitian* only if it fully satisfies the following parameter as far as possible:

$$\int \psi_1^* \hat{A}\, \psi_j\, dx = \int \psi_i^* \left(\hat{A}\psi_i\right)^* dx \qquad\qquad\text{(b)}$$

where, ψ_i and ψ_j =   Wave functions (duly representing the *physical states of the quantum system*)**

❑ *Fourth* **Postulate:** Thus, the observed values of an *'operator A'* are duly designated by the *eigen values*, a_i, in the **operator equation**; and hence, we may have the following expression:

$$\hat{A}\psi_i = a_i\, \psi_i \qquad\qquad\text{(c)}$$

Eqn. (c) is usually called the *Eigen-value Equation*. In this case A represents the 'operator A' for the observed *physical quantity*, and ψ_i designates an *1 of* $\hat{A}$ having an *eigen-value a_i*. Alternatively, one may also measure the **observed** $\hat{A}$ that eventually gives the **eigenvalue a_i**.

❑ *Fifth* **Postulate:** Importantly, the so-called **operation value** (or *average value*), <A>, of an *observed A*, with respect to the intended **operator** $\hat{A}$ – is duly accomplished by the following relationship:

$$<A> = \int\limits_{-\infty}^{+\infty} \psi^* \hat{A}\, \psi\, dx \qquad\qquad\text{(d)}$$

NOTE	In Eq. (d) the function ψ is usually predicted to be normalized to an appreciable extent as per the *Eq. (a)*

Therefore, the **average value** of the *'x-coordinate'* may be given by the following expression:

$$<x> = \int\limits_{-\infty}^{+\infty} \psi^* \hat{X}\, \psi dx \qquad\qquad\text{(e)}$$

❑ *Sixth* **Postulate:** Based on the **quantum mechanical approaches** with respect to the various intended observations, one may conveniently construct them, precisely by the help of the so-called *'classical expression'* in terms of:

---

*   That is, by replacing *'i'* with *'–i'*, wherever it occurs in the **wave function:** $\psi \cdot (i = \sqrt{-1})$.

**   That is, a minute particle e.g., **electron, proton, atom,** or a **molecule**

- **recording the variables,** and
- **converting the observed expressions to the respective** *operators*,

Table: 1.3 records the various **wave mechanical operators** for carrying out the evaluation of different **classical variables.**

Table: 1.3: Various Wave Mechanical Operators for Carrying Out the Evaluation of Different Classical Variables.

S.No.	Classical Variables	Quantum Mechanical Operator	Operator	Requisite Operation
1.	x	$\hat{x}$	x	Multiplication by 'x'.
2.	p_x	p_x	$-ih\dfrac{\partial}{\partial x}$	Taking derivative with respect to 'x' and Multiplying by $-ih$.
3.	x^2	x^2	x^2	Multiplication by x^2.
4.	t	t	t	Multiplying by 't'
5.	p_x^2	$\widehat{p_x^2}$	$-h^2\dfrac{\partial^2}{\partial x^2}$	Taking the second derivative with respect to x and then multiplying by $-h^2$.
6.	E	$\hat{E}$	$ih\dfrac{\partial}{\partial t}$	Taking derivative with respect to 't' and multiplying by ih

❑ *Seventh* **Postulate:** The wave function $\psi(x, t)$ is nothing but a suitable solution of the so-called '**time-dependent expression**' – as given under:

$$\boxed{\hat{H}\psi(x, t) = ih\,\frac{\partial\psi(x, t)}{\partial t}}$$ (f)

where, $\hat{H}$ = **Hamiltonian Operator*** of the system

2.4. Important Aspects of Wave Mechanical Approach

Let us take into consideration the following important aspects pertaining to the **wave mechanical approach**, namely:

- **The concept of Electron Charge Cloud,**
- **Probable Solutions of Schrödinger Equation to certain Simple Systems** *viz.,*
  - ➤ **Particles present in One-Dimensional Box,**
  - ➤ **Particles in a 3D Box;** and
  - ➤ **Hydrogen and Hydrogen Like systems.**

2.4.1. *The Concept of Electron Charge Cloud*

In order to understand the logistics associated with the concept of ***electron charge cloud***, we may have to consider the **H-atom** absolutely in its '**ground state**' (*i.e.,* having a **bare minimum energy level**). Let us assume that one can visualize possibly the **hydrogen nucleus** plus its **lone electron** clearly; and that:

---

* **Hamiltonian Operator:** It refers to a *function* to express the energy of a system in terms of its **momentum** and **positional coordinates**. However, in simple cases this is the sum of its *kinetic* and *potential* energies.

- the 'nucleus' occupies a *fixed place*; and
- the 'electron' ensures a *moving (dynamic) status* throughout its existence.

In an absolute hypothetical assumption if one takes it granted that an 'electron' occupies an accurate position as a 'dot' [*i.e.*, at a time interval of *one-thousandth of a second* up to *one-thousand seconds* (≡ 0.001 second)], – the optimum observed concentration of such *dots* ought to be located strategically closest to the 'nucleus'. Obviously, the *dots* might go on reducing progressively as the **distance** between the 'electron' and the 'nucleus' goes on enhancing in **all the directions** perceptively.

Comment: At this material time, one may critically observe the maximum probability of spotting the 'electrons' almost nearest to the 'nucleus'; and goes on reducing constantly as the distance from the 'nucleus' increases. However, it is next to impossible to specify precisely the so-called:

"actual status (position) of the 'electron' from the 'nucleus'".

Importantly, one may, in all its probability, ensure that a **definite region of space** might be there for the actual presence of the 'electron'.

Electron Charge Cloud – It usually refers to the **electron probability distribution pattern** as could be observed in the particular **3D–space around the 'nucleus'**, – which is termed as the '**electron charge cloud**'.

NOTE	Obviously, the specific '*zone*' that explicitly shows the presence of the *thickest electron cloud* (*i.e.*, where the number of 'dots' signifying the position of '*electron*' is maximum) identifies the '*zone*' where one would see the probability of locating the maximum number of the '*electron*'.

Fig: 1.6 shows the model of a **hydrogen atom** having the *electron charge cloud* located outside the 'nucleus'; and the '*nucleus*' predominantly contains **protons** and **neutrons** – that critically appears to be tiny *vis-a-vis* the actual size of the **electron cloud**.

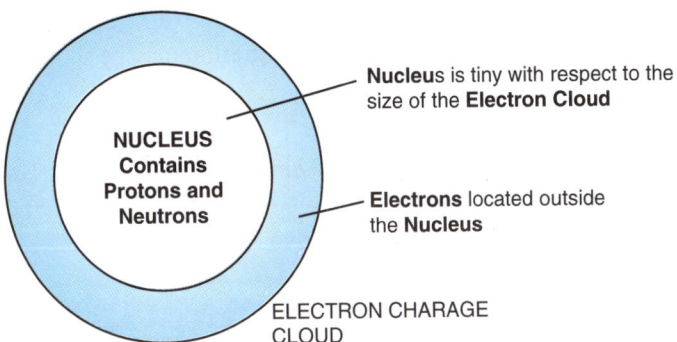

Fig. 1.6: Electron Charge Cloud in a Hydrogen Atom.

Remarks: In the above diagramatic representation of a **H-atom** the specific **region** (*zone*) located very much within the *sphere* designates in reality – '**a region of relatively high probability**'. In other words, it suggests explicitly that the '*electron*' invariably spends a large segment of its time within the *spherical region*.

NOTE	In contrast to *Bohr's Concept*, this particular region (zone) strictly as per the *wave mechanical model* does not possess any sharp boundaries.

2.4.2. *Probable Solutions of Schrödinger Equation to Certain Simple Systems*

Based on the scientific evidences and logical explanations the **Schrödinger equation** may be solved elegantly with respect to the following simple sytems, namely:

2.4.2.1. *Particle Present in One-Dimensional Box*

Let us take into consideration a specific particle (or particulate matter) with **mass (m)** – that is restricted to move in an **one-dimensional box** (Fig: 1.7) that has:

- **Length (a)**   and   • **High Walls (∞).**

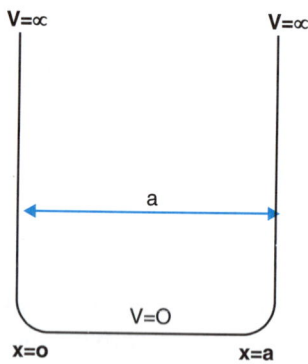

Fig. 1.7: Particle Present in one-Dimensional Box

Thus, one may assume that the inherent **potential energy** of the said particle to be '**zero**' (for the sake of *simplicity*) at every **nook and corner** inside the **one-dimensional box**.

Thus, we may have:

$$V(x) = 0 \qquad (1)$$

Therefore, very much within the '**box**', the **Schrödinger equation** may be shown in dimension '*x*', such as:

$$\left[-\frac{\hbar^2}{2m}\left[\frac{d^2}{dx^2}\right] + V(x) \right] = E\psi(x) \qquad (2)$$

Inserting the value of $V(x)$ from Eq. (1) in Eq. (2) we have:

$$-\frac{h^2}{2m}\left[\frac{d^2\psi}{dx^2}\right] = E\psi(x) \qquad (3)$$

Rewriting Eq. (3) mathematically we may have:

$$\frac{d^2\psi}{dx^2} + \left[\frac{2mE}{h^2}\right]\psi = 0 \qquad (4)$$

or

$$\frac{d^2\psi}{dx^2} + k^2\psi = 0 \qquad (5)$$

where, $k^2 = \dfrac{2mE}{h^2}$ (which being a *constant* found to be **independent** of '*x*')

Besides, **Eq. (5)** designates an ordinary **2nd order differential equation** that usually provides the '*solution*' in the following form:

$$\psi = A \cos kx + B \sin kx \tag{6}$$

where,   *A & B* = Constants.

Let us now consider a '**particle**' moving outside the '**box**', whereby $V(x) = \infty$; and hence, the **Schrödinger equation (2)** may be written as:

$$\left[-\frac{h^2}{2m}\left[\frac{d^2}{dx^2}\right] + \infty \right]\psi(x) = E\psi(x) \tag{7}$$

Thus, **Eq. (7)** may now be written **mathematically** as:

$$\frac{d^2\psi}{dx^2} = \frac{2m}{\hbar^2}(E - 0\infty)\psi = 0 \tag{8}$$

Key Observations – Eqn. (8) may only be satisfied logically if the value of ψ is '**zero**' at any point **outside the 'box'** – indicating thereby that the '**particle**' must not be seen outside the '**box**' at all; which suggest explicitly that the '**particle**' is **only confined within the 'box'**. Furthermore, it implies vehemently that 'ψ' [*i.e.*, the *amplitude of vibration for the wave in the* **wave equation**] should be '**zero**' critically along the **inside-walls of the 'box'** (*i.e.*, at $x = 0$ and $x = a$ in Fig. 1.7]. Based on the fact that '**ψ**' has got to be a *continuous function* of '*x*' throughout; and , therefore, it drives us to the '*ultimate conclusion*' that in **Eqn. (6)**, the value of '**A**' should be '*zero*'.

Therefore, one may rewrite **Eqn. (5)** in the following form:

$$\psi = B \sin kx \tag{9}$$

Because, ψ = 0 at $x = 0$ and at $x = a$; therefore, $B \sin ka = 0$ or $\sin ka = 0$ or $ka = n\pi$, so we may have:

$$k = n\pi/a \tag{10}$$

where, n = 1, 2, 3, 4, 5, ... represents the **principal quantum number**.

Therefore, the so-called **permitted solutions of Eqn. (5)** may be expressed as under:

$$\psi = \psi_n = B \sin\left[\frac{n\pi x}{a}\right] \quad (\text{where, n=1,2,3...}) \tag{11}$$

Finding the Particle within the Box – may be accomplished by looking into the probability of locating the **particle within the 'box'** in the *small length portion* **dx** which is indeed **ψ²dx**. As it has been stated earlier that the **particle should always be found within the 'box'**, therefore we may have the following expression:

$$\boxed{\int_0^a \psi^2 dx \quad \text{or} \quad \int_0^a B^2 \sin^2 \frac{n\pi x}{a} dx = 1} \tag{12}$$

[where, 'a' and 'o' are the *two* equivalent values of 'x' in Fig. 1.7]

Thus, the solution of the aforesaid equation gives precisely B as $\left[\dfrac{2}{a}\right]^{1/2}$.

Therefore, Eq. (11) may be written as follows:

$$\boxed{y = \left[\frac{2}{a}\right]^{1/2} \sin\left[\frac{n\pi x}{a}\right]} \tag{13}$$

Importantly, Eqn. (13) relates to the solution of the so-called **Schrödinger equation** for a particle duly present in a '*one-dimensional box*'.

From Eqn. (4) and Eqn. (10), we have:

$$2m \, E/h^2 = n^2 \pi^2/a^2$$

or $\qquad \boxed{E \equiv E_n = \dfrac{n^2 \hbar^2 \pi^2}{2ma^2} = \dfrac{n^2 h^2}{8ma^2}} \quad \text{(where, } \eta = 1,2,3...)$ \hfill (14)

Thus, **Eqn. (14)** vividly provides the overall expression for the inherent *energy of the particle* present in one-dimensional '**box**'.

At this point in time, it may be added that since the *energy* depends solely upon the corresponding **quantum number 'n'**, that may eventually possess any extent of **integral value**, the ensuing **energy levels** of the *particle* in the '**box**' could be **quantized accordingly**.

Problem: An '*electron*' is duly confined into a '*one-dimensional box*' with a length 1Å. Calculate the 'ground state' energy in electron volts (eV).

Solution: We know that the '*ground state*' the value of n = 1.

From **Eqn. (14)**, the '**ground state**' energy of the '*electron*' is given in the following expression:

$$E_1 = \frac{n^2 h^2}{8ma^2} = \frac{(1)^2 \, (6.626 \times 10^{-34} \, \text{Js})^2}{(8) \, (9.109 \times 10^{-31} \, \text{Kg}) \, (10^{-10} \, \text{m})^2 \, (1.602 \times 10^{-19} \, \text{J/eV})}$$

$$= 37.6 \text{ eV}$$

Variation of ψ and ψ^2 with x for x = 1, 2, 3 for a Particle in a One-Dimensional Box

The precise and accurate values of **E(energy)** for n = 1, 2, or 3 *vis-a-vis* the respective values of function ψ and ψ^2 prevailing between x = 0 and x = a (see Fig: 1.8). Thus, one may encounter **different forms of the curves** depicting the so-called observed variation(s) of ψ^2 with that of 'x' for altogether different value of 'n' indicate explicitly that:

"**the probability of spotting the particle at a given status solely rests upon 'n'**".

Hence, upon the corresponding **inherent energy** of the '*particle*'; and at x = a/2, it may become (say):

| **NOTE** | The exact '*number of nodes*' in the observed '*electron wave*' is invariably given by '$n - 1$'. |

Now, we may take into consideration **a few vital implications** by the following expression [see Section 2.4.2 (Eqn. 14)]:

$$E = \frac{n^2 h^2}{8ma^2}$$

First and foremost let us recognize the underlying fact that the **inherent energy of the particle** may eventually possess only some *critical values* with respect to $n = 1, 2, 3, \dots$ which explicitly inducts the prevalent concept of a '**quantum number**'. Because E(*i.e.*, energy) may never attain the value '**zero**', that particle located strategically in the **one-dimensional box** (Fig: 1.7) does possess a *minimum energy level* equivalent to $h^2/8\,ma^2$ even at **absolute zero value** *i.e.*, either its:

- **residual energy**; or
- **zero-point energy** (*frequently called*).

Importantly, the most crucial spacing of the *energy levels* for an **one-dimensional box** with a *given length 'a'* depends inversely upon 'm'* (which means it is obviously much wider for *electrons* than for '*nuclei*'. Hence, for the given values of a (*acceleration*) and n (*an integer*), the **E** (*energy*) differs progressively for *various isotopes*** of an element. Therefore, for a given value of m, the so-called *spacing of the energy levels* is observed to be **inversely proportional** to a^2.

Hence, in case one intelligently compares an *electron* present in a molecule having length equivalent to a few **angströms (Å)** with a molecule in a **one-dimensional box** (Fig: 1.7) *of length a few centimeters* (**cms**), it gives a clear indication that:

"**partially due to the difference in mass (m), but more big virtue of the existing difference in the actual length of the '*box*' the *energy levels* (E) of the latter are found to be quite closer by comparison of with those (such criteria) of the *former*)**"

| **NOTE** | Hence, the energy levels of the ensuing *thermal motion of molecules* are indeed found to be so **intimately close** so that the observed '*kinetic energy*' of the molecules (under investigation); and also of all **larger species** essentially appears to change continuously and perceptively. |

2.4.2.2. *Particle Present in a Cubic Box [Three-Dimensional Box; 3D-Box]*

It is, however, pertinent to state here that an almost identical treatment to the one given earlier *viz.*, **one dimensional box** (see *Section 2.4.2.1*) ultimately leads to the **Schrödinger equation** meant exclusively for a **constant-energy particle** duly present in a *field-free* **Cubic Box** – of side 'a' wherein the value of **V** (volume) is equivalent to *zero* for any value of x, y, or z existing between the *two* extremities '**0**' and 'a', but obviously remains '**infinite (∞)**' outside these limits predominantly.

Thus, we may have the following expression:

$$\frac{\partial^2 \psi}{\partial x^2} + \frac{\partial^2 \psi}{\partial y^2} + \frac{\partial^2 \psi}{\partial z^2} + \frac{8\pi^2 m}{h^2} E\psi = 0$$

---

* **m**: Mass of the '**Electron**'.

** **Isotopes**: They refer to *two* atoms having the **same atomic number, but different atomic weights**.

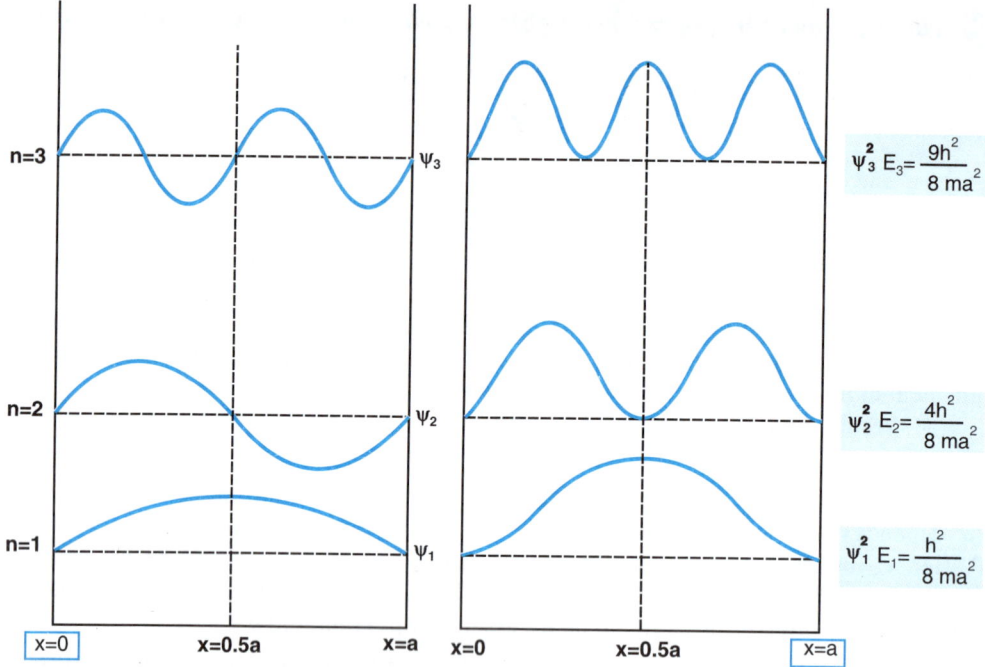

Fig. 1.8: The Observed Variations of ψ and ψ^2 with **x** for **n** = 1, 2, and 3 for a Particle duly Present in a One-dimensional Box.

In a particular instance, when ψ designates a function of *three* altogether **independent variables x, y, and z,** it may be illustrated vividly to be a product of:

*"**three** functions dependent individually upon the same available variables"*,

that is,

$$\psi(x, y, z) = \psi_x \psi_y \psi_z$$

where, ψ_x, ψ_y, and ψ_z represent the *wave functions* satisfying the following relation:

$$\frac{d^2 \psi_x}{dx^2} = -\frac{8\pi^2 mEx}{h^2} \psi_x$$

and such obviously **identical relationship** prevailing between ψ_y and E_y; and also between ψ_z and E_z. Thus, the **total energy (E)** is usually given by the following expression:

$$E = E_x + E_y + E_z$$

Interestingly; the *solution* of these equations may ultimately lead to:

$$\psi_{nxuynz} = \left(\frac{8}{a^3}\right)^{1/2} \sin\left(\frac{n_x \pi}{a}\right) x \sin\left(\frac{n_y \pi}{a}\right) y \sin\left(\frac{n_z \pi}{a}\right) z$$

and

$$E_{nx,ny,nz} = \frac{h^2}{8ma^2} (n_x^2 + n_y^2 + n_z^2)$$

Important Points – Following are a few important points pertaining to the **Schrödinger equation**, namely:

1. The need for *three* **quantum numbers** *viz.*, n_x, n_y, and n_z are an absolute necessity to specify explicitly:

   "**all the permissible solutions of the** *Schrödinger equation* **for the prevailing system**".

2. Details of various **energy (E) levels** required are:
   - **Zero-point energy [3h²/8ma²]** – that critically corresponds to the specific **wave function** $\Psi_{1, 1, 1}$;
   - **Next energy level [6h²/8ma²]** – invariably corresponds to *any of the* **wave function** *viz.*, $\Psi_{1, 1, 2}$, $\Psi_{1, 2, 1}$, $\Psi_{2, 1, 1}$ (*i.e.*, having the **nodal planes*** located strategically at: $z = a/2$, $y = a/2$, and $x = a/2$ respectively); and, therefore, this kind of an **energy level** is usually termed as – '*triply degenerate*'.
   - **Next** '*four*' **energy levels** – which may be observed easily and frequently *viz.*, $9h^2/8ma^2$, $11h^2/8ma^2$, $12h^2/8ma^2$, and $14h^2/8ma^2$ – that crucially exhibit the so-called '*degeneracies*'** **having: 3, 3, 1, and 6 respectively.**

2.4.3. *Schrödinger Wave Equation for Hydrogen Atom and Other One-Electron Species*

The **Schrödinger equation** related to the motion of a particle in *three* **dimensions** (*viz.*, x, y, and z-axes: all being perpendicular ($\perp$) to one another) can be expressed as under:

$$\frac{\partial^2 \psi}{\partial x^2} + \frac{\partial^2 \psi}{\partial y^2} + \frac{\partial^2 \psi}{\partial z^2} + \frac{8\pi^2 m}{h^2} (E - V) \psi = 0 \qquad (1)$$

where, ∂ = Partial differentiation of a variable;

         ψ = Amplitude of vibration for the *wave* in a **wave equation;**

   x, y, z = Three axis perpendicular to one another;

        m = Mass of the electron;

        h = Planck's constant;

        E = Energy; and

        V = Volume

   - *Approximation*: In order to tackle the problem in a rather *simplified manner*, one may adhere to the *first approximation* by considering:

      "**the** '*nucleus*' **to be the centre of mass and that too at rest**".

      One-electron System: In the generalized instance of a *one-electron system* the inherent '*charge*' borne by the **nucleus** stands at ± **Ze**; and hence, the overall potential energy of an '*electron*' located at a **distance 'r'** is usually given by the following expression:

---

  *   **Nodal Planes:** They relate to *two* **planes** having a zero amplitude in a system of standing waves.

 **   **Degenerecies:** These refer to the so-called **degenerative behaviours**.

$$V = \frac{Ze^2}{4\pi\varepsilon_0 r} \tag{2}$$

where, ε_0 = Permitivity of vacuum

Substituting the Value of **V** from Eqn. (2) in Eqn. (1) we have:

$$\frac{\partial^2 \psi}{\partial x^2} + \frac{\partial^2 \psi}{\partial y^2} + \frac{\partial^2 \psi}{\partial z^2} + \frac{8\pi^2 m}{h^2}\left(E + \frac{Ze^2}{4\pi\varepsilon_0 r}\right)\psi = 0 \tag{3}$$

❑ *Spherical Symmetry of System:* It suggests explicitly that it might obviously prove to be advantageous to carry out the operations in the so-called *polar coordinates viz., r,* θ, and φ as depicted in Fig: 1.9:

where, *r* = Distance of the electron from the nucleus,

θ = Angle '*r*' makes with the *z*-axis, and

φ = Angle made by the projection of '*r*' upon the *xy* **plane** that makes with the *x*-axis. Nevertheless, the **Cartesian** as well as the **polar coordinates** are duly related by:

$$x = r \sin\theta \cos\phi$$

$$y = r \sin\theta \cos\phi$$

$$z = r \cos\theta$$

$$r^2 = x^2 + y^2 + z^2$$

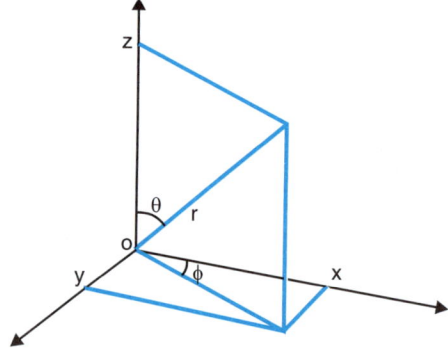

Fig. 1.9: Diagramatic Representation Showing the Relationship between Cartesian and Polar Coordinates.

2.4.3.1. *The Polar Coordinates Version of the 3D-Schrödinger Equation*

Based on the foregoing theoretical aspects supported by statement of facts, we may look at the **polar coordinates version of the three-dimensional (3D)–Schrödinger equation** as given below:

$$\frac{1}{r^2}\frac{\partial}{\partial r}\left(r^r \frac{\partial \psi}{\partial r}\right) + \frac{1}{r^2 + \sin\theta}\frac{\partial}{\partial\theta}\left(\sin\theta \frac{\partial\psi}{\partial\theta}\right) + \frac{1}{r^2 \sin^2\theta}\frac{\partial^2\psi}{\partial\phi^2} + \frac{8\pi^2 m}{h^2}\left(E + \frac{Ze^2}{4\pi\varepsilon_0 r}\right)\psi = 0$$

Explanation – Following are the explanatory notes:

1. As may be observed with a particle present in a **one-dimensional box** (see **Section 2.4.2.1**) and its subsequent '*problems*' encountered, various **limitations** have been imposed intelligently that would eventually, render it a lot easier and possible to **solve** the '**wave equation**' efficaciously.

2. In addition, the '**wave equation**' should always be either:
   - **continuous** or - **single-valued**.

3. The '**wave equation**' should attain a value '**zero**' at *infinity* (∞)*; and hence, the ensuing probability of finding the '*electron*' duly summed up over **all space must be unity.****

Points to Ponder: These essentially include:

1. Bearing in mind the various obstacles/difficulties in dealing squarely with such **unavoidable problems** do seem to be simply frightening (since difficult to manage/overcome). Hence, it might appear that in order to represent logically the so-called dependence of:

   "ψ **upon** r, θ, **and** ϕ – **one may necessarily require a 4D-graph.**"***

2. Amazingly, to our advantage – it may be amply proven that the '*mathematical form of the equation*' indeed renders it enormously possible to express explicitly:

   "ψ – **as a product of** *three* **functions** *viz.*, ψ_r, ψ_θ, **and** ψ_ϕ, **that rests exclusively upon** r, θ, **and** ϕ **respectively.**"

   Thus, we may have the following expression:

   $$\psi_{r,\theta,\phi} = \psi_r \, \psi_\theta \, \psi_\phi$$

3. Now, the '**wave equation**' may be thoughtfully segmented into *three* **individual equations** thereby expressing the vivid dependence of ψ upon r, θ, and ϕ respectively.

   Radial part of the wave function: ψ;

   Angular part of the wave function: ψ_θ and ψ_ϕ.

4. **Solutions of 'ψ':** One may critically note a portion of the **wave equation** as given by **E** for its respective values:

   $$E_n = \frac{me^4 Z^2}{8\varepsilon_0^2 h^2} \cdot \frac{1}{n^2}$$

   where, $n = 1, 2,$ and 3.

   In fact, the above '**Equation**' is found to be almost identical with the one derived for the **Bohr's theory** of the **hydrogen atom** (here 'n' also designates the '*principal quantum number*').

2.4.3.2. Hydrogen–Like Species vis-a-vis 1s, 2s, and 2p Orbitals

In fact, Table: 1.4 essentially comprises all relevant information related to the **1s, 2s, and 2p** orbitals of a *Hydrogen-like species*.

---

*   That is, the observed species should be '**finite**'.

**   That is, the ensuing '**wave function ψ**' should be *normalized*.

*** **4D-Graph:** Four-dimensional graph.

S.No.	n	l	m_1	Ψ_r	$\Psi_\theta \Psi_\phi$	Symbol
1.	1	0	0	$2\left(\dfrac{Z}{a_0 *}\right)^{1/2} e^{-Zr/a_0}$	$\left(\dfrac{1}{4\pi}\right)^{1/2}$	$1s$
2.	2	0	0	$\left(\dfrac{Z**}{2a_0}\right)^{1/2}\left(2 - \dfrac{Zr}{a_0}\right) e^{Zr/2a_0}$	$\left(\dfrac{1}{4\pi}\right)^{1/2}$	$2s$
3.	2	1	0	$\dfrac{1}{\sqrt{3}}\left(\dfrac{Z}{2a_0}\right)^{1/2}\left(\dfrac{Zr}{a_0}\right) e^{-Zr/2a_0}$	$\left(\dfrac{3}{4\pi}\right)^{1/2}\cos\theta$	$2pz$
4.	2	1	+1	$\dfrac{1}{\sqrt{3}}\left(\dfrac{Z}{2a_0}\right)^{1/2}\left(\dfrac{Zr}{a_0}\right) e^{-Zr/2a_0}$	$\left(\dfrac{3}{4\pi}\right)^{1/2}\sin\theta\cos\phi$	$2px***$
5.	2	1	−1	$\dfrac{1}{\sqrt{3}}\left(\dfrac{Z}{2a_0}\right)^{1/2}\left(\dfrac{Zr}{a_0}\right) e^{-Zr/2a_0}$	$\left(\dfrac{3}{4\pi}\right)\sin\theta\sin\phi$	$2py***$

Table: 1.4: Hydrogen-Like Species Wave Functions for $n = 1$ and $n = 2$.
[Adapted From : Sharp AG: Inorganic Chemistry, 3rd. edn., Pearson New Delhi, 2007]

2.4.4. *Mathematical Exactness of One-Electron System Vs. 1s, 2s, and 2p Orbitals of Hydrogen-Like Species*

In the recent past, there prevails an emerging trend amongst the *chemists* in terms of the following *two* critical aspects:

- **rather difficult to conceptualize the so-called mathematical statements of the *wave functions*; and**
- **preference to visualize an *orbital* possessing essentially a '*single boundary surface*'****.**

Fig: 1.10 illustrates some boundary surfaces pertaining to the specific *angular parts* with respect to **1s and 2p wave functions** for the *hydrogen atom*.

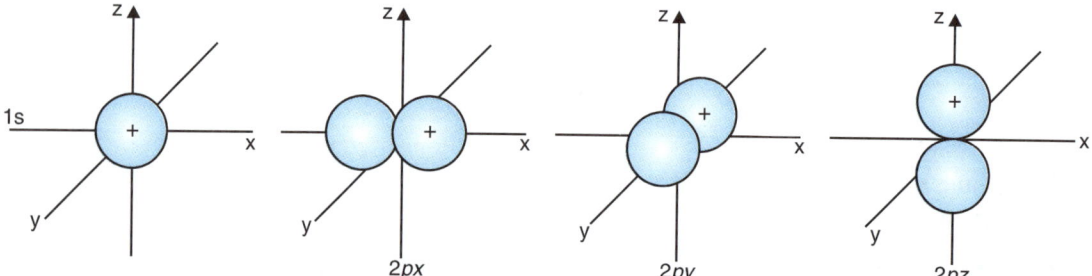

Fig. 1.10: Diagramatic Representation of the Boundary Surface for Angular Parts of 1s and 2p Wave Functions for the Hydrogen Atom.

---

* $a_0 = \varepsilon_0 h^2/\pi m e^2 = 0.529\text{Å}$

**   Z: It refers to the **Nuclear Charge.**

***   The angular portions of these **wave function** as Stated above are, in fact, normalized linear combination of other Solutions. However, this practice is being followed since it aids to emphasize that the **2p orbitals** do actually differ in orientation.

****   That is, a certain fraction (mostly ~ 90%) of the *total electronic charge* is very much held in it (*i.e., orbital*).

Therefore, in order to emphasize categorically that ψ designates a *continuous function* being the so-called *extended surfaces up to the nucleus*. However, particularly for the *p-orbitals* it may not be quite true in a situation when we intend to take into consideration the exact location of merely 90% of the total *electronic charge*. Based on this logical reasoning certain authors do prefer to designate a *p-orbital* by *two* **spheres** (see Fig: 1.10) – that, of course, **do not touch each other**.

Besides, the explicit *boundary surfaces* (in Fig: 1.10) relate particularly to **1s and 2p wave functions**, whose radial segments are devoid of '**nodes**' completely. Hence, for the **2s and 2p wave functions**, there are definite visible **changes of sign** very much within the '**envelopes**'.

2.4.5. Boundary Surfaces for the Angular Probability Functions of 1s and 2p Orbitals for Hydrogen Atom

As we have already made use of the typical **radial distribution function** to designate squarely the ensuing probability of finding an '*electron*' positioned at a distance '*r*' from the '**nucleus**', we may also employ effectively the function $\psi_\theta^2 \psi_\phi^2$ to symbolize the overall probability in terms of θ and ϕ.

It is, however, pertinent to state here that :

"**for an '*orbital*' – the squaring of the *angular segment* of the '*wave function*' virtually affords absolutely no change in the spherical symmetry, but for the p-orbitals the observed '*envelopes*' appear more elongated**".

Fig: 1.11 depicts the *specific boundary surfaces as seen for the angular probability functions of* **1s** and **2p** orbitals in the *hydrogen atom*.

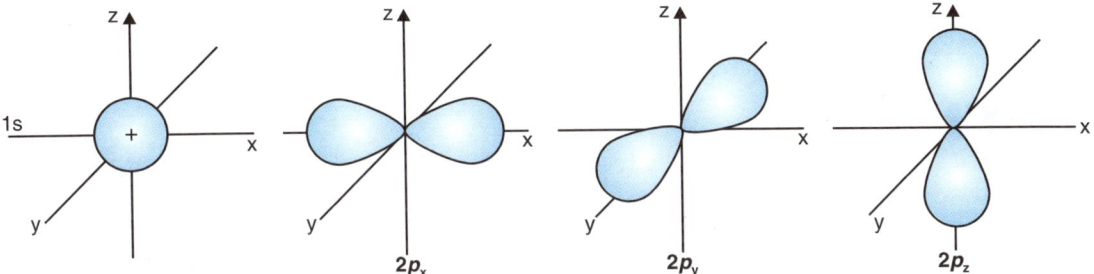

Fig. 1.11: Diagramatic Representation of the Boundary Surfaces for the Angular Probability Function of 1s and 2p Orbitals for the hydrogen Atom.

NOTE   **It may be noted clearly that positive and negative signs do disappear necessarily.**

Concept of Orbitals – In usual practice, there prevails a common perception by the '**chemists**' to construct (draw) **envelopes** (*i.e.*, the boundary of *the family of curves obtained by varying a parameter of a wave*) of varying *shapes* more closely designating the probabilities, – but scripts on them the respective **signs of the wave functions** themselves. Therefore, these '**envelopes**' and '**wave functions**' are invariably termed as the '**orbitals**'.

3D-Model: Radial and Angular Probabilities – Importantly, a full representation of the intricacies of *electron distribution probability* broadly comprises:

"**radial and angular probabilities and also needs a 3D-model wherein the observed variation in ψ^2 may be depicted by the variation in 3D – of the ensuing density of a fog designating ψ^2.**"

2.4.6. Boundary Surfaces for Angular Parts of 3D-Wave Function for the Hydrogen Atom

It has been duly proven and established that the '*d*' **radial functions** are found to be almost identical for all the *five* **members of a** group of *nd-orbitals*. However, the observed '**angular functions**' do vary perceptively as could be seen in Fig: 1.12.

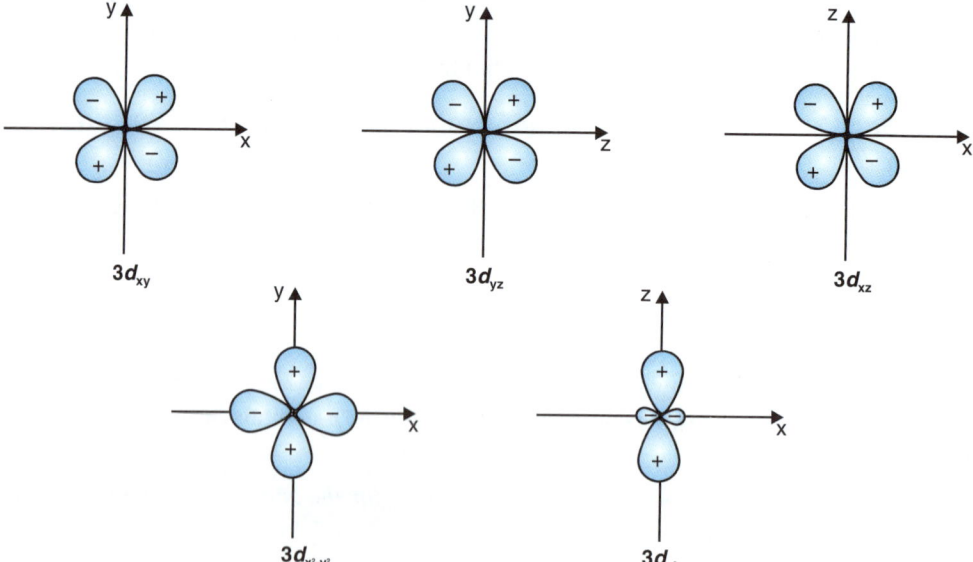

Fig. 1.12: The Boundary Surfaces for Angular Parts of the 3D-Wave Functions on the Hydrogen Atom.

Explanation

1. The d_{xy}, d_{xz} and d_{yz} orbitals do lie specifically in the *xy*, *xz*, and *yz* planes, having the 'lobes' midway between the Cartesian axes.

2. Besides, the d_{x^2} – **orbital** represents a *normalised linear combination* of:

    * $a\, d_{x^2 - x^2}$; and

    * $a\, d_{x^2 - y^2}$ **orbital,**

    i.e., each of them is analogous to the respective $d_{x^2 - y^2}$ orbital. It essentialy comprises of *two* **major lobes** that shows clearly:
    *   **directed along the z-axis; and**
    *   **having a cylindrical collar around the z-axis.**

3. The main reason attributed for allowing the combination of the ensuing 'orbitals' $d_{z^2-x^2}$ and

    $d_{z^2-y^2}$ being that: "**though there exists in all** *six* **wave functions which may be assigned to** '*orbitals*' **(having the usual 4-lobed form), there may be only '5n orbitals' with any kind of** a physical reality; and hence, based on the recognized convention that $d_{z^2-x^2}$ and $d_{z^2-y^2}$ are duly combined".

NOTE   **Obviously, the observed difference between the *three 'orbitals'* being directed between the *Cartesian axes* and the *two 'orbitals'* directed along them is certainly of critical fundamental importance in the so-called – *'chemistry of the transitional elements' viz.,* Lanthanum (La).**

2.4.7. *The Angular Momentum* and the Inner Quantum Number***

In a broader perspective, the value of the quantum number '*l*' virtually determins these important aspects, namely:

- **actual shape of an *electronic orbital*,**
- **amount of orbital angular momentum linked directly with an 'electron' present in it, usually given by the expression** $\sqrt{l(l+1)}\,(h/2\pi)$).

Importantly, the '**axis**' *via* the '**nucleus**' around which the '*electron*' may be assumed to rotate ultimately *defines*:

"the observed direction of the 'orbital angular momentum vector and the inherent magnitude of this *vector* refers to that of the respective 'orbital angular momentum'.'"

It is worthwhile to state at this point in time that –

"the resulting 'orbital angular momentum' specifically yield a 'magnetic moment', –which critically exhibits the following *two* criteria, namely":

- **direction showing the '*same sense*' as that of the *vector*; and**
- **inherent magnitude remains proportional to the respective magnitude of the *vector*.**

Points to Ponder: It has been duly ascertained that an '*electron*' present in an *s*-orbital [$l = 0$] possesses 'no orbital angular momentum'; whereas, in the corresponding *p*-orbital [$l = 1$] the resulting orbital exhibit angular momentum $\sqrt{2}(h/2\pi)$, and so on so forth. Obviously, it drives us to infer that:

"the orbital angular momentum vector essentially has ($2l + 1$) possible directions in space with respect to the ($2l + 1$) observed possible values of m_1 vis-a-vis a given value of '*l*'."

Component of the '*Orbital Angular Momentum Vector*' along the z-Axis:

Having understood the *two* vital criteria viz., **angular momentum** and **inner quantum number** in an exhaustive manner we may be specifically concerned with the so-called **component of the angular momentum vector** along the z-axis; and this would explicitly possess an altogether *divergent value* for each of the *feasible* **different orientations**, – which the aforesaid **angular momentum vector** may eventually hook-on-to.

Therefore, apparently the real prevalent magnitude of the *z*-**component** may be expressed by: $m_l(h/2\pi)$. Consequently, for an '*electron*' present in a *d*-orbital, such as: $l = 2$, may give rise to the **orbital angular momentum**:

---

*   **Angular Momentum:** It refers to the product of **moment of inertia** and **angular velocity**.

*   **Inner Quantum Number:** It relates to any one of the four integers used to describe the movement of an '*electron*' inside an '*atom*'.

$\sqrt{6(h/2\pi)}$; and the **z-component** should be +2 $(h/2\pi)$, + $(h/2\pi)$, 0, –$(h/2\pi)$, or –2$(h/2\pi)$ – **that depends exclusively upon the ultimate orientation profile observed explicitly.**

Fig: 1.13 represents clearly the *two* **typical orientations of the so-called angular vector, namely:**

- **possible orientations of the *angular momentum vector* for a, d electron ($l = 2$); and**
- **possible orientations of the *spin angular momentum vector* for an 'electron' (s = 1/2).**

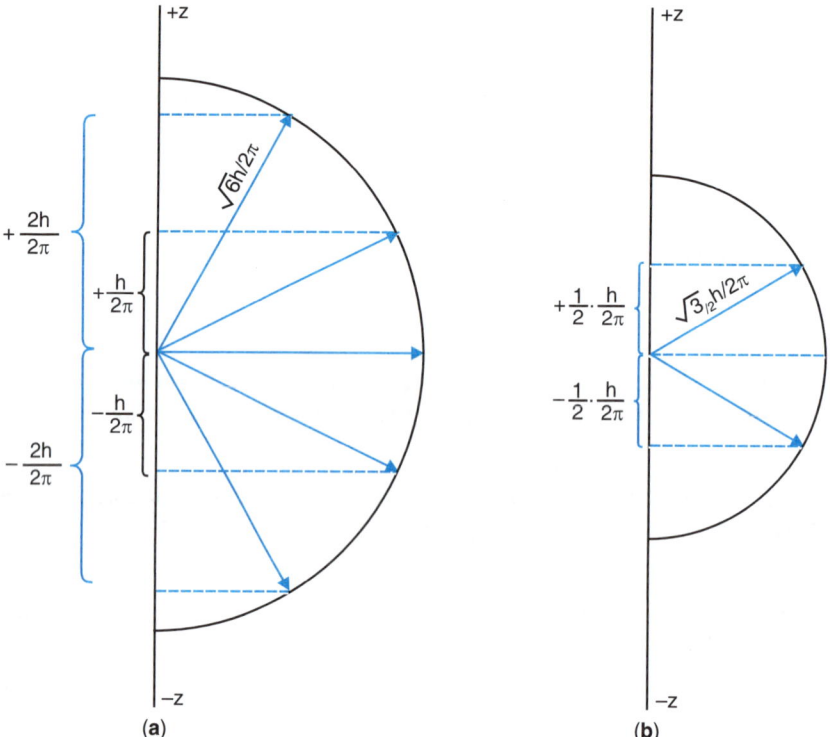

(a)   (b)

Fig. 1.13: (a) The Possible Observed Orientations of the Orbital Angular Momentum Vector for a *d* electron (*l = 2*); (b) The Possible Recorded Orientations of the Spin Angular Momentum Vector for an Electron (s = 1/2). [Adapted from: Sharpe AG: Inorganic Chemistry, 3rd edn., Pearson, New Delhi, 2007]

Explanations: These essentially include:

1. It may be seen that an '***electron***' also inherits the *spin angular momentum* that may be eventually considered as bringing into existence the critical rotation of the '***electron***' **about its own axis.**

2. Thus, the magnitude of such a *spin angular momentum* may be expressed as given below:

$$\boxed{\sqrt{s\,(s+1)\,(h/2\pi)}}$$

where, s = Spin quantum number [having a value +1/2 for all electrons]

3. The inherent **axis** (+ z to – z) defines clearly the direction of the **spin angular momentum vector**; however, the ensuing *orientation of the vector* pertaining to the respective z-axis and the underlying magnitude of the component of this *vector* having the z-**direction** would be specific point of concern to an *inorganic chemist*.

4. The resulting component (step 3) may be given by the following expression:

$$\boxed{m_s\,(h/2\pi)}$$

where, m_s = Spin magnetic quantum number [it would be $+s$ or $-s$ *i.e.*, $+1/2$ or $-1/2$]

5. Thus, there are exclusively *two* **observed possible orientations** with respect to the prevailing **spin angular momentum vector**. Hence, these factors ultimately yield the so-called prevalent **z-components** having the magnitude:

   - $+1/2\,(h/2\pi)$ and   • $-1/2\,(h/2\pi)$, as depicted in Fig: 1.13(b).

6. Importantly, for an '*electron*' possessing the following *two* **criteria**, namely:
   - **orbital angular momentum**; and
   - **spin angular momentum,**

any may arrive at the *total* **angular momentum vector** given by the following expression:

$$\boxed{\sqrt{j(j+1)}\,(h/2\pi)}$$

where, j = Inner Quantum Number [which could be $l - s$, *i.e.*, $l - 1/2$]

Thus, the component z related to the *total* **angular vector** now turns out to be $j(h/2\pi)$; and hence, there are $(2j + 1)$ *possible orientations in space*.

3. WAVE MECHANICAL CONCEPTS

Preamble: The great innovative ideas and concepts amalgamated with noticeable breakthroughs primarily introduced by **Planck** and **Einstein** suggests indeed the so-called '**particle theory of light**', besides the '**wave theory of light**' required necessarily by such glaring phenomena as:
   - **interference**; and
   - **diffraction.**

De Broglie (1924) put forward the dictum vehemently that:

"in a specific instance when '*light*' were made up of particulate matters (particles) and even then exhibited *wave-like* characteristic features; then the same must hold good for both '*electrons*' and '*particles*'."

As a result, the so-called **electromagnetic radiation** one may observe the crucial relationship occurring as a combination of: **Planck's constant** [$E = h\nu$] with **Einstein's equation** [$E = mc_0^2$] which ultimately leads to the following expression:

$$\boxed{m = \frac{h\nu}{c_0^2} = \frac{h}{c_0^\lambda}}$$

thereby establishing a *definite likage* between the following *two* properties of a **photon**, namely:

   - **Mass (m)** and   • **Wavelength (λ)**

Therefore, the **above equation** was now being suggested strongly to uphold and apply to any particulate matter having **mass (m)** if C_0, provided the velocity of the ensuing electro-magnetic radiation, was duly **replaced by the particle** itself.

Thus, by rearrangement of the *above equation,* we may have:

$$mv = \frac{h}{\lambda}$$

Interestingly, in the *above derived equation* the **left** and **right-hand sides of the said relationship** do designate the following *two* entities respectively:

- **stress particulate matter (particles)**; and
- **observed wave behavioural profile.**

Points to Ponder: These essentially include:

1. The **greater** the *mass* and *velocity of the particle*, the *shorter* would be the observed **wavelength** (λ).

2. Because, the wavelength intimately associated with any *macroscopic* **particulate mater** is found to be perceptively **smaller in dimension** *vis-a-vis* the respective dimensions of any **prevailing physical system**.

3. Besides, one must take cognizance of the glaring fact that either:
   - **diffraction phenomena**, or
   - **wave phenomena**,

   could never be observed with a '*macroscopic particulate matter*'.

4. Nevertheless, the '*electrons*' duly accelerated particularly in the following *two* instances, namely:
   - **at a velocity of 6×10^6 m 8^{-1} by a potential difference of ~100 V do exhibit a *wavelength* (λ) of ~ 1.2Å; and**
   - **recorded diffraction pattern of such electrons by in an ensuing crystal does provide a significant support for the very** *concept of the electron as a wave*.

3.1. Molecular Orbital Method

The '**molecular orbital**' refers to the resultant orbital arising from the overlapping of *atomic orbitals* of two individual *atoms* to result into the formation of a '**covalent bond**'.

It is, however, pertinent to state here that in the so-called *electron-pair theory* (or the *valence bond theory*), it is believed that a '**molecule**' is often:

"regarded to be comprised of atoms".

Obviously, the '*electrons*' embodied in the *atoms* do take possession of the '**atomic orbitals**', irrespective of the fact whether these may or may not undergo the phenomenon of *hybridization**.

❑ **Hybridized Atomic Orbitals** – In this specific situation, the atomic orbitals so formed – *duly from the same atom* do combine to give rise to the formation of the desired **hybrid orbitals**. In this manner, the resulting **hybrid orbitals** do have a tendency to overlap rather more efficiently with the *orbitals* emanated from *other atoms*, – thereby generating the '**stronger bonds**' perceptively. Hence, the **atomic orbitals** (or even the *hybrid orbitals*) are believed to remain very much intact even when the atom gets eventually **bonded chemically in a specific molecule**.

---

* **Hybridization:** It refers to a phenomenon wherein the **lower energy orbital electrons** are slightly elevated to the *next higher levels* and correspondingly the **higher energy electrons** do assume a *slightly lower level* thereby forming a **now energy level** for *all electrons involved in the shifts (ie.,* an explanation for the tetravalency of C-atom).

❑ **Formation of Molecular Orbitals** – Based, on the underlying postulates of the **molecular orbital theory** the so-called '*valency electrons*' are regarded to be closely associated with *all the nuclei* present strategically in the *entire molecule*'. In this manner, the *atomic orbitals* derived 'from **different atoms**' should be combined meticulously to produce the '**molecular orbitals**'.

In true sense, the '*electrons*' may be looked at either:

   • **as particles** or   • **as waves**

Thus, an '**electron**' critically present in an atom may be described elegantly take possession of either:

   • **an '*atomic orbital (AO)*'**, or

   • **by a '*wave function (ψ)*'**

that ultimately provides a comprehensive logic solution to the ***Schrödinger wave equation*** (see *Section 2.2*). In other words, the '*electrons*' present in a *molecule* are assumed to be taken up in the '**molecular orbitals**'.

The '**wave function**' that invariably describes a '***molecular orbital***' may be duly accomplished one of the *two* following cited procedures, namely:

   ❑ **Linear Combination of Atomic Orbitals [LCAOs]**, and

   ❑ **United Atom Method [UAM]** (*which is beyond the scope of discussion in the present context*).

3.1.1. *Linear Combination of Atomic Orbitals [LCAOs] Method*

What is the LCAO method ? The *linear combination of atomic orbitals (LCAOs) method* may be expatiated by considering two altogether separate atoms **X** and **Y** – that do inherit the '*atomic orbitals*' described duly by the *wave functions:* ψ_x and ψ_y. Now, let us consider a specific situation when the so-called *electron clouds** of these *two* atoms **X** and **Y** actually **overlap when the atoms approach each other** in the usual manner. Thus, the corresponding **wave function** for the molecule (*i.e.*, **molecular orbital ψ_{XY}**) may be duly accomplished by a *linear combination* of the ensuing *atomic orbitals* $\psi_{(x)}$ and $\psi_{(y)}$.

Thus, we may have the following expression:

$$\psi_{(XY)} = N(c_1\psi_{(X)} + c_2\psi_{(Y)})$$

where, N = Normalizing constant selected to ensure that the probability of finding an '*electron*' in the entire available space is unity; and

c_1 and c_2 = Constants so chosen to attribute a *minimum energy* level for $\psi_{(XY)}$.

Now, if the *two* atoms **X** and **Y** are perfectly *identical*, both **C_1 and C_2** should possess the *same values*. Obviously, when the two atoms **X** and **Y** are more or less exactly the same both C_1 and C_2 are also equal.

Wave Function (ψ) Squared – It has been duly ascertained that the probability of finding precisely an '*electron*' **in an assigned (given) volume of space**, *dv*, is found to be $\psi^2 dv$; and, therefore, the ensuing *probability density* profile for the observed combination of *two* **atoms X and Y** (as stated earlier) gets duly related to the respective **wave function squared**, and may be expressed as under:

$$\psi^2_{(XY)} = (c_1^2\psi^2_{(X)} + 2c_1c_2\psi_{(X)}\psi_{(Y)}) + c_2^2\psi^2_{(Y)}$$

---

* That is, one lies above the plane of the atoms, and another *electron cloud* that lies **below the plane of the aloms**,–therby giving a **π-cloud** both above and below the plane.

Explanation: The above expression of the **wave function squared** may be explained as stated under:

1. Let us critically examine the *three* **terminologies** as stated above on the *right-hand-side (RHS)* of the equation:

   ❑ **First Term** $[c_1^2 \psi_{(X)}^2]$ – is intimately related to the probability of finding an '*electron*' located on atom **(X)** **(provided 'X' designates an isolated atom)**.

   ❑ **Third Term** $[c_2^2 \psi_{(Y)}^2]$ – is closely related to the probability of finding an '*electron*' on atom **Y** if **Y** is found to be an **isolated atom**.

   ❑ **Second (Middle) Term** – actually becomes important specifically and progressively since the prevailing overlap between the *two* inherent **atomic orbitals** registers a definite increment; and hence, this terminology is mostly known as the '*overlap integral* '.

 NOTE   **Hence, the term 'overlap integral' actually designates the so-called major difference prevailing between the *electron clouds* in separate atoms and in the molecule itself, *i.e.*, the larger the value of the said '*term*' the stronger would be the bond.**

3.1.2. *Combination of Orbitals*

In usual practice, we may come across the following *five* **variants** in the *combinations of orbitals*, namely:

- *s–s* **Combinations of Orbitals,**
- *s–p* **Combinations of Orbitals,**
- *p–p* **Combinations of Orbitals,**
- *p–d* **Combinations of Orbitals,** and
- *d–d* **Combinations of Orbitals,**

which shall now be discussed individually in the sections that follows:

3.1.2.1. *s–s Combinations of Orbitals*

Let us assume that the *two* atoms **X** and **Y** are duly designated by the **H-atoms**; now the respective **wave functions** $\psi_{(X)}$ and $\psi_{(Y)}$ do represent the *1s* **atomic orbitals (AOs)** located on the *two* said atoms **X** and **Y** . In such a situation, we may observe two distinct (AOs) located on the *two* **distinct combinations** pertaining to the **wave functions** $\psi_{(X)}$ and $\psi_{(Y)}$, namely:

- **a specific instance** – where the signs of the *two* **wave functions remain the same**; and
- **a particular case** – where the signs of the *two* **wave functions are altogether different from each other.**

A Logical Assumption – In case, one of the **wave functions**, say '$\psi_{(X)}$' is being assigned arbitrarily a **+ve sign**; whereas, the other could be either +ve or –ve. At this point in time, we may come across *two* **divergent situations with respect to the 'waves'**, namely:

   ❑ **Waves that remain-in-phase** – *i.e.*, such *wave functions* that critically do possess the '*similar sign*' may be regarded as **waves that remain in phase**; and that when get combined eventually '*add up to render a larger – resultant wave*', – as shown in Fig: 1.14;

   ❑ **Waves that remain out-of-phase** – Likewise, the *wave functions* with altogether *different signs* with respect to the *waves* which are **totally out-of-phase**; and that cancel each other by means of '**destructive interference***,

---

* $\psi_{(g)}$ and $\psi_{(u)}$: They represent *a pair of* molecular orbitals.

Interference – relates to the interaction of two or more *wave motions* affecting the same segment of a medium so that the instantaneous disturbances in the *resultant wave* are equivalent to the algebraic sum of the instantaneous disturbances in the so-called *interfering waves*. Besides, when the crest of one wave meets the trough of another wave of equal amplitude, the *wave* gets destroyed at that point; conversely, the superimposition of one crest upon another leads to an increased affect ultimately.

Thus, we may have the following *two* **combinations:**

$$y_{(g)}^* = N[\psi_{(X)} + \psi_{(Y)}]$$

and

$$y_{(u)}^* = N\{\psi_{(x)} + [-\psi_{(Y)}]\} \equiv N\{\psi_{(X)} - \psi_{(Y)}\}$$

Comment: The latter equation must be considered to be the '**summation of the wave functions'**; and hence, not be regarded simply as the *mathematical difference between them*.

Formation of a Pair of Molecular Orbitals – It has been duly observed that whenever a **pair of atomic orbitals (AOs)** viz., $\psi_{(X)}$ and $\psi_{(Y)}$ combine together, they help in the formation of a pair of molecular orbitals $\psi_{(g)}$ and $\psi_{(u)}$. However, it may be noted that:

"**the number of** *molecular orbitals* **(MOs) thus generated should always be equal to the exact number of** *atomic orbitals* **(AOs) that are being involved intimately."**

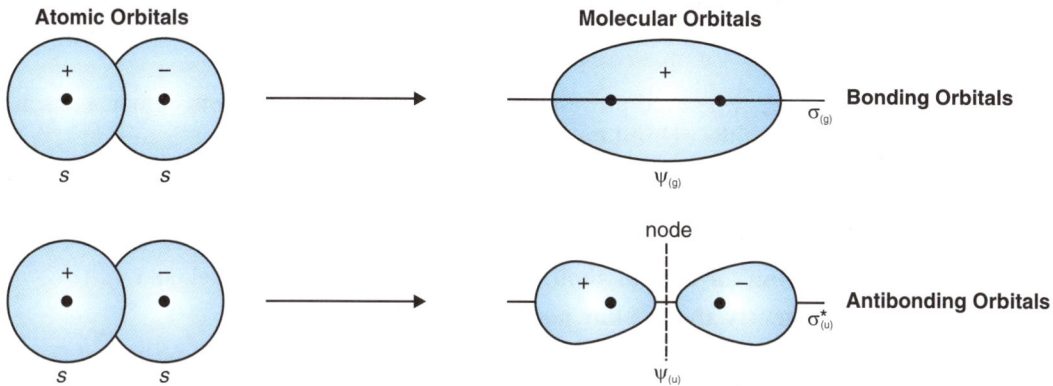

Fig. 1.14: The Diagrammatic Representation of *s-s* Combinations being Observed in Atomic Orbitals (AOs).

Bonding Molecular Orbitals – It may be observed that the ensuing **function** $\psi_{(g)}$ virtually leads to an *enhanced electron density* profile occurring between the *two* **nuclei**; and hence, designates the **bonding molecular orbital.** Obviously, it happens to be possessing a little **low-energy level** *vis-a-vis* the **actual (original) atomic orbitals (AOs).**

    Contrarily, $\psi_{(u)}$ is responsible for attributing *two* **lobes with opposite signs thereby cancelling each other**; and, therefore, giving rise to the evolution of '*zero-electron density*' prevailing in between the *two* nuclei. Amazingly, it results into an **anti-bonding molecular orbital** – that inherits a *higher energy level* as depicted in Fig: 1.15.

---

  * **That is, the signs +ve and –ve do** refer to the *signs of the wave functions*, that eventually **determine their symmetry**; and hence, have nothing to do with the inherent **electrical charges**.

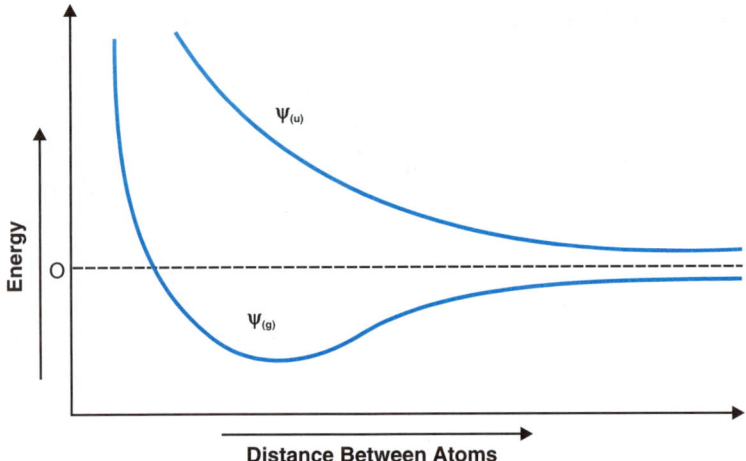

Fig. 1.15: The Energy Profile of $\psi_{(g)}$ and $\psi_{(u)}$ Molecular Orbitals (MOs).

Explanation

1. Nationally, the prevalent '**molecular orbital wave functions**' are commonly designated as: $\Psi_{(g)}$ and $\Psi_{(u)}$

   where, g = stands for '*gerade*' (means **even**); and

   u = stands for '*ungerade*' (means **odd**).

   Thus, both g and u invariably relate to the so-called ensuing **symmetry of the orbital** *about its centre.*

2. In a specific instance, when the prevailing '**sign of the wave function**' remains absolutely *unchanged* by virtue of the **orbital being reflected about its centre** (that is to say: x, y, and z are duly replaced by $-x$, $-y$, and $-z$), the orbital is said to be **gerade** (or **even**).

3. We may also use an *alternative procedure* for determining precisely the so-called *symmetry of* **the molecular orbital** by adopting the following *two* ways, namely:

   - **by rotating the orbital about the line joining the *two* nuclei**; and

   - **subsequently about a line perpendicular ($\perp$) to it.**

Inference: In case, the *sign* upheld by the *lobes do remain the same*, the resulting *orbital* is **gerade** (**even**); and if the **sign** alters, the accomplished orbital is **ungerade** (**odd**).

Energy Levels of s–s Atomic and Molecular Orbitals

It is worthwhile to state here that the observed energy level profile of the **bonding molecular orbital** $\psi(g)$ **passes** meticulously *via* a **minimum trough** (see *Fig: 1.15*).. Hence, the actual distance between the *two* atoms **X** and **Y** at this critical point duly corresponds to:

"**the internuclear distance between the atoms X and Y**",

when they ultimately gives rise to the formation of a '**bond**'. At this particular stage, let us take into considerations the   so-called inherent **energy levels of the two 1s atomic orbitals**, and also of the following *two* orbitals, such as:

   - **bonding $\psi_{(g)}$ orbitals**; and

   - **antibonding $\psi_{(u)}$ orbitals,**

which have been explicitly illustrated in Fig: 1.16.

Stabilization Energy – It relates to the ensuing energy possessed by the **binding molecular orbital** which is observed to be lower in comparison to the respective **atomic orbital** by an *amount* Δ.

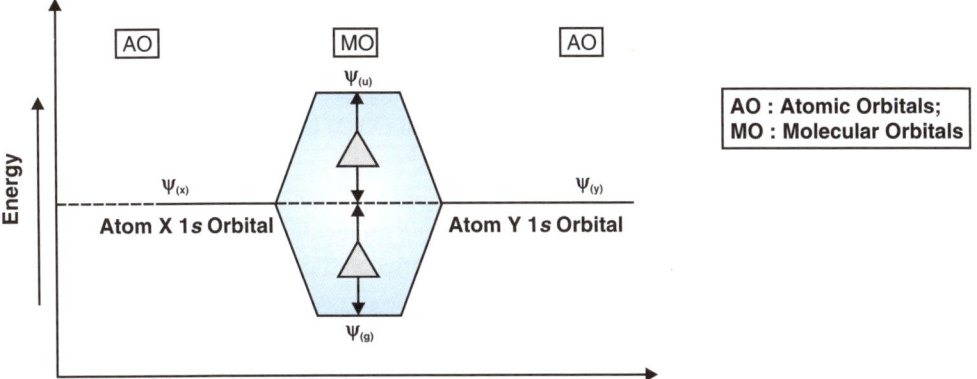

Fig. 1.16: Diagrammatic Representation of the Energy Levels of *s–s* Atomic and Molecular Orbitals.

Explanation

1. The overall observed energy derived from the so-called *antibonding molecular orbital* gets enhanced by Δ.

2. However, the **atomic orbitals (AOs)** do have a tendency to hold up to *two* **electrons***, and amazingly, the same holds good to the corresponding **molecular orbitals (MOs)**.

3. In another instance, when *two* **H-atoms get combined**, there may be *two* **electrons** that could be taken into consideration:
   - **one from the 1s– orbital of atom X**; and
   - **one from the 1s–orbital of atom Y**.

4. As and when the aforesaid **1s-orbitals** of the *two* atoms (**X** and **Y**) are combined, – these **2-electrons** together take possession of the **bonding molecular orbital** $\psi_{(g)}$. In fact, it causes an apparent saving of energy up to 2Δ, that eventually corresponds to the ultimate **bond energy profile** perceptively.

NOTE | **Perhaps it could be the reason justifying the fact how the system gets stabilized owing to the critical formation of the 'bond'.**

Remarks: Based on the foregoing scientific evidences and logical explanations one would certainly appreciate the involvement of '**certain further symbols**' in order to *expatiate* and *necessitate* the manner in which the **overlapping** of **atomic orbitals (AOs)** come into play. Thus, we may encounter the following *two* variants in the '**molecular orbitals (MOs)**', namely:

➤ **σ-Molecular Orbitals** – formed by overlapping of the orbitals along the axis joining the *two* **nuclei**; and

➤ **π-Molecular Orbitals** – generated by the so-called lateral overlapping of the atomic orbitals.

---

* That is, if they do have *opposite spins*.

3.1.2.2. *s–p Combinations of Orbitals*

Importantly, the specific combination of an **σ-molecular orbital** with a *p*-orbital takes place provided the **lobes of the *p*-orbital do point towards the axis joining along the nuclei**.

Emergence of Two Different Situations – *i.e.*, at this stage we usually observe the crucial emergence of *two* altogether different situations, namely:

❑ **First Situation** – when the emerged lobes that actually overlap do essentially exhibit the '***same sign***' – thereby giving a **bonding molecular orbital (MO)** having a distinct **enhanced electron density profile** prevailing between the nuclei; and

❑ **Second Situation** – when the observed '***overlapping lobes***' do exhibit the *opposite signs* – thereby giving rise to the formation of an ***antibonding* molecular orbital (MO)** having a definitely **minimized electron density pattern** occurring between the nuclei.

Fig: 1.17. depicts the diagramatic representation of the *s–p* **combination of the atomic orbitals (AOs)**:

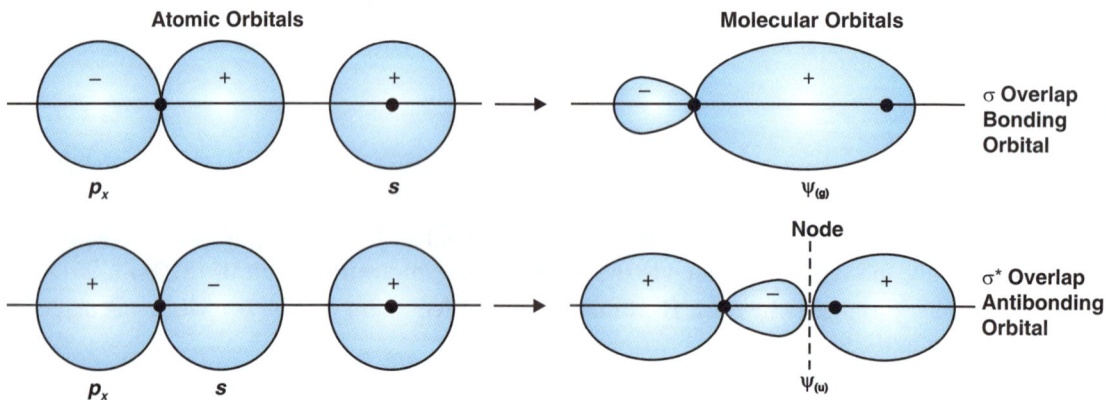

Fig.1.17: Diagrammatic Representation of the *s–p* Combinations of Atomic Orbitals (AOs).

3.1.2.3. *p–p* Combinations of Orbitals

First and foremost let us consider the combinations of *two p-orbitals*, both of them do have lobes that usually point very much along the axis linking the *two nuclei*. Interestingly, the above episode results into the production of:

• **a bonding molecular orbital (MO)**; and

• **an antibonding molecular orbital (MO)**,

explicitly which have been duly shown in Fig. 1.18.

Fig: 1.18 clearly represents the so-called *p–p* combinations of the **atomic orbitals (AOs)**.
Now, let us take into consideration the critical **combination of *two p-orbitals**, – that both essentially possess '***lobes***' which are found to be **perpendicular (⊥)** to the *axis* linking the *two nuclei*. Interestingly, such an event would cause an exclusive '***lateral overlap***' of the ensuing orbitals thereby resulting in the formation of the so-called:

• **π-bonding**   and   • **π*-Antibonding**,

molecular orbitals (MOs), – as illustrated in Fig: 1.19.

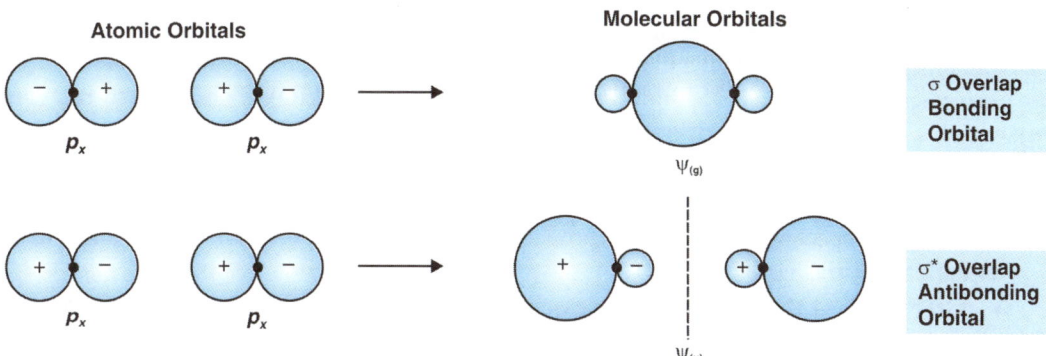

Fig. 1.18: Diagramatic Representation of *p–p* Combination of the Atomic Orbitals (AOs).

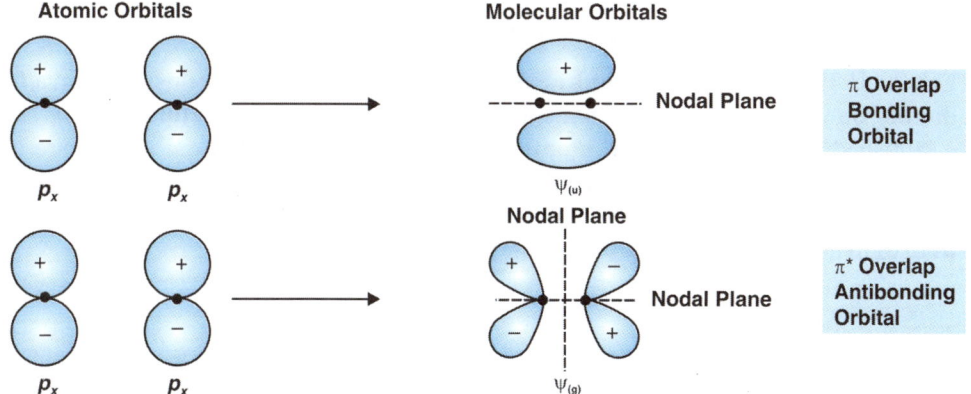

Fig. 1.19: Diagramatic Representation of *p–p* combinations Giving Rise to the Formation of π-Bonding.

Points of Difference between the *p–p* Combination MOs and the σ-Orbitals – Following are the *three* vital and important **points of differences between the *p–p* combination molecular orbitals (MOs) and the σ-orbitals, namely:**

(a) **The π-Overlap** – Importantly, the so-called **π-overlap** to come into effect the **lobes of the ensuing atomic orbitals (AOs)** (see *Fig: 1.19*) are observed to be *absolutely perpendicular* (⊥) **to the corresponding line** linking the *two* **nuclei.**

Of course, this is indeed quite contrary to the **σ-overlap** wherein the corresponding **lobes point specifically along the line linking the *two* nuclei.**

(b) **The π-Molecular Orbitals (MOs)** – In this instance, *i.e.*, for the **π-molecular orbitals, ψ*** is *zero* as could be seen along the **internuclear line** (*i.e.*, the line joining the *two* **nuclei**) and as a result the ensuing **electron density, ψ^2,** also becomes 'zero'.

NOTE   Amazingly, this is an absolute contrast to the respective σ-orbitals.

(c) **The Symmetry of π-Molecular Orbitals (MOs)** – However, it may be observed distinctly that the **symmetry of π-molecular orbitals** is quite divergent right from the one shown by the **σ-orbitals.** Now, in a typical instance when the so-called *bonding p-molecular orbital (MO)* is rotated carefully about the '**internuclear line**', one may observe a change in the '*lobes*'.

---

* **ψ:** It refers to the amplitude of vibration for the '*wave*' in '**Wave Equation**'.

Remarks: Therefore, the **π-bonding orbitals** do designate particularly the '*ungerade*' (**odd**) status; whereas, all the rest **σ-bonding molecular orbitals (MOs)** do represent specifically the '*gerade*' (**even**) status. Conversely, the so-called **antibonding π-molecular orbital (MO)** is observed to be '*gerade*' (**even**); whereas, the corresponding **σ-antibonding molecular orbitals (MOs)** as the '*ungerade*' (**odd**). **Salient Features in *p–p* Combination of Orbitals** – A few **salient features in *p–p* combinations of orbitals** are expatiated as under:

1. The overall impact of **π-bonding** has proved to be extremely important in the domain of **organic chemical entities** *viz.*,
   - **CH$_2$ = CH$_2$ (Ethene)** – which has essentially one **σ-bond** and one **π-bond** located between the *two* C-atoms;
   - **CH ≡ CH (Ethyne)** – which bears **one σ-bond** and **two-π-bonds**;
   - **CO$_2$ (Carbon dioxide) and CN (Nitrile)** – *i.e.*, an array of *inorganic compounds*; and
   - **C$_6$H$_6$ (Benzene)** – a liquid *aromatic hydrocarbon*.

2. **Ethene [CH$_2$ = CH$_2$]: Important Features** – Interestingly, **ethene** does contain a *localized* **double bond (olefinic bond)** that solely involves the **2-carbon atoms**. It has been duly established* that the **2-carbon atoms** and the **4-hydrogen atoms** in **ethene** are certainly *coplanar* (*i.e.*, they lie in the same plane); and the inherent *bond angles* are ~ **120°**. Besides, each **C-atom** makes use of its **2s and 2p orbitals** to give rise to the formation of 3 *sp^2*– **hybrid orbitals** which actually form the *σ-bonds* with:
   - **other C-atom**, and
   - *two* **H-atoms**,

thus, the remaining *p***-orbital** located strategically on each **C-atom** is found to be at *right angles* to the respective **σ-bonds** duly formed so far.

Valence-Bond Theory – Importantly, in the **valence bond theory**, (*i.e.*, the concept that the power of atoms to bind together to form a molecule is primarily based upon the exact number of orbital electrons present in their outer shells), these *two p***-orbitals** do have a tendency to overlap sideways to produce a **π-bond** (a *weak bond*). Obviously, the ensuing *sideways overlapping phenomenon* is not so prominent and prevalent *vis-a-vis* the end-to-end overlap seen in the **σ-bonds**. Perhaps this could be reason why the **C=C olefinic bond** although appears to be stronger in comparison to a C–C covalent **(single) bond**, does not prove to be *twice as strong.***

Furthermore, the molecule bearing C–C bond 'ethane' may be twisted about the C–C covalent bond easily; whereas, it may not be twisted in 'ethene' (having the **C=C olefinie bond**) because it would possibly minimize significantly the *quantum of π-overlap.*

Molecular-Orbital Theory – Hence, under this the logical explanation pertaining to the so-called **π-bonding phenomenon** appears to be slightly different. It may be further expatiated by considering the **two *p*-orbitals** that are involved intimately in accomplishing the **π-bonding combine process** to give **two π-molecular orbitals (MOs)** of which:
   - **one designates bonding phenomenon**; and
   - **one represents antibonding phenomenon.**

---

*   By means of several '**experimental measurements**'.

** That is, C-C in 'ethane' is 346 kJ. mol^{-1} and C=C in 'ethene' 598 kJ. mol$^{-1.}$

Because, there are present only *two electrons* (which may be engaged) – they do occupy the π-bonding molecular orbitals (MOs) – as they inherit the **lower energy profile**.

 NOTE   **Certainly, the explanation based upon the *molecular orbital (MO) theory* appears to be more important in such instances where there exists the non-localized π-bonding (*i.e.*, where the π-bonding covers many atoms viz., NO_3^-, CO_3^{2-}, Benzene (C_6H_6).**

3. **Ethyne [CH≡CH]: Important Features** – In the case of **ethyne (CH≡CH)** one may observe critically that **each C-atom** makes use of the *sp-hybrid orbitals* to result into the formation of **σ-bonds** to the other **C-atom** precisely and **a H-atom**. In true sense, all of these four-atoms ultimately produce a '**linear molecule**'. Obviously, **each C-atom** possesses *two p-orbitals* positioned at *right angles to one* another; and hence, they do *overlap sideways* with the equivalent *p-orbitals* located strategically on the **other C-atom**, thereby forming *two distinct π-bonds*. Therefore, it forms a **C≡C triple bond**, that eventually proves to be *much stronger* than the corresponding **C = C olefinic bond***.

 NOTE   **In a broader perspective, one may come across the majority of *strong π-bonds* that frequently occur between the elements of the *First Short Period* in the *Periodic Table* viz., C≡C, C≡N, C = C, and C = O. Perhaps this could be possible since the *atoms* are smaller in dimension; and hence, the *orbitals* being involved are fairly compact, so it is quite feasible to lay hands on the significant overlapping of orbitals.**

3.1.2.4. *p-d Combinations of Orbitals*

It may be stated at the very outset that a *p-orbital* located on one '**atom**' may eventually overlap with a *d-orbital* positioned on another atom (as depicted in *Fig: 1.20*), – thereby giving rise to both:

- *bonding* combinations; and
- *antibonding* combinations.

Because, the said *two orbitals* (*viz.*, *p-* and *d-orbitals*) *fail* to point along the line joining the *two nuclei*, – the resulting overlapping should be of the 'π-type' (see Fig: 1.20).

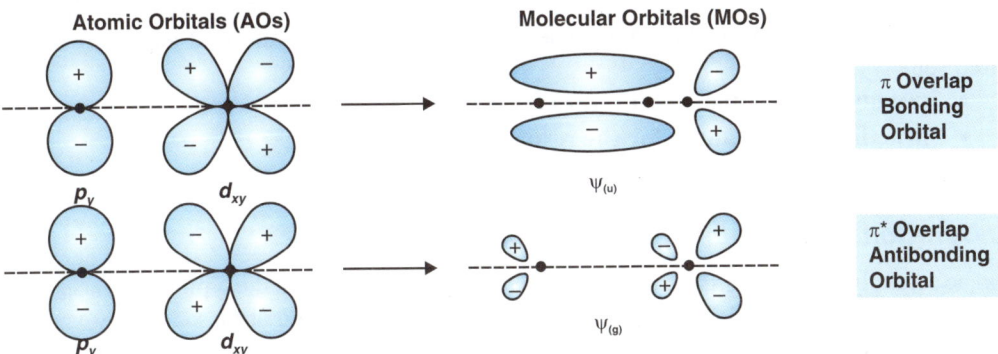

Fig. 1.20: Diagrammatic Representation of *p–d* Combinations of the Atomic Orbitals (AOs).

---

* The C≡C triple bond in '*ethyne*' is **813 kJ.mol^{-1}**.

Comment: Importantly, this kind of *bonding* is found to be solely responsible for the creation of '**short bonds**' invariably observed in the following *two* categories of elements:

- ❑ **Oxides of Phosphorus and Sulphur** *viz.*, **Phosphorus trioxide (P_3O_3); Phosphorus pentoxide (P_2O_5); Sulphur dioxide (SO_2).**
- ❑ **Transition Metal Complexes** *viz.*, **carbonyls (> C=O), and cyanides (C≡N).**

3.1.2.5. *d-d Combinations of Orbitals*

Based on wisdom, thoughtful concepts, and experimental evidences – it may be possible to:

"combine effectively two 'd'-atomic orbitals (AOs)",

thereby giving rise to the formation of both:

- • **bonding molecular orbitals (MOs); and**
- • **antibonding molecular orbitals (MOs),**

that are usually termed as **δ** and **δ*** respectively.

Fig: 1.21 shows explicitly the δ-bonding by the corresponding *d-orbitals* (*i.e.*, sideways overlapping of *two* $d_{x^2y^2}$ orbitals.

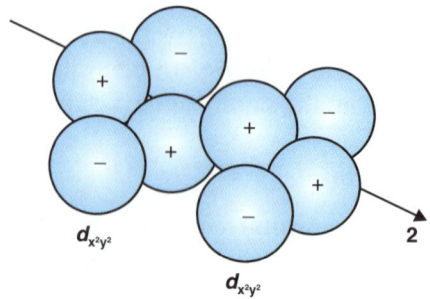

Fig. 1.21: Diagrammatic Representation of δ-Bonding by δ-Orbitals. [Showing sideways Overlap of Two $d_{x^2y^2}$ Orbitals.]

Comment: It may be observed that on *rotating these orbitals* strictly about the so-called '**internuclear axis**', the corresponding *sign* **of the 'lobes'** specifically undergo alternations at least *four* times when compared squarely with:

- • *two* **noticeable changes with π-overlap;** and
- • **practically no changes for the respective σ-overlap,**

as could be seen in Fig: 1.21.

3.1.3. *Non-Bonding Combinations of Orbitals*

It would be certainly a point of great interest to a *chemist* that practically all the typical instances exhibiting the critical overlap of the **atomic orbitals (AOs)** being taken into consideration so far duly caused either:

- • **formation of a** *bonding molecular orbital (MO)* **with a definite lower-energy profile;** or
- • **generation of an** *antibonding molecular orbital* **(MO) with a higher-energy level.**

Bonding Molecular Orbital (MO) – Therefore, to accomplish a *bonding MOs* having a distinct concentration of electron density critically observed between the *two* **nuclei,** – one may observe that:

"the signs (symmetry) of the 'lobes' – that overlap should by all means be the same."

Antibonding Molecular Orbital (MO) – Likewise, for the so-called **antibonding MOs** bearing the **signs of the overlapping lobes** should by all means be different.

Fig: 1.22 depicts certain **non-bonding combinations** of the **atomic orbitals (AOs)**.

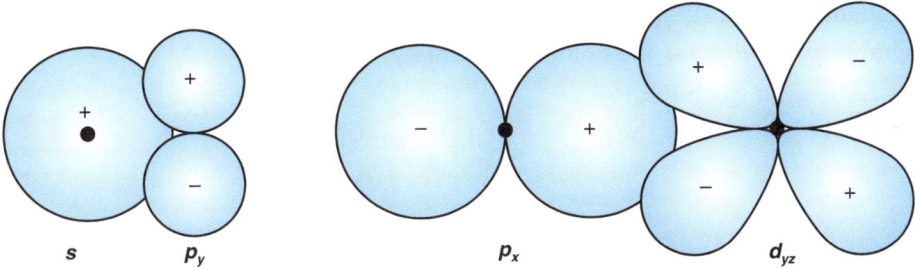

Fig. 1.22: Diagramatic Representation of Certain Non-bonding Combinations of the Atomic Orbitals (AOs).

3.1.4. *Diagrammatic Representation of the LCAO* Method*

An elaborated description related to the **diagramatic representation of the LCAO method** have been duly provided in Figs: 1.23, and Fig: 1.24 respectively.

Fig: 1.23 *i.e.*, the *first* of these descriptions we may observe that: **"the relative energies of the 1s-molecular orbitals (MOs) and their respective *constituent atomic orbitals (AOs)* are duly illustrated."**

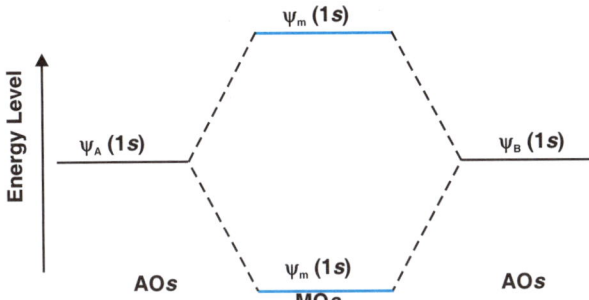

Fig. 1.23: Illustration of the Relative Energy Levels of Molecular Orbitals (MOs) and their Respective Constituent Atomic Orbitals (AOs) for H_2.

In a particular instance, when the **hydrogen molecule** is at its **ground state** (*low-energy level*), both the *electrons* essentially occupy σ_{1s} – orbital. In the *hydrogen molecule ion* (H_2^+), being duly produced on account of the action of an '*electrical discharge*' upon the **hydrogen** at low pressures. Thus, only a 'single electron' is observed duly in the orbital; and hence, the *entire bonding effect* appears to be *definitely* **smaller****.

Besides, the *bond* seen in H_2^- (*i.e.*, which has not yet been duly characterised) might be expected to be *relatively weaker* in comparison to that present in, H_2^+; whereas, there would be hardly any bonding observed at all in the so-called *hypothetical species*, H_2^{-2}.

---

* **LCAO:** Linear Combination of Atomic Orbitals, (see *Section 3.1.1*).

** That is, though the ensuing '**bond**' is still a *strong one*, the binding **energy stands at 26 kJ. cal^{-1}** *vis-a-vis* with **458 kJ. mol^{-1} for hydrogen (H_2)**

Fig: 1.24 – depicts explictly the *critical formation* of the **molecular orbitals (MOs)** for **hydrogen (H$_2$)** in *five* **different configurations,** namely:

- ψ_A and ψ_B **meant for the individual atoms;**
- ψ_m – **bonding orbital;**
- **probability function for the bonding orbital $(\psi_m)^2$;**
- ψ_m^* – **antibonding orbital; and**
- **probability function for the antibonding orbital $(\psi_m^*)^2$**

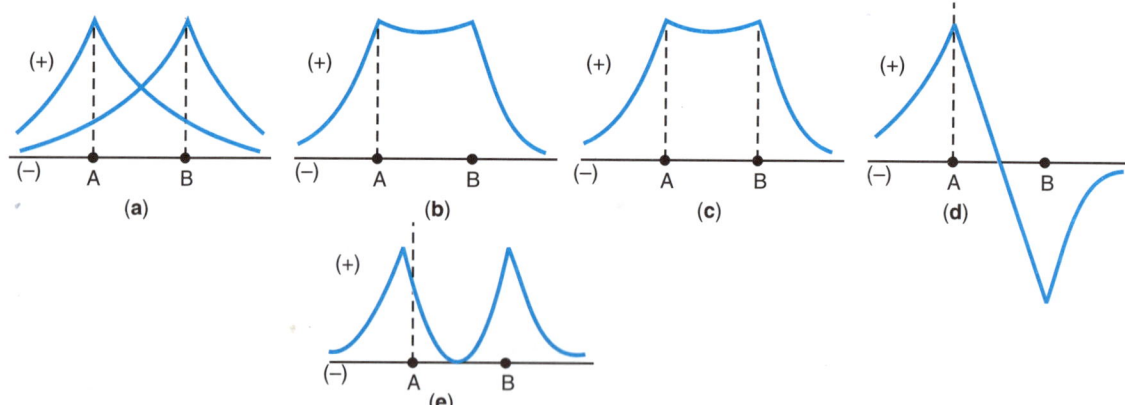

Fig. 1.24: Graphic Representation of the Formation of Molecular Orbitals (MOs) for Hydrogen (H$_2$): (a) ψ_A and ψ_B for Individual Atoms; (b) ψ_m – the Bonding Orbital; (c) Probability Functions for the Bonding Orbital $(\psi_m)^2$; (d) ψ_m^* – the Antibonding Orbital; and (e) Probability Function for the Antibonding Orbital $(\psi_m^*)^2$.

3.1.5. *Governing Rules for Linear Combination of Atomic Orbitals [LCAOs]*

There are *three* established cardinal rules that eventually govern in making a definite decision with respect to:

"which atomic orbitals (AOs) may be combined actually to form the respective molecular orbitals (MOs)", namely:

Rule – 1: The '**atomic orbitals (AOs)**' should bear approximately the *same level of energy.* It is indeed extremely important while one intends to consider the critical overlapping due between *two* **altogether divergent types of atoms.**

Rule – 2: The '**atomic orbitals (AOs)**' should by all means overlap one another to the *maximum extent.* In other words, it indicates that the *atoms* should be located reasonably close to one another to accomplish the following *two* **vital objectives:**

- **for effective overlapping; and**
- **the observed** *radial distribution functions* **pertaining to the** *two* **atoms must be absolutely identical at the ensuing distance.**

Rule – 3: To afford the generation of both '*bonding MOs*' and '*antibonding MOs*' either the *symmetry* or the *two atomic orbitals (AOs)* should **always remain unchanged** as and *when they are being rotated* either:

- **about the '*internuclear line*'; or**
- **both '*atomic*' orbitals (*AOs*) should alter their symmetry in an exactly similar manner.**

Each AO and MO Defined by Four Quantum Numbers – Interestingly in the same way we may observe vehemently that:

- each and every *atomic orbital* (*AO*) does possess a specific energy level – that may be duly defined by *four* quantum numbers; and
- each and every *molecular orbital* (*MO*) essentially has a definite energy profile – which may also be duly defined by *four* quantum numbers.

These *four* different '**quantum numbers**' are described briefly as under:

(a) **Principal Quantum Number (*n*)**– It refers to a quantum number for the so-called '*orbital electrons*', which along with the **orbital angular momentum** and **spin quantum numbers**, levels precisely the ensuing *electron-wave function*. Obviously, the prevailing *energy level profile* as well as the *average distance of an electron from the nucleus – solely depend upon the* '**principal quantum number**'.

> **NOTE** *The principal quantum number 'n' predominantly has more or less the same level of significance as could be seen in atomic orbitals (AOs).*

(b) **Subsidiary Quantum Number (*l*)** – It relates to the same degree of significance as usually observed in the **atomic orbitals (AOs)**.

(c) **Magnetic Quantum Number (m_l)** – It refers to an integer duly describing the *magnetic field* so generated by the momentum of an *electron* present in the atom. Hence, for an electron, the *quantum number* where *n* = 2, the observed **magnetic quantum number (m_l)** would be equal to: – 1, 0, or + 1.

Another school of thought advocates that the **magnetic quantum number of atomic orbitals (AOs)** be duly replaced by an altogether '*new quantum number* (λ)'. This particular analogy may be further expatiated in a **diatomic molecule**, wherin the *line joining the nuclei* is invariably considered as a '**reference direction**'; whereas, λ designates the so-called *quantization of angular momentum* in the respective *h/2π units* with reference to *this axis*. Therefore, λ actually holds the **same values** as '**m**' takes for the atoms.

Thus, we may have the following expression:

$$\lambda = -l, ..., -3, -2, -1, 0, +1, +2, +3, ..., +l$$

Here, *three* **situations** may arise, such as:

- when λ = 0: the ensuing **orbitals** are found to be *symmetrical around the axis*; and are termed as σ-**orbitals**;
- when λ = ± 1: the observed **orbitals** are known as π-**orbitals**; and
- when l = ± 2: the emanated **orbitals** are called as δ-**orbitals**.

(d) **Spin Quantum Number (m_s):** It arises due to the **spin of the electron** along its axis; and it may take up *two* **values:** $+\dfrac{1}{2}$ or $-\dfrac{1}{2}$. This is termed as the **spin quantum number (m_s)**.

Remarks: The aforesaid *four* quantum numbers viz., *n, l, m_l* and *m_s* usually render each and every electron present duly in an '**atom**' absolutely *unique* in **character and status**; and hence, one may observe predominantly that:

"**no two electrons present in an 'atom' may have the same value of all the aforesaid four quantum numbers.**"

3.2. Molecular Orbital Treatment for: (a) Homonuclear Diatomic Molecules, and (b) Heteronuclear Diatomic Molecules

In a broader perspective, we may look into certain specific examples related to the **molecular orbital treatment** for:

- **Homonuclear Diatomic Molecules**; and
- **Heteronuclear Diatomic Molecules**,

which would be discussed individually in the sections that follows:

3.2.1. Homonuclear Diatomic Molecules

In the unique process related to the meticulous build-up of the '**atoms**', the *electrons* are duly fed into the **atomic orbitals (AOs).** A survey of literature reveals that:

"**the following order of the orbitals is generally found to hold good for elements up to the inherent nuclear charge z = 20 quite satisfactorily; and beyond with a meager modifications**"

Thus, we may have the following **order of the orbitals:**

$$1s < 2s < 2p < 3s < 3p < 4s < 3d < 4p < 5s < 4d < 5p < 6s$$

$$< 4f < 5d < 6p < 7s < 5f < 6d < 7p < ... \tag{a}$$

Importantly, the following '**emperical rules**' usually known as the **(n + l) rules** may be employed approximately to remember the order stated in Eq. (a) above, namely:

(a) Amazingly, the inherent energy of the orbitals enhances with the **increasing (n + l) values.** Therefore, one may observe critically the ensuing *energies of the orbitals* in the following order:

$$n+1 = \frac{1s < 2s < 3s < 4s}{1 \quad 2 \quad 3 \quad 4} \frac{3s < 3p < 3d}{3 \quad 4} \frac{4p < 5d}{5 \quad 5} \frac{6s < 5f}{7 \quad 6 \quad 8}$$

(b) Therefore, for the same value of **n + l**, the energy gets enhanced with the **increasing n values.** Thus, the order of the *increasing energies* shall be;

$$n+l = \frac{3d < 4p < 5s}{5 \quad 5 \quad 5} \text{ and } \frac{4d < 5p < 6s}{6 \quad 6 \quad 6}$$

Interestingly, the so-called **energy levels** [*Eqn.* (a)] may be conveniently remembered with the aid of Fig: 1.25.

NOTE **Let us proceed from the '*Top*' to the '*Bottom*' as shown by the diagonal arrows in Fig: 1.25.**

According to the **Hund's rule** (see *section 1.1.6.2*) which evidently states that:

"**when several '*orbitals*' do possess the same level of energy profile*, the '*electrons*' shall be arranged in such a manner so as to yield the optimized number of the '*unpaired-spins.*'**"

---

* That is, they are duly **degenerated.**

Nevertheless, the **MO-method** usually makes use of the *whole molecule vis-a-vis* the *constituent atoms*; and, therefore, engage the **MOs,** rather than the **AOs.** In other words, in the build up of the *whole molecule* – the *total number of electrons duly derived from all the atoms in the molecule* is being fitted meticulously right inside the **moleclar orbitals (MOs).**

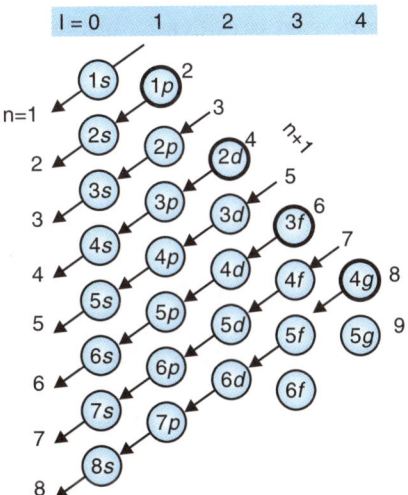

Fig. 1.25: The Mnemonic* for the Determination of the Order of the Filling of Orbitals (approximate) in Elements from the Auf Bau Principle**. [The 'orbitals' in 'Bold Circles' are the Non-existent Orbitals.]

NOTE   As stated earlier both *Auf Bau* principle and Hund's Rule are used.

Homonuclear Diatomic Molecules – The term '**homonuclear**' relates to the critical presence of only –'**one type of nucleus**' *i.e.,* only *one element* is present virtually. Likewise, the term 'diatomic explicitly means that the '**molecule**' is *actually made up of two atoms.*'

Let us now briefly discuss the following *four* types of *molecule* or *ion*, such as:

- **H_2^+ Molecule Ion;**
- **H_2 Molecule;**
- **He_2^+ Molecule Ion;** and
- **He_2 Molecule,**

in the sections that follows:

❑ **H_2^+ Molecule Ion** – Interestingly, it may be regarded as a unique combination of a *H-atom* and a H^+ *ion*; and thus, it gives rise to the formation of:

'one electron in the molecular ion',

that eventually occupies the *lowest energy* molecular orbital (MO):

$$\boxed{\sigma 1s}$$

---

* **Mnemonic:** That is, intended to help the '**memory**'.

** *Auf Bau* **Principal (Order):** It is also sometimes known as the *auf bau* (**building up**) **order** [see Eq. (a) given earlier].

❑ Obviously, the '*electron*' does occupy the σ1s bonding **molecular orbital (MO)**. The overall energy of the aforesaid is found to be certainly **lower** in comparison to that of the *constituent atom and ion* by a quantity equivalent to Δ; and hence, there is definitely certain degree of *stabilization* being attained.

NOTE	**The aforesaid species does exist prevalently, but is not quite common because H_2 is relatively *much more stable*. Nevertheless, the H_2^+ may be easily detected *spectroscopically* – as and when the H_2 – gas under vacuum is being subjected to an '*electric discharge*'.**

❑ **H_2 Molecule** – In this particular instance, there is *one electron* derived from **each atom**; and therefore, there are *two electrons* present in the molecule itself. Amazingly, these mostly occupy the *lowest energy* **molecular orbital (MO):**

$$\boxed{\sigma 1s^2}$$

The above description may be illustrated explicitly in the following Fig: 1.26.

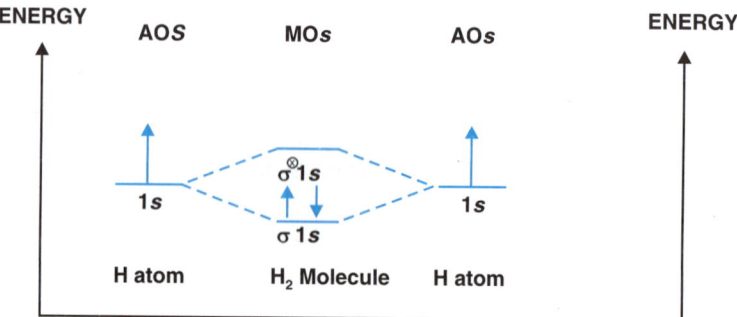

Fig. 1.26: Electronic Configuration of the Atomic Orbitals (AOs) and Molecular Orbitals (MOs) for Hydrogen.

Remarks: It may be observed that the bonding σ1s MO is almost full; and hence, the ensuing *stabilization energy* is found to be 2Δ. But a **σ-bond** is duly formed; and, therefore, the **H_2–molecule** does exist predominantly, – which is well established and recognized.

❑ **He_2^+ Molecule Ion** – It may, however, be regarded as a probable combination of **H_e atom** and a **He^+ ion**. Obviously, there exists *three electrons* that are duly present in the so-called '*molecular ion*' – that are duly arranged in the **molecular orbitals (MOs).**

$$\boxed{\sigma 1s^2, \sigma^{\ddot{A}} 1s^1}$$

Therefore, the adequately *filled* **σ1s-bond molecular orbital (MO)** eventually provides the **2Δ** stabilization perceptively; whereas, the so-called *half-filled* $\sigma 1s^\otimes$ *critically gives the Δ destabilization.* However, the overall there would be a definite **Δ stabilization**. Therefore, the helium molecule ion **[He_2^+]** can exist predominantly. It may *not be very stable*, but it has been duly observed **spectroscopically** and **confirmed**.

❑ **He^2 Molecule** – In this case, there are *two electrons* that get involved from *each atom of He*; and hence, in all *four electrons* are duly arranged in **molecular orbitals (MOs).**

$$\boxed{\sigma 1s^2, \sigma^{\otimes} 1s^2}$$

Thus, the particular **2Δ stabilization energy** duly derived from filling the so-called **σ2s-molecular orbital (MO)** is being cancelled squarely by means of the **2Δ destabilization energy** obtained from the respective filling of the $\sigma^{\otimes} 1s$ **molecular orbital (MO)**. Therefore, obviously *no* **bond formation** comes into play; and thus, the **He^2 molecule** *does not exist at all*.

3.2.2. *Heteronuclear Diatomic Molecules*

In a broader sense one may crucially observe the **identical principles** which normally is applicable whenever:

> **"the critical combination of the atomic orbitals (AOs) duly derived from *two* different atoms (as applicable when the said atoms were absolutely similar)".**

Alternatively, the above phenomenon holds good as and when the following laid down criteria are fulfilled appreciably:

(i) The *only* **atomic orbitals (AOs)** having the *similar energy level profile* may combine efficaciously.

(ii) The **AOs** must have the **optimized inherent overlap**.

(iii) The **AOs** should possess the **some level of symmetry features**.

It is, however, pertinent to state at this material time that by virtue of the fact that:

> **"the *two* atoms seem to be entirely divergent; and hence, the corresponding energy levels of their atomic orbitals (AOs) also differ slightly".**

Fig: 1.27 depicts the diagramatic representation of the so-called *electronic configuration*, *atomic orbitals (AOs)*, and *molecular orbitals* pertaining to a **heteronuclear molecule 'XY.'**

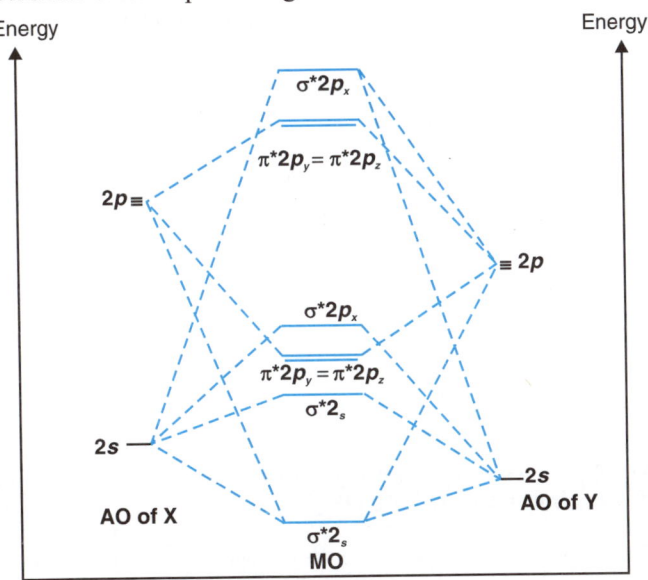

Fig. 1.27: Diagramatic Representation of the Electronic Configuration, Atomic Orbitals (AOs), and Molecular Orbitals (MOs) for a Heteronuclear Molecule 'XY'.

Fig: 1.27 explicitly shows how the *two* different atoms normally combine to give rise to the formation of the *molecular orbitals (MOs),* soonafter logically taking into consideration the typical **symmetry interaction of 2s and 2p orbitals** as could be seen in the said Figure *i.e.,* of the *two* **heteronuclear molecule 'XY', – 'Y'** is more electronegative than 'X'.

Comments: Based on the experimental results and evidences derived therefrom one may encounter a *serious problem* as could be seen in several instances that relates to:

"**the precise order of the molecular orbital (MO) energy level is not yet known with utmost certainty**".

Let us now consider *two* **typical examples (NO and CO molecule)** wherein the *two* altogether **different atoms** are found to be relatively close to each other in the '*Periodic Table*'. As a result, it is quite *logical, plausible,* and *reasonable* to assume vehemently that:

"**the '*order of energies*' for the MOs remain almost the same as for the '*homonuclear molecules*'.**"

❑ **NO Molecule** – It is well known that there exist in all *fifteen electrons* in *NO molecule*; and hence, the prevailing '**order of energy-levels**' contributed by the various **molecular orbitals (MOs) is as given under:**

$$\sigma 1s^2, \sigma 1s^2, \sigma 2s^2, \sigma 2s^2, \left(\begin{array}{c} \pi 2p_y^2, \\ \pi 2p_z^2, \end{array} \right. \sigma 2p_x^2, \left. \begin{array}{c} \pi 2p_y^1 \\ \pi 2p_z^0 \end{array} \right.$$

Observations: The presence of an '*unpaired electron*' specifically in the $\pi^\otimes$ orbital; and hence, subsequently gets *delocalized over the whole molecule* – thereby indicating that the **NO molecule** is definitely **paramagnetic*** in nature. Importantly, because the '**highest occupied molecular orbital (HOMO)**' is $\pi^\otimes$ and the ensuing energy of which is observed to be higher *vis-a-vis* the AOs of N and O atoms, the ultimate *ionization energy* of **NO** is recorded to be **much lower than that of both N and O atoms,** as given in the following available data:

Type of Species	Ionization Energy (in kJ. mol^{-1})
N	1402
O	1314
NO	894

Thus, on being subjected to *ionization* phenomenon, the bond **order of NO** gets apparently enhanced to 3.0 in **NO$^\oplus$**, as a consequence of which the prevailing 'bond length' gets reduced right from **113 pm in NO$^\oplus$** to **106 pm in NO$^\oplus$**

❑ **CO Molecule** – Interestingly, it is possible to observe certain highly specialized features introduced in **molecular orbital (MO) diagram of CO,** since the recorded *electronegativity difference profile* between **C** and **O** atom is found to be **large enough.** Fig: 1.28 shows the diagramatic representation of the **electronic configuration, atomic orbital (AO)** and **molecular orbital (MO) for carbon monoxide (CO):**

---

* **Paramagnetic:** That is, offering less resistance to the passage of the ensuing '*magnetic flux*' than does air; and hence, capable of being attracted by a magnet. the '**paramagnetic substances**' have, in other words, **magnetic permeabilities greater than '1'.** In this an **applied magnetic field** is increased by the alignment of **election orbits.** *Paramagnetion* decreases with the increase in temperature.

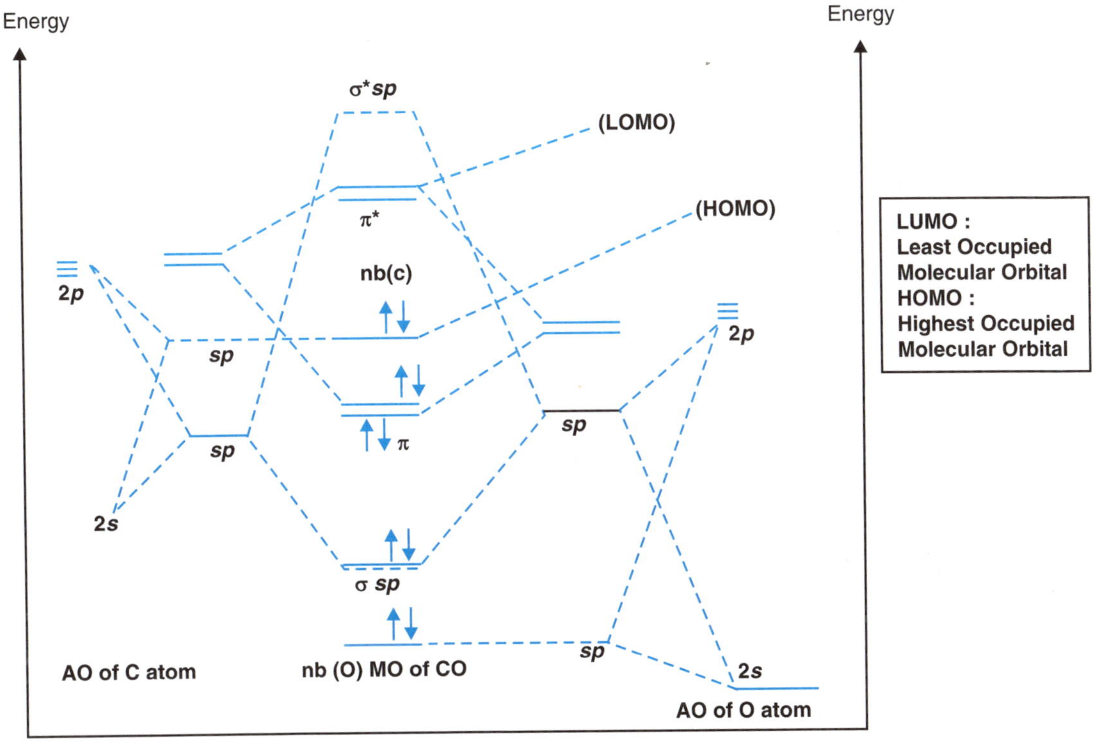

Energy

Energy

σ*sp

(LOMO)

π*

(HOMO)

nb(c)

2p

↑↓

sp

↑↓

LUMO :
Least Occupied
Molecular Orbital
HOMO :
Highest Occupied
Molecular Orbital

2p

↑↓ π

sp

sp

sp

2s

↑↓

σ sp

↑↓

sp

AO of C atom   **nb (O) MO of CO**   2s

AO of O atom

Fig. 1.28: Diagramatic Representation of the Electronic Configuration, Atomic Orbital, Molecular Orbital for Carbon Monoxide (CO). [**Adapted from: Lee JD: Concise Inorganic Chemistry, 2nd edn., Wiley (I) Pvt. Ltd. New Delhi, 2011**].

Remarks: Based on Fig: 1.28 one may take cognizance of the underlying fact that because: "the $\Delta E_{(2p-2s)}$ for C-atom (5.3 eV) is found to be significantly lower *via-a-vis* to that of $\Delta E_{(2p-2s)}$ for O-atom (15.0 eV); hence, the inherent participation of the so-called *s–p* hybrid orbitals are mostly regarded for the ultimate creation of the molecular orbital (MO) diagram of carbon monoxide (CO)".

NOTE   Besides, in the ionization of carbon monoxide (Co) *i.e.*, Co → Co⁺, the subsequent elimination of the *electron* comes into play right from the prevailing non-bonding orbital of the C-atom (HOMO). Perhaps this eventually attributes to the ensuing *interelectronic repulsion* being reduced significantly; and hence, ultimately the observed *bond length* gets reduced from *112.8 pm to 111.5 pm.*

3.3. Slater's Rules

The survey of literature reveals that **Slater** has intelligently put forward a **set of empirical rules** that, in fact, legitimately represent the flagrant overall generalization based exclusively upon the following *two* **cardinal aspects**, namely:

- **average behavioural pattern of the 'electrons' for the evaluation of the screening; and**
- **also shielding the *constant* σ (*i.e.*, the *screening* constant).**

Even though the **Slater's rule** fails to provide a *not-so-accurate* results pertaining to the *energy levels* of the *electron's* duly present in the **polyelectronic systems**; however, they are indeed found to be quite useful in the estimation of:

- **atomic dimensions (sizes)**; and
- **electronegativities**.

Slater's Rule for 'σ' Related to *ns* or *np* Orbital Electrons – Following are the *six* **important rules**, namely:

Rule – 1 :   First and foremost group together all the available *electrons* present in the following order carefully: (1*s*), (2*s*, 2*p*), (3*s*, 3*p*), (3*d*), (4*s*, 4*p*), (4*d*, 4*f*), (5*s*, 5*f*), (5*d*, 5*f*), (6*s*, 6*p*) etc.

Rule – 2 :   The *electrons to the right of the respective (ns, np)** moiety are found to be not so effective in shielding the *ns* or *np* electron; and hence, almost contribute nothing to *n*. Alternatively, the *higher energy electrons* are indeed ineffective in shielding the so-called '**lower energy electrons**' from the *nuclear charge (Z)*.

Rule – 3 :   All other '*electrons*' present in the (*ns*, *np*) **group** do contribute prevalently to the extent of **0.35** each to the respective **screening constant** except for the ensuing **1*s*-orbital electron** for which a pre-determined value equivalent to **0.30** is believed to work out in a better way.

Rule – 4 :   Importantly, all the '*electrons*' present in the (*n* – 1)th shell eventually contribute **0.85 each** to the corresponding **screening constant**.

Rule – 5 :   Likewise, all other '*electrons*' duly present in the (*n* – 2)th shell or even **lower shell**, do contribute **1.00 each to σ** (*i.e.*, the *screening or shielding constant*), thereby suggesting that the '*electrons*' actually located strategically in the: (*n* – 2)th and **lower shells** almost '*shield*' the *electrons* in the *n*th **shell** completely. Nevertheless, in the particular instance of either *ad* or *nf* electron, the aforesaid **rule (1) through (3)** practically *remain the same*, but the **rules (4) and (5)** invariably get replaced by the following '*Rule-6*':

Rule – 6 :   In this case, all the '*electrons*' present critically in the groups located left to **the (*nd*, *nf*) group** do precisely shield the so-called *d*(or *f*) **electrons** almost completely *i.e.*, they are found to contribute a value **1.00** to the ensuing **screening constant (σ)**.

Example: In order to understand fully the aforesaid conceptualized thoughts and ideas regarding the **Slater's Rules** – let us consider the following **typical example:**

For the respective '**valence-shell electrons of fluorine (Z = 9)**, the ensuing orbitals are usually grouped together as:

(1*s*)^2 (2*s*, 2*p*)^2, the **screening constant (σ)** is:

$$\sigma = 2 \times 0.85 + 6 \times 0.35 = \mathbf{3.80}$$

and, therefore, we may have:

$$\mathbf{Z}^{\otimes} = Z - \sigma = 9.00 - 3.80 = \mathbf{5.20}$$

Now, let us consider another element **Arsenic (Z = 33)**. Hence, as per '**Rule – 1**' – its respective orbitals shall be grouped together as:

$$\mathbf{(1s)^2 \ (2s, \ 2p)^8, \ (3s, \ 3p)^8, \ (3d)^{10}, \ (4s, \ 4p)^5}$$

---

*   That is, filled *s* and *p* orbitals of the outermost shell *ns*^{2} *np*^{6}

Therefore, for the corresponding **4s** or **4p electron**, we have:

$$\sigma = (10 \times 1.00) + (18 \times 0.85) + (4 \times 0.35) = \textbf{26.70}$$

and, therefore, we may have:

$$Z^\otimes = Z - \sigma = 33 - 26.70 = \textbf{6.30}$$

However, for the **3d electrons** the grouping virtually remains the same, but

$$\sigma = (18 \times 1.00) + (9 \times 0.35) = \textbf{21.15}$$

Thus, we may obtain:

$$Z^\otimes = Z - \sigma = 33 - 21.15 = \textbf{11.85}$$

Remarks: Therefore, in short the **electrons** present in different orbitals are affected indeed differently by the **same nuclear charge** which depends exclusively upon their inherited – '**proximity to the nucleus itself.**'

3.3.1. *The Limitations of Slater's Rule*

There are *three* **major limitations (drawbacks) of the Slater's Rules** – as given below:

1. It has been duly observed that the **s** and **p-orbital electrons** are believed to be largely affected by the respective *nuclear charge* exactly in a similar fashion – that is never expatiated clearly by their '**probability distribution functions**'.

2. In fact, most of the **electrons** duly present in the so-called **s, p, d, or f-orbitals** are normally taken to shield specifically the – '**higher shell electrons**' with a perfect *equal level of integrity and competence.*

 NOTE   **Obviously, this particular aspect is not-so-easily convincing since the broadly divergent (*i.e.*, both *energetically* and *stereochemically*) orbitals may not exert the same level of effect on or for the so-called '*outer-shell electrons*' perceptively.**

3. The actual observed penetration of the **higher orbital** right into the **inner orbital** (*i.e.*, particularly the **s-orbitals**) is virtually ignored totally.

3.3.2. *Slater's Orbitals*

In true sense, the timely proposed postulates under the **Slater's orbitals** were introduced so as to get: "the *hydrogens* like orbitals for *polyelectronic systems* for their judicious application in the *Quantum Mechanical Calculations*".

They have the following *characteristic features,* namely:

(a) The **Slater's orbitals** do possess essentially '**no nodes**'. Even though it play a major role in the *proper simplification of the orbitals*; however, it renders them slightly **less precise and accurate**.

(b) The observed **nuclear charge (Z)** is replaced adequately by the help of the above laid-down rules.

(c) For the **Slater's orbital number** $n > 3$, – the '**n**' is being duly replaced by $n^\otimes$, and $n - n^\otimes$ is usually termed as the *quantum defect*; whereas, the values of $n^\otimes$, for certain **n values** are found to be:

3.7 for *n* = 4; 4.0 for *n* = 5; 4.2 for *n* = 6; and the like,
thereby levelling off ultimately to **4.5 for the respective higher *n* values.**

NOTE Clement and Raimondi (1963) meticulously calculated $z^{\otimes}$ (nuclear charge) from the so-called *self-consistent field wave functions* from Helium (He) to Krypton (K_r). Eventually, they have proposed a set of rules for the calculation of σ (screening constant), when now depends on both '*n*' and '*l*'.

3.3.3. *Penetration Phenomenon of the Orbitals*

The exact and precise degree to which the *orbital of a shell* is capable of interacting with the specific '**lower-quantum shell orbitals**' is invariably termed as: **the penetration phenomenon of orbitals**. *Example:* The typical inherited energy profile of the *4s-orbital* is definitely found to be *lower* in comparison to that of the corresponding *3d–orbitals*,– which could be substantiated duly by the '**filling of the 4s – orbital**' that actually comes into play even before the *third–shell has been filled-up completely*. In fact, such a phenomenon is usually referred to as:

"**the critical penetration of the 3d-orbital by the 4s – orbital**".

Besides, one may also observe meticulously the actual **penetration phenomenon** is obviously much more as and when the ensuing **shell number gets enhanced progressively**. Hence, the **5s-orbital** does penetrate both **4f-orbital** and **4d-orbital** squarely; whereas, the particular **6s-orbital** penetrates the so-called **4f, 5d, and 5f orbitals**.

Fig: 1.29 shows the energy-level diagrammatic representation of a typical *multi-electron atom*.

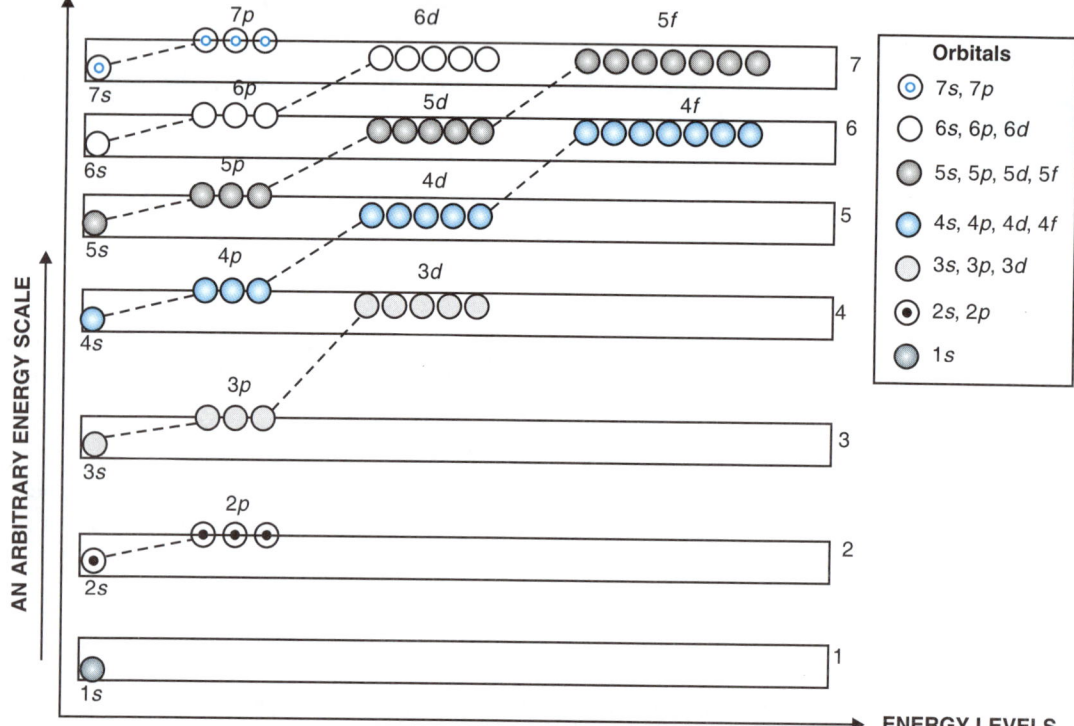

Fig. 1.29: The Diagramatic Representation of an Energy Level Profile of a Multi-electron Atom.

Explanation – The inherent *energy levels* of different 'orbitals' pertaining to various *principal energy levels* as observed in a **multi-electron atom** are duly illustrated in the **energy-level diagram** (Fig: 1.29) and the various vital aspects are stated below:

1. In **a multi-electron atom**, one may critically observe the different kinds of 'orbitals',* which perhaps could be due to the underlying fact that the inherent energies of these 'orbitals' solely depend on the following *two* **cardinal aspects**, namely:

   - **value of *n* (*i.e.*, Slater's orbital number); and**
   - **values of *l* (*i.e.*, the azimuthal number for orbital)**

2. The above analogy ascertains that the **2p-orbital** ($n = 2$; $l = 1$) possesses much *higher energy profile* *vis-a-vis* the respective **2s-orbital** ($n = 2, l = 0$).

3. Likewise, the inherent energy of the **3d-orbital** ($n = 3, l = 2$), is observed to be *significantly greater* in comparison to the corresponding **3p-orbital** ($n = 3, l = 1$), and hence, this in turn is much higher than that of the respective **3s-orbital** ($n = 3, l = 0$).

4. *Perceptible Crossing of Energy* – However, it may also be seen that there exists a *definite* and **perceptible crossing of energies**, specifically in the so-called '**higher-energy shells**' predominantly.

   Example: Energy possessed by a **4s-orbital** is observed to be obviously *much lower* in comparison to that of the **3d-orbital** (*i.e.*, even though the *former* belongs to the **higher principal energy shell**).

5. In the same vein, the inherent energy of **5s-orbital** is found to be *much lower* than the respective **5d-orbital**.

NOTE	In short it may be added that the 'energies' may be determined precisely by such critical factors, such as: • **principal quantum number (*n*); and** • **angular momentum number (*l*).**

3.3.4. *The Energy Level Diagram of Hydrogen (H_2) [or Hydrogen-like Atom]*

In true sense, the **energy-level diagram of hydrogen (H_2)** of a **hydrogen-like atom**** is observed to be distinctly divergent *vis-a-vis* the respective **energy-level** diagram of a *multi-electron atom*.

First **instance** *i.e.*, when **all the orbitals belonging to a specific principal energy level do have the same energy profile.**

Example: The **2s-orbital** and *three* **2p-orbitals** belonging to the *second* **energy level profile** essentially possess the **identical degree of energy**.

Exactly in the same manner, the following *orbital variants:*

- *3s-orbital*
- *three 3p-orbital and*
- *five 3d-orbitals,*

that critically belong to the *3rd energy level* do possess the **similar energy profiles**.

---

* That is, in the **same principal energy level** with *divergent energy profiles*.

** That is, an 'atom' having *only one electron*.

3.3.5. *Dependence of Energy of an Orbital*

A survey of literature reveals that the **dependence of energy of an '*orbital*'** solely centres around the following *three* major aspects, such as:

(i) *Nuclear charge* – It refers to the *positive* **electric charge** residing on the *nucleus of an atom*. When expressed in units equal to the charge on the nucleus, then this is numerically equal to the atomic number of the element *i.e.*, to the *number of protons to the nucleus*, and also equal to the *exact number of electrons* **surrounding the nucleus in the 'neutral atom'**.

(ii) *Principal Quantum Number* – It relates to a **quantum number** for the **orbital electrons**, which along with the **orbital angular momentum** and **spin-quantum number**, levels suitably the so-called '**electron-wave function**'. Thus, the ensuing **energy level** and the *average distance of an electron from the nucleus depend solely upon the* '**principal quantum number (n)**'.

(iii) *Presence of Other Electrons* – It usually refers to the presence of other electrons particularly in the *lower energy orbitals* with altogether **divergent 'l'** values (*i.e.*, the **azimuthal number for orbital**).

Special Remarks

(1) Because of the critical presence of *only one electron* in either **hydrogen (H_2)** or **hydrogen-like atoms** there prevails a commendable '*attractive effect*' of the **nuclear charge (e^+)** upon the *electron* remains totally undisturbed. Hence, the inherent *energy of an orbital** is mostly determined by the **principal quantum number (n) only**.

(2) Nevertheless, in the so-called **multi-electron atoms** one may observe an array of *electrons i.e.*, a few of them do essentially occupy the *lower-energy orbitals* which actually:

"tend to minimize the overall effect of nuclear charge (Z) located on the *electron* present specially in *higher-energy* orbitals."

How does this take place? Perhaps it may be by virtue of the fact that these aforesaid *orbitals* do intervene critically between the *nucleus* and the *electrons* exclusively in the higher energy orbitals. Therefore, such an effect is usually termed as the '**screening effect**'.

(3) Because, the *electrons* particularly present in – '**divergent types of orbitals' (*viz.*, s, p, d, or f)** **get screened differently these are obviously being attracted to various extents by the nucleus.**

Consequently, it may be observed explicitly that – **"different orbitals are invariably held at different distances right from the nucleus; and hence, do possess different energy levels prominently."**

(4) Thus, the above logical explanations do expatiate the ensuing variance in the inherent energies of *s, p, d,* and *f* **orbitals** even though they actually possess the **similar principal energy level** present in a *multi-electron atom*.

> The '*Slater's Rules*' (*section 3.4*) helps to record the extent of screening of an *electron* by other *electrons* in the atom *via* calculation using certain *empirical rules*.

3.3.6. *Important Applications of Slater's Rules*

In addition to the various meritorious advantages of the **Slater's Rules** already discussed earlier there are quite a few **important applications** which are enumerated as under:

---

*   That is, the *solitary electron* usually occupies at any point in time.

1. **Transition metal atoms usually lose 'ns' – electrons much earlier in the 'ionization' process –** Let us consider the **manganese (Mn) atom** (*Atomic Number: 25*) having an **electronic configuration***:

$$1s^2\ 2s^2\ 2p^6\ 3s^2\ 3p^6\ 3d^5\ 4s^2$$

Now, let us calculate the *effective nuclear charge*, $Z_{effective}$ duly registered by both **3d** and **4s electron**. Thus, we may have the following expressions:

$$Z_{effective} = \text{Registered by } \textbf{3d-electron}$$

$$= 25 - (0.35 \times 4 + 1.0 \times 18) = \textbf{5.60}$$

$$Z_{effective} = \text{Registered by } \textbf{4s-electron}$$

$$= 25 - (1 \times 0.35 + 13 \times 0.85 + 1.0 \times 10) = \textbf{3.60}$$

Based on the above data, it may be inferred explicitly that – 'the attraction between the **3d-electron and the nucleus of *Mn* atom is greater than the attraction between the 4s-electron and the nucleus of *Mn* atom**'. Therefore, it clearly suggests that the critical removal of the **4s-electron** right from an *isolated gaseous atom of Mn* should be obviously a lot easier *vis-a-vis* the removal of the so-called **3d-electron** from it.

2. **Filling of a 4s-Orbital is Faster than a 3d-Orbital** – For this let us consider a **potassium (K) atom** (**Z** = 19) in which *18 out of a total of 19 electrons* do occupy electrons having the following **electronic configuration***:

$$1s^2\ 2s^2\ 2p^6\ 3s^2\ 3p^6$$

Interestingly, the **19th electron** may easily enter either the **4s-orbital** or the **3d-orbital**. Hence, one may calculate *effective nuclear charge* (*viz.*, $Z_{effective}$) as given in the following expression:

$$Z_{effective} = \text{Registered for the } \textbf{19th electron} \text{ as it gains entry into the } \textbf{4s-orbital}$$

$$= 19 - 0 \times 0.35 - 8 \times 0.85 - 10 \times 1.0$$

$$= 19 - 16.80 = \textbf{2.20}$$

$$Z_{effective} = \text{Registered by the } \textbf{19th electron} \text{ as it enters the } \textbf{3d-orbital}$$

$$= 19 - 0 \times 0.35 - 18\ 0 \times 1.0$$

$$= 19 - 18 = \textbf{1.00}$$

Therefore, looking at the aforesaid *calculated data* one may gainfully draw a conclusion that – '**a 4s-electron could be eventually subjected to a definite *increased electrostatic attraction*****' in comparison to a 3d-electron**'.

---

* That is, duly arranged as per the **Slater's Rules**.

** **Z** = **Nuclear charge** or **Atomic Number**.

*** That is, it resembles to that of **Argon (Ar)** [$Z = 18$].

**** That is, perhaps on account of a **higher-effective nuclear charge 2.30**.

Remarks: As a result, the **K-atom** having the specific **outer electronic configuration** viz., $3s^2p^64s^1$ might be intimately associated with typical *lower energy profile* vis-a-vis with the particular **outer electronic configuration** viz., $3s^2p^6d^1$. Thus, the *former configuration, $3s^2p^64s^1$,* should definitely be rather 'more stable' vis-a-vis the latter configuration, $3s^2p^6d^1$. In this manner, the so-called *4s-orbital* gets invariably registered even much before than the respective *3d-orbital*.

3. **Usual Observed Trends in the Registered Successive Ionization Energies** – The various experimental data and scientific evidences do suggest vehemently that the very *first-ionization energy* is mostly at *lower ebb* than the *second-ionization energy*; whereas, the **2nd ionization energy** is invariably at a *lower ebb* vis-a-vis the **3rd ionization energy**, and so on so forth.

4. *Anion* **Bears a Bigger Dimension and** *Cation* **Bears a Smaller Size** *vis-a-vis* **the Parent Atom** – For this we may take into consideration the particular instance of **Chlorine (Cl) atom** (**Z = 17**) having the following **electronic configuration**:

$$1s^2 \; 2s^2 p^6 \; 3s^2 \; 3p^5$$

Therefore, we may calculate the *effective nuclear charge (e.g., $Z_{effective}$)* – as stated in the following expressions (*i.e.*, individually for the **chlorine atom** and **chlorine ion** respectively:)

$$Z_{effective} = \text{Registered by the 'last electron' of } \textbf{chlorine atom}$$
$$= 17 - (6 \times 0.35 + 8 \times 0.85 + 2 \times 1.0)$$
$$= \textbf{6.10}$$

$$Z_{effective} = \text{Registered by the 'last electron' of } \textbf{chlorine ion}$$
$$= 17 - (7 \times 0.35 + 8 \times 0.85 + 2 \times 1.0)$$
$$= \textbf{5.75}$$

Important Observations – **These essentially include**: In the typical instance, $Z_{effective}$, as could be seen in the **Cl⁻-ion** appears to be less *vis-a-vis* the **Cl-atom**. Hence, the prevailing *electrostatic attraction* between the **electrons and the nucleus** is found to be **greater** in the **Cl-atom** than in the **Cl⁻-ion**. Consequently, the **'nucleus'** should have apparently *much lower hold* upon the **electrons** duly present in the **Cl⁻-ion** in comparison to the **Cl-atom**. Thus, the ultimate dimension of the **Cl⁻-ion** must be *bigger* than the dimension of the **Cl-atom**.

5. **Progressive Decrease in the Atomic Size within a Period with Increase in Atomic Number** – In order to understand the progressive decrease in the atomic size occurring in a period having an increase in **atomic-number (Z)** – we may take into consideration the so-called:

"elements of the *First*-transition series from Selenium (Se) to Zinc **(Zn)"**.

It is however, pertinent to state here that the calculated **effective nuclear charge, $Z_{effective}$,** is found to increase in a progressive manner with the enhancement in the atomic number. Perhaps this could be the reason why the respective *'hold of the nucleus upon the ensuing electrons'* appears to go on increasing right from **Se** to **Zn**. As a result, the overall effect would appear:

"in terms of a regular decrease in the atomic size with the subsequent increase in the atomic number (Z) – as we move along the series of elements in the Periodic Table".

FURTHER READING REFERENCES

Abei Ew *et. al.* [Eds.]	:	**Comprehensive Organometallic Chemistry,** Elsevier, Amsterdam, 1982.
Bockhoff FJ	:	**Elements of Quantum Theory**, 2nd edn., Addison Wisley, Reading (UK), 1976.
Buckinghan J [Ed.]	:	**Dictionary of Organometallic Compounds,** Chapman and Hall, London (UK), 1984.
Buttler LS and Harrod JF	:	**Inorganic Chemistry**, Benjamin/Cummings, Redwood City, California (USA), 1989.
Cotton FA *et al.*	:	**Advanced Inorganic Chemistry**, 6th ed., Wiley India Pvt. Ltd., New Dehi, 2011.
De Kock RL and Gray HB	:	**Chemical Structure and Bonding**, Benjamin/Cummings, Manio Park, California (USA), 1980.
Ebsworth EAV *et al.*	:	**Structural Methods in Inorganic Chemistry**, Blackwell, Oxford (UK), 1987.
Guha S	:	**JD Lee Concise Inorganic Chemistry**, 2nd ed., Wiley India, Pvt. Ltd., New Delhi, 2013.
King RB	:	**Encyclopedia of Inorganic Chemistry**, John Wiley and Sons, New York, 1994.
Macintyre JE [Ed.]	:	**Dictionary of Inorganic Compounds,** Chapman and Hall, London (UK), 1992.
Murrell JN *et al.*	:	**The Chemical Bond**, 2nd ed., Wiley, Chichester (UK), 1985.
Smith D	:	**Inorganic Substances**, Cambridge University Press, Cambridge (UK), 1990.
Sharpe AG	:	**Inorganic Chemistry**, 3rd ed. Pearson Education Ltd. New Delhi, 2007.
Webster B	:	**Chemical Bonding Theory**, Blackwell, Oxford (UK), 1990.
Wilkinson G. *et al.* [Eds]	:	**Comprehensive Coordination Chemistry**, Pergamon Press, Oxford (UK), 1987.

PROBLEMS WITH SOLUTIONS

[A] Problems Based on Bohr's Sommerfeld's Atomic Model

Q.1. **The spectrum of Helium (He) atom is very much similar to that of Lithium ion (Li$^+$ion). Explain.**

Solution: Because both the He atom and Li$^+$ion do have exactly the similar number of electrons (*e.g.*, He = 1s^2; Li$^+$ = 1s^2 2s^0), the said two species essentially possess the same spectrum.

Q.2. **In the *Bohr series of lines of the H spectrum*-specify which of the *inter-orbit jump* the electron usually gives *third* line from the *red-end*?**

Solution: Since it is known that the spectral lines at the so-called *red-end* of H-spectrum critically belong to the corresponding *Balmer series*; hence, the lines of the series are duly accomplished by virtue of the following *electronic jumps*, such as:

$n_3 \rightarrow n_1$ (*First* line); $n_4 \rightarrow n_1$ (*Second* line); $n_5 \rightarrow n_1$ (*Third* line) etc., Therefore, *third* line is obtained solely due to $n_5 \rightarrow n_1$, or 5 → 1 electronic jump.

Q.3. **If the ionization potential (IP) of H atom is 13.6e V, determine the IP of He$^+$ ion.**

Solution: Because both **H atom** and **He$^+$ ion** essentially have one electron, the ionization potential (IP) of He$^+$ is obviously given by the following expression:

$$(IP)_{He^+} = (IP)_H \times (\text{Atomic Number of He})^2 \text{ eV}$$

$$= 13.6 \times 2^2 \text{ eV}$$

$$= \textbf{54.4 } \textbf{\textit{eV}}$$

i.e., $(IP)_{He^+} = \textbf{54.4 } \textbf{\textit{eV}}$

Q.4. **How would you determine the value of frequency (v) of radiation duly emitted when an electron falls from $n = 4$ to $n = 1$ orbit in H-atom? Given: Ionization Energy (IE) of H atom = 2.18 × 10^{-18} J atom^{-1}; and h = 6.625 × 10Js.**

Solution: The ΔE for $n_4 \rightarrow n_1$ jump = $E_4 - E_1$

$$= -2.18 \times 10^{-18} \left[\frac{11}{n_4^2} - \frac{11}{n_1^2} \right] \text{J atom}^{-1}$$

$$= -2.18 \times 10^{-18} \left[\frac{11}{4\hat{1}} - \frac{11}{2} \right] \text{J atom}^{-1}$$

$$= \textbf{+ 2.044 ×10}^{-18} \textbf{ J atom}^{-1}$$

We know that: $\boxed{\Delta E = hv}$ (1)

[ΔE = Energy difference; v = frequency; h = Planck's constant]
Rearranging Eq. (1) we have:

$$v = \frac{\Delta E}{h}$$

$$= \frac{2.044 \times 10^{-18} \text{ J atom}^{-1}}{6.625 \times 10^{-34} \text{ Js}}$$

$$= 3.085 \times 10^{15} s^{-1}$$

Q.5. **Calculate precisely the wavelength of the electron emitted when in a H-atom the electron falls from infinity to the respective stationary state [Given: $R = 1.097 \times 10^7 m^{-1}$]**

Solution: The **frequency (v)** in terms of the **wave number ($\bar{v}$)** is given by the following exprssion:

$$\bar{v} = R \left[\frac{1}{n_1^2} - \frac{1}{n_2^2} \right]$$

$$= 1.097 \times 10^7 \left[\frac{11}{1^2} - \frac{1}{\infty^2} \right] m^{-1}$$

$$= 1.097 \times 10^7 m^{-1}$$

since we know that

Wavelength = 1/wave number

or

$$\boxed{\lambda = 1/\bar{v}}$$

$$= \frac{1}{1.097 \times 10^7 m^{-1}}$$

$$= 0.91 \times 10^{-7} m$$

Q.6. **Which of the following one-electron ions has two ionization energy (IE) = 54.4e V?**
- **B^{4+}** • **Li^{2+}** • **Be^{2+} and** • **He^{2+}.**

Solution: Thus, we may have the following **ionization energies (IE)** of the aforesaid *one-electron ions*:

$$(IE)_{B^{4+}} \ (Z = 5) = 13.6 \times Z^2 eV = 13.6 \times 5^2 \ eV = \textbf{340 } \textit{eV}$$

$$(IE)_{Li^{2+}} \ (Z = 3) = 13.6 \times Z^2 eV = 13.6 \times 3^2 \ eV = \textbf{122.4 } \textit{eV}$$

$$(IE)_{Be^{2+}} \ (Z = 4) = 13.6 \times Z^2 eV = 13.6 \times 4^2 \ eV = \textbf{217.6 } \textit{eV}$$

$$(IE)_{He^{2+}} \ (Z = 2) = 13.6 \times Z^2 \ eV = 13.6 \times 2^2 \ eV = \textbf{54.4 } \textit{eV}$$

Based on the above results, obviously **He^{2+}** ion does possess the *ionization energy:* IE = **54.4 eV.**

Q.7. **Of the following electronic transitions which one would show the minimum wavelength:**
$n_4 \rightarrow n_1$; $n_2 \rightarrow n_1$; $n_4 \rightarrow n_2$; and $n_3 \rightarrow n_1$

Solution: Since we know the following relationship that

$$\boxed{\Delta E = hv} \hspace{3cm} \text{(a)}$$

Because, $v = c/\lambda$; putting this value in Eq. (a), we get

Hence,

$$\Delta E = \frac{hc}{\lambda}$$

(b)

or

$$\lambda = \frac{hc}{\Delta E}$$

(c)

Since in *Eq. (c)*, **hc** is a constant; and hence, *Eq. (c)* becomes

$$\lambda = \frac{1}{\Delta E}$$

(d)

Furthermore, since $(\Delta E)_{n_4 \to n_1}$ corresponds to $n_4 \to n_1$, the *transition is maximum i.e., the wave length of this specific transition* (λ) is observed to be *minimum*.

Q.8. **How will you determine the value of the most probable radius (in pm) for the electron duly present in He$^+$ion ?**

Solution: Since it is known that the radius of the *first* orbital ($n = 1$) for the respective **He$^+$** ion (**Z = 2**), we have:

$$= \frac{0.529 \times n^2}{Z} \text{Å}$$

$$= \frac{0.529 \times n^2}{Z} \times 10^2 \text{pm}$$

$$[\because 1\text{Å} = 10^2 \text{pm}]$$

$$= \frac{52.9 \times n^2}{Z} \text{pm} = \frac{52.9 \times 1^2}{2} \text{pm}$$

$$= \textbf{26.5 pm}$$

Q.9. **In a H-atom, in case the inherent energy of an electron in the *ground state* is found to be 13.6 eV (electron volt), determine the value of energy of this particular atom in the *second excited state*.**

Solution: Since we know that in the **second excited state** of the *H-atom electron* is duly located in the *third orbit* ($n = 3$).

Therefore, we may have the following expression:

$$E_3 = -\frac{eV}{n^2} \times Z^2 eV$$

or

$$E_3 = -\frac{13.6}{(3)^2} \times (1)^2 \text{ eV}$$

$$= -\frac{13.6}{9} \text{ eV} = \textbf{-1.51eV}$$

Q.10. **[In SI Units*]** Applying *de Broglie's equation*, calculate the wavelength associated with the motion of the earth, a stone, and an electron. The respective masses and velocities of these objects are as stated under:

Mass	Velocity
Earth = 6×10^{24} kg	3×10^4 ms^{-1}
Stone = 0.1 kg	1.0 ms^{-1}
Electron = 10^{-30} kg	6×10^5 ms^{-1}

Solution: Let us calculate the wavelength associated with the motion of the **Earth, Stone,** and **Electron** separately:

(a) *For Earth:*

$$\lambda = \frac{h}{mv} = \frac{6.6 \times 10^{-34}}{(6 \times 10^{24}) \times (3 \times 10^4)}$$

$$= 3.68 \times 10^{-63} \text{ m}$$

(b) *For Stone:*

$$l = \frac{h}{mv} = \frac{6.6 \times 10^{-34}}{(0.1) \times (1.0)}$$

$$= 66 \times 10^{-34} \text{ m}$$

(c) *For Electron:*

$$\lambda = \frac{h}{mv} = \frac{6.6 \times 10^{-34}}{(10^{-30}) \times (6 \times 10^{-5})}$$

$$= 1.1 \times 10^{-9} \text{ m}$$

NOTE The wavelength (λ) for an *'electron'* is found to be maximum; and, therefore, it obviously is a measurable quantum.

Q.11. **[In SI Units]** How would you calculate the wavelength (λ) of a body of mass 1 mg moving with a constant velocity of 3 ms^{-1}?

Solution: Let us apply **de Broglie's Equation** as follows:

$$\lambda = \frac{h}{mv}$$

$$= \frac{6.63 \times 10^{-34}}{(9.11 \times 10^{-31}) \times 2.5 \times 10^7} \text{ m}$$

$$= \frac{6.63 \times 10^{-10}}{9.11 \times 2.5} \text{ m} = 0.2911 \times 10^{-10} \text{ m}$$

---

* **The SI Units:** There is no '*v*' an internationally accepted set of units for the physical sciences based on the *metric system* which are termed as the **SI units** (based on the *system internationale*.)

Q.12. **[In CGS Units*] Calculate the wavelength of matter wave associated with an electron moving with a velocity of 1.20×10^7 cm.s^{-1}. [Mass of electron $= 91 \times 10^{-28}$g; Planck's Constant (h) $= 6.626 \times 10^{-27}$ erg.s].**

Solution: In the above particular example:

$$v = 1.20 \times 10^7 \text{ cm.s}^{-1}; m = 9.1 \times 10^{-28} g; \text{ and}$$

$$h = 6.626 \times 10^{-27} \text{ erg.s}$$

Now, applying the **De Broglie's Equation**

$$\lambda = \frac{h}{mv} = \frac{6.626 \times 10^{-27} \text{ erg.s}}{(9.1 \times 10^{-28} \text{ g}) \times (1.20 \times 10^7 \text{ cm.s}^{-1})}$$

$$= \frac{6.626 \times 10^{-6} \text{ g.cm}^2.s^{-2}.s}{9.1 \times 1.20 \text{ g.cm.s}^{-1}}$$

$$= \mathbf{0.6057 \times 10^{-6} \text{ cm}}$$

Q.13. **[In SI units] Calculate the uncertainty in the position (Δx) of an asbestos particle, if 0.1% error in the measurement of velocity is permitted. Given that: Velocity of particle $= 10^{-6}$ ms^{-1}; Mass of particle $= 10^{-14}$ kg; Planck's constant (h) $= 6.6 \times 10^{-34}$ Js; and Diameter of particle $= 10^{-6}$ m. Comment on the result.**

Solution: Thus, we have: v $= 10^{-6}$ms^{-1}, $m = 10^{-14}$kg, $h = 6.6 \times 10^{-34}$ Js, and $\Delta v = 0.1\%$ of the velocity of particle.

$$= \frac{0.1 \times 10^{-6}}{100} \text{ m.s}^{-1} = 10^{-9} \text{ m.s}^{-1}$$

∴ $$\Delta x^{**} = \frac{h}{2\pi \times \Delta v \times m}$$

∴ $$= \frac{6.6 \times 10^{-34} \text{ Js}}{2 \times 3.14 \times (10^{-9} \text{ m.s}^{-1}) \times (10^{-14} \text{ kg})}$$

$$= \frac{6.6 \times 10^{-11}}{6.28} m$$

$$= 1.05095 \times 10^{-11} m$$

$$= \frac{1.05095 \times 10^{-6}}{10^5} m$$

$$= \mathbf{0.0000105 \times 10^{-6} m}$$

---

*   **CG's Units: Centimeter-gram second** system of measurements.
** That is, uncertainty in the position.

B] Problems Based on Wave-Mechanical Approach

Q.12. **[In SI Units] A sophisticated microscope making use of the appropriate 'photons' is being employed to locate strategically an 'electron' present in an atom within a distance of 0.1 Å. What will be the degree of uncertainty involved in the measurement of its velocity.**

[Mass of electron (m) = 9.1 × 10⁻³¹kg; and Planck's constant (h) = 6.626 × 10⁻³⁴Js]

Solution: We have,
$$\Delta x = 0.1 \text{ Å}$$
$$= 0.1 \times 10^{-10} m$$

But, m = 9.1 × 10⁻³¹ kg; and h = 6.626 × 10⁻³⁴ Js

Now, according to **Heisenburg's uncertainty principle**

$$\boxed{\Delta x . m . \Delta v = \frac{h}{2p}}$$ (a)

From Eq. (a) we have:

$$\Delta v = \frac{h}{2\pi \times Dx \times m}$$ $\boxed{\Delta v = \text{Velocity of the particle}}$

$$= \frac{6.626 \times 10^{-34} \text{ Js}}{2 \times 3.14 \times (0.1 \times 10^{-10} \ m) \times (9.1 \times 10^{-31} \text{ kg})}$$

$$= \frac{6.626 \times 10^{7}}{2 \times 3.14 \times 0.1 \times 9.1} \text{ m.s}^{-1}$$

$$= \textbf{1.159} \times \textbf{10}^{7} \textbf{ ms}^{-1}$$

Q.13. **[In SI Units]. How would you calculate the prevailing uncertainty in the exact position (Δx) of an 'electron', if velocity of electron (Δv) is 0.1%?**

[Velocity of electron = 2.2 × 10⁶ ms⁻¹; Mass of electron = 9.1 × 10⁻³¹ kg; and Plancks' constant (h) = 6.6 × 10⁻³⁴ Js.]

Solution: We have
$$\Delta v = 0.1\% \text{ of the velocity of electron}$$

$$= \frac{0.1 \times 2.2 \times 10^{6}}{100} \text{ m.s}^{-1}$$

$$= \textbf{2.2} \times \textbf{10}^{3} \textbf{ m.s}^{-1}$$

Now, the uncertainty of the exact position (Δx) is given by the following expression:

$$\Delta v = \frac{h}{2\pi \times m \times \Delta v}$$

or
$$= \frac{6.6 \times 10^{-34} \text{ Js}}{2 \times 3.14 \times (9.1 \times 10^{-31} \text{kg}) \times (2.2 \times 10^{6} \text{ ms}^{-1})}$$

$$= \frac{6.6 \times 10^{-6}}{2 \times 3.14 \times 9.1 \times 2.2} m$$

$$= 0.0524953 \times 10^{-6} \, m$$

$$= \textbf{524.953} \times \textbf{10}^{-10} \, \textbf{\textit{m}}$$

Q.14. **[In CGs Units] An electron has a speed of 30,000 cm.s^{-1} accurate upto 0.001%. What would be the extent of uncertainty in locating its precise and exact position?**
[Given: Mass of an electron = 9.1 × 10^{-28} g; and Planck's constant (h) = 6.626 × 10^{-2} erg.s]

Solution: We have the following given parameters: h = 6.626 × 10^{-27} erg.s; m = 9.1 × 10^{-28} g; and the velocity of electron = 30,000 cm.s^{-1}.

$$\therefore \qquad \text{Velocity of electron } \Delta v = \frac{30,000 \times 0.001}{100} \text{ cm.s}^{-1}$$

$$= \textbf{0.3 cm.s}^{-1}$$

Now, as per **Heisenberg's uncertainty principle**, we have the following expression:

$$\boxed{\Delta x \,.\, m \,.\, \Delta v = \frac{h}{2\pi}}$$

or
$$\Delta x = \frac{h}{2\pi \times m \times \Delta v}$$

$$= \frac{6.626 \times 10^{-27} \text{ erg.s}}{2 \times 3.14 \times (9.1 \times 10^{-28} \text{ g}) \times (0.3 \text{ cm.s}^{-1})}$$

$$= \textbf{3.86.48 cm}$$

[C] Problems Based on Quantum Numbers

Q.15. **An *'electron'* is present in 4f orbital. What possible values for the Quantum Number of n, l, m and s it can have?**   **[B.Sc., 1981; Punjab University]**

Solution: Since we are aware that for **4f orbital** the respective *orbital numbers* for n = 4 and l = 3, m = 0, ± 1, ± 2, ± 3 (*seven values*). Hence, seven values of m indicate that the electron may be present in any one of the **seven 4f orbitals**; and thus, this electron m any possess any of the above *seven values*. Therefore, for this electron, we have: $s = +\dfrac{1}{2}$ **or** $-\dfrac{1}{2}$.

Hence, the n, l, m, and s values for an electron located in the **4f orbital** are:

n = 4, l = 3, m = any *one of the seven values* namely: 0, ± 1, ± 2, ± 3, and $s = +\dfrac{1}{2}$ or $-\dfrac{1}{2}$.

Q.16. **Write the name of the orbital for which the Quantum Numbers are: $n = 2$ and $l = 1$.**

                                          [Himachal, 1982; Jabalpur, 1981]

Solution: It is known that $n = 2$ indicates explicitly the second main shell. Thus, when $n = 2$, $l = 0$, 1, then $l = 0$ stands for a *sub-shell*; where as, $l = 1$ implies **p sub-shell**. Thus, $p(l = 1)$ **sub-shell** essentially belongs to the **2nd shell**; and, therefore, $n = 2$ and $l = 1$ actually represents the **2p sub-shell**.

Q.17. **Write the correct orbital notation for each of the following set of Quantum Numbers:**

      (i) $n = 1$; $l = 0$; $m = 0$

      (ii) $n = 2$; $l = 1$; $m = 1$

      (iii) $n = 3$; $l = 2$; $m = -1$

                                          [Meerut (B.Sc.,): 1984]

Solution: Please refer to the *Reference Table* in a textbook and note that:

      (i) The orbital for which: $n = 1$, $l = 0$, and $m = 0$ is '*1s*'.

      (ii) The orbital for which $n = 2$, $l = 1$, and $m = 1$ is '*1s*'.

      (iii) The orbital for which: $n = 3$, $l = 2$, and $m = -1$ is '*$3d_{yz}$*'.

Q.18. **What is maximum number of electron in (i) all the 4d orbitals; (ii) all the orbitals with $l = 1$.**   **[Kurukshetra (B. Sc.): 1983]**

Solution: We are aware that the *maximum number of electrons* present in an orbital with a given value of l is equal to $2(2l + 1)$; and hence, we have:

      (i) Maximum number of electrons in **4d orbitals ($l = 2$)**

$$= 2 (2 \times 2 + 1) = 10$$

      (ii) Maximum number of electrons in all the orbitals with ($l = 1$)

$$= 2(2 \times 1 + 1)$$

$$= 6$$

Q.19. **What are the n, l, and m values for $3s$, $3p_x$, and $3d_{xy}$ electrons?**

                                          [Meerut (B. Sc.,): 1984]

Solution: Please refer to the *Reference Table* in a Textbook and note that:

      (i) For *3s* electrons: $n = 3$; $l = 0$; $m = 0$.

      (ii) For *$3p_x$* electrons: $n = 3$; $l = 1$; $m = 0$.

      (iii) For *$3d_{xy}$* electrons: $n = 3$; $l = 2$; $m = -2$

Q.20.   **(i)** **How do the 'electrons' get duly distributed in different sub-shell for $n = 3$?**

      (ii) **Give the *quantum numbers* for the electron in the *first sub-shell* appearing in the $n = 3$ shell.**

Solution: (i) When the quantum number $n = 3$, the shell is the **3rd shell**; and $l = 0$ [*i.e.*, *3s* **sub-shells**, which means 1 (*3p* sub-shell) and 2(*3d*-sub-shell)]. Obviously, the *third* **main shell** critically contains *three* **sub-shells**–that are *3s*, *3p*, and *3d*.

Since we know that the **total number of electrons** present duly in the *third* **main shell** $= 2 \times 3^2 = 18$; and hence, these **18 electrons** present duly in *3s, 3p,* and *3d* **sub-shells** are usually distributed as under:

Electron in *3s* sub-shell ($l = 0$) $= 2(2l + 1) = 2(2 \times 0 + 1) = \mathbf{2}$

Electrons in *3p* sub-shell ($l = 1$) $= 2(2 \times 1 + 1) = \mathbf{6}$

Electrons in *3d* sub-shell ($l = 2$) $= 2(2 \times 2 + 1) = \mathbf{10}$

(ii) It is, however, quite evident that the *1st sub-shell* located in the *third* **main shell** is the sub-shell that essentially comprises *two* **electrons** as shown above. Hence, the *four* **quantum numbers** for the *two* **electrons** present duly in the *3s* **sub-shell** are as given under:

Quantum Numbers

		n	l	m	s
	1st electron	3	0	0	$+1/2$
Electron Number					
	2nd electron	3	0	0	$-1/2$

Q.21. **What would be the values of other Quantum Numbers of an electron for which the** *principal quantum number* **is 3?**

Solution: The respective values of other **Quantum Numbers**, such as: *l*, *m*, and *s* are as given under:

n	l	m	s
	0	0	$+1/2, -1/2$
		0	$+1/2, -1/2$
	1	$+1$	$+1/2, -1/2$
		-1	$+1/2, -1/2$
3		0	$+1/2, -1/2$
		$+1$	$+1/2, -1/2$
	2	-1	$+1/2, -1/2$
		$+2$	$+1/2, -1/2$
		-2	$+1/2, -1/2$

Q.22. **Given** $n = 4$ **principal, find out the following:**

(i) **Name of the shell;**

(ii) **Number of sub-shells present in it;**

(iii) **Total number of orbitals present in each sub-shell; and**

(iv) **Number of orbitals present in the principal shell.**

Solution: (i) $n = 4$ is the **4th shell** (*i.e.*, N-shell).

(ii) Number of sub-shells in the main shell with a given value of $n = 4$.

(iii) Since it is known that for $n = 4$ shell, $l = 0$ (*4s* sub-shell), 1(*4p* sub-shell), 2(*4d* sub-shell), 3(*4f* sub-shell). Now, the *total number of orbitals* present in a *sub-shell* having a given value of $l = (2l + 1)$, we have:

Number of orbitals in *4s* sub-shell ($l = 0$) $= (2 \times 0 + 1) = \mathbf{1}$

Number of orbitals in *4p* sub-shell ($l = 1$) $= (2 \times 1 + 1) = \mathbf{3}$

Number of orbitals in *4d* sub-shell ($l = 2$) = $(2 \times 2 + 1) = $ **5**

Number of orbitals in *4f* sub-shell ($l = 3$) = $(2 \times 3 + 1) = $ **7**

(iv) Number of orbitals present in the **principal shell** *i.e.,* **4th shell** ($n = 4$) = $4^2 = $ **16**

Q.23. **Determine the number of electrons in atoms that have the following levels in the ground state:**

(a) **K-shell, L-shell, *3s* sub-shell, and half filled *3p* sub-shell; and**

(b) **K, L, and M shells, and *4s*, *4p*, and *4d* sub-shells. What would be the *two* probable elements?**

Solution: (a) Number of electrons present in the **K-shell ($n = 1$)** = $2n^2 = 2 \times 1^2 = $ **2**

Number of electrons present in the **L-shell ($n = 2$)** = $2 \times 2^2 = $ **8**

Number of electrons present in the **sub-shell ($l = 0$)** = $2(2l + 1) = 2(2 \times 0 + 1) = $ **2**

Half the number of electrons present in the *3p* **sub-shell ($l = 1$)** = $\dfrac{2(2 \times 1 + 1)}{2} = 3$

∴   The total number of electrons present = $2 + 8 + 2 + 3 = $ **15**

Thus, the above number (15) explicitly indicates the so-called atomic number of the element, which as per the **Periodic Table of the Elements** stands for *Phosphorus* [P].

(b) Number of electrons present in the **K-shell ($n = 1$)** = $2n^2 = 2 \times 1^2 = $ **2**

Number of electrons present in **L-shell ($n = 2$)** = $2 \times 2^2 = $ **8**

Number of electrons present in **M-shell ($n = 3$)** = $2 \times 3^2 = $ **18**

Number of electrons present in *4s* **sub-shell ($l = 0$)** = $2(2l + 1) = 2(2 \times 0 + 1) = $ **2**

Number of electrons present in *4p* **sub-shell ($l = 1$)** = $2(2 \times 1 + 1) = $ **6**

Number of electrons present in *4d* **sub-shell ($l = 2$)** = $2(2 \times 2 + 1) = $ **10**

∴   Total number of electrons = $2 + 8 + 18 + 2 + 6 + 10 = $ **46**

Thus, the above number (**46**) implies clearly the atomic number of the element, that according to the **Periodic Table of the Elements Stands** for **Palladium [Pd].**

Q.24. **How will you determine the exact number of orbitals duly present in the sub-shell having $n = 4$ and $l = 3$?**

Solution: The sub-shell having $n = 4$ and $l = 3$ is *4f* sub-shell. Obviously, the *4f* sub-shell does possess *seven* orbitals since when $l = 3$, $m = 0, +1, +2, +3, +4$ (Seven values).

Q.25. **How would you arrange the *'orbitals'* having the following *quantum numbers* in the increasing order of their ensuing energy?**

(a) $n = 3, l = 0, m = 0, s = +1/2$

(b) $n = 3, l = 1, m = 1, s = \pm1/2$

(c) $n = 3, l = 2, m = 1, s = \pm1/2$

(d) $n = 4, l = 0, m = 0, s = +1/2$

Solution: Obviously, the *orbitals* duly designated by the given *quantum numbers* are: (a) *3s*; (b)*3p*; (c) *3d*; (d) *4s*. Thus, according to the **Auf Bau principle** the ensuing energy of *3s*, *3p*, *3d*, and *4s* orbitals increases progressively as given under:

$$3s < 3p < 4s < 3d.$$

Q.26. **Explain with logistics when the following sets of quantum numbers are not practically possible.**

	n	l	m	s
(a)	2	2	1	$+1/2$
(b)	1	0	-1	$-1/2$
(c)	3	2	3	$+1/2$

Solution: (a) The $n = 2$ (*2nd* **shell**) and $l = 2$ (*d*-**sub shell**) duly designates the **2d-sub shell**. Because, the *2nd* **shell** fails to contain the *d*-**sub shell**, the **2d-sub shell** does not exist at all. As a result, this particular **set of quantum number is not at all possible practically**.

(b) The $n = 1$ and $l = 0$ designates the *1s* **sub-shell**. Since for this specific **sub-shell** m must have **zero ($m = 0$) value** and certainly not -1 (as given in *question above*), the **given set of quantum numbers is not possible at all**.

(c) The $n = 3$ and $l = 2$ represents the *3d* **sub-shell**. Since the *d* **sub-shell** must have any of the *five* values of m. *e.g.*, $\pm 2, \pm 1, 0$, **and not 3**, as given in the above question, the **given set of quantum numbers is not possible practically**.

Q.27. **How would you find out the maximum number of electrons in an atom having the** *quantum number* **as: $n = 4$ and $m = \pm 1$?**

Solution: The $n = 4$ designates explicitly the *4th* **shell**, which eventually comprises *4s*, *4p*, *4d*, and *4f* **sub-shells**. Hence, the observed values of m for these **sub-shells** are for: *4s*, $m = 0$ (*one 4s* **orbital**); *4p*, $m = 0, +1, -1$ (*three 4p orbitals*); *4d*, $m = +2, -2, +1, -1, 0$ (*five 4d orbitals*) and *4f*, $m = +3, -3, +2, -2, +1, -1, 0$ (*seven 4f orbitals*). Therefore, the **orbitals** for which $m = \pm 1$ may be: $4p_x, 4d_{xy}, 4f_z$. Since practically each of these **orbitals** does contain *two* **electrons**, the *maximum number of electrons present duly in the atom* $= 2 + 2 + 2 = 6$.

Q.28. **The set of quantum numbers: $n = 3, l = 2, m = \pm 2$ for an electron specifically in the** *ground state* **of an atom (X) having atomic number (Z) = 19 is not possible at all. Explain.**

Solution: The values $n = 3$ and $l = 2$ designates the *3d* **sub-shell**. The complete **electron cloud (EC)** of **atom (X)** with *atomic number* (Z) $= 19$ is $1s^2, 2s^2, 2p^6, 3s^2, 3p^6$, and $4s^1$. Since this **electron cloud (EC)** fails to contain any *sub-shell*, the given **set of quantum numbers is not possible practically**.

Q.29. **We have three different orbitals having** *quantum number values*: **$n = 5, l = 0; n = 4, l = 1$; and $n = 3, n = 2$. In which orbital will the electron gain entry at first instance?**

Solution: Obviously, the various **orbitals** duly designated by the given values of the **quantum numbers** are given as: $5s(n = 5, l = 0)$; $4p (n = 4, l = 1)$; and $3d (n = 3, l = 2)$ respectively. However, the corresponding inherent energies of these *orbitals* is found to be in the order: $3d < 4p < 5s$. Since the *3d* **orbital** critically possesses the *lowest energy level*, the electron will enter *3d* **orbital** for which the values of: $n = 3$ and $l = 2$.

Q.30. **How will you determine the exact number of orbitals for which the following set of quantum number is possible: $n = 3, l = 2$, and $m = +2$?**

Solution: The given values $n = 3$ and $l = 2$ designates precisely the *3d* **sub-shell**. Since this particular **sub-shell** essentially possesses *five* different values of m, such as: $+2, -2, +1, -1$, and 0, this specific **sub-shell** has *five 3d* **orbitals**, for instance: $3d_{xy}, 3d_{yz}, 3d_{zx}, 3d_{x^2-y^2}$ and $3d_{z^2}$. Besides, $m = +2$ designates one of the *five* **orbitals** already stated above. Hence, we have **one orbital** (*3d* **orbital**) for which: $n = 3, l = 2$, and $m = +2$.

Q.31. Arrange systematically *ns, np, (n –1) d; (n – 1) f* orbitals in the increasing order of their inherent energy.

Solution: The energy of orbitals increases with the simultaneous increment of their respective $(n + l)$ **values**. In a situation, when the *two orbitals* do possess the same value of $(n + l)$, the orbital with *higher values of n* definitely has *higher energy level* $(n + l)$ **values** for the given orbitals are as stated under:

Orbitals:	ns	np	$(n-1)d$	$(n-1)f$
$(n + l)$ values:	$n + 0$	$n + 1$	$n - 1 + 2$	$n - 1 + 3$
	$= n$	$= n + 1$	$= n + 1$	$= n + 2$

Based on the aforesaid results one may observe that np and $(n - 1)d$ orbitals do have the identical value of $(n + l)$. Since *np orbital* possesses a relatively *higher* value of $n(= n)$ *vis-a-vis* $(n - 1)d$ **orbital** $(= n - 1)$ the *np* **orbital** definitely shows higher energy in comparison to $(n - 1)d$ **orbital**.

Therefore, the *increasing order of energy* is as given by: $ns < (n - 1)d < np < (n - 1)f$.

Q.32. Which of the orbitals with the following set of *quantum numbers* have the *same energy profile*?

(a) $n = 2, l = 1, m = +1$,

(b) $n = 2, l = 1, m = -1$

(c) $n = 2, l = 1, m = 0$, and

(d) $n = 2, l = 0, m = 0$

Solution: The various orbitals that are duly designated by the given the **sets of quantum numbers** are:

$2p_x$ for $(n = 2, l = 1, m = +1)$;

$2p_y$ for $(n = 2, l = 1, m = -1)$;

$2p_z$ for $(n = 2, l = 1, m = 0)$, and

$2s$ for $(n = 2, l = 0, m = 0)$

Thus, all the *three 2p orbitals* usually degenerate; and hence, they do have the same energy profile, but the **2p orbital** has the *different energy level*.

Q.33. What would be the atomic number (Z) of an atom having quantum numbers as: $n = 3$, $l = 0$, and $m = 0$?

Solution: Since $n = 3, l = 0$, and $m = 0$ explicitly designates the *3s orbital*, the *electron cloud (EC)* of the atom having *3s orbital* is found to be: $1s^2, 2s^2, 2p^6, 3s^1$ $(Z = 11)$ or $1s^2, 2s^2, 2p^2, 3s^2$ $(Z = 12)$.

Therefore, the '*atom*' having $n = 3, l = 0$, and $m = 0$ may have either **atomic number 11 (Na)** or **atomic number 12 (Mg)** 1(according to the **Periodic Table of the Elements**).

Q.34. Determine the exact number of *radial nodes* duly present in 3s and 2p orbitals.

Solution: The number of *radial-nodes* in a given orbital $= n - l - 1$. Therefore, we may have:

In *3s* orbital the number of radial nodes $= 3 - 0 - 1 = 2$

In *2p* orbital the number of radial nodes $= 2 - 1 - 1 = 0$.

UNIVERSITY QUESTIONS

1. How is **e/m ratio of the electron determined?** [Madras '85, Madurai '85, Poona '86]
2. Write a short note on '**Discovery of Neutron**' [Kanpur' 82, Garhwal '88]
3. Describe **Rutherford's Model** of an atom. How was it improved by Bohr?
   [Banaras'80, Poona '81]
4. Write a note on '**Photoelectric Effect**'. [Nagpur' 81, Baroda' 82, Madras '85]
5. Write a note on '**Planck's Quantum Theory of Radiation**'. [Kanpur' 82, '83 and '84]
6. Write a note on '**Compton Effect**'. [Nagpur' 81, Baroda '82]
7. Give a brief account on **Bohr's Model of Hydrogen Atom** covering the following points:
   (a) Postulates; (b) Energy and Radius; (c) Short-comings. [Allahabad '89]
8. Explain how **Bohr's Theory** accounts for the line spectrum of H-atom.
   [Calcut' 80, Osmania' 81, Himachal '82, Rajasthan '82, '86, Banaras '80]
9. How did Sommerfeld modify the **Bohr's Model of atom?** [Rajasthan '87]
10. Deduce the following relation from **Bohr's Theory**.

$$E_n = \frac{2\pi^2 Z^2 me^4}{n^2 h^2}$$

[Gauhati 2000]

11. Write the demerits of **Bohr's atomic model**. [Nagpur, 2002]
12. (a) Discuss **Bohr's Model** of the atom. How does it account for the **Hydrogen Spectra?** What are its **limitations?**
    (b) Calculate the frequency of spectral line emitted in case of **Hydrogen atom** when the *electron* jumps from level 2 to level 1. [Lucknow 2001]
13. K.E. of an electron is 5.76×10^{-15} J. Calculate the wavelength associated with the electron. [Mass of electron = 9.1×10^{-31} kg and $h = 6.626 \times 10^{-34}$ Js] [Delhi '99]
14. Describe **Mosley's work on the X-ray spectra of elements** with respect to instrument and equation. Bring out its theoretical importance. [Poona '86, Madras '85]
15. (a) Derive **Schrödinger Wave Equation** in three dimensions and what is the justification for this equation.
    (b) Draw the **shape of d-orbitals.**
    (c) Place the electrons identified by **quantum numbers 'n'** and 'l': (i) $n = 4, l = 1$; (ii) n = 4, 1 = 0; (iii) $n = 3, l = 2$; (iv) $n = 3, l = 1$ – in the order of their decreasing energy.
    [Himachal 2000]
16. (a) Derive **Schrödinger Wave Equation**. How does it keep in determining position of electrons in an atom?
    (b) Write a note on **Hamiltonian Operator**. [Kanpur 2000]
17. Write a note on '**Magnetic Quantum Number**' [Kanpur 2000]
18. (a) Write a note on **significance of ψ.**
    (b) What is the **significance of *four* Quantum Numbers?** [Delhi 2002]
19. (a) **Define Orbital**
    (b) What are *n*, l and *m* values for $2p_x$ and $3p_y$ electrons?
    (c) Explain briefly time independent **Schrödinger Wave Equation**. [Delhi 2003]

20. (a) What are **Quantum Numbers**? Explain them clearly.
    (b) Mention number of unpaired electrons in a **sub-shell** of the element whose atomic number is 29.
    (c) Give all the **Quantum Numbers** of the 15th electron of the atom with atomic number 19.
    (d) Designate the orbital for which:
         (i) $n = 3$, $l = 1$, and $m = 0$; and
         (ii) $n = 2$, $l = 1$, and $m = 0$.
    (e) Write down the general form of **Schrödinger's Wave Equation** and define each of its term.
    (f) Explain **Heisenberg Uncertainty Principle**. How does it influence the concept of the electron?
    (g) A moving cricket ball, weighing 200 g, is to be located within 0.2 Å. What is the uncertainty in the velocity? Comment on your result. Given: h = 6.626×10^{-34} Js   [Lucknow 2001]

21. (a) Define an **orbital**.
    (b) What are n, l, and m values for $2p_x$ and $3p_y$ electrons?
    (c) Explain briefly time independent **Schrödinger Wave Equation**.

22. (a) Calculate the **Screening constant** for the elements with atomic numbers 11 and 17.
    (b) State **Paulie's Exclusion Principles**.

[Nagpur 2003]

23. The **1st, 2nd, 3rd, and 4th shells** of an atom may contain maximum 2, 8, 18, and 32 electrons respectively. Explain this arrangement in terms of **Quantum Numbers**.   [Rohilkhand 2003]

24. (a) Which of the following orbitals is nearest to the nucleus: *Af, 5d, 6s,* and *6p*?

    (b) Name the rule according to the **V.S.E.C of C-atom** is: $2s^2, 2p_{x^1}, 2p_{y^1}, 2p_{z^0}$; and not

         $2s^2, 2p_{x^2}, 2p_{y^0}, 2p_{z^0}$.   [HN Bahuguna 2006]

25. What is the values of n, m, l for $2p_{x^1}$ **electron**?   [Meerut 2007]

26. (a) Which of the following orbitals are not possible: **1p, 2s, 2p, 2d,** and **3f**?
    (b) What are the values of **quantum number** of *10th electron* of K-atom?
    (c) What are n, l, and m values for $3p_x$ **electron**?   [Meerut 2006]

27. (a) Arrange the following *orbitals* in the increasing order of their energy: **5p, 3s, 4d,** and **6s**.
    (b) What are the **values of n and l for $5d$ orbital**?   [Purvanchal 2007]

28. (a) Write a note on **Heisenburg's Uncertainty Principle**.
    (b) What are **Quantum Numbers**? Explain their significance.   [HN Bahuguna, 2007]

29. (a) What is **Schrödinger Wave Equation**? Give the **derivation and significance of ψ and ψ^2**.
    (b) Give a brief account of the **Wave nature** of electron. Derive the **Broglie equation**.
    (c) Write short notes on:
         (i) **Heisenburg's Uncertainty Equation**, and
         (ii) **Pauli's Exclusion Principle**.   [Meerut 2008]

30. (a) Explain **Pauli's Exclusion Principle**.

    (b) Derive **de Broglie's Equation** for a particle of mass *m* and moving with a velocity of *v*.
    [HN Bahuguna 2008]

31. Discuss **Heisenberg's Uncertainty Principle** in details alongwith its experimental verification.
    [Agra 2008]

32. (a) Explain **Quantum Numbers** and their significance in characterising an electron in an atom.

    (b) What is **Heisenberg's Uncertainty Principle?**   [Meerut 2009]

33. (a) Explain the concept of **Atomic Orbital** and explain the differences between *orbit* and orbital with suitable examples.

    (b) What are **Quantum Numbers?** Explain **Hund's Rule of maximum multiplicity**.
    [Gurukul Kangri 2008]

PROBABLE QUESTIONS

1. Even though the *Hydrogen atoms* do have only one electron, their respective spectra have several spectral lines. Why?

2. Explain the ensuing relationship between energy levels and the resulting frequencies of the spectral lines. What does the position of the spectral line imply with respect to the exact location of the '**electron in an atom**'?

3. How would you differentiate the spectra of *ionized* **helium (He) and lithium (Li)** from that of **hydrogen (H)**?

4. What are the articulated assumptions of **Bohr's Theory** of the **Hydrogen Spectra?**

5. Calculate the *circumference, radius of the orbit, velocity, kinetic energy*, and *total energy* of an '**electron**' in:

    (a) **second, third, fifth, and sixth orbit of H-atom**;

    (b) **first, second, third, and fourth orbit of He$^+$ and Li^{2+} ions**.

    Comment on the results thus obtained.

6. Why do we find the energy of the **2nd orbit** greater than the **1st orbit?**

7. Calculate the *Ionization Potential* of **H, Li, and B**.

8. Based on the *Bohr's Model*, calculate the velocity, kinetic energy and radius of the **M shell of Li^{2+} and He$^+$ ions**.

9. State briefly **Bohr's criterion** for a *stationary orbit*. Calculate the frequency of light radiated when an electron in an excited state falls duly from the **2nd orbit to the 1st orbit**.

10. Enumerate the various underlying facts in support of the belief that electrons are the **essential constituents of atoms?**

11. The energy in the **Bohr's orbit** is usually given by $-A/n^2$ (where, $A = 2.179 \times 10^{-18}$ J. Calculate the *frequency of the radiation* and the *wavenumber of the compounding radiation*, when the electron moves from **3rd orbit** to the **2nd orbit. Use $h = 6.62 \times 10^{-34}$ Js**.

12. Consider *two* **H-atoms**. The electron in the **1st atom** is the $n = 1$ Bohr orbit, and that in the **2nd atom** is in the 4th orbit:

    (a) which atom has the **ground state configuration?**

    (b) in which atom, is the **electron moving faster?**

(c) Which atom has **greater ionization energy?**

(d) Which atom has **lower potential energy?**

13. What is the relationship prevailing between **Planck's constant (h)** and the following:

(i) **Back body radiation,**

(ii) **Photoelectric effect,** and

(iii) **Computer Scattering of X-rays.**

14. How would you show that **Planck's constant (h)** has the units of an **angular momentum?**

15. In case, an *electron* and a *photon* are travelling with almost the *same velocity*, which one has a *greater wavelength* and why?

16. Calculate the energy of an *electron* whose **De Broglie** *wavelength* is 125 pm.

17. How does the **Heisenberg's uncertainty principle** apply to the solar systems? Explain.

18. Write **Schrödinger's equation** and explain each term involved.

19. What is **Pauli's Exclusion Principle** and **Hund's Rule?**

20. How many *electrons* are actually present in the **Valence Shell** of V (Z = 23), CO(27), Kr(36), Ag(47), Te(52), Hg(80), Gd(64), and U(92)?

21. What do you understand by the consequence of the **stability of the** *half-filled* and *completely filled* **shells?** Why should these shells be associated intimately with higher stability profile?

22. How do the electrons get duly distributed in the *M and L shells*? What are their actual **Quantum Designations?**

23. What do you understand by the **penetration of the orbitals?** What would be the ultimate consequences of the **orbital penetration?**

24. What **atoms** are duly indicated by the following **configurations:**
    $1s^2 2s^2 2p^1$; [Ar] $4s^2 3d^1$; [Kr] $4d^4 5s^3$; and [Kr] $5s^2 5p^1$? Give suitable reasons.

25. **De Broglie** proposed that all matter has a *dual behaviour*. How does a *matter wave* differ from an *electromagnetic wave?*

26. How will you differentiate between *1s and 2s*; and between $p_x, p_y,$ and p_z orbitals?

27. How will you show that the **De Broglie's equation** also holds good for the **matter waves?**

28. How would you actually account for and explain duly the various divergent shapes of the so-called–**degeneration of orbitals?**

29. Name the **different orbital variants commonly known in elements?**

30. In the **heavier atoms,** the orbitals do have a tendency to behave in a manner very much similar to the '**orbitals of hydrogen**'. Comment.

MODERN PERIODIC TABLE: ELECTRONIC CONFIGURATION OF ATOM AND ATOMIC PROPERTIES

1. INTRODUCTION

The underlying concept of '**The Periodic Table**' was initially put forward by the so-called:

"**the magnificent widely accepted generalization stated by Mendeleev in 1869 and by Lothar Meyer in the following year,**"–that the *characteristic features (properties) of the elements may be duly represented as the periodic functionalities of their respective atomic weights,*–which was soonafter put forward by the said to academic genium in the form of a **Periodic Table**. Interestingly, the '**Original version of the Periodic Table**' has virtually undergone a sea change in terms of relevant modification perhaps due to:

- **discovery of the** *Noble Gases* (*viz.*, **Neon, Argon, Krypton, Xenon, Radon**);
- **actinides** (*actinism–refers to the production of chemical change in a substance upon which the electromagnetic radiation is incident*); and
- **Subsequent elucidation of the structure of atoms**.

In addition, it has been duly established and recognized that the so-called *Modern Periodic Law* is essentially the dire consequence of:

"**the periodic variation duly observed in electronic configuration**".

1.1. Chronological Events Related to Historical Development of Periodic Table

These chronological events are stated as under:

- **Aristotle [384–322 BC]**–The observed *atomic nature of matter (Greek:* α γ 0 μ 0 σ = incapable of further subdivision) had been duly accepted since **Aristotle (384–322 BC)***.
- **Paracelsus [ca AD 1495–1551]**–made a fruitful attempt to correlate the ensuing **biological functioning** of the so-called:

    "*triaprima* of Albert Magnus [Mercury (Hg), Sulphur (S), and Salt]."

- **Boyle's Theory (1677) of Atoms**–It relates to the *tiny individual particulate matters* of various **shapes** and **masses**; and hence, it enabled the *Inorganic Chemistry* to correlate vehemently the

---

*   Although Aristotle's ideas were more of a '**philosophical character only**'.

characteristic to correlate vehemently the *characteristic features of the elements* together with their respective atomic masses.

Obviously, the earlier attempts to establish a due correlation between the mass and the properties of the ensuing *elements* **by making use of** *Dalton's Law (1808)* **were found to be constrained severally by the following glaring facts, namely:**

➢ **ambiguity seen between the** *'atomic weights (A)'* **and** *'equivalent weights'*,

➢ **non-availability of consistent set of atomic or combining weights;**

➢ **a good number of elements not even been discovered; and**

➢ **combining the** *ratios of elements* **in already known chemical entities (compounds) were found to be** *erroneous perceptively*.

Comment: However, the advent of a progressive systematic approach in the domain of '*inorganic chemistry*' has proved to be *synonymous* with the simultaneous aggressive investigative study pertaining to the '*periodic relationships*' of the **elements** and their corresponding **compounds**.

- **Dobereiner JW (1929)**–He solemnly suggested that the prevailing '**triads of elements**' possessing *identical characteristic features,* such as:
  - **Ca, Sr, Ba** • **Li, Na, K** • **Cl, Br, I** • **S, Se, Te**.

  wherein the **atomic weight** of the '*middle element*' was the mean of the other *two* **elements**.
- **Pattenkofer (1850)**–He showed that the **atomic weights** of the identical elements differed particularly by the **integral multiples of '8'**, such as:

  Li(7), Na(23), K(39), Mg(24), Ca(40), Sr(88), Cu(40), S(32), Se(80), Te(28).
- **Chancourtois (1862)**–He demonstrated vividly that the elements do differ from each other by '**16**' or its *multiple in atomic weight* invariably had a similar behaviour; and, therefore, concluded that the characteristic features of the **elements** of the corresponding *numbers*.
- **Nieuwland JA (1932)**–He observed that on arranging the *known* elements in the order of increasing **atomic weights**, every '**8th element**' did possess *similar characteristic features* (usually known as the **Law of Octaves**).

Though *Nieuwland* **was virtually ridiculed for providing** *'blank spaces'* **for the yet-to-be discovered elements, he was considered to be the coveted forerunner of the today's** *'Periodic Table'*; **and hence, was awarded the most deserved the** *Davy Medal (1887)* **in proper and marked recognition of his invaluable contribution.**

1.2. Mendeleef's Periodic Law and Periodic Table

Let us first discuss the '**periodic law**' at length in order to understand appreciably the so-called '**Modern Periodic Table**'.

The Periodic Law It was **Medeleev** who forward the famous **Periodic Law** which states that:

"**the characteristic features of the elements do represent the periodic functions of their** *atomic weights* **respectively**".

Now, based on this dictum, if the ensuing elements were arranged meticulously according to the *increasing* **atomic weights**, identical elements must take place at regular intervals. Amazingly, exactly the similar observations were duly recorded by a *three* noted scientists, namely:

- **Dobereiner** • **Pathenkofar** or • **Chancourtois,**

Critically on a rather '*smaller scale*'; however, **Mendeleev** extended the same to practically all the *known elements* at his era.

Since it is well known that:

"the characteristic features of the *elements* solely depend upon the so-called *electronic configuration* of the ensuing *elements*"; and, therefore, it would be worth while to remember the dictum duly proposed by **Mosley** with respect to the **Periodic Law**–that states as under:

"the properties of elements do designate the *periodic function* of their respective *Atomic Numbers*".

Therefore, if the elements are carefully put in order in a sequence intelligently as per their prevailing 'atomic numbers', identical elements are most likely to take place at *regular intervals*.

Example: Obviously, one may critically take cognizance of the underlying fact that when one takes a closer look at the so-called *long form* of the Periodic Table of Elements,–as depicted in Table: 2.1.

Table : 2.1: The Long Form of the Periodic Table or The Bohr's Periodic Table of Elements.																		
ns^x						$(n-1)d^x ns^2$									$ns^2 np^x$			
n	s^1 s^2	d^1	d^2	d^3	d^4	d^5	d^6	d^7	d^7	d^8	d^9	d^{10}		p^1	p^2	p^3	p^4	p^5 p^6
1	H He																	
2	Li Be													B	C	N	O	F Ne
3	Na Mg													Al	Si	P	S	Cl Ar
4	K Ca	Sc	Ti	V	Cr	Mn	Fe	Co	Ni	Cu	Zn			Ga	Ge	As	Se	Br Kr
5	Rb Sr	Y	Zr	Nb	Mo	Tc	Ru	Rh	Pd	Ag	Cd			In	Sn	Sb	Te	I Xe
6	Cs Ba	La*	Hf	Ta	W	Re	Os	Ir	Pt	Au	Hg			Tl	Pb	Bi	Po	At Rn
7	Fr Ra	Ac**																

	f^1	f^2	f^3	f^4	f^5	f^6	f^7	f^8	f^9	f^{10}	f^{11}	f^{12}	f^{13}	f^{14}
*4f^x	Ce	Pr	Nd	Pm	Sm	Eu	Gd	Tb	Dy	Ho	Er	Tm	Yb	Lu
**5f^x	Th	Pa	U	Np	Pu	Am	Cm	Bk	Cf	Es	Fm	Md	No	Lr

Limitations of the Long Form of the Periodic Table–It is, however, pertinent to state here that the so-called '**Periodic Table of Elements**' does represent categorically:

"*not* a relevant source of information in itself; and hence, may project the principles upon which the classification is based predominantly, and nothing else whatsoever".

Comment: Therefore, we may critically observe the fact that: "the underlying limitations pertaining to the *modern form of the Periodic Table* do designate the so-called certain crucial limitations of the *Auf Bau principle*"

Following are the *eight* cardinal observed limitations of the long form of the Periodic Table, such as:

1. It is not only tedious but also cumbersome to draw effectively since it has: $2(s) + 6(p) + 10(d) + 14(f) = 32$ **columns,** for each and every probable type of *electronic configuration*. Therefore, on an

average, a division of nearly **100 elements** into several groups might leave ultimately **3 to 4 elements per group**.

In this may, not many elements are prone to exhibit the similar behavioural pattern.

2. Importantly, the so-called **inner transition elements** are virtually thrown out of the *main-body* of the *Periodic Table*. Hence, these are usually shown at the *foot of the Periodic Table–as* **two** altogether **individual periods** in such a manner that they do not seem to belong to the **main congregation of elements** or as if they fail to behave in the expected way in the **main Periodic Table**.

3. Even though the '**isotopes***' do essentially possess **identical chemical characteristics** they invariably do inherit altogether *divergent characteristics* (that are mostly physical in nature). Besides, the **isotopes of hydrogen** are found to be quite different, that the so-called *heavier-isotopes* are invariably known by altogether having '**different names with different symbols**', namely:

> **It resembles all these groups of elements in its properties.**
> * **Deuterium (D)** – for the **isotope** with *atomic mass number* **2**; and
> * **Tritium (T)** – for the **isotope** with *atomic mass number* **3**.

Comment: In true sense, the aforesaid '*problem*' does persist quite often in the *Periodic Table*.

4. The exact, and precise strategical position of '**hydrogen**' still remains a '**matter of controversy**':
   * whether it is an *one-electron element (Group I)*, one electron shorter than the *inert gas element (Group VII)*; or
   * exist as *half-the valence shell 'filled element* (**Group IV**)'

5. Even though the Periodic Table possesses '**groups**' that do belong to almost all the **electronic configuration**; and hence, there exist practically '**no such group**' having the following *electronic configuration of the element as:*

$$(n-1)d^4ns^2$$

i.e., it relates to a configuration found specifically for **Tungsten [W] with** $n = 6$.

Important Points–These essentially include:

   (i) Almost identical situations may exist for the '**4d**' and inner transition elements (or *f*-block elements) *viz.*, elements from **Cerium (Ce) to Lutetium (Lu), and their homologues from Thorium (Th) to Lawrenthum (Lr)**.

   (ii) The aforesaid critical observation relates directly to the consequence of the **Hund's Rule** (see **Chapter-1 section 1.1.6.2**) of *maximum multiplicity*; and, therefore, is certainly not a *major drawback* of the **Periodic Table**.

   (iii) Nevertheless, it is certainly not yet clear from Table: 2.1 as to why the *elements* having either:
   * **half-filled '*d*' subshell**; and
   * **completely filled subshells**,

---

   * **Isotopes:** They refer to **elements** that do have *same number of protons* but **different number of protons**; and hence, **different masses**.

- *viz.*, **Chromium (Cr), Molybdenum (Mo) [having $(n-1)d^5 ns^1$ configuration];** whereas, **Copper (Cu), Silver (Ag), and Gold (Au) [having $(n-1)d^{10} ns^1$ configuration]** fail to exhibit the so-called characteristics usually expected of their corresponding *ground state* **electronic configuration.** *

6. Even through one may critically observe the **4th period 3d elements**, the *electronic configurations* are duly expressed in the respective **5th (4d)** and **6th period** elements **(5d)**, the respective **Auf Bau principle** is violated grossly by a large segment of the elements.

Perhaps, based on these glaring facts the *elements with different* valence shell configurations are meticulously placed together in the same group of the *d block elements* since quite a few of the inherent characteristics present in these elements categorically exhibit close resemblences to one another. Table: 2.2 provides explicitly the electronic configuration of the various elements.

		Table 2.2: Ground State Electronic Configuration of Elements	
Z		**Elements**	**Configuration**
1	H	Hydrogen	$1s^1$
2	He	Helium	$1s^2$
3	Li	Lithium	(He) $2s^1$
4	Be	Beryllium	(He) $2s^2$
5	B	Boron	(He) $2s^2 2p^1$
6	C	Carbon	(He) $2s^2 2p^2$
7	N	Nitrogen	(He) $2s^2 2p^3$
8	O	Oxygen	(He) $2s^2 2p^4$
9	F	Fluorine	(He) $2s^2 3p^5$
10	Ne	Neon	(He) $2s^2 2p^6$
11	Na	Sodium	(Ne) $3s^1$
12	Mg	Magnesium	(Ne) $3s^2$
13	Al	Aluminium	(Ne) $3s^2 3p^1$
14	Si	Silicon	(Ne) $3s^2 3p^2$
15	P	Phosphorous	(Ne) $3s^2 3p^2$
16	S	Sulphur	(Ne) $3s^2 3p^4$
17	Cl	Chlorine	(Ne) $3s^2 3p^5$
18	Ar	Argon	(Ne) $3s^2 3p^6$
19	K	Potassium	(Ar) $4s^1$
20	Ca	Calcium	(Ar) $4s^2$
21	Sc	Scandium	(Ar) $3d^1 4s^2$
22	Ti	Titanium	(Ar) $3d^2 4s^2$
23	V	Vanadium	(Ar) $3d^3 4s^2$
24	Cr	Chromium	(Ar) $3d^5 4s^1$
25	Mn	Manganese	(Ar) $3d^5 4s^2$
26	Fe	Iron	(Ar) $3d^{96} 4s^2$
27	Co	Cobalt	(Ar) $3d^7 4s^2$
28	Ni	Nickel	(Ar) $3d^6 4s^2$

(Contd..........)

* That is, the observed **stability profile** appears to be virtually vanishing for all the intended *chemical interactions* of most of these elements (except in the particular case of Ag); thus, all the elements actually lose the *d* electrons in the course of chemical reactions yielding the **higher valent** ions perceptively.

29	Cu	Copper	(Ar) $3d^{10} 4s^1$
30	Zn	Zinc	(Ar) $3d^{10} 4s^2$
31	Ga	Gallium	(Ar) $3d^{10} 4s^2 4p^1$
32	Ge	Germanium	(Ar) $3d^{10} 4s^2 4p^2$
33	As	Arsenic	(Ar) $3d^{10} 4s^2 4p^3$
34	Se	Selenium	(Ar) $3d^{10} 4s^2 4p^4$
35	Br	Bromine	(Ar) $3d^{10} 4s^2 4p^5$
36	Kr	Krypton	(Ar) $3d^{10} 4s^2 4p^6$
37	Rb	Rubidium	(Kr) $5s^1$
38	Sr	Strontium	(Kr) $5s^2$
39	Y	Yttrium	(Kr) $4d^1 5s^2$
40	Zr	Zirconium	(Kr) $4d^2 5s^2$
41	Nb	Niobium	(Kr) $4d^4 5s^1$
42	Mo	Molybdenum	(Kr) $4d^5 5s^1$
43	Tc*	Technetium	(Kr) $4d^6 5s^1$
44	Ru	Ruthenium	(Kr) $4d^7 5s^1$
45	Rh	Rhodium	(Kr) $4d^8 5s^1$
46	Pd	Palladium	(Kr) $4d^{10}$
47	Ag	Silver	(Kr) $4d^{10} 5s^1$
48	Cd	Cadmium	(Kr) $4d^{10} 5s^2$
49	In	Indium	(Kr) $4d^{10} 5s^2 5p^1$
50	Sn	Tin	(Kr) $4d^{10} 5s^2 5p^2$
51	Sb	Antimony	(Kr) $4d^{10} 5s^2 5p^3$
52	Te	Tellurium	(Kr) $4d^{10} 5s^2 5p^4$
53	I	Iodine	(Kr) $4d^{10} 5s^2 5p^5$
54	Xe	Xenon	(Kr) $4d^{10} 5s^2 5p^6$
55	Cs	Casium	(Xe) $6s^1$
56	Ba	Barium	(Xe) $6s^2$
57	La	Lanthanum	(Xe) $5d^1 6s^2$
58	Ce*	Cerium	(Xe) $4f^2 6s^2$
59	Pr*	Praseodymium	(Xe) $4f^3 6s^2$
60	Nd*	Neodymium	(Xe) $4f^4 6s^2$
61	Pm*	Promethium	(Xe) $4f^5 6s^2$
62	Sm	Samarium	(Xe) $4f^6 6s^2$
63	Eu	Europium	(Xe) $4f^7 6s^2$
64	Gd*	Gadolinium	(Xe) $4f^7 5d^1 6s^2$
65	Tb*	Terbium	(Xe) $4f^9 6s^2$
66	Dy*	Dysprosium	(Xe) $4f^{10} 6s^2$
67	Ho*	Holmium	(Xe) $4f^{11} 6s^2$
68	Er*	Erbium	(Xe) $4f^{12} 6s^2$
69	Tm*	Thullium	(Xe) $4f^{13} 6s^2$
70	Yb	Ytterbium	(Xe) $4f^{14} 6s^2$
71	Lu	Lutetium	(Xe) $4f^{14} 5d^1 6s^2$
72	Hf	Hafnium	(Xe) $4f^{14} 5d^2 6s^2$
73	Ta	Tantalum	(Xe) $4f^{14} 5d^3 6s^2$
74	W	Tungsten	(Xe) $4f^{14} 5d^4 6s^2$
75	Re	Rhenium	(Xe) $4f^{14} 5d^5 6s^2$
76	Os	Osmium	(Xe) $4f^{14} 5d^6 6s^2$

(Contd..........)

77	Ir*	Iridium	(Xe) $4f^4 5d^7 6s^2$
78	Pt*	Platinum	(Xe) $4f^{14} 5d^9 6s^1$
79	Au	Gold	(Xe) $4f^{14} 5d^{10} 6s^1$
80	Hg	Mercury	(Xe) $4f^{14} 5d^{10} 6s^2$
81	Tl	Thallium	(Xe) $4f^{14} 5d^{10} 6s^2 6p^1$
82	Pb	Lead	(Xe) $4f^{14} 5d^{10} 6s^2 6p^2$
83	Bi	Bismuth	(Xe) $4f^{14} 5d^{10} 6s^2 6p^3$
84	Po	Polonium	(Xe) $4f^{14} 5d^{10} 6s^2 6p^4$
85	At	Astatine	(Xe) $4f^{11} 5d^{10} 6s^2 6p^5$
86	Rn	Radon	(Xe) $4f^{14} 5d^{10} 6s^2 6p^6$
87	Fr	Francium	(Rn) $7s^1$
88	Rs	Radium	(Rn) $7s^2$
89	Ac*	Actinium	(Rn) $6d^1 7s^2$
90	Th*	Thorium	(Rn) $6d^2 7s^2$
91	Pa*	Protoactinium	(Rn) $5f^2 6d^1 7s^2$
92	U*	Uranium	(Rn) $5f^3 6d^1 7s^2$
93	Np*	Neptunium	(Rn) $5f^5 7s^2$
94	Pu*	Plutonium	(Rn) $5f^6 7s^2$
95	Am*	Amercium	(Rn) $5f^7 7s^2$
96	Cm*	Curium	(Rn) $5f^7 6d^1 7s^2$
97	Bk*	Berkelium	(Rn) $5f^7 6d^2 7s^2$
98	Cf*	Californium	(Rn) $5f^9 6d^1 7s^2$
99	Es*	Einstenium	(Rn) $5f^{11} 7s^2$
100	Fm*	Fermium	(Rn) $5f^{12} 7s^2$
101	Md*	Mendelevium	(Rn) $5f^{13} 7s^2$
102	No*	Nobelium	(Rn) $5f^{14} 7s^2$
103	Lr*	Lawrencium	(Rn) $5f^{14} 6d^1 7s^2$

* Configurations are uncertain

7. Likewise, all the **fourteen lanthanons** do not possess the anticipated $(n-2) f^x (n-1) d^1 ns^2$ configuration in the *ground state*. Hence, these are simply placed strictly as per the **increasing atomic numbers**.

8. The so-called *borderline elements* e.g., **Zinc (Zn)**, **Cadmium (Cd)**, and **Mercury (Hg)** for the respective *d* and *p* elements; and **Lutetium (Lu)** for the corresponding *d* and *l* elements. Besides, the **first elements of the d series** viz., **Scandium (Sc)**, **Yttrium (Y)**, and **Lanthanum (La)** are obviously found to be *not too different vis-a-vis the preceeding elements*, such as: Cu, Ag, Au, Y, Ca, Sr, and Ba respectively.

NOTE   **Note Importantly, their characterisitcs features may be duly accomplished by the *simple extrapolation* in a period. Hence, in the *long table*, they have been drastically separated and the differences in their properties have been expressed in a dramatic way.**

2. SYSTEMATIC APPROACH TO MODERN PERIODIC TABLE

Having acquired the fundamentals of the **Periodic Table** it would be absolutely necessary to explore a systematic approach to the so-called **Modern Periodic Table** with particular reference to the following relevant aspects, namely:

  (i) **Moseley's Modern Periodic Law,**

 (ii) **Long Form Periodic Table**

    (a) **Merits of Long Form of Periodic Table** *vis-a-vis* **Mendeleef's Periodic Table; and**

    (b) **Demerits of Long Form of Periodic Table,**

2.1. Moseley's Modern Periodic Law

Moseley (1919)* carried out a quantitative investigative study of the *characteristic X-rays*** from **thirty eight elements** by employing a crystal of potassium ferrocyanide [K_4 Fe(CN)$_6$] as the so-called:

- **diffraction grating**, and
- **a photographic plate***.

Later on, Barkla and Sadler duly observed that each ray duly comprised of *two* **spectral lines** (*viz.*, K_a and K_β for **K rays**). However, the plots of *square root of the ensuing frequency (v)* of the characteristic **X-ray line** *Vs* atomic number (Z)**** of the corresponding *element* critically yielded in:

- **straight lines for Aluminium (Al) (Z = 13) to silver (Ag) (Z = 47) for the respective K-line;** and
- **Zirconium (Zr) (Z = 40) to Gold (Au) (Z = 79) for the respective L-lines respectively.** L-lines respectively.

Fig: 2.1 depicts the plots of the frequency (V) related to the *characteristic* X-rays Vs the corresponding **atomic number (Z):**

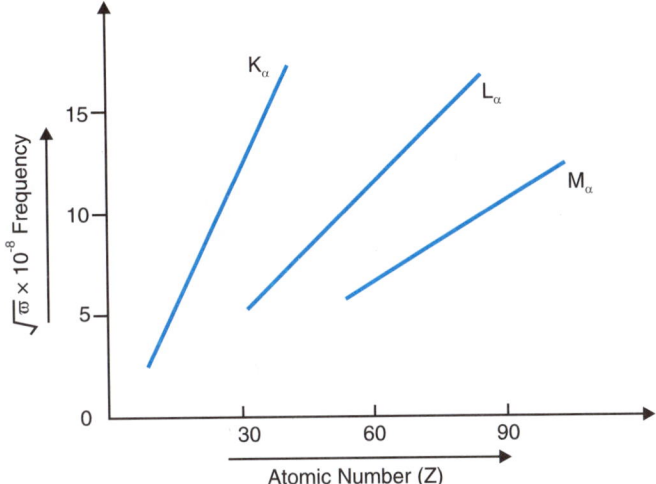

Fig.2.1: The Plots of the Frequency (V) of the Characteristic X-Rays *Vs* the Atomic Number (Z).

Thus, based on the aforesaid findings and observations **Moseley** solemnly proposed that:

$$\sqrt{V} = a\ (Z - b)$$

---

    *   Moseley HGJ (1919)

  ** That is, X-ray crystallographic studies.

 *** As a '**Screen**' duly developed by Bragg.

**** As determined by **Rutherford's experiment.**

where,

a = Proportionality constant (1.0 for K-line) and (7.4 for L-line); and

b = Same for all the lines of a given series.

Remarks: These essentially include:

1. No such relationship could be observed when the **atomic number (Z)** was duly replaced by the **atomic weight (A)** of the element.

2. Thus, **Moseley** vehemently concluded that the so-called **nuclear charge** (Z) or a characteristic feature related intimately to Z serves as a **fundamental property of the elements**.

Importantly, Moseley (1919) opined strongly that since the vital and important *physical* and *chemical* **characteristic features** of an '*element*' dependent exclusively upon the precise number of the **electrons** and their respective arrangements in **various orbitals of the atom**, the so-called *systematic classification of the elements* must be based meticulously upon the following *two* **guiding aspects**, namely:

- **number of the electrons (*i.e.*, Atomic Number Z)**; and
- **their respective arrangement in various *orbitals***.

2.2. Long Term Periodic Table

The major objective of the **Periodic Table** relates to the *systematic* and *articulated*:

"**organization and systematizing the so-called chemistry of the Elements.**"

In a broader perspective, the **Long Term Periodic Table** critically fulfils and satisfies the aforesaid *objective* quite elegantly and satisfactorily. As an date, the total number of **known elements at present** exceeds 100 (*i.e.*, including those elements that are prepared exclusively *via* the **Nuclear Reactions** in the recent part). However, a meticulous systematic investigative study reveals the characteristic features of various **moieties** thereby enlightening a fairly **good insight** with respect to the overall.

Table 2.3: illustrates the Long Form of Periodic Table:

Table 2.3: The Long From of the Periodic Table.																	
ns^x					$(n-1)d^x ns^2$								$ns^2 np^x$				
n	s^1 s^2	d^1	d^2	d^3	d^4 d^5 d^6	d^7	d^7	d^8	d^9	d^{10}		p^1	p^2	p^3	p^4	p^5	p^6
1	H He																
2	Li Be											B	C	N	O	F	Ne
3	Na Mg											Al	Si	P	S	Cl	Ar
4	K Ca	Sc	Ti	V	Cr Mn Fe	Co	Ni	Cu	Zn			Ga	Ge	As	Se	Br	Kr
5	Rb Sr	Y	Zr	Nb	Mo Tc Ru	Rh	Pd	Ag	Cd			In	Sn	Sb	Te	I	Xe
6	Cs Ba	La*	Hf	Ta	W Re Os	Ir	Pt	Au	Hg			Tl	Pb	Bi	Po	At	Rn
7	Fr Ra	Ac**															

	f^1	f^2	f^3	f^4	f^5	f^6	f^7	f^8	f^9	f^{10}	f^{11}	f^{12}	f^{13}	f^{14}
*4f^x	Ce	Pr	Nd	Pm	Sm	Eu	Gd	Tb	Dy	Ho	Er	Tm	Yb	Lu
**5f^x	Th	Pa	U	Np	Pu	Am	Cm	Bk	Cf	Es	Fm	Md	No	Lr

- **behavioural pattern**, and
- **vital characteristics**,

pertaining to a majority of the elements.

Salient Features– The **Long Form Periodic Table** possesses the following *two* **salient features**, such as:

1. It broadly helps us to understand the basic reason as to why some elements resemble with one another; whereas, in some instances they do differ from the other elements with respect to their inherent properties. Thus, it helps us to arrange the elements in the **long Form Periodic Table** in such a manner bearing in mind the **precise electronic configuration**.

2. It significantly helps one to-**understand comprehensively the fundamental cause of** *periodicity* **of properties and also the reason why similar characteristic features invariably recur at** *certain regular intervals* **viz., as could be observed after: 2, 8, 8, 18, 18, and 32 elements**.

> **NOTE**   The aforesaid numbers: **2, 8, 8, and 32** are sometimes termed as the *Magic Numbers*.

3. The **Long Form Periodic Table** is provided at the back of little page of the book. It has been constructed intelligently by arranging the *various* **known elements:** 'specially in the order of their increasing Atomic Number'. Besides, these elements are arranged in *two* distinct modes:
    - **Horizontal Rows: constitute the 'Periods'**; and
    - **Vertical Rows: constitute the 'Groups'**.

☐ **PERIODS**–It may be observed that there are *seven* **periods** or *horizontal rows* in the so-called **Long Form Periodic Table**, which shall now be treated individually in the sections that follows:

   ➢ *First Period:* It is duly made up of only *two* **elements,** such as:
      - **Hydrogen** and • **Helium**

Properties of Hydrogen–are somewhat very much akin to those of the **Group–1: alkali metals** (*viz.* **Li, Na, K, Rb, Cs, and Fr**); and also to those of the **halogens: Group-17** (*viz.,* **F, Cl, Br, I, At, and Uus**) in certain critical aspects perceptively.

Comments:

1. Though it is more reasonable but not absolutely logical to place it in Group-1
2. Helium–designates a typical 'Nobel Gas'; and hence, belongs to **Group–18**.
   ➢ *Second Period:* The **second period** comprises in all *eight* **elements**. It commences with **Lithium (Li) (an** *alkali metal* **of Group-1)**. The other elements duly placed in the *succeeding moieties* are: **Beryllium (Be), Boron (B), Carbon (C), Nitrogen (N), Oxygen (O), Fluorine (F), and Neon (Ne)**–a noble gas.
   Comments: The *second* **period** also ends with a **noble gas** duly positioned in **Group-18**.
   ➢ *Third Period:* It critically starts with **Sodium (Na)** and also comprises *eight elements* viz., **Sodium (Na), Magnesium (Mg), Aluminium (Al), Silicon (Si), Phosphorus (P), Sulphur (S), Chlorine (Cl),** and **Argon (Ar)**–the noble gas.

Comments: The first *seven* elements are duly placed in **Groups: 1, 2, 13, 14, 15, 16, and 17** respectively; whereas, **Argon (Ar)** is placed in **Group-18**.

> **NOTE**   Amazingly, the *third period*, just like the *first* and *second periods* usually terminates with a 'Noble Gas'.

➤   *Fourth Period:* As usually it also commences with an **alkali metal Potassium (K)**. Importantly, one may have to *cross-over* almost **seventeen** elements before one comes across a **Noble Gas: Krypton (Kr)**. Obviously, the *fourth* period comprises essentially of **Eighteen Elements** starting from: **Potassium (K)** [$Z = 19$] to the **Noble Gas: Krypton (Kr)** [$Z = 36$]. Interestingly, the **first two elements** *viz.*, Potassium (K) and **Calcium (Ca)**–do exhibit explicitly the so-called **recurrence of characteristic features**.

Thus, we may take cognizance of the fact that: **Potassium (K)** resembles **Lithium (Li)** and **Sodium (Na)** of **Group-1**, wherein it very much falls; whereas, **Calcium (ca)** resembles both **Beryllium (Be)** and **Magnesium (Mg)** belonging to **Group-2** wherein it falls.

Important Points: These essentially comprise:

1. A difficult situation arises soonafter the element **Calcium (Ca)**, since the very next element **Scandium (Sc)**–does possess very little in common with **Aluminium (Al)**.

2. Further, one may have to cross over *ten* elements *i.e.*, from **Scandium (Se)** to **Zinc (Zn)** before one arrives at an element **Gallium (Ga)**,–that eventually shows a close resemblance with **Aluminium (Al)** belonging to **Group-13**.

3. **Gallium (Ga)**, therefore, is duly positioned just below **Aluminium (Al)**, in **Group-13**.

4. The suceeding *five* elements, such as: **Germanium (Ge), Arsenic (As), Selenium (Se), Bromine (Br)**, and **Krypton (Kr)** are usually positioned strategically in **Groups: 14, 15, 16, 17, and 18** respectively.

5. Out of the *eighteen* elements belonging to the *fourth period*, there are *eight* elements, namely: **Potassium (K), Calcium (Ca), Gallium (Ga), Germanium (Ge), Arsenic (As), Selenium (Se), Bromine (Br)**, and **Krypton (Kr)** [Z: 19, 20, 31, 32, 33, 34, 35, 36] are invariably called as:
   - **Normal** • **Main Group or** • **Representative Elements,**

   since they resemble intimately to the **elements** of the *first-three periods* under which they are placed.

6. However, the remaining *term* elements from **Scandium (Se)** to **Zinc (Zn)** are usually termed as: **Transition Elements**. Interestingly, the prevailing *differences* between the **Normal Elements** and the **Transition Elements** rests prevalently:

   'in the way the *entrance of electrons occurs*'.

7. Since the *inherent* chemical characteristic features are mostly determined by the *actual number of electrons present duly in the outermost shell.*, the so-called properties of the **transition elements**, *do not change radically.*

8. Besides, in the specific instance of the **Normal Elements**,–the so-called gradation of properties upon *moving across a period* is found to well marked and prouounced perceptively.

9. The **Transition Elements** do differ from the **Normal Elements** in an array of **specific features** *viz.*,
   - exhibition of *variable valency*,
   - forming **coloured ions**, and
   - showing *paramagnetic aspects*.

Hence, the *fourth period* is normally called as the **First Long Period**.

> **Fifth Period:** Just like the **fourth Period**, it also comprises **eighteen elements** [$Z = 37$ to 54]. It starts from **Rubidium (Rb)** [$Z = 37$] and continues up to **Xenon (Xe)** [$Zn = 54$],–that is also a **Noble Gas**. However, out of the **18 elements** only *eight* are found to be the so-called **Normal Elements** *viz.*, $Z = 37, 38, 49, 50, 51, 52, 53$ and 54; whereas, the rest of the *ten elements viz.*, $Z = 39$ to 48 belongs to the class of the **Transition Elements**.

> **Sixth Period:** It essentially consists of **thirty two elements** *i.e.*, $Z = 55$ to 86, which comprises *three* **distinct types of elements,** namely:

> ❑ **Normal Elements (Nos: 8):** $Z = 55, 56, 81, 82, 83, 84, 85$ and 86;

> ❑ **Transition Elements (Nos: 10):** $Z = 57, 72$ to 80; and

> ❑ **Rare Earth Elements (Nos: 14):** Starting from **Cerium (Ce)** ($Z = 58$) **and stretched over to Lutium (Lu)** ($Z = 71$).

NOTE

1. The *rare earth elements* are so called since they occur *quite seldomly*. Besides, these are so *identical to one another*, and also to Lanthanum (La); and hence, these are termed collectively as 'Lanthanides'.

2. In case, these elements are arranged in a *horizontal manner* in the same moiety, one would certainly observe an *under expansion* of the Periodic Table. Hence, in order to avoid this situation they are duly arranged as a *separate series* below the Periodic Table as shown.

> **Seventh Period:** Importantly, it solely comprises the typical **Radioactive Elements** having the **Atomic Numbers** commencing from **87 and beyond**.

Points to Ponder:

1. As to date, the elements upto **Atomic Number '108'** only have been duly discovered.

2. The elements **beyond the Atomic Number '92'** have been meticulously prepared *via* the *Nuclear Reactions artificially*.

3. Very much akin to the elements **beyond the Atomic Number '92'** are called as *Lanthanides* and in the same vein the elements just after **Aetinium (Ac)** *i.e.*, right from **Thorium (Th)** to **Lawrencium (Lr)** [$Z = 90$ to 103] are invariably known as: *Actinides*.

4. These elements are arranged systematically as an altogether *'separate series' i.e.*, just below the **Periodic Table** (as shown vividly).

5. **Elements Beyond Uranium (U)**–The elements beyond **Uranium (U)** that have been prepared artificially by means of the aforesaid **Nuclear Reactions** are usually known as–'**Transuran Elements'**. Thus, the so-called **common names** of such elements having the **Atomic numbers** ranging between **93 to 105** are as stated under:

  - **Neptunium (Np)** • **Plutonium (Pu)** • **Americium (Am)** • **Curium (Cm)**
  - **Berkeylium (Bk)** • **Californium (Cf)** • **Einstenium (Es)** • **Fermium (Fm)**
  - **Mendelevium (Md)** • **Nobelium (No)** • **Lawrencium (Lr)** • **Kurchatavium** and • **Hahnium**.

6. **IUPAC convention of 1977**–It has suggested the **New-Nomenclatures** for all such elements having the **Atomic Number beyond 100**. These names are as given under:

- Un-nil (Unium) (Unu, 101) • Un-nil-Hium (Unh, 102) • Un-nil-Trium (Unt, 103) • Un-nil-Quadium (unq, 104) • Un-nil-pentium (Unp, 105)
- Un-nil–Hexium (Unh, 106) • Un-nil-Septium (Uns, 107)
- Un-nil-Octium (Uno, 108) • Un-nil Enium (Une, 109)
- Un-un-Nilium (Unn, 110) • Un-un-Unium (Unu, 111) and
- Un-un-Hium (Unh, 112).

Table: 2.4 depicts the various number of elements in different periods:

Table : 2.4: The Various Number of Elements in Different Periods.			
Period	Orbitals filled before $ns^2 np^6$ configuration is achieved	Number of orbitals that should be filled up	Number of the elements
First	$1s$	1 = 1	2
Second	$2s, 2p$	$1 + 3$ = 4	8
Third	$3s, 3p$	$1 + 3$ = 4	8
Fourth	$4s, 3d, 4p$	$1 + 5 + 3$ = 9	18
Fifth	$5s, 4d, 5p$	$1 + 5 + 3$ = 9	18
Sixth	$6s, 4f, 5d, 6p$	$1 + 7 + 5 + 3$ = 16	32
Seventh	$7s, 5f, 6d, 7p$	$1 + 7 + 5 + 3$ = 16	32

❑ **GROUPS** – The various elements that are arranged systematically in the **Vertical Columns** of the **Periodic Table** are termed as the **Groups**. Nevertheless, in the latest **Long Form Periodic Table** [see section (B)], the corresponding **Transition Elements**–are invariably taken into account whenever we intend to carry out the **numbering of the Groups**. In actual practice, all the **Groups** are usually numbered in the so-called **Arabic** numerals (viz., 1, 2, 3, 4, …). Further numbered as **Groups 1 and 2 respectively**.

Besides, the **Groups** that are usually **headed by an element are numbered** as a specific Group, such as:

Element	Group	Element	Group
• Scandium (Sc)	3	• Iron (Fe)	8
• Titanium (Ti)	4	• Cobalt (Co)	9
• Vanadium (V)	5	• Nickel (Ni)	10
• Chromium (Cr)	6	• Copper (Cu)	11
• Manganese (Mn)	7	• Zinc (Zn)	12

Likewise, the **Groups** which are normally headed by another set of **elements** are numbered respectively as a **specific Group,** for instance:

Element	Group	Element	Group
• Boron (B)	13	• Oxygen (O)	16
• Carbon (C)	14	• Fluorine (F)	17
• Nitrogen (N)	15	• Helium (He)	18

In a situation, when the **Transition Elements** are set aside, the *rest of elements (i.e.)* the so-called **Normal Elements)** *i.e.*, excluding **Hydrogen (H)** and **Helium (He)** do arrange themselves in various **Groups** as depicted under:

Group	1	2	13	14	15	16	17	18
Element	Li	Be	B	C	N	O	F	Ne
	Na	Mg	Al	Si	P	S	Cl	Ar
	K	Ca	Ga	Ge	As	Se	Br	Kr
	Rb	Sr	In	Sn	Sb	Te	I	Xe
	Cs	Ba	Tl	Pb	Bi	Po	At	Rn
	Fr	Ra						

Remarks: The various **elements** belonging to **each Group** are critically found to be analogous to one another.

Examples: These essentially include:

- **Lithium (Li)** and **Sodium (Na)** belonging to **Group 1** do largely resemble the **elements** *viz.*, **Calcium (Ca)**, **Strontium (Sr)**, **Barium (Ba)** and **Radium (Ra)**.
- **Likewise, Fluorine (F)** and **Chlorine (Cl)** belonging to **Group 17** have characteristic features very much in common with: **Bromine (Br), Iodine (I), Astatine (At)**, and so on 80 forth.
- Elements of **Group 18** (*i.e.*, the **Nobel Gases**) do lie in between the following *two* **entities**, namely:
  - ➢ **Electronegative Hologons**; and
  - ➢ **Electropositive Alkali Metals**.

2.3. Cause of Periodicity

The **Periodic Table** reveals explicitly that all such elements that essentially possess almost **identical characteristic features** do have a tendenty to occur at *regular intervals* **in the periodic table**. Obviously, such elements invariably occur in the *same* **vertical column** *i.e.*, the **similar group**. Thus, we may observe that–

   "**the occurrence of similar characteristic features of elements after the specific regular intervals *i.e.*, when they are arranged meticulously in the order of enhancing** *Atomic Numbers* **is usually known as periodicity.**"

   At this point in time, one may raise an interesting querry as to the fundamental *cause of Periodicity*, Therefore, in order to put forward a possible and plausible answer to the querry let us take into consideration the various elements that are critically present in **Groups: 1, 17, and 18**.

   Table 2.5 records the observed **Electronic Configurations of the Groups: 1, 17, and 18** in an elaborated manner:

Salient Features–They essentially include:

1. Elements belonging to **each Group** do exhibit *identical properties*; however, they usually *differ from the elements* that belong to the **other Groups**.

Table : 2.5: The Observed Electronic Configurations of Various Groups: 1, 17, and 18.		
Elements of Group-1	**Elements of Group-17**	**Elements of Group-18**
Li: $1s^2\ 2s^1$		He: $1s^2$
Na: $1s^2\ 2s^2p^6\ 3s^1$	F: $1s^22s^2p^5$	Ne: $1s^22s^2p^6$
K: $1s^22s^2p^6\ 4s^1$	Cl: $1s^22s^2p^63s^2p^5$	Ar: $1s^22s^2p^63s^2p^6$
Rb: $1s^22s^2p^63s^2p^6d^{10}4s^2p^6\ 5s^1$	Br: $1s^2\ 2s^2\ p^63s^2p^6d^{10}4s^2p^5$	Kr: $1s^2\ 2s^2p^6\ 3s^2p^6d^{10}\ 4s^2p^6$
Cs: 2, 8, 18, 18, $5s^2p^66s^1$	I: 2, 8, 18, 18, $5s^2p^5$	Xe: 2, 8, 18, 18, $5s^2p^6$
Fr: 2, 8, 18, 32, 18, $6s^2p^6\ 7s^1$	At: 2, 8, 18, 32, 18, $6s^2p^5$	Rn: 2, 8, 18, 32, 18, $6s^2p^6$

Examples:
1. All the '**Alkali Metals**' duly present in **Group-1** are found to be:
    - **Strongly metallic in character**; and
    - **electropositive in nature**,

Since they promptly give rise to the formation of catious (*viz.*, Li$^+$ and Na$^+$) almost instantly.

2. Besides, the **Halogens** belonging to **Group-17** are, in fact, found to be both:
    - **strongly non-metallic**; and
    - **electronegative in nature**,

since they do form **anions** (*viz.*, **F, Cl$^-$**) quite promptly.

3. **Elements of Group-18** that eventually connect the *two* so called *extreme kinds of elements* are found to be:

'**chemically quite inert (*i.e.*, least reactive in nature)**'.

Explanation of Table: 2.4–They essentially include:

1. Let us first and foremost look at the status of their **electronic configurations**. Because the **electrons** located in the *inner shell* fail to participate normally in any sort of *chemical combination*,–one may hence to examine exclusively the so-called:

"**outermost electronic configurations of the elements**".

Hence, one may critically observe that:

'all the *elements* belonging to a specific *family* (*or Group*) do have identical outer electronic configuration.'

In this way, we may encounter the following *three* divergent situations, namely:

1. All the **elements belonging to Group 1** essentially possess at least **one electron** present duly in the **outermost s orbitals** preceeded by the so-called **Noble Gas configuation** perceptively (*viz*, *ns^{2}p^6: for Li 1s^2*). However, their *outer electronic configuration* may eventually be written *ns^1*.

2. Practically all the **elements of Group 17** do possess essentially *seven* **electrons** located strategically in the **outermost orbital**. In fact, *two of* these electrons are duly present in the **S orbitals**. Thus, their actual **outer electronic configuration** is observed to be *ns^{2}p^2*.

3. Thus, all the **elements of Group 18** [*with Helium (He) as an exception*] do have *eight* **electrons** located in the **outer orbital**,–whereby we have:

- *two* of these electrons are duly accomodated in the *S* orbitals, and
- *six* remaining electrons do fill up the *p*-orbitals completely.

Example: However, in **Helium (He)** there are present only *two* **electrons** and amazingly, both of which are located strategically in the *1s*-**orbital**; and hence, found to be **filled up completely**.

Comment: In this manner, practically **all the orbitals** are more or less **fully occupied** in the aforesaid elements,–that as mentioned earlier are found to be *inert* **chemically**. In addition, their **outer elect** configuration is found to be: ns^2p^6, except for **Helium (He)** which is $1s^2$.

2.4. THE COVALENT RADII [ATOMIC RADII]

Types of Radii– In true sense, it is indeed a difficult task to define precisely the *radius of an isolated atom*; and thus, it leads to a rather more difficult proposition to measure accurately:

- **internuclear distances in compounds (*chemical entities*), or**
- **gaseous molecules,**

that are determined; and thus, the *radius of the individual atoms* duly calculated therefrom.

The **radius of an atom**, therefore, solely depends upon the following *five* **cardinal factors,** namely:

- **bond order,**
- **degree of covalent or ionic nature of bond,**
- **metallic characteristic feature,**
- **oxidation status *vis-a-vis* size of neighbouring atoms,** and
- **crystalline/molecular structure.**

Therefore, an attempt has been duly made to *define a large number of radii for the elements*. Following are a few *typical and classical radii*, such as:

(i) **Covalent Radii,**
(ii) **Orbital Radii,**
(iii) **Bragg-Slatter Radii**, and
(iv) **van der Waal's Radii,**

which shall now be discussed separately in the sections that follows.

❑   **Covalent Radii**–The **covalent Radius (γ)** of an atom may be defined as:

"**half of the distance between the nuclei of two similar atoms thereby forming a covalent bond.**"

Besides, the *intermolecular distance* between *two* **C-atoms** present in a **diamond crystal** stands at **154 pm**. Therefore, the *single-bond radius* of carbon is **77 pm** since the *two atoms* present in a **diamond crystal** are meticulously held together by a **single bond** perceptively.

Importantly, the so-called **isotonic Radii** are in deed **additive in nature**; and hence, provide the **covalent-bond radii** only if the *bonds so formed are single bonds* and the ensuing *electronegativity* of the *bond atoms* fails to differ to an *appreciable extent*.

Example: Let us assume X_A and Y_B designate the so-called **electronegativities** of the *two* **elements A and B** possessing the **covalent radii** γ_A and γ_B, the observed bond length γ_{A-B} prevailing between **A and B** is usually provided by the following **Eq. (1):**

Thus, we may have the following expression:

$$r_{A-B} = r_A + r_B - g[X_A - X_B]$$

...Eq. (1)

Remarks: These essentially comprise:

1. The **Eq. (1)** holds good if the **bond (A – B)** does not exhibit the so-called *pi-character*, as could be observed by the various **bond lengths** *viz.*,

- C–I = 214 • C–Br = 194 • C–Cl = 176 • Si–I = 244 • S–Cl = 199 (pm),

It is worthwhile to state here that the **covalent Radii** (*Atomic Radii*) of the respective atoms being: C=77, Si = 157, S = 104, F = 72, Cl = 100, Br = 114, and I = 135 (pm). Nevertheless, wherever the *pi-bonding* comes into play the observed **bond lengths** do critically deviate from the values duly accomplished by the help of **Eq. (1)**; and hence, the *extent of deviation* being *directly proportional* to the corresponding *pi-character of the bond*. In this manner, the so-called: **observed** and **calculated bond lengths** related to such bonds are as given below:

- Si–F = 156 (182),
- Si–Cl = 202 (217), and
- Si–F = 156 (176) [in pm].

Multiple Bonded Atoms: In this particular instance, the respective radii are invariably known as the: **multiple bond radii**. Therefore, the **carbon** essentially possesses:

- **double-bonded radius of 66 pm in Ethylene ($CH_2 = CH_2$), and**
- **triple bonded radius of 60 pm in Acetylene (CH º CH).**

Obviously, as the **atoms** do approach distinctly closer to form the respective **multiple bonds**,–that eventually leads to *greater orbital overlap* necessarily; and hence, the **covalent bond length** gets reduced along with the **bond order**.

NOTE Neverthess, the term '*radius*' bears no physical significance pertaining to the so-called–*multiple-bonded atoms*–as they are being removed from sphericity.

Additional Types of Covalent Radii: These are usually prevalent in **tetrahedral** and **octahedral radii** forms; and hence, are certainly vital and important for:

- **Ionic Crystals**, and
- **Coordination Compounds**.

Importantly, the observed **tetrahedral radii** are indeed *smaller* in comparison to the **octahedral radii**.

Following are typical values of some of the so-called **Metallic and Covalent Radii** of certain elements from the **Periodic Table**:

Radius	K	Ba	La	Cr	In	
Metallic	231	217	188	159	162	*pm*
Covalent	203	198	169	145	150	*pm*

❑ **Orbital Radii**–It is normally determined by the aid of **principle quantum number**. It is found to be optimum in the so-called: '**radical probability distribution curve (RPDC) of the outermost orbital**'.

For Cations Obtained–by carefully stripping the **atom** of the ensuing *outermost quantum shell electrons*; and thus, the **orbital radius** gets duly reduced by a factor ranging between **1.5–5.0** on being compared with the **atomic radius**. Hence, the orbital radii **(in pm)** for the respective *outer orbitals* for: **Sodium (Na), Magnesium (Mg)**, and **Aluminium (Al)** are found to be:

Na : [*2p*: 25, *3s*: 171];

Mg : [*2p*: 25, *3s*: 128]; and

Al : [22 for *2p*, 104 for *3s* and 131 for *3p*].

Comments: However, the observed radii for the **2p-orbitals** duly correspond to the **ions**; whereas, those for the corresponding **3s-orbitals** for **Na** and **Mg**, and for the respective **3p-orbitals** for **Al** correspond to the **atomic radii**:

For *Anions Formed*–by the actual gain of electrons in the outermost orbitals, the respective **radii** fail to alter significantly.

❑ **Bragg-Slater Radii**–The **Bragg–Slatter Radii** was initially proposed by **Bragg (1920)** in terms of:

"**a set of radii for such *elements* whose sum provided the observed internuclear distances in a few hundred crystals both *ionic* and *covalent* falling with 6 *pm* only.**"

Slater (1964) put forward an *empirical rule* that

"**the ionic radii of the *catious* are nearly 85 pm lower and those of the anions are 85** *pm* **higher *vis-a-vis* the corresponding *atomic radii***"

Furthermore, **Slater** suggested a '**unique radius**' for an element in **multiples of 5 pm**,–that ultimately provided critically the so-called **observed intermolecular distances** in the *solid crystals* occurring within 6 *pm*.

Examples: Following are some of the **typical values of the Bragg-Slater radii**, such as:

- **For P, S, Cl = 100** *pm* • **For As, Se, Br = 115** *pm*
- **For Co, Ni, Zn, Cu = 135** *pm* • **For Cr, Ti, V, I, M$_n$ = 140** *pm* and
- **For Mg, Hg = 150** *pm*.

❑ **van der Waals Radii**–In a particular instance when the **molecules** with an *inherent* **intermolecular attraction** in terms of the ensuing *chemical forces viz.*,

- **Noble Gases** • **Saturated Molecules** (*e.g.*, Methane, Carbon Monoxide, Chlorine)– that shows a tendency to condense to form a solid, weak dispersion forces.

Amazingly, the so-called prevailing **internuclear distances** in most of these cases are found to be intimately related to the **van der Waals radii**. Thus, if the *two atoms* are absolutely *identical* and only the **weak-dispersion forces** do critically exist between them, we may have:

'**half of the internuclear distance being called as the** *van der Walls radius*'.

Example: It has been observed that we may have:

In Solid	Internuclear Distance (*pm*)
Helium (He)	360
Neon (Ne)	320
Argon (Ar)	380

which suggests that the **van der Waals radii** for: **He, Ne,** and **Ar** are **180, 160,** and **190** pm respectively*.
Usefulness of Lenard–Jones Radii–In case, the molecules do possess **kinetic energy** (*i.e.*, at a temperature other than *OK*),–the so-called **Lenard-Jones radii** are certainly proved to be more useful. Hence, the **Lenard-Jones radii** may be defined as:

> "**half of the distance of closest approach of *two* molecules having no kinetic energy other than that ultimately arrived at from the *van der Waals attraction forces* predominantly**".

Points to Ponder: Interestingly, the **vander Waals radius** for an element is observed to absolutely *variant in nature*. It exclusively rests upon:

> '**compression by the prevalent External Forces perceptively**'.

Following are a few *typical values* for the **van der Waals radii:**

- **H** = 120–145 *pm*
- **N** = 155 *pm*
- **F** = 150–160 *pm*
- **Mg** = 170 *pm*

- **C** = 165 – 170 *pm*
- **O** = 150 *pm*
- **Na** = 230 *pm*
- **Ni** = 160 *pm*

2.5. THE IONIC RADII

Based on the fundamentals of **Wave Mechanics** (*Chapter-1*), we may infer quite precisely that:

> '**radius of the *isolated ion or atom* shall be an infinite entity**'.

Since the **electrons** are invariably transparent to the **X-rays**, the exact position of the **electrons** located very much around the *nucleus* may not be determined experimentally either:

- **in an ionic compound**, or
- **ionic lattice**.

In order to understand the actual **existence of ions** or the so-called **Electron Density (ED) Map radii**, we may have to make **certain assumptions**; and hence, to ensure a *reliable and gainful concept of the ionic radii*. These essentially comprise the following *three* vital and important aspects, namely:

(a) **critical existence of the ions,**

(b) **accurate and precise apportioning of *internuclear distances* between the ions,** and

(c) **observed consistency of *ionic radii* (*i.e.*, additivity of radii).**

2.5.1. *The ED Map Radii*

Importantly, the '*ions*' do exist either in **molten form** or **aqueous solutions**; and hence, the calculations based upon the important *assumption* that the **ions** are duly present in:

- the so-called *Ionic solids* that invariably provide significantly precise and accurate description of the **ionic compounds** (*chemical entities*),–whereby the direct evidence results critically from the *Electron Density Maps* (*ED maps*); and
- that relates for a *restricted number of ionic bonds* by making use of the sophisticated **X-Ray crystallographic technique**,–which explicitly provides the following *two* cardinal characteristics:

---

* That is, the **kinetic energy** of the molecules is assumed to be '**zero**'.

> ➤ **internuclear distances**, and
> ➤ **electron densities (EDs)**.

Salient Features: These essentially include:

1. The ensuing **charge density** profile extends *right from one atom to the other* in an almost **continuous fashion**. However, it eventually sets down to an *extremely low value of*:

$$\text{`200 electron nm}^{-3} \text{ (0.2 electron Å}^{-3}\text{)'},$$

present duly at the **outer edges of the ion**.

2. The *integration* of the so-called observed *electron density* towards such a point does specifically provide approximately:
   - **10.05 electrons around Na;** and
   - **17.70 electrons around Cl nucleus,**

whereas, presumably **0.25 electrons** duly located strategically in the so-called **internuclear space** not included obviously in the usual **arbitrary integration** episode.

3. It critically substantiates a reasonable **direct evidence** for the absolute presence of **ions** in the *solid state*. Besides, the regions of **minimum (ED)** may be adequately considered to:
   "**define the so-called extremeties of the *spherical* ions that essentially comprise a large segement of the *electrons* for which the *ED* map radii may be calculated easily**".

4. Importantly, the actual difference being observed between the **ED map** and the **Ionic radii of Pauling**,–that is being regarded to be an *accurate measure* of the ensuing **Covalent Bonding** occuring in the so-called **Ionic Chemical Entities** (or **compounds**); and hence, enhances in the following order perceptively.

$$\boxed{K^+ < Na^+ < Li^+ < Cu^+ \text{ and } Cl^- < Br^-}$$

5. In true sense, a precise and correct *apportioning* of the **internuclear distance** could be definitely possible and feasible since *these radii* are found to be *'additive in nature'*.

6. **Construction of a Permanent Table of Ionic Radii**–It may be constructed rather conveniently right from the so-called **internuclear distances** found prevalently in the '**Ionic Crystals**' perceptively, *if the ionic radius of one of the ions is known*.

7. Even though logically the radii fail to maintain a **constant level** throughout a particle series, the ensuing observed **variations** –are small in quantum; and, therefore, for the *halides of metals*–the respective radius (ρ) in pm is as stated under:

$$\boxed{\begin{array}{llll}
r(Na) & = & r(Li) + 25 + 3 & \qquad r(Cl) \quad = \quad r(F) + (50 + 4) \\
r(K) & = & r(Na) + 32 + 2 & \qquad r(Br) \quad = \quad r(Cl) + (16 + 1) \\
r(Rb) & = & g(K) + 14 + 1 & \qquad r(1) \quad = \quad g(Br) + (25 + 1)
\end{array}}$$

❑ **Ionic Radii Determined Experimentally**–Based on the literature survey a good number of both:
   - **Empirical Method** and • **Semiemperical Method**,

have been duly put forward so as to carry out the calculation of the **ionic radii**. Since, the observed *internuclear I-I distance* present in **Lithium Iodide (LiI)** was found to be the **shortest (4.26 *pm*)** exceptionally.

Later on, **Lande (1920)** vehemently assumed that the *'smallest ion'*, which may be duly observed particularly in the **ionic crystals** happens to be the **Lithium Ion [Li⁺]**. Thus, when **Li⁺** gets duly combined with relatively *bigger-in-size* **Halide Ions**,–it would too small to keep the **halide ions** *from even touching each other i.e.,* **Li⁺ and X⁻**.

Points to Ponder: These essentially comprise:

1. **Bigger the size of the halide ion (X⁻)**, there is always a greater possibility for this to be true.
2. Besides, the so-called observed **internuclear distance** explicitly demonstrated that the **I–I distance** present in the *LiI-crystal*–happens to be the **shortest (426 pm)**.
3. Let us assume that in the **LiI**–the *two* **I⁻** ions almost touch each other, the ultimate observed **ionic radius** of the corresponding **I⁻ ion** was eventually taken as:

   '**half-of-the internuclear distance = 213 pm**'.

4. Thus, starting from the **internuclear distance**, as could be seen in *other iodides*, we may eventually arrive at the so-called **radii of *cations* by difference only** *e.g.*, one may obtain in **KI (Potassium Iodide)**–the actual internuclear distance of **353 pm**, which ultimately provides:

   '**the so-called Ionic Radius for K⁺ = 353–213 = 140 pm**'.

Likewise, we may obtain certain other values as stated under:

- **Na⁺ = 99** • **Rb⁺ = 144** • **Li⁺ = 74** • **F⁻ = 132** • **Cl⁻ = 172** • **Br⁻ = 188 (in pm)**

NOTE   **Braggard West (1926) based on the aforesaid principle accomplished the *radius* for the oxide ion (O⁻) from the silicates (SiO_2); and thus, calculated the *ionic radii* of more than 80 ions.**

❑   **Molar Refractivity** *vis-a-vis* **Ionic Radii**–The **molar fraction (R)** of a substance may be determined experimentally by means of the **Lorenz-Lorentz (Eq. 2)**, as given under:

$$R = \frac{n^2 - 1}{n^2 + 2} \frac{M}{\rho}$$

...Eq. (2)

where,

n = Refractive index of the material (substance)

M = Molecular weight of material

ρ = Density of material

Remarks: Importantly, the **molar fraction (R)** has been observed to be:

- **additive** and • **refraction equivalents**,

pertaining to the *'**atoms**'* and *'**bonds**'* may be assigned accordingly. Since the **molar fraction (R)** essentially has be dimensions of m/ρ (*i.e.*, molecular weight/Density of substance)–the **volume**; and hence, the **ionic refraction for each ion** may be placed proportional to **r³** [*i.e.* **the radius (r) of the ion**]; and hence, we may have:

$$r_t = kR^{1/3}$$

Thus, if '**k**' is duly assumed to be constant for various ions, we may have the following expression:

$$\boxed{\frac{r_1}{r_2} = \left(\frac{R_1}{R_2}\right)^{1/3}}$$
...Eq. (3)

Therefore, based on **Eq. (3)**, as the **internuclear distance** in NaF is **231** *pm*; and the *ionic refractions* related to **Na⁺** and **F⁻** are 0.5 and 2.5 respectively, we may have the so-called **Ionic Radius for Na**, –as given under:

$$231 \times \frac{3\sqrt{0.5}}{3\sqrt{0.5} + 3\sqrt{2.5}} = \textbf{85.25 } pm$$

NOTE | Likewise, the so-called *Molar Polarization* of the ions may also be employed for the precise determination of the *ionic radii* (r^3) using the equations very much akin to "**Eq. (2) and Eq. (3).**

❑ **Ionic Radii Vs ED-Maps**–As on date, the advent of having reference to the accurate and precise **electron density maps (ED-maps)** exclusively based on the so-called **high-precision X-Ray Diffraction (XRD) methods**, it have really become a lot easier and convenient to observe rather more realistically:

"**how ED really varies along the line between the *adjacent nuclei* located strategically in an Ionic Crystal**".

Example: A classical example is duly given by **Lithium Fluoride (LiF)**–as illustrated in Fig: 2.2.

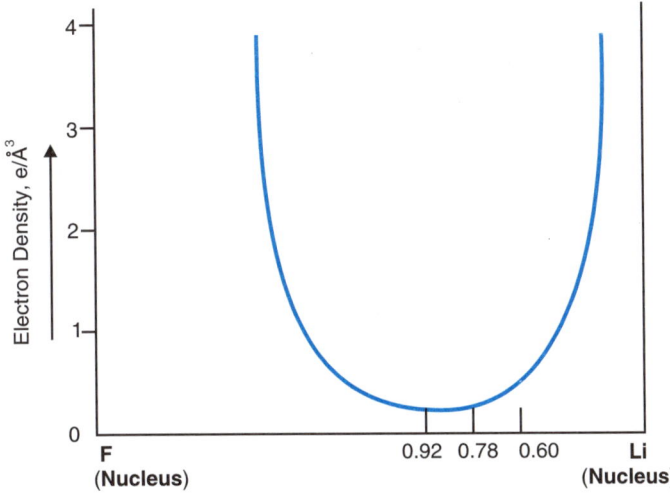

Figure 2.2: Graphic Representation of a Plot of Experimentally Measured Electron Density Between Adjacent Fluorine (F) and Lithium (Li) Nuclei in LiF. (The Pauling and Goldschmidt Radii of Li⁺ are duly indicated by 'P' and 'G', while 'M' denotes the actual minimum).
[Adapted From: Krebs H: Fundamentals of Inorganic Crystal Chemistry, McGraw Hill, New York, 1968].

Explanation: The various aspects in Fig: 2.2 may be explained as under:

1. It has been duly observed that neither the so-called:
   - **Gold Schmidt (G) radius,** nor   • **Pauling (P) radius (a *fortiori*),**

pertaining to Li^+ are at the **minimum (M) ebb of electron density (ED)**.

2. Therefore, these aforesaid **radii** might be considered to be "**wrong**" *in an absolute sense*, even for sure the complete sets to which they eventually belong and explicitly show **adequate interval consistency.**

Shannon and Prewitt (1976) proposed the most successful and established **ED maps** having an extensively voluminous **base-of-data** duly supported, altogether new and vigorous efforts were made to establish the so-called–**Tables of Radii**.

Table: 2.6 records a selected list of these radii, and in its *footnotes* duly substantiated with *additional values* and an elaborated discussion pertaining to how these radii were derived.

Table: 2.6: The Ionic Radii[1,2] (Å)					
Ion	**C.N.**	**Radius**	**Ion**	**C.N.**	**Radius**
A. *Alkali and Alkaline Earth Cations*					
Li^+	4	0.73	Fr^+	6	1.94
	6	0.90	Be^{2+}	4	0.41
	8	1.06		6	0.59
Na^+	4	1.13	Mg^{2+}	4	0.71
	6	1.16		6	0.86
	8	1.32		8	1.03
	12	1.53	Ca^{2+}	6	1.14
K^+	4	1.51		8	1.26
	6	1.52		10	1.37
	8	1.65		12	1.48
	10	1.73	Sr^{2+}	6	1.32
	12	1.78		8	1.40
Rb^+	6	1.66		10	1.50
	8	1.75		12	1.58
	10	1.80	Ba^{2+}	6	1.49
	12	1.86		8	1.56
	14	1.97		10	1.66
Cs^+	6	1.81		12	1.75
	8	1.88	Ra^{2+}	8	1.62
	10	1.95		12	1.84
	12	2.02			
B. *Group IB (11)*					
Cu^+	2	0.60	Ag^+	8	1.42
	4	0.74	Au^+	6	1.51
	6	0.91	Cu^{2+}	4	0.71
Ag^+	2	0.81		4(sq)	0.71
	4	1.14		6	0.87
	4 (sq)	1.16	Au^{3+}	4(sq)	0.82
	6	1.29		6	0.99

(Contd...........)

C. Group IIB (12)

Zn^{2+}	4	0.74	Cd^{2+}	12	1.45
	6	0.88	Hg^{2+}	2	0.83
	8	1.04		4	1.10
Cd^{2+}	4	0.92		6	1.16
	6	1.09		8	1.16
	8	1.24			

D. Other Non-Transition Metal Ions

NH^{+4}	6	1.61 (?)	Tl^{3+}	8	1.12
Ti^{+}	6	1.64	Sb^{3+}	6	0.90
	8	1.73	Bi^{3+}	6	1.17
	12	1.84	Sc^{3+}	6	0.89
Pb^{2+}	6	1.33		8	1.01
	8	1.43	Y^{3+}	6	1.04
	10	1.54	C^{4+}	4	0.29
	12	1.63	Si^{4+}	4	0.90
B^{3+}	4	0.25		6	0.54
Al^{3+}	4	0.53	Ge^{4+}	4	0.53
	6	0.68		6	0.67
Ga^{3+}	4	0.61	Sn^{4+}	4	0.69
	6	0.76		6	0.83
In^{3+}	4	0.76		8	0.95
	6	0.94	Pb^{4+}	4	0.79
	8	1.06		6	0.92
Tl^{3+}	4	0.89		8	1.08
	6	1.03			

E. First Transition Series Metals

Ti^{2+}	6	1.00	Ni^{2+}	6	0.83
V^{2+}	6	0.93	Ti^{3+}	6	0.81
Cr^{2+}	6 (LS)	0.87	V^{3+}	6	0.78
	6(HS)	0.94	Cr^{3+}	6	0.76
Mn^{2+}	4(HS)	0.80	Mn^{3+}	6 (LS)	0.72
	6 (LS)	0.81		6 (HS)	0.79
	6 (HS)	0.97	Fe^{3+}	4 (HS)	0.63
Fe^{2+}	4 (HS)	0.77		6 (LS)	0.69
	6 (LS)	0.75		6 (HS)	0.79
	6 (HS)	0.92	Co^{3+}	6 (LS)	0.69
Co^{2+}	4 (HS)	0.72		6 (HS)	0.75
	6 (LS)	0.79	Ni^{3+}	6 (LS)	0.70
	6 (HS)	0.89		6 (HS)	0.74
Ni^{2+}	4	0.69	Ti^{4*}	6	0.75
	4 (sq)	0.63			

(Contd...........)

F. Second Transition Series Elements

Ion	C.N.	Radius	Ion	C.N.	Radius
Pd^{2+}	4 (Sq)	0.78	Rh^{3+}	6	0.81
	6	1.00	Nb^{4+}	6	0.82
Nb^{3+}	6	0.86	Mo^{4+}	6	0.79
Mo^{3+}	6	0.83	Ru^{4+}	6	0.76
Ru^{3+}	6	0.82	Rh^{4+}	6	0.74

G. Third Transition Series Elements

Ion	C.N.	Radius	Ion	C.N.	Radius
Pt^{2+}	4 (Sq)	0.74	W^{4+}	6	0.80
	6	0.94	Re^{4+}	6	0.77
Ta^{2+}	6	0.86	Os^{4+}	6	0.78
Ir^{3+}	6	0.82	Ir^{4+}	6	0.77
Hf^{4+}	6	0.85	Pt^{4+}	6	0.77
Ta^{4+}	6	0.82	Th^{4+}	6	1.08

H. Anions

Ion	C.N.	Radius	Ion	C.N.	Radius
F^-	2	1.15	O^{2-}	8	1.28
	4	1.17	S^{2-}	6	1.70
	6	1.19	Se^{2-}	6	1.84
Cl^-	6	1.67	Te^{2-}	6	2.07
Br^-	6	1.82	OH^-	2	1.18
I^-	6	2.06		3	1.20
O^{2-}	2	1.21		4	1.21
	3	1.22		6	1.23
	4	1.24	N^{3-}	4	1.32

Important Trends and Correlations–There are several **important trends and corelations** that may be observed meticulously among these results obtained so far. However, following *six are the principal ones*, namely:

1. For a **cation** which critically shows *several* **coordination numbers (C.N.)**, the respective *radius* gets enhanced with **increasing C.N.**

2. For a particular '**Vertical Group**' (*viz.*, **Alkalis** or **Alkaline Earths**) the **radii** usually gets enhanced with **Z(Atomic Number)** for a given **Coordination Number** (*i.e.*, *number of ligands attached to a central metal in a complex*).

3. For an '**isoelectronic series**' *viz.*, Na^+, Mg^{2+}, Al^{3+}, and Si^{4+}–the **radii** usually get decreased with an **increasing charge**.

4. For any **element** essentially possessing **several oxidation numbers** *viz.*, Ti^{2+}, Ti^{3+}, and Ti^{4+}, – the ensuing radius gets **decreased with an increasing charge.**

---

a: Shannon *RD et al.* (1976)–This research paper provides the radii for several other ions and also for other **coordination number**.

b: For **Lanthanide** and **Actimide radii** (refer to relevant Tables elsewhere).

c: The unannctated 6 means '**octahedral**'; for which no specific **geometry** is implied for **other numbers** unless stated clearly [*viz.*, 4(sq.) means square]; Low spin (LS) and High spin (HS).

5. For each of the **different transition series**,–the ions of the **identical charge** do display an overall typical trend towards **decreasing radii** having an **increasing Atomic Number (Z)**. Nevertheless, in the **d block**, these trends are not found to be smooth*. Thus, the behaviour of the **Lanthanide** and **Actinide radii** could be explored eleswhere in the text.

6. For the **transition metals** which may have both **high and low spin stated**, for example:

   'In Fe^{2+} (ferrous)–the radius is layer for the high-spin ion'.

2.5.2. *Pauling's Univalent Radii*

Let us assume that the **ionic radius of an ion** must be:

'**inversely proportional to the effective Nuclear Charge Z* of the Ion**'.

Pauling strongly proposed a set of **ionic radii**,–that are regarded to be the most widely used ones. Furthermore, **Pauling** vehemently taken into logistic consideration the so-called:

"**isoelectronic ion pairs** *viz.* Na^+F^-, K^+Cl^-, Rb^+Br^-, and Cs^+I^- having the similar '*ionic radius rating*'; and also the same percentage of the inherent ionic characteristic feature."

Pauling promulgated the assumption that:

"**in the solid state the ions virtually touch each other**"

However, the following *two* important characteristics were calculated precisely as stated below:

❑ **Internuclear Distance [r]**–was determined with the help of **X-Ray crystallographic (XRC) studies**, and

❑ **Nuclear Charge [Z*]**–was arrived at by making use of:

   • **Slater's Rules**–as applicable for the respective *lighter elements*; and

   • **Molar Refractivity and X-Ray Studies**–as applicable for the respective *heavier elements*.

Calculation–Based on **Eq. (3)**, the internuclear distance in **NaF** is **231** *pm* (as stated earlier). Thus, if r_{Na} and r_F designate the **ionic** radii of the said *two ions*, we may have:

$$\boxed{r_{Na} + r_F = 231}$$ ...Eq. (4)

Thus, the **electronic rearrangement** of **anion (F^-)** and **cation (Na^+)** is $1s^2\ 2s^2\ 2p^6$. Therefore, the ensuing *screening constant** for the said *two ions* is given as under:

$$8 \times 0.35 + 2 \times 0.85 = \mathbf{4.50}$$

Hence, we may have:

$$Z^*_{Na} = 11.00 - 4.50 = \mathbf{6.50},$$

$$Z^*_F = 9.00 - 4.50 = \mathbf{4.50}$$

Since the **ionic radii** are found to be proportional to Z^{*-1}, we may have the folllwing expression:

$$\boxed{r_{Na} = \left[\frac{C}{Z_{Na}}\right] \quad \text{and} \quad r_F = \left[\frac{C}{Z_F}\right]^*}$$ ...Eq.(5)

---

* It is also known as the '**Shielding Constant**'

where, C = Proportionality depnding upon the *electronic structure of the ion*.

Let us now insert the respective values from **Eq. (5)** into **Eq. (4)**, we have:

$$C/6.50 + C/4.5 = \textbf{231}$$

thereby giving:

C = 614 [**for ions of Neon configuration**]

Therefore, we have:

$$\boxed{\begin{array}{l} r_{NA} = \textbf{614/6.50 = 94.5 } pm \\ r_F = \textbf{614/4.5 = 136 } pm \end{array}}$$

Because of the **Isoelectronic systems**, the so-called **screening constant** shall remain exactly th same; and , therefore, the precise *univalent radius* for the following *two ions* is found to be:

➢ **For Oxide Ion [O²⁻]**

$$r_{o^{2-}} = \frac{.614}{(8.0 - 4.5)} = \frac{614}{3.5} = \textbf{175.4 } pm$$

➢ **For Magnesium Ion [Mg²⁺]**

$$r_{Mg^{2+}} = \frac{614}{(12.0 - 4.5)} = \frac{614}{7.5} = \textbf{81.3 } pm$$

❑ **Variations of Ionic Radii: Observed in Iso-electronic Ions**

The iso-**electronic ions** are those that do possess the **same number of electrons**; however, they essentially show a glaring difference being held on their nuclei,–as recorded in Table: 2.7.

Importantly, the observed overall effect of **nuclear charge** related to the respective **ionic size** may be expatiated in an explicit manner by:

"**taking into consideration the exact radii of these iso-electronic species**".

It is, however, pertinent to state here that almost all of them do consist of the **some number o electrons**. Nevertheless, they actually display their variance predominantly only in the charge located strategically upon the nucleus. Therefore, it is quite evident that since the **inherent nuclear charge increases**, the corresponding electrons are being held more rigidly by the said **nucleus**; and hence, the so-called *ionic radius* deceases perceptively.

S.no.	Ion	N^{3-}	O^{2-}	F^-	Na^+	Mg^{2+}
	Table: 2.7: The Radii of Iso-electronic Ions.					
1	**Number of Electrons**	10	10	10	10	10
2	**Charge on the Nucleus**	+7	+8	+9	+11	+12
3	**Radius (Å)**	1.71	1.45	1.33	0.95	0.60

2.5.3. Observed Periodic Trends in Ionic Radii

In a broader perspective, it has been duly observed that:

the *Ionic Radii* do increase appreciably in moving from the top-segment to the bottom one in any given *Group*; and hence, decrease perceptively while moving along a *period*".

Table: 2.8 records the so-called **Ionic Radii** of certain common ions:

Table: 2.8: The Ionic Radii (Å) of Certain Common Ions.

Group 1	Group 2	Group 13	Group 16	Group 17
Li^+	Be^{2+}	B^{3+}	O^{2-}	F^-
0.60	0.31	0.20	1.45	1.33
Na^+	Mg^{2+}	Al^{3+}	S^{2-}	Cl^-
0.98	0.65	0.50	1.84	1.81
K^+	Ca^{2+}	Ga^{3+}	Se^{2-}	Br^-
1.33	0.99	0.61	2.02	1.96
Rb^+	Sr^{2+}	In^{3+}	Te^{2-}	I^-
1.48	1.10	0.81	2.22	2.12
Cs^+	Ba^{2+}	Tl^{3+}		
1.67	1.29	0.91		

Let us now briefly look into the various cardinal aspects pertaining to these *two* **characteristics**:
- **Variation in a *Group*,** and
- **Variation in a *Period*.**

➤ **Variation in a Group**–Based on the available data in Table 2.8 it is quite evident:

'**the ionic radii critically increase progressively with a simultaneous increase in the Atomic Number (Z) present within a Group**'.

Therefore, we may take cognizance of the fact that with the so-called **increase in the atomic size**, while moving down the **Group**, the respective **ionic radius (r) also increases** accordingly. However, one may further observe the following important aspects, namely:

- Within a specific moiety there prevails a sudden and rapid increase in the **ionic radius (r)**–as one proceeds from one element to another in the segment of *first few elements specifically*;
- However, the observed increase in the above episode is not found to be so rapid as one visualizes amongst the last elements categorically.

Examples:

1. An increase in the **Ionic Radius (r)** is observed to be very fast and rapid as one critically proceeds from:
- **Li^+ to Na^+** and   • **Na^+ to K^+,**

however, the said increase is certainly **not that rapid** as we proceed from
- **K^+ to Rb^+ and Rb^+ to Cs^+**

Explanation–A logical explanation to the above particular behavioural pattern could be solely due to the fact:

'**between K^+ and Rb^+ ions there is a crucial entry of elements belonging to the *First Transition Series*.**'

Since in these elements the specific **last-occupied shell** happens to be the same, enhancing **Nuclear Charge (Z)** with the so-called enhancement in the **Atomic Number (Z)**:

"**it vehemently tends to afford** *contraction of electron charge clouds* **of the** *atoms* **and the** *ions.*"

 NOTE   **In fact, it tends to oppose the so-called expected enhancement in the** *Ionic Radius (r)* **as one moves from K^+ to Rb^+ and from Rb^+ to Cs^+.**

> **Variation in a Period–** It has been observed that very much: "**within a period the respective radii of 'cations' of the normal elements of Groups: 1, 2, and 13 decrease with an increase in** *Atomic Number***; and the same could be ascertained with regard to the radii of 'anions' of Groups: 16 and 17**".

Importantly, the above dictum is fairly vivid and understandable, because the so-called *atomic size* decreases as one moves along a period; and thus, the ensuing *ionic radius* also decreases accordingly.

2.6. The Ionization Energy or Ionization Potential [IE or I or Δ_iH]

Based on the scientific evidences, logical explanations, and intelligent interpretations we may refer to the **Ionization Energy** or **Ionization Potential** as:

"**the most characteristic '***atomic property***' that is duly exhibited both critically and elegantly by the atoms of all elements.**"

Importantly, the aforesaid characteristic features obviously provides us with a fairly concrete idea and concept pertaining to the virtual tendency of the elements to get rid of the '**electron**' present strategically in the *outermost shell* to yield a '**cation**' perceptively. Therefore, with the aid of such a crucial property one may lay hands on to:

"**an array of other characteristic features of the ensuing elements** *viz.***, metallic character (electropositive charge), reducing power, and the like,–that may be expatiated**".

What do we understand by Ionization Energy?

The **Ionization Energy** (or *Ionization Potential*) of an element may be defined as:

"**Amount of energy required critically to remove the highest, loosely bound electron* from the respective isolated '***neutral gaseous atom***' present duly in its lowest energy state** so as to bring forth its conversion right into a '***unipositive gaseous cation***'.**"

In this manner, the so-called **Ionisation Energy** of an atom **M(g)** may be duly expatiated by an accepted process as illustrated by the following equation:

$$M(g) + \text{Energy supplied} (+I_1) - e^- \longrightarrow M^+(g) + e^-$$

An Isolated Gaseous Atom   **A Gaseous Cation**

Observation: Importantly, the *Ionization Energy* as stated above is broadly termed as: *First* **Ionization Energy**, because it exclusively corresponds to the careful **removal of only one 'electron'**. Therefore, it is mostly designated as: I_1 **or** $+ I_1$, – wherein the **positive (+) sign** duly represents the absorption of energy the corresponding atom **M(g)**. It may also be represented as: Δ_i**H**.

---

*   That is, the *outermost shell electron*.

**   That is, the *ground state*.

Endothermic Process: In reality, the critical, removal of an electron from a gaseous atom in the aforesaid reaction, clearly shows an **endothermic process**–because the removal of the electron needs a certain quantum of energy.

Measurement of Ionization Energy: In an '**electron**' the ionization energy is usually measured in terms of: **Volts per atom [eV. atom^{-1}]** and **kilocalories per mole [kcal. mol^{-1}]** or **kilo Joules per mole [kJ. mole^{-1}]**. Interestingly, these *quantities* are intimately related, –as could be seen in the following expressions

$$1\,eV.atom^{-1} = 96.4705\ kJ.mole^{-1} = 23.06\ kcal.mole^{-1}\ 1\ kcal \times mole^{-1} = 4.185\ kJ.mole^{-1}$$

Thus, we may take note of the following *two* important points:

- **The resulting values of Ionization Potential may be depicted vividly by** *placing a positive* (+) *sign before them*; and
- **The positive (+) sign designates explicitly the so-called 'absorption of energy an atom M(g)'.**
- **The values of** *First-Ionization Potential* **(expressed as kJ. mole^{-1}) of the representative elements** *viz.*, **the elements of s–and p–block elements–are duly recorded in Table: 2.9.**

Table: 2.9: The First Ionization Potentials [kJ.mole^{-1}] for Some Representative Elements (s-and p-Block Elements.)*

| Group→ Period↓ | s-Block Elements | | p-Block Elements | | | | | |
	IA	IIA	IIIA	IVA	VA	VIA	VIIA	ZERO
1	1 H 1312.0							2 He 2372.3
2	3 Li 520.2	4 Be 899.5	5 H 800.6	6 C 1086.4	7 N 1402.3	8 O 1314.0	9 F 1881.0	10 Ne 2080.7
3	11 Na 495.8	12 Mg 737.7	13 Al 577.6	14 Si 786.5	15 P 1011.8	16 S 999.6	17 Cl 1251.1	18 Ar 1520.5
4	19 K 118.9	20 Ca 589.8	31 Ga 578.8	32 Ce 762.2	33 As 944.0	34 Se 940.9	35 Br 1139.9	36 Kr 1350.7
5	37 Rb 403.0	38 Sr 549.5	40 In 558.3	50 Sn 708.6	51 Sb 531.6	52 Te 869.3	53 I 1008.4	54 Xe 1170.4
6	55 Cs 375.7	56 Ba 502.9	81 Tl 589.3	82 Pb 715.5	83 Bi 703.3	84 Po 812.0	85 At —	86 Rn 1037.0

* The Positive (+) sign has not been shown in the above Table before the values.

The Higher Successive Ionization Potentials [I_1, I_2, I_3, I_4, I_5, ...]

Besides, the *First* **Ionization Potential (I_1)** as stated above, we are also well aware of the subsequent: Second (I_2), **third (I_3), Fourth (I_4),** ans so on so forth **Ionization Potentials.**

❑ **Second Ionization Potential (I_2)**–It may be duly defined as:

"the energy needed essentially to remove elegantly one more electron right from the *gaseous cation* [M$^+$(g)] to accomplished the desired *double positively charged geaseous catio.* [M^{2+}(g)]."

Thus, we may express the **Second Ionization Potential (I$_2$)** as given below:

$$M^+(g) + I_2 \xrightarrow{\ -e^-\ } M^{2+}(g) + e^-$$

❑ *Third* **Ionization Potential (I$_3$)**–In the same vein, it designates the energy required to remove 'one additional electron' from the respective **double positively charged gaseous cation [M^{2+}(g)]** to yield the corresponding **[M^{3+}(g) cation]**; and hence, we may have the following expression

$$M^{2+}(g) + I_3 \xrightarrow{\ -e^-\ } M^{3+}(g) + e^-$$

The above episode obviously suggests that:

'one may accomplish the *ionization* of higher grades Ionization Potentials'.

Interestingly, it may be achieved by the **removal of one electron at a time** in a step-wise sequentia manner (as shown above).

Thus, the values of *light successive ionization potentials* with regard to the *first ten elements* of th **Periodic Table** are duly provided in Table: 2.8.

Based on the documented **First Ionization Potentials**, expressed duly in **kJ.mole^{-1}**, pertaining to the representatives elements **(s-and p-Block Elements)**,–as shown in Table: 2.10, that for a *given element* the ensuing *magnitudes of its successive ionization potentials* are obtained in the order, –as given under

$$I_1 < I_2 < I_3 < I_4 < I_5 \text{ L and so on so forth.}$$

Besides, we may also categorically visualize the so-called **systematic increment** in the values o **successive Ionization Potentials**,–that could be expatiated as given below based upon:

'the underlying Idea and Concept of Effective Nuclear Charge'

At this point in time, it is absolutely clear that the '**cations**' are duly formed by the *careful removal o electrons* right from the **atom/cation** one after the other systematically.

Thus, we may have the following expression:

$$M(g) \xrightarrow{\ -e^-\ } M^+(g) \xrightarrow{\ -e^-\ } M^{2+}(g) \xrightarrow{\ -e^-\ } M^{3+}(g) \cdots \longrightarrow M^{n+}(g)$$

Table: 2.10: The Values of I$_1$, I$_2$, I$_3$ ... I$_n$ (in kJ.mole^{-1}) of First Ten Elements of Periodic Table.

Elements with Atomic Number and Electronic Configuration	Values of Successive Ionization Potentials							
	I_1	I_2	I_3	I_4	I_5	I_6	I_7	I_n
H(1) ($1s^1$)	1312.0							
He(2) ($1s^2$)	2372.3	5250.4						
Li(3) ($1s^2$, $2s^1$)	520.3	7298.1	11814.9					

(Contd..........)

Element								
Be(4) $(1s^2, 2s^2)$	899.5	1757.1	14848.7	21006.5				
B(5) $(1s^2, 2s^2p^1)$	800.6	2427.0	3659.8	25025.7	32826.6			
C(6) $(1s^2, 2s^2p^2)$	1086.4	9352.6	4620.5	6222.6	37830.4	47276.9		
N(7) $(1s^2, 2s^2p^3)$	1402.3	2856.1	4578.1	7475.1	9444.9	58266.4	64359.8	
O(8) $(1s^2, 2s^2p4)$	1314.0	3388.2	5300.4	7469.3	10989.5	13326.4	71334.5	84077.4
F(9) $(1s^2, 2s^2p^5)$	1681.0	3371.2	6050.4	8407.7	11022.7	15164.0	17867.7	92037.8
Ne(10) $(1s^2, 2s^2p^6)$	2080.7	3952.3	6122.0	9370.0	42178.0	15238.0	19999.0	13069.0

Based on the above expression, we may observe explicitly that a given '**cation**' does possess a definitely **lesser number of electrons** *vis-a-vis* **its parent atom or cation**. Furthermore, with the progressive *reduction in the number of electrons*, the overall magnitude of the so-called *effective nuclear charge* enhances from **M(g) to M^{n+}(g)**. Thus, with the observed increase in the **effective nuclear charge**, the resulting *magnitude of the force of attraction* prevailing between the **Nucleus** and **Outermost electron cloud shell** gets also enhanced as stated under:

Thus, we may have the following expressions:

- ❑ **Atom/Cation**   : $M(g)\ M^+(g)\ M^{2+}(g)\ M^{3+}(g) \dots M^{n+}(g)$
- ❑ **Effective Nuclear Charge**   : **Increasing Progressively →**
- ❑ **Force of Attraction**   : **Increasing Progressively →**

NOTE

It is, however, pertinent to state here that an enhancement in the *observed force of attraction*–the ultimate magnitude of energy to *remove the outer-most shell electron* (or the so-called **Ionization Potential**) appears to increase progressively from **M(g)** to **M^{2+}(g); and hence, we may have:**

$$I_1 < I_2 < I_3 < I_4 < \dots \text{ and so on so forth.}$$

Points to Ponder: Amazingly, the **total energy** actually needed to dislodge *two* **electrons** right from a *gaseous* **M(g)** atom to accomplish a respective *gaseous* **cation M^{2+}(g)**, is the sum of:

"First (I$_1$) and Second (I$_2$) Ionization Potentials".

Thus, we may have the following expression:

$$\boxed{M(g) + (I_1 + I_2) \xrightarrow{\ -2e^-\ } M^{2+}(g) + 2e^-}$$

2.7. GOVERNING FACTORS THAT INFLUENCE THE IMPORTANCE OF IONIZATION POTENTIAL

Following are various cardinal governing factors that influence the importance of **Ionization Potentials** of *atoms of elements* elegantly:

- **Effective Nuclear Charge (Z),**
- **Size of an Atom (γ),**
- **Principal Quantum Number (*n*),**
- **Shielding Effect,**
- **Concept of Half-filled and completely-filled Orbitals,** and
- **The Precise Nature of Orbitals.**

2.7.1. *Effective Nuclear Charge (Z)*

It is an universal fact that–'**greater being the magnitude of effective nuclear charges (Z), greater would be the electrostatic force, of attraction exerted by the nucleus upon the outer electrons perceptively.**'

From the above dictum one may safely infer that it would be certainly a rather difficult task to remove the so-called **electron from the outermost shell of an atom** having specifically the *higher effective nuclear charge*.

In other words, it means explicitly that–

"**greater would be the magnitude of effective nuclear charge, higher would be the quantum of energy (*i.e.*, the Ionization Potential) required to eliminate the outermost-shell electron**".

Based on the above glaring factor one may logically explain the underlying variation in the **Ionization Potential values** of the elements of a specific **Period** of the *representative elements* of the **Periodic Table**, as given under:

Elements of *Second* Period:

    Li Be B C N O F Ne

Effective Nuclear Charge

1.30 1.95 2.60 3.25 3.55 4.55 5.20 5.85 [Increasing Mode]

Ionization Potential [kJ.mole^{-1}]

520.3 899.5 800.6 1086.4 1402.3 1314.0 1681.0 2080.7

2.7.2. *Size of the Atom (r)*

In this case, the greater being the magnitude of '**g**' in an atom, more away would be the **outer-most shell electron** from the **Nucleus**; and hence, lesser shall be the ultimate *force of attraction* being exerted predominantly by the said Nucleus upon the **outer-most shell electron**.

Therefore, an **increase in size of the atom (*r*)**, will evidently far more convenient and easy to get away an **electron** right from the **outer-most shell**, which ascertains predominantly that:

'**higher being the value of '*r*' for an atom,–lower would be its Ionization Energy**'.

Thus, based on the factual evidences as discussed under '**size of the atom**' above, one may specifically expatiate the variation in the *ionization potential values* of the elements present in a *Group* as well as a *Period* in the **Periodic Table**,–as explained under:

(a) Variation in Ionisation Potential of the Elements in a Group

It has already been expatiated abundantly earlier that as one proceeds from the **top-end to the bottom-end in a *Group***:

"the value in size of the atom (γ) enhances; and hence, the ensuing magnitude of ionization potential gets decreased significantly,–as given below for the elements belonging to Group–IA,":

Elements of Group IA	Atomic (Covalent) Radius (A)	Ionization Potential (kJ.mol^{-1})
H (1)	0.32	1312.0
Li (3)	1.23	520.3
Na (11)	1.54	495.8
K (19)	2.03	118.9
Rb (37)	2.16	403.4
Cs (55)	2.35	375.7

(b) Variation in Ionisation Potential of the Elements in a Group

Based on the fact that on proceeding from **left-hand side (LHS) to right-hand side (RHS)** in a *Period*:

"the observed value in size of the atom (*r*) decreases; and hence, the magnitude of ionization potential gets duly increased appreciably as depicted under for the element pertaining to the *Third Period*."

Thus, we may particularly observe the **abnormal values** of the *two* specific elements: **Magnesium (Mg)** and **Phosphorus (P)** Elements of *Third* Period:

Na   Mg   Al   Si   P   S   Cl   Ar

Atomic (Covalent Radius (Å))

1.54   1.36   1.18   1.11   1.06   1.02   0.99   0.98

Ionization Potential [kJ.mole^{-1}]

495.8   737.7   577.6   786.5   1011.8   999.6   1251.1   1520.5

2.7.3. *Principal Quantum Number [n]*

As discussed in **Chapter-1**, the so-called *three* **Quantum Numbers (*n, l, m*)** do represent the *direct consequence* of the critical solution of the **Schrödinger Wave Equation** put forward so as to *elaborate the behavioural pattern of electron as a wave*. Nevertheless, we may recall that the **principal quantum number (*n*)** either:

- **designates the major energy level**, or
- **energy shell '*n*' wherein the electron is present**.

In true sense, it specifically determines the so-called **average distance of an *electron* from the** *nucleus*. Since an increment in the **value of '*n*'** shows the electron to move further away right from the nucleus; and hence, its energy enhances perceptively.

It may be further observed that the greater value of '*n*' for the respective **valence shell electron of an atom**,–it would push the electron farther away from the nucleus; and, therefore, smaller would be the force of attraction influenced by the nucleus on it. Alternatively, it suggests that relatively **lesser quantum of energy** shall be needed to get rid of the **valence-shell electron**. In this way the increase in the corresponding value of '*n*' of the *valence shell electron*, the ensuing ionization potential gets lowered and *vice-versa*.

In short, one may also express the stream of thoughts as:

"with the increase of the principal number '*n*' of the respective orbital from which the electron is intended to be removed, the magnitude of ionization potential gets decreased significantly".

- **Variation of Ionization Potential of Elements Present in a Group,**
- **Successive Ionization Potentials of an Element,** and
- **Ionization Potentials of Inert gases** *vis-a-vis* **the Alkali Metals.**

(a) Variation of Ionization Potential of Elements Present in a Group

It can be seen explicitly that as we move down a **group** in the **Periodic Table**, the corresponding value of the **Principal Quantum Number** (*n*) of the orbital where from the *electron* is intended to be removed gets enhanced appreciably; and hence, the respective magnitude of the ensuing **Ionization Potential [kJ.mole^{-1}]** gets decreased as depicted below related to the elements of **Group II A**:

Elements of Group II A	Configuration of Valence Shell Electrons	Quantum Number (*n*)	Ionization Potential (kJ.mole^{-1})
Be	$2s^2$	2	899.5
Mg	$3s^2$	3	737.7
Ca	$4s^2$	4	589.8
Sr	$5s^2$	5	549.6
Ba	$6s^2$	6	502.9

(b) The Successive Ionization Potential of an Element

In order to understand the various steps that are duly involved in the *successive ionization potential* of an element one may have to look into the following *three* **cardinal sequences** perceptively:

Stage – 1: The observed values of I_1 and I_2 (*i.e.*, **Ionization Potentials**) of **Lithium (Li)** atom is found to be: $1s^2, 2s^2$ that eventually stands at: **520.3 kJ.mol^{-1} and 7298.1 kJ.mole^{-1}** respectively; and hence, the same may be duly represented by the following *two* **set of equations:**

$$\text{Li} (1s^2, 2s^1) \xrightarrow[\;I_1 = 520.3 \,\text{kJ.mole}^{-1}\;]{-e^- \,[\text{from } 2s \text{ orbital with } n = 2]} \text{Li}^+ (1s^2)$$

$$\xrightarrow[\;I_2 = 7298.1 \,\text{kJ.mole}^{-1}\;]{-e^- \,[\text{from } 1s \text{ orbital with } n = 1]} \text{Li}^{2+} (1s^1)$$

Remarks: The reason why I_2 is higher in comparison to I_1 may be expatiated based on the underlying fact that in the specific instance of I_1–the **electron** gets out of the *orbital* that essentially has greater value of *n*(= 2); whereas, in the specific instance of I_2 the **electron** gets removed from the respective **1s orbital** that critically possesses a **lower value of *n* (=1).**

Stage-2: The 3rd ionization potential (I_3) of **magnesium (Mg)** is found to be fairly high *vis-a-vis* to I_1 and I_2 [*i.e.*, I_1 = 737.7 kJ.mole^{-1}; I_2 = 1450.7 kJ.mole^{-1}; and I_3 = 7732.8 kJ.mole^{-1}]. The above glaring observed variations in the *three* **successive ionization potentials** may be duly expatiated by taking into consideration: $I_1, I_2,$ and I_3–as represented by the following *three* **distinct equations,** namely:

$$\text{Mg}[1s^2, 2s^2p^6, 3s^2] \xrightarrow[\substack{I_1 = 737.7 \text{ kJ.mole}^{-1}}]{-e^- \text{ (from 3s orbital with } n = 3)} \text{Mg}^{\text{Å}}(1s^2, 2s^2p^6, 3s^1)$$

$$\xrightarrow[\substack{I_2 = 1450.7 \text{ kJ.mole}^{-1}}]{-e^- \text{ (from 3s orbital with } n = 3)} \text{Mg}^{2+}(1s^2, 2s^2p^6)$$

$$\xrightarrow[\substack{I_3 = 7732.8 \text{ kJ.mole}^{-1}}]{-e^- \text{ (from 2p orbital with } n = 2)} \text{Mg}^{3+}(1s^2, 2s^2p^5)$$

Important Observations: From the above elaborated equations it is critical and important to observe that in the particular instance of I_1 and I_2 the **electron** is being knocked out (removed) from an *orbital with* $n = 3$; whereas, in the case of I_3–the said **electron** is actually removed right from an *orbital possessing a lower value of n* (=2). Therefore, one may safely infer that I_3 has an **extremely high ionization potential** *vis-a-vis* to both I_1 and I_2.

Stage–3: The actual observed values of I_1, I_2 and I_3 (**in kJ. mole^{-1}**) of the *two* elements **Berilium (Be)** and **Boron (B)** are as stated under:

	I1	I2	I3
Be (Berilium):	899.5	1757.1	14848.7
B (Boron):	800.6	2427.0	3659.8

Observations: We may observe the following cardinal aspects:

1. There exists a big-gap in the **Ionization Potentials of Berilium (Be)** right from I_2 (= 1757.1 kJ. mole^{-1}) to I_3 (= 14848.7 kJ.mol^{-1}) *vis-a-vis* with **Boron (B):** I_2 (= 2427.0 kJ.mole^{-1}) to I_3 (= 3659.8 kJ.mole^{-1}).

How do we account for the 'big gap' in the aforesaid Ionization Potential?

The above anomaly may be explained vividly as under:

$$\text{Be}(1s^2, 2s^2) \xrightarrow[\substack{I_1 = 899.5 \text{ kJ.mole}^{-1}}]{-e^- \text{ (from 2s orbital with } n = 2)} \text{Be}^+(1s^2, 2s^1)$$

$$\xrightarrow[\substack{I_2 = 1757.1 \text{ kJ.mole}^{-1}}]{-e^- \text{ (from 2s orbital with } n = 2)} \text{Be}^{2+}(1s^2)$$

$$\xrightarrow[\substack{I_3 = 14848.7 \text{ kJ.mole}^{-1}}]{-e^- \text{ (from 1s orbital with } n = 1)} \text{Be}^{3+}(1s^1)$$

$$\text{B}(1s^2, 2s^2, 2p^1) \xrightarrow[\substack{I_1 = 800.6 \text{ kJ.mole}^{-1}}]{-e^- \text{ (from 2p orbital with } n = 2)} \text{B}^+(1s^1, 2s^2)$$

$$\xrightarrow[\substack{I_2 = 2427.0 \text{ kJ.mole}^{-1}}]{-e^- \text{ (from 2s orbital with } n = 2)} \text{B}^{2+}(1s^1, 2s^1)$$

$$\xrightarrow[\substack{I_3 = 3659.7 \text{ kJ.mole}^{-1}}]{-e^- \text{ (from 2s orbital with } n = 2)} \text{B}^{3+}(1s^1)$$

Remarks: From the above deliberations it may be observed that in the specific instance when the so-called **Ionization Potential, I_3,** of the B_e *atom* an *electron is removed* right from the ensuing **Is orbital** with critically **lower value of** n (=1); whereas, in the particular case of I_3 **of B atom the electron is knocked out** from the **2s orbital** with **higher value of** n (=2).

(c) The Ionization Potentials of Inert Gases (He, Ne, Ar, Kr, Xe) *vis-a-vis* to those of the Alkali Metals (Li, Na, K, Rb, Cs)

As we proceed through an **Inert Gas (*Zero Group*)** to an **Alkali Metal (*IA Group*)**, we may critically take cognizance of the fact that there prevails a *significantly large decrease* in the value of **ionization potential** as depicted under:

Inert Gas [Zero Group]	Alkali Metal [IA Group]
1. **He** [Z = 2] = 2372.3 kJ.mole^{-1} $1s^2$	1. **Li** [Z = 3] = 520.3 kJ.mole–1 $1s^2$, $2s^1$
2. **Ne** [Z = 10] = 2080.7 kJ. mole^{-1} $1s^2$, $2s^2p^6$	2. **Na**[Z = 11] = 495.8 kJ mole^{-1} $1s^2$, $2s^2p^6$,$3s^1$
3. **Ar** [Z = 18] = 1520.7 kJ.mole^{-1} $1s^2$, $2s^2p^6$, $3s^2p^6$	3. **K**[Z = 19] = 418.9 kJ. mole^{-1} $1s^2$, $2s^2p^6$, $3s^2p^6$,$4s^1$
4. **Kr** [Z = 36] = 1350.7 kJ.mole^{-1} $1s^2$, $2s^2p^6$, $3s^2p^6$ d^{10}, $4s^2p^6$	4. **Rb** [Z = 37] = 403.0 kJ.mole^{-1} $1s^2$, $2s^2p^6$, $3s^2p^6d^{10}$,$4s^2$ p^6, $5s^1$
5. **Xe** [Z = 54] = 1170 kJ.mole^{-1} $1s^2$, $2s^2p^6$, $3s^2p^6d^{10}$, $5s^2p^6$	5. **Cs**[Z = 35] = 375.7 kJ.mole^{-1} $1s^2$, $2s^2p^6$, $3s^2p^6d^{10}$, $4s^2p^6d^{10}$, $5s^2p^6$, $6s^1$

Explanation–The observed appreciably large decrease may be explained on the basis of the underlying fact that:

"**in the specific instance of an alkali metal it would be a lot easier and convenient to get rid of an 'electron' right from an orbital belonging to a higher shell; whereas, in the typical case of an *inert gas*, it is rather more difficult to eliminate (remove), the 'electron' from the orbital which belongs to the lower shell.**"

2.7.4. *The Shielding Effect*

In a multi-electron atom, the electrons that are found to be lying in between:

- **Nucleus and**   • **Valence-Shell*,**

that eventually exert a shielding effect particularly upon the so-called '**outermost–shell electron**' from the *nucleus*. Thus, the overall effect of shielding the **valence-shell electron from the nucleus**–that is prevalently brought into effect due to the *intervening electrons* (**Valence Shell**) is invariably termed as the:

- **Shielding Effect**   or   • **Screening Effect.**

Another school of thought advocates that the magnitude of the **shielding effect** in actual practice determines the inherent *quantum of the force of attraction* existing between the **nucleus** and the **valence shell electron**. Hence, it may be duly inferred that:

'**higher is the magnitude of shielding effect showing its effect upon the *valence-shell electron*, lower would be the magnitude of the force of attraction prevailing between the *nucleus* and the *valence shell electron*.**'

Furthermore, we may conclude that lower would be the **quantum of energy** (or *Ionization Potential*) to affect the removal of the valence shell electron. Hence, it may be derived explicitly that:

'**with the progressive increase of the shielding effect–the ionization potential gets decreased accordingly**'.

---

*   Also known as–**Intervening Electrons.**

Determination of the Magnitude of Shielding Effect [Screening Effect]

The magnitude of **Shielding Effect** is usually determined by the actual quantum of **shielding constant sigma (σ)**. It has been observed that **higher is the ensuing amount of shielding effect** duly produced by the so-called *intervening electrons* (**valence-shell**); and hence, lower shall be the ultimate value of **ionization potential**.

In short, it may be added that–

"**with the increase of shielding constant (σ), the magnitude of ionization potential decreases proportionally**".

Importance of Shielding Constant (σ)–In a broader sense, the above vital factors may be utilized effectively to explain:

'the variations of *Ionization Potentials* of the elements of a given *Group* in the Periodic Table'.

Besides, as we slowly move down a *Group* vertically, the critical number of *inner-shell electrons* (or the **intervening electrons**) gets enhanced progressively; and therefore, the respective magnitude of **shielding effect (σ) duly caused by these electrons located strategically upon the so-called valence - shell electron** increases. Alternatively, an **increases in the magnitude of shielding effect (σ)** results into a **decrease in the ionization potential**. However, the above analogy may be shown explicitly by taking into consideration the corresponding **Ionization Values** of the various elements listed in Group IA, –as given under:

Elements of Group IA	Electronic Configuration	Number of Inner Shell Electrons		Value of (σ)		Ionization Potential (kJ.mole^{-1})	
Li (3)	2, 1	2		1.70		520.3	
			I		I		D
Na (11)	2, 8, 1	10	N	8.80	N	495.8	E
			C		C		C
K (19)	2, 8, 8, 1	18	R	16.80	R	418.9	R
			E		E		E
Rb (37)	2, 8, 18, 8, 1	36	A	34.80	A	403.0	A
			S		S		S
Ca (55)	2, 8, 18, 18, 8, 1	54	I	52.80	I	375.7	I
			N		N		N
			G		G		G

Remarks: These essentially include:

1. Based on the above facts one may obviously expatiate the **substantial reduction** in the observed value of **Ionization Potential** as one moves from

'**an Inert Gas [Zero Group] to an Alkali Metal [IA Group]'**.

Example: As we move from **Helium (He)** to **Lithium (Li)** there prevails a relatively *large reduction in the so-called Ionization Potential* [*viz.*, He = 2372.3 kJ.mole^{-1} and Li = 520.3 kJ. mole^{-1}]. Importantly, the approximate **4.5 folds decrease** in the observed **Ionization Potential** may be explained as under:

❏   In the **Li atom [1s², 2s¹]**, the respective electron that need to be virtually removed from the corresponding **2s-orbital** gets either *shielded or screened* effectively by the aid of the **2-electrons** duly present in the **1s-orbital** perceptively.

❏   **Besides,** in the **Helium (He) atom [1s²]**–the respective **electron** that requires to be removed from the corresponding **1s-orbital** is *not shielded or screened* effectively by the presence of the **other electron** in the *1s-orbital*. Perhaps, it could be on account of the fact that-both the electrons present in **He-atom** do precisely reside in the same **1s-orbital** predominantly.

❏   Based on the same **logical explanations supported by theoretical arguments** it would be easy and convenient to put forward a convincing argument with regard to the reduction in the **Ionization Potential** in the following instances from the **Periodic Table**:

- **From Ne[Z = 10] to Na [Z = 11]**;
- **From Ar [Z = 18] to K [Z = 19]**;
- **From Kr [Z = 36] to Rb [Z = 37]**; and
- **From Xe [Z = 54] to Cs [Z = 55]**.

2.7.5. Concept of *Half-Filled* and *Completely-Filled* Orbitals

As per the **Hund's Rule** (see **section 1.1.6.2, Chapter-1**) we have already seen that:

"**half-filled (ns^1, np^3, nd^5) or completely-filled (ns^2, np^6, nd^{10}) orbitals are indeed observed to be comparatively much more stable**";

and, therefore, certainly much more energy is critically needed to get rid of an **electron** from such kind of *orbitals*.

It obviously suggests that the so-called **Ionization Potential** of an atom essentially possessing **half-filled or completely-filled orbitals** occurring prevalently in its respective **electronic configuration** is found to be comparatively higher *vis-a-vis* that expected usually right from its strategic position in the **Periodic Table**.

Importantly, the above mentioned, criterion has been employed as a strong point so as to explain the underlying fact that:

"**higher value exhibited by the *Ionization Potential* of a few of the elements *vis-a-vis* to that of the elements present on the left-hand side (LHS) in a specific Period of the Periodic Table**".

Example: As one proceeds from **LHS to RHS** in a **Period** pertaining to the long-form of the **Periodic Table**, normally the **Ionization Potential** gets enhanced perceptively on account of the successive enhancement in the respective *nuclear charge* (or **Atomic Number**); and hence, get reduced in **atomic size**.

Exceptions to the Above Observations: It has been observed critically that there are quite a few **elements** that do exhibit certain typical **irregular trends**.

Examples: Following are the *two* **specific examples**:

(i) **Berilium [Be] (Z = 4)** present in the **2nd Period** does possess a definte *higher* Ionization Potential in comparison to **Boron (B) (Z = 5)**. Likewise, **Nitorgen (N) (Z = 7)** possesses *higher value of* Ionization Potential *vis-a-vis* Oxygen [O] (z = 8).

(ii) **Magnesium [Mg] (Z = 12)** present duly in the **3rd Period** possesses a definite *higher* **Ionization Potential** *vis-a-vis* **Aluminium [Al] (Z = 13)**. Likewise, **Phosphorous [P] (Z = 15)** does possess a **higher Ionization Potential** in comparison to **Sulphur [S] (Z = 8)**.

Comments: The above mentioned **irregulations** may be expatiated vividly based on the following *two* **cardinal aspects, namely:**

- The **Ionization Potential** of *Berilium (Be)* compared to that of *Boron (B)*; and of Magnesium (Mg) compared to that of *Aluminium (Al)*, and
- The **Ionization Potential** of *Nitrogen (N)* compared to that of *Oxygen (O)*; and of *Phosphorus (P)* compared to that of Sulphur (S).

(i) **The Ionization Potential for Berilium (Be) compared to that of Boron (B); and of Magnesium (Mg) compared to that of**

Aluminium (Al)–Following are the *two* **equations** that duly represent the **Ionization Potential** of **Berilium (Be)** and **Boron (B)**,–which may be expressed as under.

$$\text{Berilium [Be] } (1s^2, 2s^2) \xrightarrow[\text{--}e^-\text{ (from completely filled 2s orbital)}]{} Be^\oplus (1s^2, 2s^2)$$

$$\text{Boron [B] } (1s^2, 2s^2, 2p^1) \xrightarrow[\text{--}e^-\text{ (from Partially filled 2p orbital)}]{} B^\oplus (1s^2, 2s^2)$$

Remarks: Thus, we may observe keenly that:

- ❑ In **Be**–it is rather **more difficult** to knock out an electron right from the so-called **completely-filled 2s-orbital**.
- ❑ In **B**–it is indeed a lot easier to get rid of an *electron* crucially from the **partially–filled 2p-orbital**.

Hence, we may infer that the **Ionization Potential** of *Be* is found to be defiantly **higher** *vis-a-vis* that of **B**.

 NOTE   It is however, pertinent to state here that both Be and *b* do possess exactly the *identical value of Principal Quantum Number (N = 2).*

Likewise, **Mg (magnesium)** having the inherent *higher value of* **Ionization Potential** *vis-a-vis* Al (aluminium), which may be solely attributed due to the underlying fact that:

"in the specific instance of Mg the '*electron*' has got to be removed right from the ensuing completely-filled 3s-orbital; whereas, in the case of *Al* the said '*electron*' is being duly removed right from the partially–filled 3p-orbital".

We may have the following *two* **expressions** each for **Mg** and **Al** respectively:

$$\text{Magnesium [Mg] } (1s^2, 2s^2, p^6, 3s^2) \xrightarrow[I_1 = 737.7 \text{ kJ.mole}^{-1}]{-e^-\text{ (completely-filled 3s-orbital)}} Mg^\oplus (1s^2, 2s^2 p^6, 3s^1)$$

$$\text{Aluminium [Al] } (1s^2, 2s^2 p^6, 3s^2 p^1) \xrightarrow[I_1 = 577.6 \text{ kJ.mole}^{-1} \text{ mole}^{-1}]{-e^-\text{ (Partially-filled 3p-orbital)}} Al^\oplus (1s^2, 2s^2 p^6, 3s^2)$$

(ii) The Ionization Potential of Nitrogen (N) Compared to that of Oxygen (O), and of Phosphorus (P) Compared to that of Sulphur (S)–The following *two* equations do explicitly represent the so-called **Ionization Potential** of **Nitrogen (N)** and **Oxygen (O)**; and hence, may be expressed as under:

$$\text{Nitrogen [N] } (1s^2, 2s^2, 2p^3) \xrightarrow[\text{I}_1 = 1402.3. \text{ kJ. mole}^{-1}]{-e^{\ominus}\text{(From Half-filled 2p Orbital)}} \text{N}^{\oplus} (1s^2, 2s^2, 2p^2)$$

$$\text{Oxygen [O] } (1s^2, 2s^2, 2p^4) \xrightarrow[\text{I}_1 = 1314.0. \text{ kJ. mole}^{-1}]{-e^{\ominus}\text{(From Partially-filled 2p Orbital)}} \text{O}^{\oplus} (1s^2, 2s^2, 2p^2)$$

Remarks: These essentially include:

1. In **Nitrogen (N)** it is found to be rather more difficult to get rid of an *electron* right from the ensuing **half-filled 2p-orbital**; whereas, in the specific instance of **Oxygen (O)**–it is definitely a lot easier to knock out the *electron* from **a partially-filled 2p-orbital**.

Inference–Hence, the **Ionization Potential** of **Nitrogen (N)** is observed to be higher *vis-a-vis* that of **Oxygen (O)**.

Similarly, based on the aforesaid dictum we may account for the *higher* **Ionization Potential** of **Phosphorous (P)** *vis-a-vis* to that of **Sulphur (S)**,–as expressed under:

$$\text{Phosphorus [P] } (2, 8, 3s^2, 3p^3) \xrightarrow[\text{I}_1 = 1011.8. \text{ kJ. mole}^{-1}]{-e^{\ominus}\text{(From Half-filled 3p-Orbital)}} \text{P}^{+} (2, 8, 3s^2, 3p^2)$$

$$\text{Sulphur [S] } (2, 8, 3s^2, 3p^4) \xrightarrow[\text{I}_1 = 999.6 \text{ kJ. mole}^{-1}]{-e^{\ominus}\text{(From Partially-filled 3p Orbital)}} \text{S}^{+} (2, 8, 3s^2, 3p^3)$$

Special Remark: It is worthwhile to state here that the so-called **Noble Gases** [*viz.,* He, Ne, Ar, Kr, Xe, Rn, Uuo] $(ns^2 p^6)$; **Zn** $(3d^{10}, As^2)$; **Cd** $(4d, 5s^2)$, and **Hg** $(4f^{10}, 5d^{10}, 6s^2)$, they essentially possess the so-called *completely-filled orbitals* do predominantly possess an **extremely high-Ionization Potential**.

2.7.6. *The Precise Nature of Orbitals*

It has been proven and established that the so-called **precise nature of orbitals** related to the **valence-shell** (or *intervening electrons*) from which the *electron* is intended to be removed invariably **influences grossly** the obvious:

'**magnitude of the Ionization Potential**'

Nevertheless, the prevailing-'*relative order of energy*' pertaining to the ensuing *orbitals: s, p, d,* and *f* belonging to a given *n*th **shell** is duly expressed as under:

$ns < np < nd < nf$

Importantly, the aforesaid order explicitly indicates that the ensuing '**relative order of energy**' by means of which the *electron* available in these *orbitals* may be removed quite easily and conveniently as stated below:

Relative order of energy *vis-a-vis* **the case of removal of electrons increasing** ®

$ns < np < nd < nf$

Remarks: Based on the above observed '**relative order of energy**' it may be understood vividly that to affect the removal of an *electron* from the *f*-orbital shall be accomplished in the **easiest manner; whereas,** to knock out the same from the corresponding *s*-orbital would eventually term out to the **most difficult and cumbersome task*.**

Inference: Hence, the rest of the factors remaining the same, the observed **Ionization Potentials** of an '*electron*' present in the respective *s, p, d,* and *f* orbitals are found to be in the following order:

$$ns > np > nd > nf$$

NOTE   **Based on these factual delebrations it may be rather easy to explain the inherent** *successive ionization potentials of an element*; **and also the** *higher values of Ionization Potential* **pertaining to Berilium (Be) and Magnesium (Mg)** *vis-a-vis* **to that of Boron (B) and Aluminium (Al) respectively.**

Some Important Aspects for Nature of Orbitals

Having understand certain vital and important aspects related to the **Nature of Orbitals**, it would be rather an appropriate opportune time to have a brief exposure to the following *two* **cardinal aspects**, namely:

- **The Successive Ionization Potentials of an Element [I_1 and I_2];** and
- **First Ionization Potentials of Two Different Elements belonging to the *Same Period*,**

which shall now be discussed individually in the sections that follows:

❑   **The Successive Ionization Potentials of an Element [I_1 and I_2]**–the **Ionization Potential of Boron (B) atom ($1s^2, 2s^2, 2p^1$)** are found to be: $I_1 = 800.6$ **kJ.mole^{-1}** and $I_2 = 2427.0$ **kJ.mole^{-1}** (*as mentioned earlier*) may be duly represented by the following *two* **equations:**

$$\text{Boron [B] } (1S^2, 2s^2, 2p^1) \xrightarrow[\text{I_1 = 800.6 kJ. mole}^{-1}]{-e^{\ominus}(2p^1)} B^{\oplus}(1S^2, 2s^2)$$

$$B^{\oplus}(1S^2, 2s^2) \xrightarrow[\text{I_2 = 2427.0 kJ. mole}^{-1}]{-e^{\ominus}(2s^1)} B^{2+}(1S^2, 2s^1)$$

Remarks: From the above *two* **equations** it may be observed explicitly that the value of I_1 is lower *vis-a-vis* that of I_2,–that could be duly explained based on the underlying fact:

"**it is indeed a lot easier to get rid of an** *electron* **from the 2p-orbital than to remove the same from the respective 2s-orbital, I_1, which eventually corresponds to the removal of a 2p-electron is found to be lower than I_2, that actually corresponds to the critical removal of a 2s-electron.**"

Besides, one may also take cognizance of the fact that both these **orbitals** *viz.* 2p and 2s do belong to the same shell, which relates to the **glaring observations:**

'**value of Principal Qunatum Number (*n*) pertaining to 2p and 2s-orbitals remains absolutely identical *viz., n* = 2.'**

❑   *First* **Ionization Potentials (IPs) of Two Different Elements Belonging to the Same Period**– In a broader perspective, as we proceed from *Left* to *Right* in a '**Period**',– the respective **Ionization Potentials** do increase accordingly: Nevertheless, quite a few of the **elements** do exhibit almost '*irregular trends*'.

---

*   That is, the quantum of energy needed actually to remove an '**electron**' from the *f*-orbital is minimum, whereas, that required to remove the same from *s*-orbital is maximum.

Example: The so-called *First* Ionization Potential (IP) of Berilium (Be) is found to be equal to 899.5 kJ.mole^{-1},–which being obviously higher *vis-a-vis* that of Boron (B) having the *First* Ionization Potential (IP) equal to **800.6 kJ.mole^{-1}** [*see under section 3.(b)*]. However, one may express these IPs as stated under:

$$\text{Berilium [Be] } (1s^2, 2s^2) \xrightarrow[I_1 = 899.5\,k.J.mole^{-1}]{-e^{-1}\,(2s^1)} \text{Be}^{\oplus}\,(1s^2, 2s^1)$$

$$\text{Boron [B] } (1s^2, 2s^2, 2p^1) \xrightarrow[I_2 = 1800.6\,k.J.mole^{-1}]{-e^{-1}\,(2p^1)} \text{B}^{\oplus}\,(1s^2, 2s^2)$$

Remarks: These essentially comprise:

1. The observed **IP of Be** is higher in comparison to that of **B,** because it would be rather more difficult to get rid of an electron from a **2s-orbital of Be**–that certainly needs much more energy in comparison to the same circumstance from a **2p-orbital of B**.

2. Thus, on the similar stream of logical explanation it may also be expatiated vividly the underlying fact that:

"the observed Ionization Potential (IP) of Magnesium [Mg] is found to be at a higher ebb than that of Aluminium [Al]."

The above observations may be expressed as under:

$$\text{Magnesium [Mg] } (2, 8, 3s^2) \xrightarrow[I_1 = 737.7\,k.J.mole^{-1}]{-e^-\,(3s^1)} \text{Mg}^{\oplus}\,(2, 8, 3s^1)$$

$$\text{Aluminium [Al] } (2, 8, 3s^2, 3p^2) \xrightarrow[I_2 = 577.6\,k.J.mole^{-1}]{-e^-\,(3p^1)} \text{Al}^{\oplus}\,(2, 8, 3s^2)$$

Remarks: These essentially include:

1. The various divergent aspects stipulated above are indeed found to be absolutely independent of each other.

2. Hence, it is quite impossible to predict the precise and exact extent to which they would be able to exert their ultimate effect upon the so-called ensuing magnitude of **Ionization Potential (IP)**.

❑ **Variation of Ionization Energies**

Ionization Energy–may be defined as–'**the energy essentially required to remove the abundantly and loosely bound electron right from an isolated gaseous atom**'.

In other words, when a small quantum of energy is duly provided to an atom,–an *electron* may be subsequently promoted to a **higher energy level** perceptively; whereas, if the quantum of energy supplied is significantly large, the said *electron* may be removed completely.

Determination of Ionization Energies: In true sense, these are duly determined right from the *spectra itself*; and measured precisely in **kJ.mole^{-1}**. Thus, we may conveniently get rid of more than **one electron** from a **good number of atoms**. Besides, we may observe the following aspects:

➤ *First* Ionization Energy–relates to the energy needed to remove the *first electron* and subsequently convert **M to M$^{\oplus}$**;

➤ *Second* Ionization Energy–refers to the energy so required to remove the *second electron* and also convert **M$^+$ to M^{2+}**; and

> *Third* **Ionization Energy**–helps in the conversion of M^{2+} to M^{3+} and so on so forth.

Factors Influencing the Ionization Energy–Following are the *four* cardinal factors that largely influence the **Ionization Energy**, namely:

(i) **Size of the atom**;

(ii) **Charge on the nucleus,**

(iii) **Efficiency with which the *Inner Electron Shells* usually screen the nuclear charge**; and

(iv) Type of electron being involved precisely (s, p, d, or f).

Remarks: These essentially include:

1. The above mentioned factors are invariably interrelated.

2. However, in a relatively **small atom** the *electrons* are being held together **rather tightly**; whereas; in a **large atom** the said electrons are usually held together less strongly.

3. Consequently, the **ionization energy (IE)** *gets duly reduced as the size of atoms increases.*

Table: 2.11 records precisely the so-called critically observed trend of the **Group 1** and **Group 2** elements; and also by the *other main groups*:

Table: 2.11: The Ionization Energies of Group 1 and Group 2 Elements (kJ. mole⁻¹)						
	First	*Second*		*First*	*Second*	*Third*
Li	520	7296	Be	899	1757	14847
Na	496	4563	Mg	737	1450	7731
K	419	3069	Ca	590	1145	4910
Rb	403	3650	Sr	549	1064	4207
Cs	476	2420	Ba	503	965	—
Fr	—	—	Ra	509	979	3281*

*An (stumated value)

Comparison of First and Second Ionization Energies for Group 1 Elements

It may be observed precisely that the critical removal of a *second* electron predominantly involves a substantial amount of energy that usually ranges between **7 to 14 folds more** *vis-a-vis* the so-called *First* **Ionization Energy**. Since, the observed *Second Ionization Energy* appears to be too exhorbitant that, **a second electron is not removed**. Nevertheless, the so-called large difference prevailed between:

'**the ensuing *First* and *Second* Ionization Energies**',

is found to be intimately related to the so-called structure of the **Group 1 atoms** Obviously, most of these atoms do possess just only **one electron** located strategically in their **outer shell**. Thus, it would be a lot easy task to get rid of the *single* outer electron, **it needs much more energy to remove the second electron**; as it critically involves the **breaking into a filled shell of electrons**.

Ionization Energies for the Group 2 Elements–It depicts explicitly that the so-called *first* **Ionization Energy** is found to be almost *double the value vis-a-vis* the respective **Group 1 element**.

Explanation: Perhaps it could be solely due to the *inherent* **enhanced nuclear charge** given out in the form of a *smaller size* **of the Group 2 elements**. Thus, once the *first-electron* is knocked out, the ensuing ratio of:

'charges held on to the nucleus to the precise number of orbital electrons* gets enhanced perceptively',

and hence, it helps in the reduction of the original size of the **Group 2 elements**.

Example: Obviously, the **magnesium ion [Mg$^\oplus$]** is found to be definitely **smaller in size** *vis-a-vis* the 'Mg-atom'. In this way, the remaining *electrons* duly present in **Mg$^\oplus$** are observed to be even **more strongly held together**; and as a result the *second* **Ionization Energy** certainly is greater *vis-a-vis* the **First-Ionization Energy**.

Points to Ponder: **(1)** Interestingly, the subsequent removal of a *third* **electron** right from a **Group 2 element** has proved to be a really **herculian (difficult) task** based on the following *two* **major reasons**, namely:

- there exists a definite increment in the *effective nuclear charge*; and, therefore, the *residual electrons* **are being held more strongly**, and
- **further removal of another electron would critically involve the cleavage of a completed shell of electrons perceptively**.

(2) Amazingly, the **Ionization Energy (IE)** virtually depends upon the exact '**type of electron**'–that is removed ultimately. Besides, the *r, p, d,* and *f* electrons do possess *orbitals* essentially having altogether *different shapes*.

❏ the *s*-electron is found to penetrate very close to the *nucleus*; and hence, is more *strongly held* than a *p*-electron,

❏ likewise, the *p*-electron is more tightly held in comparison to the *d*-electron, and

❏ the *d*-electron is more intimately held together *vis-a-vis* an *f*-electron.

(3) Having an assumption that all other factors remain equal, the observed **Ionization Energies (IEs)** are found to follow the order:

$$\boxed{s > p > d > f}$$

(4) It may also be seen that an increase in **IEs** is certainly not so *smooth* and *convenient* as one proceeds from LHS to RHS in the **Periodic Table**.

Example: The *First* **IE** for a **Group 13 element**** is actually less *vis-a-vis* that for the adjacent **Group 2 element.*****

Salient Features of Ionization Energy (IE)–following are the salient features of IE:

1. **The Ionization Energy (IE)** gets reduced progressively while **descending a** *specific* **Group**; and gets enhanced while **crossing a Period**,–as given in Table: 2.12.

Table: 2.12: The Comparison of Certain first Ionization Energies (IEs) (kJ. mole^{-1})							
Li	Be	B	C	N	O	F	Ne
520	899	801	1086	1403	1410	1681	2080
Na	Mg	Al	Si	P	S	Cl	Ar
496	737	577	786	1012	999	1255	1521

  * That is, the **effective nuclear charge**.
 ** That is, where a *p*-electron is being removed.
*** That is, where a *s*-electron is being removed.

Remarks: The removal of the so-called successive electrons eventually becomes rather more difficult and cumbersome as depicted under:

Ist Ionization Energy < 2nd Ionization Energy < 3rd Ionization Energy

---

Likewise, it is indeed an important and critical observation wherein one may observe an array of deviations from the so-called aforesaid **generalizations**. Thus, the **Ist Ionization Energies (IGs)** of the *elements* present in the **Periodic Table** may be given in Table: 2.13.

Table: 2.13: The First Ionization Energies [IEs] of Elements in the Periodic Table																	
Group→1	**2**										**13**	**14**	**15**	**16**	**17**	**18**
Period																	
1 H • 1311																	He • 2372
2 Li • 520	Be • 899											B • 801	C • 1086	N • 1403	O • 1410	F • 1681	Ne • 2080
3 Na • 496	Mg • 737	**3**	**4**	**5**	**6**	**7**	**8**	**9**	**10**	**11**	**12**	Al • 577	Si • 786	P • 1012	S • 999	Cl • 1255	Ar • 1521
4 K • 419	Ca • 590	Sc • 631	Ti • 656	V • 650	Cr • 52	Mn • 717	Fe • 762	Co • 458	Ni • 736	Cu • 745	Zn • 906	Ga • 579	Ge • 760	As • 947	Se • 941	Br • 1142	Kr • 1351
5 Rb • 403	Sr • 549	y • 616	Zr • 674	Nb • 664	Mo • 685	Tc • 703	Ru • 711	Rh • 720	Pd • 804	Ag • 731	Cd • 876	in • 558	Sn • 708	Sb • 834	Tc • 869	I • 1191	Xe • 1170
6 Cs • 376	Ba • 503	La • 541	Hf • 760	Ta • 760	W • 770	Re • 759	Os • 840	Ir • 900	Pt • 870	Au • 889	Hg • 1007	Tl • 589	Pb • 715	Bi • 703	Po • 813	At • 912	Rn • 1037
7 Fr	Ra	Ac															

First ionization energies of the elements (Numerical values are given in kJ mol⁻¹. Large circles indicate high values and small circles low values.) After Sanderson, R.T., Chemical Periodicity, Reinhold, New York.

The critical and important observed variations based upon the *First* Ionization Energies [IEs] of the elements [*viz.*, He, Li, Be, B, N, O, Ne, Na, Mg, Al, P, Ar, K, Zn, Ca, Kr, Rb, Cd, In, Xe, Cs, Hg, Tl, and Rn] are explicitly illustrated in Fig. 2.3.

Explanations: The graph represented in Fig: 2.2 clearly, shows *three* **spectacular features**, namely:

1. The **Noble Gases** *viz.*, **He, Ne, Ar, Kr, Xe,** and **Rn** [*Group 18; Table: 2.11*] do vehemently possess the *highest* **Ionization Energy levels** in their so-called respective *periods*.

2. Likewise, the **Group 1 metals** *viz.*, **Na, K, and Rb** [see **Table: 2.11**] do have the *lowest* **Ionization Energy levels** in their corresponding **periods**.

3. Obviously, there prevails a generalized *upward trend* specifically in the observed 'horizontal period', *viz.*,
   - **From Li to Ne (Period 2)**, or
   - **From Na to Ar (Period 3).**

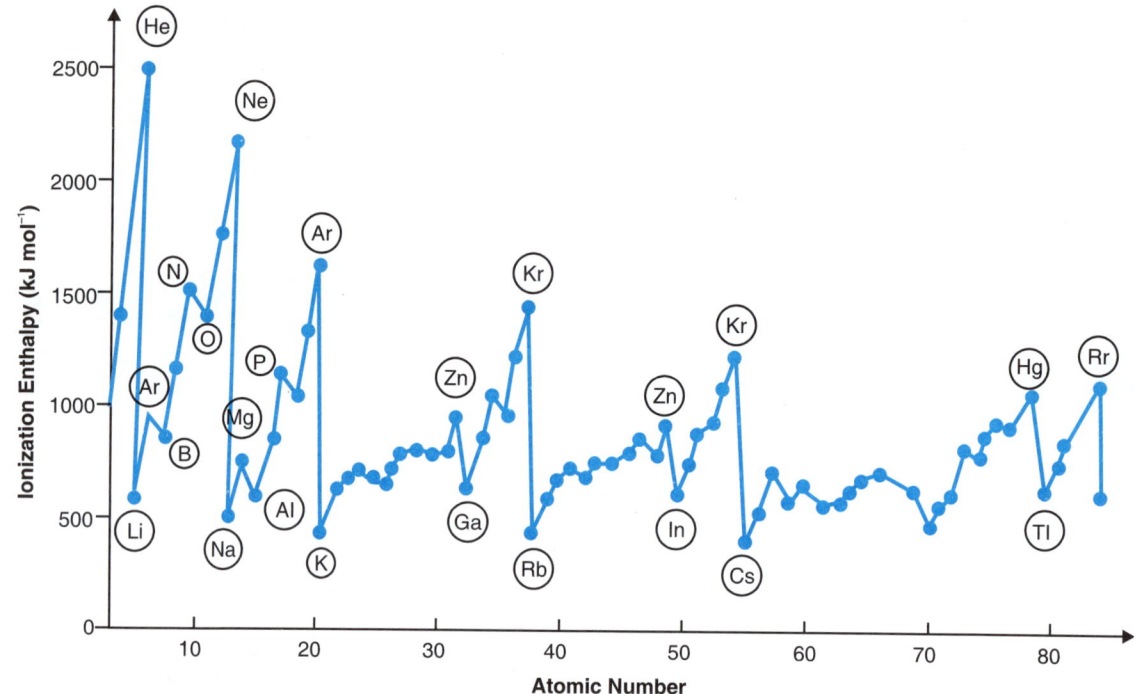

Fig.2.3: The Observed Variations in First Ionization Energies [IEs] of the Elements.

Besides, the observed values for both **Ne** and **Ar** are found to be the highest in their respective periods (*viz., 2 and 3*) since a **large quantum of energy** is needed to get rid of an *electron* right from:

'**an available stable filled shell of electrons**'.

Interestingly, the pattern of '**Graph**' (see Fig. 2.2) fails to show a smooth and progressive increment in a perfect smooth pattern. Perhaps it results into the so-called **high-values for Be and Mg**; and hence, it is solely attributed due to:

"**the overall stability of a filled *s*-shell**"

Importantly, the **IE-values** of N and P are also relatively at a *higher level*; and, therefore, it vividly shows that:

"**a half-filled *p*-level is stable specifically**".

The **IE-values** for **B** and **Al** are found to be perceptively *lower level* since the systematic removal of *one* **electron** leaves behind a-**stable filled s-shell**; and exactly in the same vein with **O and S-a stable half-filled s-shell is accomplished ultimately**.

Following are *three* important '**electronic arrangement**' exhibiting an '***extra-stability***' profile.

Remarks: These essentially comprise:

1. The *First* **Ionization Energy (IE)** gets lowered in a systematic manner while descending the **main groups**.

2. Nevertheless, an apparent deviation from the aforesaid trend does take place in **Group 13** [*viz.*, **B, Al, Ga, In, Tl, Uut**],–wherein the so-called anticipated decrease duly comes into play between **B**

and **Al**; however, the **IE-values** for the rest of the elements *viz.*, **Ga**, **In** and **Tl** fail to maintain the *same* trend rigidly,–and are indeed **irregular in nature** (as depcited in Table: 2.14).

- Fixed s-level
- Half-Filled p-level
- Completely-Filled Noble Gas Structure

Table: 2.14: The Ionization Energies [IGs] for Group 13 Elements [kJ.mole^{-1}]			
	First	Second	Third
B	801	2427	3656
Al	577	1816	2744
Ga	579	1979	2962
In	558	1820	2704
Tl	589	1971	2877

Comments: The major possible reason for the change occurring at **Ga** is perhaps owing to the underlying fact that:

'it is invariably precoded by *ten (10) elements* of the very 1st transition series (where the so-called 3d-shell is being filled)'

Eventually, it critically renders **Ga particularly smaller** *vis-a-vis* what it would otherwise be. Besides, a rather relatively smaller overall effect may be observed with the so-called:

"2nd and 3rd transition series; and also the critical presence of three transition series exert a marked and pronounced effect upon the values for Ga, In, and Tl plus the overall effect still unhibits in Group 14 and 15 perceptively".

NOTE Interestingly, the so-called Transition Energies [TEs] pertaining to the transition elements are indeed slightly irregular; however, the 3rd row elements commencing at Hf do have lower values *vis-a-vis* the expected ones due to the interpolation of the 14-lanthanide elements lying between La and Hf precisely.

3. THE ELECTRON AFFINITY [EA]

The **electron affinity (EA)** is referred to as:

"the energy released when an extra electron is being incorporated to a neutral gaseous atom."

Since only '*one electron*' is being added invariably that ultimately results into the formation of a uni-negative ion,–that critically evolves energy. Amazingly, the **Electron Affinity (EA)** solely depends on the **size** and the **effective nuclear charge**.

However, **EA** cannot be determined directly, but may be accomplished indirectly from the so-called **Born-Haber Cycle**. The important **electron affinity (EA) values** for an array of selected elements are duly provided in Table: 2.15.

Table: 2.15: A Few Important Electron Affinity (EA) Values [kJ.mole^{-1}]												

$$H \rightarrow H^- -72$$
$$He \rightarrow He^- -54$$

Li	$\rightarrow$	Li$^-$	-57	Na	$\rightarrow$	Na$^-$	-21					
Be	$\rightarrow$	Be$^-$	-66	Mg	$\rightarrow$	Mg$^-$	-67					
B	$\rightarrow$	B$^-$	-15	Al	$\rightarrow$	Al$^-$	-26					
C	$\rightarrow$	C$^-$	-121	Si	$\rightarrow$	Si$^-$	-135					
N	$\rightarrow$	N$^-$	-31	P	$\rightarrow$	P$^-$	-60					
O	$\rightarrow$	O$^-$	-142	S	$\rightarrow$	S$^-$	-200					
O	$\rightarrow$	O^{2-}	-702	S	$\rightarrow$	S^{2-}	-332					
F	$\rightarrow$	F$^-$	-333	Cl	$\rightarrow$	Cl$^-$	-348	Br	$\rightarrow$	Br$^-$	-324 I	I$^-$ -295
Ne	$\rightarrow$	Ne$^-$	-99									

Remarks: These essentially comprise.

1. The **negative electron affinity [–ve EA]** values do indicate that the energy is usually evolved when an atom **accepts an electron.** Hence, the above values do indicate explicitly that the **halogens** mostly evolve a huge quantum of energy while producing a –ve **halide ion**; and hence, it is not aetonishing at all that these ions normally occur in a good number of **chemical entities (compounds).**

2. Besides, the evolution of energy takes place as and when **one electron** is being added to either an **O or S-atom** thereby giving rise to the generation of such species as: $O^{\ominus}$ and $S^{\ominus}$; however, a substantial quantum of energy gets absorbed when **2-electrons** are being incorporated to form: O^{2-} and S^2 **ions.** Hence, the ultimate **electron affinities [EAs] for** O^{2-} and S^{2-} do possess a +ve sign:

3.1. THE BORN-HABER CYCLE

The Cycle–It is a well-known fact the *experimental determination of lattice energies* seems to be rather difficult; and hence, these are duly determined by means of so-called: **Indirect calculations** by making use of:

     'a cyclic phenomenon usually termed as the Born-Haber Cycle or Born-Haber Fajans Cycle'.

     First-law of Thermodynamics–Based on its *necessary consequence* one may observe that:

     'the so-called **enthalpy of a reaction would always be, the same whether the reaction occurs in single-or multi-steps sequence**'.

     The above dictum relates to the *Hess's Law*, which categorically is a statement of the law pertaining to the respective **Conservation of Energy.** Thus, the **Born-Haber Cycle** relates the so-called **Standard Molar Enthalpy** of the typical formation of ΔH_t (*i.e.,*) when mole of the respective **Ionic Compound:** $M^{\oplus}X^{\ominus}$ gets generated right from the elements in their rather *more* stable physical states at **298.15 K** and **101.3 kPA** (1 *atmospheric pressure*).

Example: Let us take into account the typical instance with regard to the formation of $M^{\oplus}X^{\ominus}$ (an *ionic chemical entity*) derived from **M** and X_2 (*viz.,* **NaCl from Na and Cl$_2$**) followed usually in *two* paths (Path I and Path II)–as stated under:

$$M(S) + \frac{1}{2} X_2(g) \xrightarrow{\boxed{\text{Path I}}} M^+X^-(S)$$

$$\downarrow$$

$$X(g) \xrightarrow{} X^-(g)$$

$$\boxed{\text{Path I}}$$

$$M(g) \xrightarrow{} M^+(g)$$

Explanation–Both these paths may be explained individually as under:

Path I

Thus, one mole of **M** and half-a mole of X_2 is found to combine almost directly in their so-called 'standard states' so as to yield the crystals of $M^\oplus X^\ominus$ very much in the standard state. In case, the *heat released* is **QkJ**, it may be expressed as under:

$$\boxed{M(s) + \frac{1}{2} X_2(g) + M^+ \ X^-(s) \,||\, \Delta H_1 = -Q \, kJ. \, mole^{-1}}$$

Path II

Thus, we may perform the same reaction with the help of the following *five* **sequential steps**:

Step-1: The solid element **M** is duly vaporized (*atomized*) by the aid of a precise input of **S kJ.mole^{-1}** belonging to the sublimation energy of **M(s)**:

Thus, we may have the following expression:

$$\boxed{M(g) \longrightarrow M^\oplus(g) + e^\ominus \,||\, \Delta H = SkJ. \, mole^{-1}}$$

Step-2: **M(g)** is subsequently converted into the corresponding **gaseous cation** $M^\oplus$ by means of the input of the **ionization energy (I)** thereby releasing *one electron per atom* perceptively. Hence, we may have the following reaction:

$$\boxed{\frac{1}{2} X_2(g) + e^\ominus \longrightarrow X(g) \,||\, \Delta H = I \, kJ.mole^{-1}}$$

Step-3: In this case, 1/2 a mole of the **diatomic gas** X_2 gets duly dissociated into atoms by the input of 1/2D, where Δ represents the ensuing **dissociation energy** of **gases** X_2 into the respective atoms:

Thus, we may have the following expression:

$$\boxed{\frac{1}{2} X_2(g) \longrightarrow X(g) \,||\, \Delta H = \frac{1}{2} D \, kJ. \, mole^{-1}}$$

Step-4: At this point the **electrons** are incorporated to the atomic **X** to produce the respective anions $X^{\ominus}-$ thereby releasing the corresponding electron affinity *X*:

Hence, we may have the following reaction:

$$\boxed{X(g) + e^\ominus \longrightarrow X^\ominus(g) \,||\, \Delta H = -XkJ. \, mole^{-1}}$$

Step-5: At the final stage the **gaseous cations** and the gaseous anions virtually combine to form the desired crystals of $M^\ominus X^\oplus$ along with the critical release of the **lattice energy,** *Uc*, of the crystal.

Thus, we may have the following expression:

$$\boxed{M^+(g) + X^-(g) \longrightarrow M\overset{\oplus}{X}{}^{\ominus}(s)\,||\,\Delta H = +U_c \text{ kJ. mole}^{-1}}$$

The **total enthalpy change** observed for the **Path II** with respect to the critical formation of the crystals of $M^{\oplus}X^{\ominus}$ obtained from the standard states of **M** and X_2, ΔH_{II} is given as under:

$$\Delta H_{II} = s + I + 1/2\,D - X + U_c$$

But from the **Hess's Law**, we have: $\Delta H_I = \Delta H_{II}$, so that:

$$-Q = S + I + 1/2\,D - X + Q$$

Therefore, the may have:

$$\boxed{U_c = S + I + \frac{1}{2}D - X + Q}$$

Since, the various terms **S, I, D, X,** and **Q** may be determined experimentally, the so-called **Lattice Energy**, U_c, of the resulting **Ionic Crystals** may be calculated with great ease and fervour.

Another school of thought proclaims the **Electron Affinity** as the **Electron Affinity Energy**. It may be duly measured in various ways, namely:

- **Electron Volts (eV)** • **Kilocalories (k cal) and** • **Kilojoules (kJ).**

Based on the *scientific evidences* and *logistic explanations* the values of **Electron Affinity** are invariably shown by a **negative (minus) sign**–that is usually placed before them. Interestingly, the **negative sign** vividly designates the so-called –'**release of energy**' in the addition of an electron to the **M(g)** atom.

Table: 2.16 records the corresponding values of *First* **Electron Affinity** (E_1) [*expressed commonly as:* **kJ. mole^{-1}**] of both *s*-and *p*-block elements.

Table: 2.16: The Electron Affinities [EAs] of s-and p-Block Elements in kJ. mple^{-1}.]							
Group→ **Period** **IA**	**IIA**	**IIIA**	**IVA**	**VA**	**VIA**	**VIIA**	**Zero**
1 $+e^-$ $H \rightarrow H^-$ -72.9							$+e^-$ $He \rightarrow He^-$ -0
2 $+e^-$ $Li \rightarrow Li^-$ -59.8	$+e^-$ $Be \rightarrow Be^-$ -0	$+e^{-1}$ $B \rightarrow B^-$ -23	$+e^-$ $C \rightarrow C^-$ -122	$+e^-$ $N \rightarrow N^-$ $+20.1$	$+e^-$ $O \rightarrow O^-$ $+780$	$+e^-$ $F \rightarrow F^-$ -322	$+e^-$ $H \rightarrow H^-$ -0
3 $+e^-$ $Na \rightarrow Na^-$ -52.9	$+e^-$ $Mg \rightarrow Mg^-$ -0	$+e^-$ $Al \rightarrow Al^-$ -44	$+e^-$ $Si \rightarrow Si^-$ -120	$+e^-$ $P \rightarrow P^-$ -74	$+e^-$ $S \rightarrow S^-$ -200.4 $S \rightarrow S^{-2}$ $S \rightarrow S$ 590	$+e^-$ $cl \rightarrow cl^-$ -348.7	$+e^-$ $Ar \rightarrow Ar^-$ -0
4 $+e^-$ $K \rightarrow K^-$ -48.9	$+e^-$ $Ca \rightarrow Ca^-$ -0	$+e^-$ $Ga \rightarrow Ga^-$ -36	$+e^-$ $Ge \rightarrow Ge^-$ -116	$+e^-$ $As \rightarrow As^-$ -77	$+e^-$ $Se \rightarrow Se^-$ -195 $-2e^-$ $Se \rightarrow Se^{2-}$ $+420$	$+e^-$ $Br \rightarrow Br^-$ -324.5	$+e^-$ $Kr \rightarrow Kr^-$ -0
5 $+e^-$ $Rb \rightarrow Rb^{-1}$ -46.9	$+e^-$ $Sr \rightarrow Sr^-$ -0	$+e^-$ $In \rightarrow In^-$ -34	$+e^-$ $Sn \rightarrow Sn^-$ -121	$+e^-$ $Sb \rightarrow Sb^-$ -101	$+e^-$ $Te \rightarrow Te^-$ -190.1	$+e^-$ $I \rightarrow I^-$ -2.96	$+e^-$ $Xe \rightarrow Xe^-$ -0

* The –ve sign shows the *release of energy* (**exothermic reaction**) when an *electron* is being added to an atom to convert it into an anion; whereas, the +ve sign indicates the *absorption of energy* (**endothermic reaction**) for the same phenomenon to occur.

3.2. The Second Electron Affinity (E_2)

Besides, the *First* **Electron Affinity (E_1)**, as defined above, the corresponding *second* **Electron Affinity (E_2)** with respect to certain elements *viz.*, **O, S** and **Se** is also duly recognized.

Thus, the **Second Electron Affinity (E_2)** may be defined for the element, **M(g)** as:

"**the exact and precise quantum of energy required necessarily to add one more electron to the respective mono-negative anion, "$M^-(g)$ to convert it into the corresponding di-negative anion, $M^{2-}(g)$.**"

Hence, we may have the following expression:

$$\underbrace{M^-(g)}_{\text{Mono-negative Anion}} \longrightarrow +e(g) + \textbf{Energy Supplied } (= +E_2) \longrightarrow \underbrace{M^2(g)}_{\text{Di-negative Anion}}$$

Explanations–They essentially include:

1. While adding one additional electron to the **mono-negative anion** [$M^-(g)$] against the so-called prevailing *electrostatic repulsion* between the **extra electron** which is duly added on to $M^-(g)$. Thus, the negative charge resides upon $M^-(g)$ anion; and hence, the energy instead of being released (as in the particular instance of *First* **Electron Affinity**) is being duly supplied to the $M^-(g)$ **anion** to convert it into $M^{2-}(g)$ **ion**.

2. In this manner, the phenomenon of incorporating a **2nd electron** to $M^-(g)$ **anion** is indeed an '*endothermic process*'.

3. Nevertheless, the **2nd electron affinity (E_2)** is being designated by a **+ve sign**, since it duly represents the *quantum of energy absorbed* (*i.e.*, **supplied**) and not released as in the specific instance of the *First* **Electron Affinity (E_1)**.

4. Furthermore, the subsequent formation of a **di-negative ion** [$M^{2-}(g)$] right from the *neutral* **M(g)** **atom**, wherein the *electrons* are usually being incorporated one by one. Thus, the observed **energy** that equals to the respective *First* **Electron Affinity (E_1)** is generated by the subsequent addition of one electron to the corresponding *neutral* **M(g) atom** so as to afford its conversion into $M^-(g)$ (*i.e.*, **mono-negative anion**), and the overall phenomenon is found to be *exothermic* in nature. Thus, we may have the following expression:

$$M(g) + e^-(g) \rightarrow M^-(g) + \text{Liberated Energy } (= -E_1)\ldots\text{an } \textbf{\textit{Exothermic Process}} \qquad \text{(a)}$$

> **NOTE**   The **–ve sign** appearing before E_1 shows the liberation of energy.

5. Besides, the energy equivalent to *second* **Electron Affinity (E_2)** needs to be provided duly by the crucial **addition of one more electron** to $M^-(g)$ for its conversion into the respective $M^{2-}(g)$ *i.e.*, a **di-negative anion**; and the overall process is observed to be *endothermics* in nature. Hence, we may have the following reaction:

$$M^-(g) + e^- + \text{Energy Supplied } (= +E_2) \rightarrow M^{2-}(g) \ldots\text{an } \textbf{\textit{Endothermic Process}} \qquad \text{(b)}$$

> **NOTE**   The **+ve sign** placed prior E_2 indicates clearly the absorption of energy.

6. Since $E_2 > E_1$, one may predominantly observe the **net absorption of energy** in the so-called formation of the **di-negative ion** $M^{2-}(g)$ from the respective *neutral* **M(g) atom**. Thus, the precise

quantum of energy being absorbed in this *typical phenomenon* equals to $E_2 - E_1$, that may be accomplished by the addition of Eqs. (a) and (b) stated above.

Hence, we have the following expression:

$$\boxed{\underset{\text{A neutral atom}}{M(g)} + 2e^-(g) + \text{Energy Absorbed} (= E_2 - E_1) \longrightarrow \underset{\text{A di-negative ion}}{M^{2-}(g)}}$$

Typical Illustrations to Expatiate the concept of Second Electron Affinity (E_2)–Following are the *three* typical examples that may be used for the gross illustration of the underlying concept of **Second Electron Affinity (E_2)**:

(a) **Formation $O^{2-}(g)$ from Neutral $O(g)$ Atom**– In reality, the critical formation of $O^{2-}(g)$ starting from $O(g)$ may be illustrated by the help of following sequence of equations:

$$O(g) + e^-(g) \mapsto O^-(g) + E_1 \; (= -141 \text{ kJ.mole}^{-1}) \dots \textbf{Exothermic Phenomenon}$$

$$O^-(g) + e^-(g) + E_2 \; (= +921 \text{ kJ.mole}^{-1}) \mapsto O^{2-}(g) \dots \textbf{Endothermic Phenomenon}$$

Let us now **add** the above *two* equations:

$$O(g) + 2e^{-1}(g) + \text{Energy} \; (= +921 - 141 = 1780 \text{ kJ.mole}^{-1}) \longrightarrow O^{2-}(g)$$

Inference: Thus, the overall (total) **Electron Affinity (EA)** for the *two* electrons pertaining to the respective O-atom is equal to **1780 kJ.mole^{-1}**.

(b) **Formation of $S^{2-}(g)$ from $S(g)$**–In this case, we may also obtain in a similar manner the following *two* sequential reactions:

$$S(g) + e^-(g) \longrightarrow S(g) + E_1 \; (= -200.4 \text{ kJ.mole}^{-1}) \dots \textbf{Exothermic Phenomenon}$$

$$S(g) + e^-(g) + E_2 \; (= +1790.4 \text{ kJ.mole}^{-1}) \longrightarrow S^{2-}(g) \dots \textbf{Endothermic Phenomenon}$$

Adding the above two equations, we have:

$$S(g) + 2e^-(g) + \text{Energy} \; (= +790.4 - 200.4 = 590 \text{ kJ.mole}^{-1}) \longrightarrow S^{2-}(g)$$

Inference: Hence, the total overall **Electron Affinity (EA)** for *two* electrons pertaining to **Sulphur atom** = **+ 580 kJ. mole^{-1}**.

(c) **Formation of $Se^{2-}(g)$ from $Se(g)$**–In this specific instance, we may also accomplish on the same lines the following *two* sequential reactions:

$$Se(g) + e^-(g) + Se^-(g) + E_1 \; (= -195 \text{ kJ.mole}^{-1}) \dots \textbf{Exothermic Process}$$

$$Se^-(g) + e^-(g) + E_2(= +615 \text{ kJ.mole}^{-1}) \longrightarrow Se^{2-}(g) \dots \textbf{Endothermic Process}$$

Let us now add the above *two* equations:

$$Se(g) + 2e^-(g) + \text{Energy} \; (= +615 - 195 = +420 \text{ kJ.mole}^{-1}) \longrightarrow Se^{2-}(g)$$

Inference: Therefore, we may infer that the *total* electron affinity (EA) for the said *two* electrons for Se atom = + 420 kJ. mole^{-1}.

3.3. Various Important Factors that Influences the Magnitude of Electron Affinity (EA)

Based upon the scientific evidences and logical explanations it has been proven adequately that:

'**there are several factors upon which both the magnitude as well as the inherent sign of the prevailing** *Electron Affinity* (*EA*) **solely depends**'.

These factors, in general, are very much akin to all those that virtually determine the so-called **Ionization Energies (IEs)** but certainly in the **reverse order**.

Following are the *three* **cardinal factors** that actually influence the ultimate magnitude of the **Electron Affinity (EA)**, namely:

   (a) **Atomic Radius**,

   (b) **Effective Nuclear Charge**, and

   (c) **Electronic Configuration**,

which shall now be treated individually in the sections that follows:

3.3.1. *Atomic Radius*

It has been established beyond any reasonable doubt that **greater the atomic radius**, lower would be the inherent tendency of the atom to attract the *additional electrons* very much towards itself. Therefore, *lesser would be the force of attraction* being exerted by the **nucleus** particularly upon the *extra electron* **being added on to the outer-shell of the atom**. Perhaps it would permit the release of *smaller quantum of energy* as and when the **additional electron** is being added to the atom to result into the formation of an **atom**.

Inference: **Hence, we may infer that the-smaller atoms do possess higher electron affinities perceptively**.

3.3.2. *Effective Nuclear Charge*

It is proven that higher the **effective nuclear charge (Z)**, greater would be the inherent tendency of the atom to enable the due **attraction of the additional electron towards itself**. Hence, higher would be the force of attraction being exerted by the nucleus upon the *extra electron* that eventually adds on to the so-called '**outer-shell of the atom**'. Thus, amazingly a higher quantum of energy gets released as and when the additional is being added on to the **atom**.

Inference: Therefore, the atoms having *higher effective nuclear charge (Z)* do exhibit explicitly higher *electron affinities (EAs)*.

3.3.3. *Electronic Configuration*

Importantly, the actual effect of **electronic configuration** upon the ensuing *magnitude of* **electron affinity (EA)** may be adequately expatiated by the following *four* typical examples:

3.3.3.1. *The Electron Affinity Values of Elements of Group IIA [Be, Mg, Ca, Sr, Ba, Ra]*

The **ns-orbital** of the **valence-shell** of the atoms of these elements has been found to be **filled completely**; and, therefore, the addition of an *additional electron* from outside to this **ns-orbital** is completely not feasible at all.

Inference: As a result the *elements of Group IIA do* possess almost *zero* **electron affinity (EA)**.

3.3.2. *The Electron Affinity (EA) Values of N and P*

The **valence shell electron configuration** of N and P as given under:

$$N = 2s^2p^3 \quad \text{and} \quad P = 3s^2p^3$$

Interestingly, the status of **2p and 3p orbitals** present in N and P respectively are found to be duly **half-filled**; and, therefore, are found to be **stable extra ordinarily**. Perhaps that could be the viable reason that:

'**the addition of any extra (additional) electron from outside to such orbitals is not feasible at all.**'

Inference: As a result, N and P both elements do essentially possess extremely low **electron affinity (EA) values** *viz.*,

$$[N = + 20.1 \text{ kJ.mole}^{-1} \text{ and } P = - 74 \text{ kJ. mole}^{-1}]$$

3.3.3.3. *Electron Affinity (EA) Values of Halogens*

It may be observed that the **valence shell electron configuration** of the *halogen atoms viz.*, **Group VII A: F, Cl, Br, I, At, Uus**–(ns^2p^5) critically shows a dire need for *one electron* so as to stabilize its configuration by attaining the so-called **Noble Gas configuration** (ns^2p^6)–that is found to be **extremely stable**.

Therefore, the **halogen atoms** (stated above) do have really a reasonably strong inherent tendency to accept elegantly an *additional electron* from *outside source*, thereby rendering the **ultimate configuration** of the *resultant halide ion* very much akin to that of the **Noble Gas**; and, therefore, gives them a **reasonably stable status**.

Inference: Consequently, one may infer that the **halogen atoms** do exhibit an **extremely high value of the electron affinity (EA)**.

3.3.3.4. *The Electron Affinity Values of Noble Gases*

In general, the **Noble Gases** do possess **extra-ordinary stable configuration** (ns^2p^6)–that eventually exhibit no tendency whatsoever to accept the *additional electron* obtained from an outside source.

Inference: Therefore, it may be infered that the observed **electron affinity (EA) values** of the *Inert Gases* are found to be almost **zero**.

3.4. The Periodic Trends of Electron Affinity (EA)

Preamble: The **periodic trends of electron affinity (EA)** specifically relates to the observed variation of **Electron Affinity** either:

- **in a Period**, or
- **in a Group**.

Nevertheless, based upon the theoretical considerations and subsequent interpretations one really finds it difficult to pronounce *general views* pertaining to the so-called 'Periodic Variations' of the **electron affinity (EA)** of the array of elements present in the **Periodic Table** for the following *three* good valid reasons, namely:

(i) The exact and precise **Electron Affinities (EAs)** pertaining to only a very few atoms are known.

(ii) Besides, the elements *viz.*, **Be Mg** and **Noble Gases (He, Ne, Ar, Kr, Xe, Rn, Uuo)** do invariably possess– '**zero electron affinity (EA) Values**'; whereas, N and P have extremely **small EA Values.**

(iii) Importantly, the respective elements of the second period in the **Periodic Table** fail to adopt the so-called generalized trend of *reduction in the EA-values* while proceeding from *top* to **bottom** in a particular **Group in the Periodic Table**.

Points to Ponder: In a broader perspective, the observed **general trend of variations in EA** for various elements duly present in the **Periodic Table** may be discussed briefly with particular reference to the exact status:

- **Variations in a 'Period'**, and
- **Variations in a Group**.

❑   **Variations in a '*Period*':** It is quite evident that as we move from **LHS to RHS in a Period,** we may critically take cognjzance of the fact that:

'**the size of the atoms gets decreased progressively, –while the corresponding effective Nuclear Charge (Z)* increases**'.

In fact, the said *two* factors vehemently favour a definite increase in the **force of attraction** caused by the **nucleus upon the electron** perceptively. As a result, the **atom** shows a higher tendency to:

"**critically attract an extra electron from outside towards itself; and, therefore, the electron affinity (EA) enhances from LHS to RHS**".

In this manner, the so-called **alkali metals** (*viz.*, **Li, Na, K, Rb, Cs, Fr**) which usually occur at **LHS of the Periodic Table** do possess essentially **lower EA-values**; whereas, the corresponding **non-metals** (*viz.*, **halogens**) that normally lie at **RHS** of the **Periodic Table** do exhibit **higher EA values**.

Important Exceptions–There are quite a few important exceptions with regard to the *elements* present in *each* Period–that do possess the so-called '**abnormal EA-values**'.

Examples: These essentially include:

1. Amazingly, in the *Second Period* elements *viz.*, **Li, Be, N, and Ne** do possess **abnormal EA values**.

2. Likewise, in the *Third* Period elements *viz.*, **Na, Mg, P, and Ar** do have also the **abnormal EA-values**.

3. Importantly, the **abnormal EA-values** of **Li** and **Na** cannot be logically explained by the help of *simple mechanism*; whereas, those of **Be, N, Mg, P, Ne, and Ar**–have already been explained earlier.

---

*   That is, the **Atomic Number**.

❑ **Variations in a '*Group*':** It is, however, pertinent to mention here that as we proceed *down a Group*; one may critically take cognizance of the underlying fact that:

"**both the *size of the Atoms* and the effective *Nuclear Charge* (Z) gets enhanced progressively**".

Obviously, the increment in the so-called *Atomic Size* has an overwhelming tendency to *lower* the **EA-values** progressively; whereas, the corresonding **increment in the Nuclear Charge (Z)** does exhibit a tendency to enhance **the respective EA-values** perceptively.

Thus, the overall *net result* shows that the actual effect caused by the so-called **progressive increment** of:

"**the atomic size seems to be more influential than the effect exhibited by the progressive increase in the Nuclear Charge (Z)**".

As a result, we may critically observe that the ensuing–'**electron affinity (EA) does decrease in a progressive manner as one proceeds from top to bottom in a Group**'

Examples: The aforesaid analogy may be explicitly expatiated from the respective **Electron Affinity (EA) values** of the following classical examples:

❑ *Group IA*: **alkali metals** *viz.*, **Li, Na, K, Rb, Cs, Fr;**
❑ *Group IV A*: *viz.*, **C, Si,** and **Ge,**
❑ *Group VIA*: *viz.*, **S, Se,** and **Te;**
❑ *Group VI A*: *viz.*, **O, S,** and **Se;** and
❑ *Group VII A*: *viz.*, **Cl, Br,** and **I**

Important Exceptions–Following are a few **important exceptions**, namely:

1. Even through the *elements belonging* to the *second* **Period** of the **Periodic Table** do possess rather **smaller size** *vis-a-vis* the elements of the respective *Third* **Period**; however, the corresponding **EA-values** of the *elements* occurring first below them in the *Third* **Period**.

    Examples: (1) These essentially include: **B < Al, N < P, O < S, and F < Cl.**

2. Besides, the so-called *lower* **EA-values** for the elements belonging to the *Second* **Period** may be duly supported by the fact that by virtue of their inherent *smaller size of the atoms* of the **elements of the *Second* Period**, the crucial incorporation of an **additional electron** to their *atoms* does generate perceptively:

    "**a high electron density all around the resulting anions**".

NOTE   The aforesaid higher electron density enhances the so-called *repulsion profile* prevailing between the *electrons*, invariably termed as: *electron-electron repulsion,* thereby the atoms of the elements belonging to the *Second Period* exhibit much lower tendency to attract the '*additional electron*' from outside, thus, relatively *lower EA-values for all such elements.*

Fig: 2.4 vividly depicts the so-called observed variation in the **Electron Affinity (EA)** profiles of *s*-and *p*-**Block Elements** duly present in a particular *Group* and *Period*.

3.4. Classical Applications of Ionization Potential (IP) and Electron Affinity (EA)

The classical applications of the **Ionization Potential (IP)** and the **Electron Affinity (EA)** may be explored further in a comprehensive manner under the following *four* **categories**, namely:

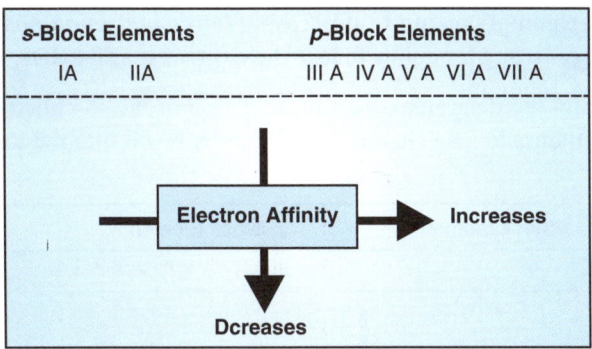

Fig.2.4: Diagrammatic Representation of Observed Variation in the Electron Affinity (EA) of *s*-and *p*-Block Elements in the Periodic Table.

- **Inherent Tendency of Elements to Produce Cations and Anions;**
- **The Reducing and Oxidizing Power of Various Elements.**
- **Metallic and Non-metallic Character of Elements**; and
- **The Relative Stability and Activity of Elements,**

which shall now be described separately in the sections that follows:

3.5.1. *Inherent Tendency of Elements to Produce Cations and Anions*

It has already been proven and ascertained that the magnitude of **Ionization Potential (IP)** of an *element* provides an ample idea pertaining to the so-called inherent tendency of the *element* to knock out an *electron*; and thereby forming a **cation**. From this dictum one may safely infer that:

'**greater the IP-value of an element lesser would be its inherent, tendency to get rid of the** *electron*; **and hence, to produce a** *cation*.'

It ultimately leads to the underlying fact that the elements possessing relatively **lower IP-values**, such as:

- **Alkali Metals (Group IA)** *viz.*, **Li, Na, K, Rb, Cs**, and **Fr**; and
- **Alkaline Earth Metals (Group IIA)** *viz.*, **Be, Mg, Ca, Sr, Ba**, and **Ra**,

do exhibit a strong tendency to *lose their outermost shell electron* perceptively. Thus, they virtually form **cations** rather easily and conveniently; whereas, those possessing critically **higher IP-values**, such as: **Halogens (Group VIIA)** are found to be incapable of *losing the electron*; and, therefore, exhibit little tendency to produce the respective **cations**.

In fact, these type of *elements*, based upon their **inherent higher electron affinity (EA) values** invariably display a strong tendency to **accept the 'electrons'** elegantly so as to produce the **Anions**. Thus, **greater the EA-values** of an *element* – leads to a higher inherent tendency to *accept the electrons*; and hence, to produce the **anions**. Perhaps this could be the possible reasons that:

"**the elements having** *higher EA-values*" *viz.*, **Halogens do have a rather strong tendency to accept an electron**",

and hence, produce an anion very easily; whereas, those elements having **lesser EA-values** (*viz.*, **Group IA and Group IIA elements**) are not in a state to–'**accept an electron elegantly**'; and, therefore, do not show any tendency to produce an **Anion**.

NOTE   Instead such elements owing to possessing *lower ionization potential (IP) values* do exhibit a rather strong tendency to lose the electrons; and hence, generate the cations.

Fig: 2.5 illustrates explicitly the diagrammatic representation of the so-called inherent tendency of the entire *s*-and *p*-**Block Elements** to lose electrons and thereby result into the formation of *cations* in the **Periodic Table**.

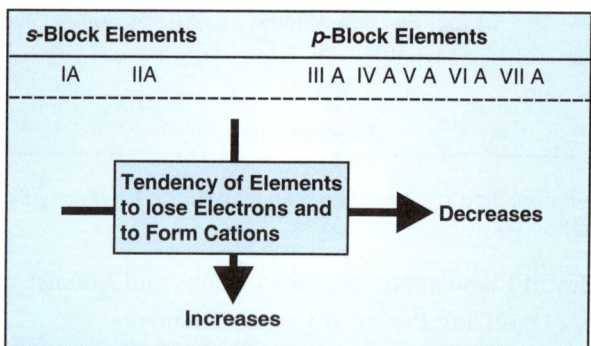

Fig.2.5: The Observed Variation for the Inherent Tendency of *s*- and *p*-Block Elements to Lose Electrons and Subsequently Result Cations in the Periodic Table.

Explanation: Likewise, with the progressive increase of **EA-values** of the respective elements from **LHS to RHS** in a specific **Period**, one may critically notice the inherent tendency of the *elements* to accept the *electrons*; and hence, to produce the respective **Anions** that enhances perceptively.

Thus, the corresponding *decrease of electrons affinity (EA) values* right from **top to bottom** in a **Group and the** tendency gets reduced appreciably, – as depicted in Fig: 2.6.

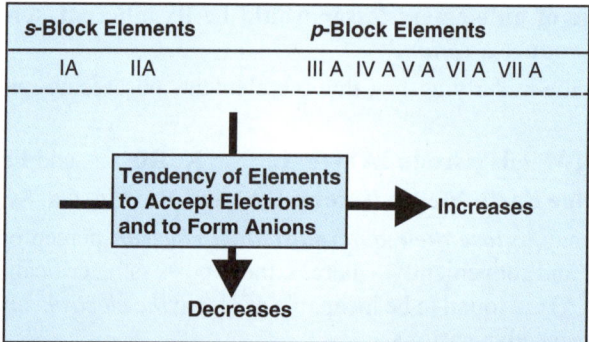

Fig.2.6: The Observed Variations in the Tendency of *s*- and *p*-Block Elements to Accept Electrons so as to Produce Anions in the Periodic Table.

3.4.2. *The Reducing and Oxidizing Power of Elements*

Obviously, a **reducing entity** (*viz.*, an **Atom** or **Ion** or **Molecule**) is the one that evnetually loses either *one* or *more* electrons; and thereby gets duly **oxidised** to a *higher* **valencey state**.

Example: Let us consider a particular instance when an *element*, duly present in its **gaseous state, M(g)**, happens to *lose one electron*; and thereby gets converted to **M$^+$(g) cation**, then the so-called element **M(g)** is regarded to be acting as a **Reducing Agent**.

We may express the above reactions as follows:

$$M(g) - e^- \longrightarrow M^+(g)$$

A Reducing Agent	An Oxidized State
[Having a Lower valency State = 0]	[Having a Higher valency state = +1]

Remarks: In may be observed that the **Reducing Power** of an *element*, present in its **gaseous state**, is invariably determined due to its inherent tendency to *lose the electrons*. Besides, one may take cognizance of the underlying fact that:

'**greater being its inherent tendency to lose the** *electrons,*–**greater would be its ultimate Reducing Power**'.

Besides, it has also been proven that '**lesser being the** *IP–value* **of an element, present duly in the gaseous state, greater**' would be its inherent tendency to lose the electrons perceptively.'

NOTE	In short, it may be added that the *elements* that are usually present in the *gaseous state* and also possessing significantly *lower IP-values* will be certainly much stronger Reducing Agents.

Furthermore, the observed **Ionization Potential (IP)** of the *elements* gets decreased progressively upon descending squarely a **Group**; and also increases accordingly as one proceeds from **LHS to RHS in a Period**. In other words, the ensuing, *reducing power of the elements* gets duly *enhanced* while proceeding down a **Group**; and hence, gets *reduced* as one moves from **LHS to RHS** in a specific **Period**,–as depicted in Fig: 2.7.

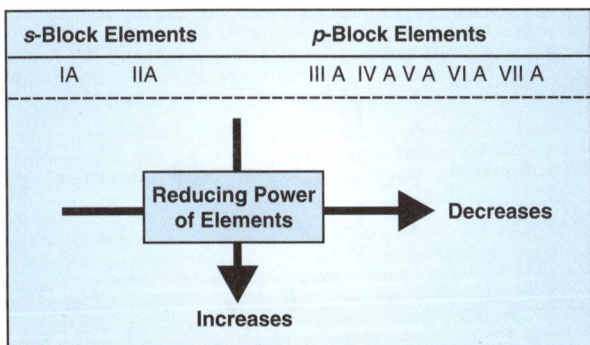

Fig.2.7: The Observed Variation in the Reducing Power of *s*-and *p*-Block Elements in the Periodic Table.

Remarks: In this way, the *elements* that are located strategically at the **extreme LHS of the Periodic Table** (*viz.*, **Alkali Metals**) that essentially possess perceptively **low IP-values** are found to be **stronger reducing agents**; whereas, the ones lying at the **extreme RHS of the Periodic Table** (*viz.*, **Halogens**)– predominantly possess **high IP-values** are known to be **weaker reducing agents** (see *Fig: 2.6*)

Explanations–It is, however, important to mention at this point in time that an **oxidizing substance** (*viz.*, **Atom or Ion or Molecule**) usually being designated as the one which:

"**critically gains either one or even more** *electrons*; **and thereby gets duly reduced to the** *corresponding lower valency state* **ultimately.**"

Example: Let us consider **M(g)**–an *element* present in the **gaseous form**, which eventually *gains one electron*; and thus, gets duly converted into the respective . **M⁻(g) anion**. Hence, the **element M(g)** is believed to be acting as an **oxidizing agent**, and may be expressed as under:

$$M(g) + e^- \longrightarrow M^-(g)$$

<table>
<tr><td>An Oxidising Agent
[Having a Higher Valence
State = 0]</td><td>A Reduced State
[Having a Lower Valence
State = –1]</td></tr>
</table>

Remarks: Based on the aforesaid statement of facts supported by logical explanations–it may be stated vehemently that:

'**the inherent oxidising power of an element its respective** *gaseous state* **could be determined precisely by its tendency to accept the electrons perceptively'.**

Besides, the **greater** being its inherent tendency to accept duly the number of electron, –the greater would be its ultimate **oxidising power**.

 NOTE   Importantly, the *elements* adequately available in the *gaseous state*–that possess critically *high-electron affinity (EA) values* do evolve as the so-called stronger Oxidizing Agents.

Fig: 2.8 illustrates explicitly the so-called observed variations with respect to the **Oxidising power of** the *s*-**and** *p*-**block elements** in a **Periodic Table** predominantly.

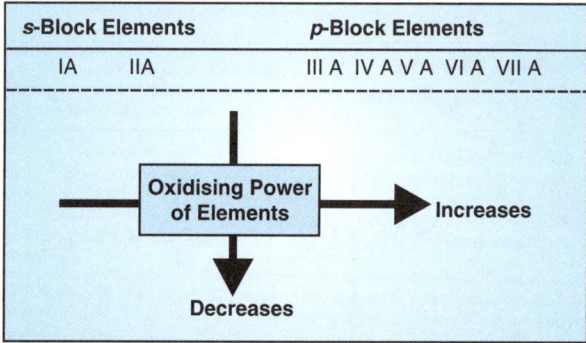

Fig.2.8: The Observed Variation of the Oxidising Power of *s*-and *p*-Block Elements in the Periodic Table.

Explanations: Because the **Electron Affinity (EA)** of the various elements invariably gets decreased as we critically descend a **Group**: and hence, increases progressively from **LHS to RHS** in a **specific Period**. Thus, the overall **oxidising power** of the elements gets decreased articulately upon descending a **Group**; and hence, gets **increased from LHS to RHS in a Period** (see Fig: 2.7).

Besides, the various **elements** located strategically at the **extreme LHS of the Periodic Table** (*viz.*, the **Alkali Metals**) – that essentially possess **lower Electron Affinity (EA) values** are observed to be relatively *poor (week) oxidising agents*; whereas, those positioned at the **extreme RHS of the Periodic Table** (*viz.*, **Halogens**)–that do have definitely **higher Electron Affinity (EA) values** *i.e.*, **Strong Oxidising Agents.**

❑ **The Inherent Oxidizing Properties of Halogens**—It has already been observed that the **Halogens** do possess the **optimized EA-values** in the **Periodic Table**; and, therefore, behave as the **most powerful oxidising agents** *i.e.*, they have the ability to **retain** (*accept*) the **maximum number of electrons perceptively**.

It is worthwhile to mention here that the so-called:

"**oxidising power of halogens gets enhanced progressively with the ensuing increment of their EA-values.**"

Based on such a valid factual observation one would be able to ascertain that the –'**oxidising power of Halogens**' should be in the **following order*:**

$$\boxed{\text{Cl} \, (-348.7) > \text{Br} \, (-324.5) > \text{F} (322.0) > \text{I} \, (-295.0)}$$

In the above expression, fluorine (F) has been observed to be the most effective as well as *powerful* oxidising agent; and hence, the **oxidising power of Halogens** gets reduced from **F** to **I** as stated under:

$$\boxed{\text{F} > \text{Cl} > \text{Br} > \text{I}}$$

Comments: In the above expression one may critically take cognizance of the fact that the **oxidising power of the Halogens** does follow an '*unprecedented order*' perceptively–that eventually is caused by virtue of the fact:

"**their ensuing Electron Affinity (EA) alone fails to control the oxidising power of elements**".

Nevertheless, the oxidising power of a given **Halogen** solely depends and being guided by an array of *divergent* energy terms, for instance:

- **Heat of fusion,**
- **Heat of evaporation,**
- **Heat of dissociation,** and
- **Heat of hydration.**

In short, their *aggregate sum* depends entirely upon the inherent *oxidising power* of a given **Halogen atom**.

Thus, one may infer that in a situation when most of these so-called '**energy terminologies**' pertaining to various **Halogen atoms** are summed up with the aid of **Born-Haber Cycle** (see *Section 3.1*), the resulting sum is found to be in the **following order** perceptively:

$$\boxed{\text{F} < \text{Cl} < \text{Br} < \text{I}}$$

NOTE	The above '*order*' indeed designates the actual order of the –'Oxidising Power of Halogens'.

Important Points: Following are the *two* **important points** with regard to the **oxidising characteristics of Halogens**, namely:

❑ **Fluoride (F⊖)**–is considered to be the **extremely powerful oxidising agent** amongst all the **Halogens** (*viz.*, **Cl, Br, I**), which being quite evident based upon the underlying fact that:

---

* Here, the **Electron Affinity (EA) values** in **kJ.mole^{-1}** are also provided in parenthesis.

"**Fluorine (F_2) has the potential and ability to *oxidise* or displace any of the other *Halide* Ions (*viz.*, Cl^-, Br^-, I^-) both in *solution* and *dry state*".**

Thus, we may have the following expression:

$$[X^\ominus = Cl^-, Br^-, I^-] \quad \underset{\text{Fluorine}}{F_2} + \underset{\text{Halide Ion}}{X^\ominus} \longrightarrow \underset{\text{Fluoride Ion}}{2F^\ominus} + \underset{\text{Halogen}}{X_2}$$

❏   Likewise, the **chlorine gas (Cl_2)** would be able to displace both **Bromide (Br^-) and Iodide (I)** **ions** from their respective solutions; and hence, **Bromine (Br_2)** shall be able to displace only the **Iodide (I) ions**,–as we may express below:

$$Cl_2 + 2X^- \longrightarrow 3Cl^- + X_2 \, [X^- = Br^-, I^-]$$

$$Br_2 + 2I^- \longrightarrow 2Br^- + I_2$$

NOTE   In a broader sense, any halogens of low atomic number will be able to *oxidise or displace halide ions* of the corresponding higher atomic number from their solutions.

3.4.3. *Metallic and Non-metallic Character of Elements*

In true sense, the **metallic and non-metallic character of elements** relates to the corresponding *electropositive or basic* and respective *electronegative or acidic* characteristic feature of elements.

Importantly, we may define the **metallic character of an element** as:

'**its inherent tendency to lose either one or more electrons to produce a cation**'.

Thus, we may have the following reaction:

$$\underset{\text{Element}}{M} - ne \longrightarrow \underset{\text{cations}}{M^{n+}}$$

Remarks: It has been established beyond any reasonable doubt that the greater being the *inherent tendency* of an element to *get rid of the electron*. To produce the *cation*, the greater would be the so-called metallic (or *electropositive character*) of the element.

Besides, it is also known that:

'**the lower being the *Ionization Potential (IP)–value* of a specific element, higher would be its** *inherent tendency* to lose the respective *electron* to produce the *cation*'.

It also endoreses the underlying fact that **smaller being IP-value of an element, greater would be its corresponding metallic character.**

Fig: 2.9 explicitly illustrates the observed variation of **metallic or electropositive character (or basic character)** of certain typical representative **elements** (*viz.*, *s*-and *p*-block Elements) present in the **Periodic Table**.

Explanation: As we may keenly observe that the **Ionization Potential (IP)** does decrease from *top to bottom* in a group, the respective **metallic character** (or the **basic character**) of the elements does increase in the **same direction** perceptively. Exactly, in the same fashion having the critical and specific increment in **IP from LHS to RHS** in a **Period**, we may obviously observe a decrease in the so-called **metallic character** of the ensuing elements in the *same direction*.

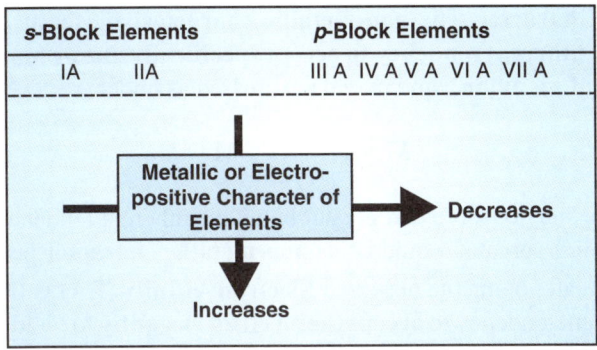

Fig. 2.9: The Illustration of the Observed Variation of Metallic (or Electro-positive Character) of Representative Elements (*viz.*, *s*–and *p*–Block Elements) of the Periodic Table.

Some Important Aspects– It is, however, pertinent to state here that we may critically observe the successive and progressive increment with regard to the **metallic character of the *elements* upon descending a *Group*** remains specifically apparent as well as important in the *three* **Groups,** namely: **IVA, VA, and VIA,** –that actually *begin with* the '*Non-Metals*' *viz.*, **C, N,** and **O**; and **subsequently,** end with '*Metals*' *viz.*, **Pb, Bi,** and **Po** respectively.

❑ **General Increase in the Metallic Characteristic Feature**–Thus, in **going down** (*descending*) a *Group* gets apparently confirmed by virtue of the underlying fact that:

"**the oxides of the respective elements of a *Group* turn out to be** *progressively basic in character*,–as we proceed from top to the bottom (descending) in a particular *Group*".

Examples: The **oxides of M_2O_3** kind of the *elements* of **Group VA** virtually become progressively **more basic in nature** (*i.e.*, become *less* **acidic**) as may be observed right from **Nitrogen trioxide (N_2O_3)** to **Bismuth trioxide (Bi_2O_3),** as given under:

$$\underbrace{N_2O_3, P_2O_3}_{\textbf{Acidic}} \quad \underbrace{As_2O_3, Sb_2O_3}_{\textbf{Amphoteric}} \quad \underbrace{Bi_2O_3}_{\textbf{Basic}}$$

$\xrightarrow{\hspace{3cm}}$
Basic Character Increases Progressively

❑ **General Decrease in the Metallic Characteristic Feature**–In this instance, we may take cognizance of the fact that the *elements* in going from **LHS** to **RHS** in a specific **Period** actually become:

"**progressively** *less basic* or successively *more acidic* as we proceed from *LHS* to *RHS*".

Example: The corresponding '**oxides**' of the *elements* belonging to the *third* **Period** invariably exhibit this kind of behavioural pattern,–as depicted under:

Na_2O Sodium Oxide [Strongly Basic]	MgO Magnesium Oxide [Basic]	Al_2O_3 Aluminium Trioxide [Amphoteric]	SiO_2 Silicon Dioxide [Feebly Acidic]	P_2O_5 Phosphorus Pantoxide [Acidic]	SO_3 Sulphur Trioxide [More Acidic]	Cl_2O_7 Chlorine Heptoxide [Most Acidic]

$\xrightarrow{\hspace{6cm}}$
Basic Character Decreasing Progressively

❑ **Non-metallic Character**–The **non-metallic characteristic feature** of an **element** may be defined –'**as its inherent tendency to accept specifically one or more electron to result into the formation of an anion**', and may be expressed as under:

$$M + ne^- \longrightarrow M^{n-}$$

Remarks: Amazingly, one may observe that greater being the inherent tendency of an element to accept an electron to yield an **Anion**, greater would be its **non-metallic character** perceptively.

In other words, the greater being the observed **Electron Affinity (EA) value** of a particular *element* greater would be its inherent tendency to **accept the electron elegantly** to yield the **Anion**. Alternatively greater being the **EA-value of an element, greater would be its non-metallic or electro negative character** predominantly.

Fig: 2.10 explicitly shows the observed variation of non-metallic (or *electronegative*) character of *s*–and *p*-Block elements in the **Periodic Table**.

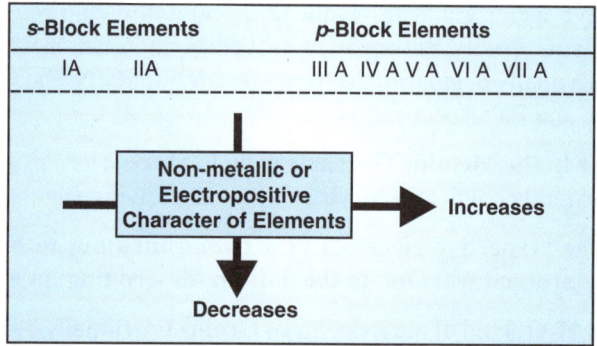

Fig. 2.10: The Observed Variation of Non-metallic (or Electronegative Character of *s*-and *p*-Block Elements in the Periodic Table).

Explanation–As the **Electron Affinity (EA) values** get decreased right from **top to bottom** (*vertically-down*) in a **Group**, the **non-metallic** characteristic feature of the *elements* also gets decreased in the same direction. Likewise, the respective increase in the **EA-values** from **LHS to RHS** in a **Period** renders the **non-metallic character** of the ensuing *element* to increase progressively.

The Metals–Non-metals and Metalloids– In true sense, the *observed trend of variation* in the **Metallic** and **Non-metallic** character of the *solvents* as stated earlier, has been fully utilized as the fundamentals so as to classify the so-called *s*– and *p*-**Block Elements** lying strategically at the **lower LHS** of the *Periodic Table*. The elements explicitly exhibit the **optimum electropositive character**; and, therefore, termed as the–'**Metals**'; whereas, the *elements* lying at the **upper RHS** of the **Periodic Table** do exhibit the **least electropositive character**; and hence, are called as the **Non-metals**.

Importantly, the *elements* that invariably located in between the Metals and Non-Metals* categorically exhibit the *dual characteristic features of Metals and Non-metals*; and hence, are known as the **Metalloids**.

Fig: 2.11 vividly shows the division of *s*-and *p*-**Block Elements** right into the **Metals–Non-metals–and Metalloid**.

---

* That is, the **elements** lying in a ***diagonal strip*** (see Fig: 2.9).

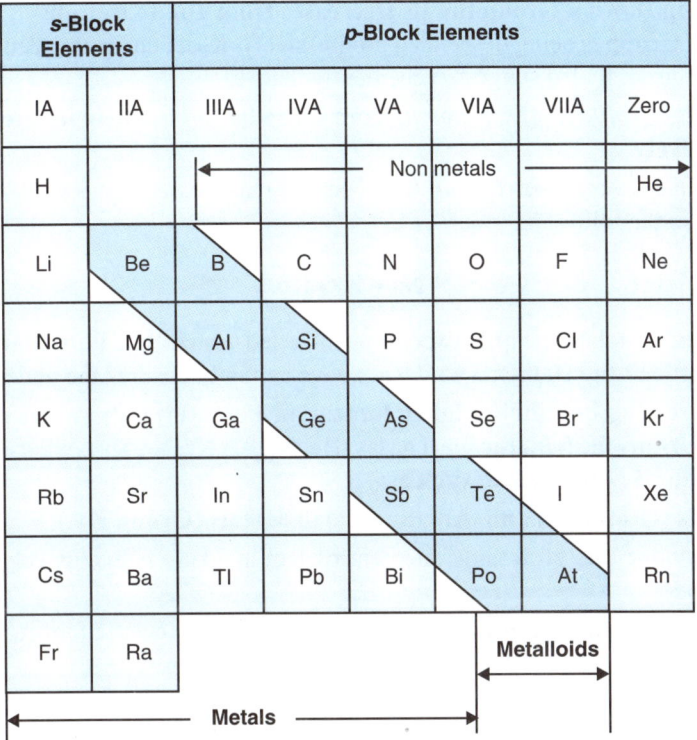

s-Block Elements		p-Block Elements					
IA	IIA	IIIA	IVA	VA	VIA	VIIA	Zero
H				Non metals			He
Li	Be	B	C	N	O	F	Ne
Na	Mg	Al	Si	P	S	Cl	Ar
K	Ca	Ga	Ge	As	Se	Br	Kr
Rb	Sr	In	Sn	Sb	Te	I	Xe
Cs	Ba	Tl	Pb	Bi	Po	At	Rn
Fr	Ra					Metalloids	

Metals

Fig. 2.11: The Explicit Division of *s*- and *p*- Block Elements into the Metallic–Non-metals –and Metalloids:

3.4.4. The Relative Stability and Activity of Elements

In a broader perspective, the underlying fundamental *concept* of **Ionization Potential (IP)** provides us with an *absolute quantitative measure of*:

- **Stability of different elements**; and
- **Activity profile of various elements**,

which ascertains the fact that such a *vivile concept* may be duly exploited so as to expatiate the so-called:

"**the relative stability and activity of various elements**".

Examples: Following are the *two* classical examples:

(a) **In a Period the IP Enhances from LHS to RHS**–It is already known that in a **Period** the observed **Ionization Potential (IP)** gets enhanced progressively from **LHS to RHS** *i.e.*, when we happen to **move from left to right in a Period**. Therefore, certainly more energy is needed to *knock out the electron*,–that eventually indicates that:

"**stability of the elements increases progressively or their inherent activity gets decreased accordingly**".

In this manner, the so-called **Third Period** ensures the observed relative activity of: **Na**, **Mg**, and **Al** to be prevail in the following order:

Na > Mg > Al

(b) In Proceeding Down a Group the IE Decreases from Top to Bottom– Thus, it ascertains the dictum that when the **Group** is being descended progressively–lesser energy is needed predominantly so as to get rid of an *electron*. It also suggests vehemently that:

"**the ensuing *Stability Profile* of the *elements* goes on reducing or their inherent activity goes on increasing progressively**".

In this fashion, in the very *First* **Group** the observed *relative activity profile* of: **Na, K,** and **Rb** is found to be existing in the following order:

$$\boxed{\textbf{Na} < \textbf{K} > \textbf{Rb}}$$

Remarks: The prevailing relationship between the so-called **Ionization Potential (IP)** as well as the ensuing **activity profile of the elements** as stated above certainly ensures the underlying fact that:

"**such elements possessing definitely higher Ionization Potential (IP) values are observed to be having lesser reactivity profile (*viz.*, the inert gases: He, Ne, Ar, Kr, Xe, Rn); whereas, those possessing the IP-values are indeed *very reactive*, such as:**"

- **Alkal Metals (Group IA) and Alkaline Earth Metals (Group IIA)**

An Important Exception: The aforesaid statement of fact and rule is exclusively applicable to such *elements* as stated under:

➢   **that essentially possess dominant electropositive characteristic feature *viz.*, Alkali Metals (Group IA, and Alkaline Earth Metals Group IIA), and**

➢   **certainly not to those which are indeed highly electronegative in nature *viz.*, Ionization Potential (IP) of Fluorine (F) is 1681.0 kJ. mole^{-1}; –that being extremely high; and hence, clearly exhibits that F must be *inert*, but F turns out to be *quite reactive* perceptively.**

4. THE ELECTRO NEGATIVITY [EN]

Electronegativity is defined as–

"**the inherent power of an atom in a molecule to attract electrons to itself**".

Alternatively, it may also be defined as per Pauling (1932) as:

"**the relative inherent tendency/ability/power of a bonded atom in a molecule to attract the shared-electron pair towards itself.**"

From the above *two* defines it is quite obvious that various investigators vehemently suggested and endorsed a specific method of his ownself in order to calculate precisely and accurately the so-called **electronegativities** of various elements according.

In general, a good number of *approaches* or *scales* have been duly suggested and put forward to measure accurately the **electronegativity** of the *atoms*. In reality, these proposed *approaches* (or *scales*) are indeed quite arbitrary; and hence, are solely based upon the **different types of experimental data** *viz.*,

- **Bond Energy** • **Dipole Moment** • **Ionization Potential (IP) and** • **Electron Affinity (EA).**

Following are the four widely accepted, recognized, and practised approaches of **electronegativity**, namely:

(a) **Pauling's Approach (1932),**

(b) **Mulliken's Approach (1934).**

(c) **Sanderson's Approach (1955)**, and

(d) **Alfred and Rochow's Approach** (1985).

which shall now be discussed briefly and individually in the sections that follows:

4.1. Pauling's Approach (1932)

It is also known as **Pauling's Scale (1932)**, which is based solely upon the so-called:

"**an emperical relation prevailing between the inherent energy of a bond (usually known as** *Bond Energy*) **as well as the electronegativities of the bonded atoms**".

Explanation: We may now consider a **A–B** bond exisiting between *two* **dissimilar atoms:** A and B belonging to a molecule AB. For this, let us assume that the ensuing *bond energies* of the following *three* **bonds** *viz.*, **A–A, B–B,** and **A–B** be designated respectively as:

$$E_{A-A}, E_{B-B}, \text{and } E_{A-B}$$

However, it has been observed critically that the inherent energy of **A–B** bond [E_{A-B}] is invariably **higher** *vis-a-vis* the so called **geometric mean** of the existing energies of **A–A** [E_{A-A}] and **B–B** [E_{B-B}] bonds.

Thus, we may have the following expression:

$$E_{A-B} > \sqrt{E_{A-A} \, ' E_{B-B}}$$

Ionic-Resonance Energy– The precise difference existing between E_{AB} and $\sqrt{E_{A-A} \times E_{B-B}}$ is usually termed as the **Ionic Resonance Energy** of the **A–B bond**; and hence, is represented by Δ_{AB}.

Thus, it is given by:

$$D_{AB} = E_{AB} - \sqrt{E_{A-A} \, ' E_{B-B}} \qquad \text{...(a)}$$

Importantly, the ensuing Δ_{A-B} **values** fail to possess any sort of an *additive* **characteristic feature,** which means that in case *three* **covalent bonds** *e.g.*, **A–B, B–C,** and **C–A,** –then the sum of Δ_{A-B} and Δ_{B-C} is never equivalent to: Δ_{C-A}. Thus, it obviously means that:

$$D_{A-B} + D_{B-C} \, ^1 D_{C-A} \qquad \text{...(b)}$$

Validity of Relation (a)– From the aforesaid expression of thoughts the **relation (a)** is *true* only in a situation when the prevailing **electronegativities** of: **A, B,** and **C** are found in the following order:

$$X_A > X_B > X_C$$

Therefore, the **validity of relation (a)** as stated above may be duly *verified* and *ascertained* by meticulously calculating the values of:

$$D_{FO}, D_{o-Si}, \text{and } \Delta_{F\,Si}$$

for the respective **F–O, O–Si,** and **F–Si** covalent bonds (*i.e.*, the ensuing **electronegativities** of **F, O, and Si** exist in the order $X_F > x_O > x_{Si}$)

Hence, by making use of the *given* **Bond Energy** values of: **F–O, O–Si, F–Si, F–F, O–O,** and **Si–Si bonds** as stated below:

$$E_{F-O} = 184.14 \text{ kJ. mole}^{-1} \qquad E_{O-Si} = 372.46 \text{ kJ.mole}^{-1}$$
$$E_{F-Si} = 535.58 \text{ kJ.mole}^{-1} \qquad E_{F-F} = 154.84 \text{ kJ.mole}^{-1}$$
$$E_{O-O} = 138.10 \text{ kJ.mole}^{-1} \qquad E_{Si-Si} = 179.95 \text{ kJ.mole}^{-1}$$

In short it may be added that by making use of **relation (a)**, one may easily calculate the values of:

$$\boxed{D_{F-O}, D_{O-Si}, \text{ and } D_{F-Si}}$$

Thus, we may have the bonds as stated under:

$$\Delta_{F-O} = E_{F-O} - \sqrt{E_{F-F} \times E_{O-O}}$$
$$= 184.14 - \sqrt{154.84 \times 138.10}$$
$$= 184.14 - 146.23$$
$$= \mathbf{37.91 \text{ kJ.mole}^{-1}}$$

$$\Delta_{O-Si} = E_{O-Si} - \sqrt{E_{O-O} \times E_{Si-Si}}$$
$$= 372.46 - \sqrt{138.10 \times 179.55}$$
$$= 372.46 - 157.64$$
$$= \mathbf{214.82 \text{ kJ. mole}^{-1}}$$

$$\Delta_{F-Si} = E_{F-Si} - \sqrt{E_{F-F} \times E_{Si-Si}}$$
$$= 535.58 - \sqrt{154.84 \times 179.95}$$
$$= 535.58 - 166.92$$
$$= \mathbf{368.66 \text{ kJ.mole}^{-1}}$$

Therefore, with the help of the values of Δ_{F-O}, Δ_{O-Si}, and Δ_{F-Si} – as calculated above, it may be depicted clearly that:

$$\Delta_{F-O} + \Delta_{O-Si} \neq \Delta_{F-Si}$$

or $\qquad\qquad 37.91 + 214.82 \neq 368.66$

or $\qquad\qquad\qquad \mathbf{252.73 \neq 368.66}$

Importantly, the observed values of **Ionic Resonance Energy (Δ_{A-B} value)** fail to exhibit the so-called 'additive characteristics' perceptively. Besides, it has been duly proven and demonstrated that:

"**the square roots ($\sqrt{\ }$) of such Δ_{A-B} values vehemently exhibit approximately the *additive characteristics*".

Alternatively, if one considers the **bonds** *e.g.*, **A–B, B–C,** and **C–A,** the resulting sum of:

$\boxed{\sqrt{\Delta_{A-B}} \text{ and } \sqrt{\Delta_{B-C}}}$ is almost equivalent to $\sqrt{\Delta_{C-A}}$, which may be duly expressed as under:

$$\boxed{\sqrt{D_{A-B}} + \sqrt{D_{B-C}} = \sqrt{D_{C-A}}} \qquad\qquad ...(c)$$

Validity of Relation (c)–Obviously, it may be duly verified by the underlying fact that:

> ➤ **for F–O, O–Si, and F–Si Bonds,** and

> ➤ **sum of** $\sqrt{\Delta_{F-O}}$ **and** $\sqrt{\Delta_{O-Si}}$ **nearly equals to** $\sqrt{\Delta_{F-Si}}$;

and hence, it may be expressed as under:

$$\sqrt{\Delta_{F-O}} + \sqrt{\Delta_{O-Si}} = \sqrt{\Delta_{F-Si}}$$

or

$$\sqrt{37.91} + \sqrt{214.82} = \sqrt{368.66}$$

or

$$6.15 + 14.65 = 19.20$$

or

$$20.80 = 19.20$$

Importance of Electronegativity (EN) Values: It is important to meation at this critical point that:

"the resulting *square root* of the **Ionic-Resonance Energy** $(\sqrt{\Delta_{A-B}}$ **values**) relates to a definite measure of the ensuing **Partial Ionic character of the so-called A–B** *covalent bond*".

Importnatly, with the progressive increment in the so-called *ionic* **characteristic features** of the *A–B covalent bond,*–the observed magnitude of $\sqrt{\Delta_{A-B}}$ get also duly increased perceptively. The actual difference in the *prevailing* **electronegativity** occuring between the ensuing:

'**bonded atoms (A and B)** [*i.e.,* $x_A - x_B$ ($x_A > x_B$)]',

also found to get enhanced with the increase in the **Partial** *Ionic Characteristics* of the **A–B bond.**

Thus, we may have the following expression:

$$\sqrt{\Delta_{A-B}} = \text{Precise Quantum of Ionic characteristics in A–B bond.}$$

and $\quad\quad x_A - x_B = $ Quantum of the Ionic characteristics in A–B bond

As a result, we may have the following *two* **expressions:**

$$\boxed{x_A - x_B = \sqrt{D_{A\text{-}B}}}$$

$$\boxed{x_A - x_B = K\sqrt{D_{A\text{-}B}}}$$

where, **K** is a constant whose value is **0.208**.

Substituting the value of Δ_{A-B} from **Eq. (a),** we have:

$$\boxed{x_A - x_B = K[E_{A-B} - \sqrt{E_{A-A} \cdot E_{B-B}}]^{1/2}}$$

In fact, the actual value of the constant, K, that derives from the conversion of the *experimental values of* E_{A-B} duly measured in **k cals . mole**$^{-1}$ into the respective **electron-volt (eV).**

Thus, we may eventually have the following expression.

$$\boxed{x_A - x_B = 0.208 \, [E_{A-B} - \sqrt{E_{A-A} \cdot E_{B-B}}]^{1/2}} \quad\quad\quad …(d)$$

Salient Feautres of Eq. (d)–These essentially include:

1. **Eq. (d)** provides exclusively the fifferences in the so-called **electronegativity (EN) values**.
2. **Eq. (d)** may also be employed intelligently in order to calculate precisely, **'x'-value of an atom** which is solely based on the fact that:

'the exact 'x'-*value* of the other *element* is known'

3. **Pauling** specifically assigned an **arbitrary value 4.0** for the **electronegativity (EN)** of *Fluorine (F)*; and hence, calculated such values pertaining to *other elements* by making use of this equation.
4. Interestingly, the particular calculation of the **EN-values, Pauling** vehemently expressed the inherent

Ionic Resonance Energies ($\sqrt{\Delta_{AB}}$ values) not in *k cal*, but instead in **electron volts (eV)***.

Table: 2.17 records the precise and accurate values of **electronegativities (ENs)** for the atoms of such '*elements*' that **pauling** has actually calculated. However, these **EN-values** are found to be **absolutely arbitrary** and exclusively relative in nature perceptively.

NOTE	The major demerit of the *Pauling's scale* being that the inherent *Bond Energies* (*viz.*, E_{A-A}, E_{B-B}) are not yet fully known for several solid elements.

Table: 2.17: The Electronegativity (EN) Values of Elements Being Estimated Precisely by Pauling's Approach [Scale used: F = 4.0]

Group→ Period	IA	IIA	IIIB	IVB	VB	VIB	VIIB	←VIII→			IB	IIB	IIIA	IVA	VA	VIA	VIIA	Zero
1	1 H 2.3																	2 He –
2	3 Li 1.0	4 Be 1.5											5 B 2.0	6 C 2.5	7 N 3.0	8 O 3.5	9 F 4.0	10 Ne –
3	11 Na 0.9	12 Mg 1.2											13 Al 1.5	14 Si 1.8	16 P 2.1	16 S 2.5	17 Cl 3.0	18 Ar –
4	19 K 0.8	20 Ca 1.0	21 Sc 1.3	22 Ti 1.5	23 V 1.6	24 Cr 1.6	25 Mn 1.5	26 Fe 1	27 Co 1.8	28 Ni 1.8	29 Cu 1.9	30 Zn 1.6	31 Ga 1.6	32 Ge 1.8	33 As 2.0	34 Se 2.4	35 Br 2.8	36 Kr –
5	37 Rb 0.8	38 Sr 1.0	39 Y 1.2	40 Zr 1.4	41 Nb 1.6	42 Mo 1.8	43 Tc 1.9	44 Ru 2.2	45 Rh 2.2	46 Pd 2.2	47 Ag 1.9	48 Cd 1.7	49 In 1.7	50 Sn 1.8	51 Sb 1.9	52 Te 2.1	53 I 2.5	54 Xe –
6	55 Ca 0.7	56 Ba 0.9	57 71 La–Lu 1.1 1.2	72 Hf 1.3	73 Ta 1.5	74 W 1.7	75 Re 1.6	76 Os 2.2	77 Ir 2.2	78 Pt 2.2	79 Au 2.4	80 Hg 1.9	81 Tl 1.8	82 Pb 1.8	83 Bi 1.9	84 Po 2.0	85 At 2.2	86 Rn –
7	87 Fr 0.7	88 Ra 0.9	89–95 Ac–Am 1.1 1.3															

Adapted From: Madan RD: **Modern Inorganic Chemistry**, 3rd ed., S Chand & Co. Pvt Ltd., New Delhi, 2011.

* That is, **1ev.bond^{-1} = 23 kcal.g^{-1} bond.**

4.2. Mulliken's Approach (1934)

Mulliken suggested vehemently that:

"**the observed calculated mean (***average***) of the Ionization Energy (IE) as well as the Electron Affinity (EA) of an atom must be an obvious and reliable measure of the *electronegativity (EN)* (also designated as '***X**m*') of the atom**".

Thus, we may have the following expression:

$$X_m \text{ (or EN)} = \frac{1}{2}(IE + EA)$$

It is, however, important to mention here that the so-called **Ionization Energy (IE)** and the **Electron Affinity (EA)** pertaining to an atom in the **Mulliken's definition** do not show any relevance to the so-called '*experimentally derived values*' for the isolated **gaseous atom** invariably present in the **ground state** (*i.e.*, the *low energy state*). In fact, they usually represent the calculated values for the atom in its respective **valence state**, which means that the *atom* does form an *integral part of a molecule.*

Example: The above episode may be further expatiated by considering the typical example of **Berylium difluoride (BeF$_2$)**, wherein the so-called **IE of Be atom** employed in the **Mulliken's approach** (or *derived formula*) which is not the one that is being observed experimentally for:

'**an isolated *Be atom* with its configuration as $1s^2\ 2s^2$; however, it designates the *hypothetical energy* essentially needed to knock out an electron right from an sp-*hybrid orbital* of the said Be atom perceptively**'.

Alternatively, it represents the *average of the* **Ionization Energies (IEs)** and **Ionization Potentials (IPs)** that are critically pressed into service to remove **2s and 2p electrons** right from the promoted '*valence state*' of the **Be atom** duly present in the **BeF$_2$ molecule itself.**

Based on the above logical explanations and scientific evidences, one may specifically glance at the following *articulated sequential steps:*

$$1s^2 2s^2 \longrightarrow 1s^2 2s^1 2p^1 \longrightarrow 1s^2 2s^0 2p^1$$

[Be in a Ground State]   [Promoted Valence State]

$$\overset{IE(p)}{\big|} \longrightarrow 1s^2 2s^1 2p^0$$

Remarks: The **Ionization Energy (IE)** and **Electron Affinity (EA)** being employed in the **Mulliken's approach** may be calculated by the help of a *complicated mathematical methodology*–that does not carry any concrete concern at this point in time.

Mulliken's Values (M) at Par with Pauling's Values (X)

Interestingly, a plethora of chemists put forward an array of *Empirical Equations* with an attempt to provide '**XM**' almost at par with the inherent values of '**X**' upon the so-called **Pauling's values**,–as given under:

$$X_M = 0.336\left[\frac{IE + EA}{2} - 0.615\right]$$ | **Both IE and EA are expressed in Electron Volts (eV)**

Merits of Mulliken's Approach: The most prevalent merit of the **Mulliken's approach** being related to the *electronegativities* of an *element*,–which is invariably present in altogether divergent **hybridization** or **oxidation** forms; and hence, may be calculated conveniently.

Example: The observed values of **XM of carbon (C)** in such *hybridized states* as: sp^3, sp^2, and sp, duly calculated by the so-called **Mulliken's approach** are found to be: **2.48, 2.66,** and **2.99** respectively; whereas, the corresponding **Pauling's Values of Carbon (C)** in all its states was found to be **2.55** precisely.

4.3. Sanderson's Approach (1955)

The **Sanderson's approach (1955)** refers to an **electronegativity scale** which is entirely based upon an altogether *new quantity*, usually termed as–**Stability Ratio (SB)** for an atom.

Thus, **Stability Ratio (SB)** may be defined as: **'the ratio of the average electron density (ED)'**.

Importantly, the **Electron Density (ED)** actually differs from point to point since the *electrons* are not observed to be distributed evenly very much around the nucleus itself. Therefore, its average **hypothetical electron density (ED_h)**,–that the '*atom*' would have if it were an *inert* gaseous atom (*viz.,* **He, Ne, Ar, Kr, Xe, Rn**)

Hence, we may have the following relationship:

$$\text{Stability Ratio (SR)} = \frac{\text{Electron Density (ED)}}{\text{Hypothetical Electron Density } (ED_h)}$$

Sanderson suggested that the **Stability Ratio (SR)** of an atom, A, explicitly refers to a measure of the **electronegativity (x_A);** and we may have the following expression:

$$(x_A)_{Sanderson} = SR = \frac{ED}{ED_h} \qquad \text{...(e)}$$

Hence, **Eq. (e)** is invariably known as the –'**Sanderson's Equation**'.

Remarks: The Sanderson **Electron Density (ED)** relates to a critical and specific measure of the so-called:

"comparative compactness of the atom",

and hence, is duly accomplished by *dividing the number of electrons* [*i.e.,* **the ensuing Nuclear Charge (Z)**] located strategically *round the nucleus* by an important factor $4/3\ \pi r^3$,–where:

r = Non-polar covalent radius of an '*atom*'.

Thus, we may arrive at the following expression:

$$ED = \frac{Z}{4/3\ \pi r^3} = \frac{3Z}{4pr^3} = \frac{Z}{4.19r^3} \qquad \text{...(f)}$$

Furthermore, the resultant value of **ED** pertaining to a *specific* **Atomic Number** is usually determined from the so-called:

'**interpolation obtained by plotting the ensuing Electron Densities (EDs) of the respective Inert Gas Atoms *Vs* the atomic numbers**'.

Inference: Based on the above **Eq. (f)**–one may easily lay hands on to the **Sanderson's Electronegativity** values and the same may be meticulously converted empirically to the Pauling's values by making use of the following expression:

$$X_{Pauling} = 0.21\, X_{Sanderson} + 0.77$$ ...(g)

4.4. Alfred and Rochow's Approach (1958)

Alfred and rochow (1958) intelligently proposed an altogether different approach of **electronegativity** profile–that is based upon:

"the crucial presence of the so-called 'Covalent Radii'."

Importantly, as per the above proposed approach we may critically take cognizance of the underlying fact that:

'the Electronegativity (EN) of an atom designates the inherent force of attraction prevailing between the nucleus of I *first atom* and the electron of an adjacent *second atom* being bonded to it recuredly; and hence, separated from the *nucleus* by the so-called covalent Radius (*r*)'.

Explanation: In this manner, if the **effective nuclear charge (Z_{eff})**, the respective **effective nuclear charge (Z^*)** duly felt by the *electron* under the influence of the nucelus, $Z_{eff}\, e$,.

Therefore, the *force of attraction* existing between the **Nucleus** and the **Electron** is found to be as stated under:

$$(Z_{eff}\ e)\ \acute{}\ 1/r^2$$ ...(h)

where, r = Distance (in cms) between Nucleus and Electron.

Hence, the resulting observed **force of attraction** very much equals to observed **electronegativity (x_A)** of an *atom*.

We may have the following expression:

$$x_A = \frac{(Z_{eff}\ e)}{r^2} = \frac{Z_{eff}\ e^2}{r^2}$$ ...(i)

Correlation of Alfred and Rochow's Electronegativity (x_A) Values Vs Pauling's Values

In order to establish a possible correlation of the **Alfred-Rochow's electronegativity (x_A) values**, (that are eventually found to be the absolute values), as obtained from **Eq. (i)** with the **Pauling's Values**, (which are observed to be **arbitrary** and **merely relative in nature**); and hence, the **Alfred and Rochow's Values** are plotted meticulously against the **Pauling's Values**,–so as to accomplish the **Best Plausible Straight Line** drawn through the various points carefully.

Now, right from the **crucial slope** and the **intercept of this line** the aforesaid *two* scientists meticulously derived the following relationship prevailing **Electronegativity Values** [*as denoted by* $(x_A)_{AR}$]; and hence, arrive at the following cardinal expression:

$$(x_A)_{AR} = 0.359\frac{Z_{eff}}{r^2} + 0.744$$ ...(j)

Merits of Alfred and Rochow's Approach–(1) As on date, the **Alfred and Rochow's approach** has elegantly mustered a tremendous acceptance and meritorious plus points that:

"the *Effective Nuclear Charge (Z$_{eff}$)* **may be determined by using the** *Slattor's Rules*; **whereas,** 'γ' *i.e., non-polar covalent radius of an atom,* **may be estimated experimentally for a majority of the** *elements* **in the Periodic Table.**"

(2) It shows absolutely little dependence upon the following *two* **critical factors**, such as:

- **electron affinities (EAs)**; and
- **bond-dissociation energies (D)**,

that are virtually known for only a small segment of *elements* in the **Periodic Table** perceptively.

Demerit of AR-Approach–The most serious **demerit** of this approach that the inherent observed values of the so-called *covalent radii (r)* are indeed not-so reliable at all.

Special Note: Drago (1960) after examining and evaluating critically the different suggested approaches pertaining to **Electronegativity (EN)** came to the ultimate conclusion and inference that:

'**The Pauling's Approach**'

having practically all its *limitations* is found to be the **Best Approach**; and hence, being practiced overwhelmingly in the domain of **Inorganic Chemistry**.

4.5. Various Factors Governing Electronegativity (EN)

There are several important factors that govern the **electronegativity (EN)** of an '*element*' in the **Periodic Table**, namely:

- **Size of the Atom,**
- **Charge on the Atom,**
- **Number of Inner Shells,**
- **Charge on the Ion,**
- **Nature and Number of Atoms wherein the Atom gets Bonded,**
- **Hybridization,** and
- **Effect of the Precise Nature of Constituents.**

4.5.1. *Size of the Atom*

It has been ascertained that the *smaller* **the size of the atom**, higher would be the inherent tendency of atom to get attracted duly towards the so-called **shared pair of electrons**.

Based on this concept and dictum one may infer that:

"the smaller atoms do essentially possess significantly higher electronegativity (EN) profile *vis-a-vis* the larger atoms perceptively".

Examples: Following are the *two* **classical examles**, namely:

(1) **The electronegativity (EN) values of Group IA elements (viz., H, Li, Na, K, Rb, Cs) duly get reduced from H(Z = 1) to Cs (Z = 55).**

As the **covalent radii** (or *atomic radii*) of the aforesaid elements invariably get duly enhanced in the same order.

(2) The electronegativity (EN) values of elements of *Second* Group enhances from Li (Z = 3) to F(Z = 9)

Since the **covalent radii** (or *atomic radii*) of these elements usually get reduced in the *same order,* – as given under:

Elements of Second Period	:	Li	Be	B	C	N	O	F
Covalent (Atomic) Radius (A)	:	1.23	0.90	0.82	0.77	0.75	0.73	0.72

——— DECREASING ———→

Electronegativity	:	1.0	1.5	2.0	2.5	3.0	3.5	4.0

——— INCREASING ———→

4.5.2. *Charge on the Atom*

Amazingly, whenever an **atom** acquires a **positive charge,**–which may be *integral* or *partial,*–it would exhibit a tendency to attract the *electrons* in a much stronger state *vis-a-vis* a *Neutral Atom* perceptively.

Thus, one may critically observe the following aspects elegantly:

- a **'cation'** shall indeed be more electronegative in comparison to the **parent atom,**–that in turn would be defintely more electronegative than its respective **'anion'**.

- *greater* the so-called inherent *positive*-**oxidation form** of an **atom** in a given species,–the higher would be its inherent **electron-attracting ability** (or profile); and, therefore, the greater would be its **electronegativity characteristics**.

Examples: The observed oxidation status of the *central chlorine atom* is found to be:

➢ **+1 in HClO, and**
➢ **+5 in HClO$_3$**

Hence, in the above instance the **chlorine atom** would certainly be **more electronegative in HClO$_3$ in comparison to HClO**. Furthermore, the observed release pattern of **hydrogen as H$^+$ in HClO$_3$** should be certainly much more easier than an identical *alteration in HClO.*

Inference: Thus, ultimately **HClO$_3$** explicitly behaves as a *much more stronger acid* in comparison to HClO.

4.5.3. *Number of Inner Shells*

It has been proven scientifically and ascertained duly that:

'**the atom having higher number of *Inner Shells** certainly possesses lesser value of electronegativity (EN) *vis-a-vis* the atom having smaller number of Inner Shells'**.

Example: This fact may be further expatiated be considering the **EN-values** of the *Halognes* that eventually decrease from **F(Z = 9)** to **At(Z = 85)**. Because the precise number of the **Inner Shells** gets increased progressively in the same order as vividly depicted in the following critical data:

---

* That is, the *shell* prevailing between the **nucleus** and the *outer-most shells.*

The Halogens [Elements of VIIA Group]	Number of Inner Shells [Complete Electronic Configuration Shown in Parenthesis]	Electronegativity (EN)
F (9)	1 (2, 7)	4.0
Cl (17)	2 (2, 8, 7)	3.0
Br (35)	3 (2, 8, 18, 7)	2.8
I (53)	4 (2, 8, 18, 18, 7)	2.5
At (18)	5 (2, 8, 18, 32, 18, 7)	2.2

4.5.4. *Charge on the Ion*

It also refers to the actual **type of ion**. Thus, a **cation** virtually attracts the ensuing *'electron pair'* more rapidly to itself *vis-a-vis* the atom from which it has been obtained ultimately. Perhaps it could be on account of the relatively **smaller size of the cation** in comparison to the so-called **parent atom** $[M^+ < M]$. In this manner, the **cation** $[M^+]$ possesses *higher values of electronegativity (EN)* vis-a-vis its parent atoms $M(M^+ > M)$, as could be observed critically in Table: 2.18*.

Table: 2.18: The Electronetativity (EN) Values of Certain Specific Cations vis-a-vis to their Respective Parent Atoms.

S.No.	Atom and Cation	Atomic (Covalent) Radius (Å)	Oxidation State of Atom and Cation	Electronegativity (EN)
1	Li	1.23	0	1.0
	Li^+	0.60	+1	2.5
2	Mo	1.30	0	1.8
	Mo^{4+}	0.68	+4	2.24
	Mo^{6+}	0.62	+6	2.35
3	Fe	1.17	0	1.80
	Fe^{2+}	0.76	+2	1.83
	Fe^{3+}	0.64	+3	1.96
4	Sn	1.41	0	1.80
	Sn^{2+}	1.12	+2	1.81
	Sn^{4+}	0.71	+4	1.96
5	Pb	1.47	0	1.80
	Pb^{2+}	1.20	+2	1.87
	Pb^{4+}	0.84	+4	2.33

Salient Features: Based on the EN-values of *some cations* when being compared to their respective *parent atoms*, one may vehemently observe the following **salient features**, namely:

1. The **electronegativity (EN) values** of: **M, M^{2+}, and M^{4+}** species are found to exist in the following order:

$$M^{2+} > M^+ > M$$

---

* That is, **atoms** from which the **cation** has bean derived also forms an integral part of the **Table**.

2. Whereas, the prevalent **size of these species** is observed to be in the **reverse order** perceptively, – as given under:

$$M^{2+} < M^+ < M$$

3. Interestingly, the **Anion** particular attracts the so-called '*Electron–Pair*' rather less promptly in comparison to its **Parent Atom**. Besides, it could also be by virtue of the underlying fact that:

'an *Anion* is definitely bigger in size *vis-a-vis* its *Parent Atom i.e.,* $X^- > X$ predominantly'.

4. It eventually leads to the glaring fact that:

'an *Anion* [X] certainly possesses lesser electronegativity (EN) in comparison to its *Parent Atom [X]*';

and hence, we may have the following expression:

- **Order of Electronegativity (EN): $X^- < X$; and**
- **Order of size of Atoms: $X^- > X$**

The aforesaid factual statement and findings further expatiates the analogy that the *Fluoride Ion (F^-)* possesses definitely *lesser electronegativity* (*EN*) value than the F atom itself [*i.e.,* $F^- = 0.8$, and $F = 4.0$].

4.5.5. *Nature and Number of Atoms wherein the Atom Gets Bonded*

It has already been established adequately that:

'since the electronegativity (EN) of an atom fails to be associated with the so-called characteristic feature of this *Atom* in its isolated form, it rests exclusively upon the nature and number of *Atoms* to which the some (*Atom*) gets duly bonded'.

Remakrs: Therefore, it could be the most plausible reason that the observed **EN-value of an Atom** is never found to be '*constant*'.

Example: Following is a classical example:

❑ The **electronegativity (EN)** value of the **Phosphorus (P) Atom** present in **phosphorus trichloride (PCl_3) molecule** is observed to be altogether divergent from that seen in **phosphorus pentafluride (PF_5) molecule**; and hence, the **nature and number of the Atoms** both to which the **phosphorus (P) Atom** is bonded gets altered predominantly.

4.5.6. *Hybridization*

What do we mean by 'Hybridization'?

In an attempt to explain and expatiate the so-called inherent **stabilities** (and **shpaes**) of the various existing '**Covalent Molecules**', the novel idea and concept of **hybridization** came into being.

Therefore, we may define **hybridization as–**

'the mathematical approach for the significant improvement of the *Wave Functions*' of the resulting combined atoms by means of an absolute *linear combination* of the authentic and pure *s, p,* and *d* orbitals thereby forming altogether newer orbitals',

–as given in the following [**Eq. (k)**]:

$$y_i = a_i \psi_s + b_i \psi_{px} + c_i \psi_{py} + d_i \psi_{pz}$$ ....(k)

and, $\int_0^\infty \psi_i \psi_i^* \, d_\tau = 1$ for the normalization of ψ_i

Nevertheless, the **newer orbitals** are indeed found to be very ***strongly directional;*** and, therefore, do exhibit distinctly higher **electron densities (EDs)** in specified regions perceptively. Thus, it would certainly give rise to a **greater overlap profile** that may ultimately lead to the generation of much **stronger bonds**.

Importantly, the so-called radial segment of the '***wave function***' of the **orbitals** that finally need to be hybridized specially must be *almost identical,*–otherwise the so-called **hybridized orbitals** will critically inherit such typical characteristics as:

- **possess only a low level of electron density (ED)**; and
- **certainly become quite unsuitable for the bond formation**.

Comments: In short, one may safely take cognizance of the fact that those orbitals which critically belong to either:

➢ **having the same principal quantum number**, or
➢ **occur perceptively in the adjacent energy level for the *d*-orbitals,**

virtually do combine to yield the desired **Hybrid Orbitals**.

The Hybridization Phenomenon: It is, however, pertinent to state here that the **hybridization phenomenon** may be usually regarded to involve the following ***three* cardinal sequential steps**:

(a) creation of the '**excited states**' that may crucially involve these ***two* processes**, such as:

- **unpairing of the electrons**; and
- **promotion of the electrons,**

to the corresponding next available orbital having higher inherent energy.

(b) The observed **hybridization phenomenon** of the *orbitals* in the typical *excited state (higher energy state)* [**see Eq. (k)**] providing the *hybrid orbitals* possessing **higher electron density** very much along the definite directions in the **spatial arrangement** (*i.e.*, also referred to as the –'**stereospecificity**') **of the hybridized orbitals**'.

(c) The observed *overlapping* of the **hybridized orbitals** having the appropriate orbitals of other atoms. Interestingly, the overlap is definitely found to be much greater in the particular instance of the '**hybrid orbitals**' *vis-a-vis* the so-called '**pure orbitals**'. Amazingly, such an event ultimately gives rise to the formation of:

'much stronger bonds having lower energy profile'.

Table 2.19 records the various forms of the commonly encountered **hybrid orbitals, together with** their *specific* **orientation profiles** in their **spatial arrangements**.

Fig: 2.12 illustrates diagrammatically the various recognized shapes of the **hybrid orbitals** that perceptively is an indicative of the so-called stereochemistry of the central atoms present critically in the **covalent compounds**.

Table: 2.19: the Stereochemistry of 'Hybridized Orbitals'*			
No. of Orbitals	Atomic Orbitals Employed	Orientation	Relative Bond Strength
1.	s	*Nondirectional*	1.000
	p	Mutually perpendicular three axes	1.732
2	$s\,p_z$	Linear	1.932
3	s, p_x, p_y	Equilateral triangle	1.991
	d_z, p_x, p_y		
4	s, p_x, p_y, p_z	Tetrahedral	2.000
	$s, d_{xy}, d_{yz}, d_{xx}$		
	$d_{x^2-y^2}, s, p_x, p_y$	Square	2.694
	$d_{x^2-y^2}, d_{z^2}, p_x, p_y$		–
5.	$s, p_x, p_y, p_z, d_{z^2}$	Trigonal bipyramid	–
	$s, p_x, p_y, d_{x^2-y^2}$	Square pyramid	–
6.	$s, p_x, p_y, p_z, d_{x^2-y^2}, d_{z^2}$	Octahedral	2.923
7.	sp^3d^3	Pentagonal bipyramid	–
8.	$d_{z^2}, d_{xy}, d_{yz}, d_{zx}, p_x, p_y, p_z, s$	Dodecahedron	–
	$d_{x^2-y^2}, d_{xy}, d_{xz}, d_{yz}, s, p_x, p_y, p_z$	Square antiprism	–
	$f_{xyz}, d_{xy}, d_{yz}, d_{zx}, s, p_x, p_y, p_z$	Cubic	–

* Adapted from: Manku GS: **Theoretical Principles of Inorganic Chemistry**, Tata Mc Graw Hill Publishing Co., Ltd., 1985.

NOTE   For the *s* and *p* hybrid orbitals– the *relative* bond energies **are those given by** Pauling.

General Remarks– These essentially comprise:

1. Obviously, the so-called '**hybrid orbitals**' are found to be:
   - **neither *s*-orbitals;**
   - **nor *p*-orbitals,**

however, they do vividly exhibit characteristic features *altogether divergent* from either aforesaid types of orbitals,–as illustrated in Fig: 2.13, wherein the observed **Electron Density (ED)** contours for:

   (a) **shows an *sp*-hybrid orbital explicitly**; and
   (b) **depicts an sp^3-hybrid orbital**.

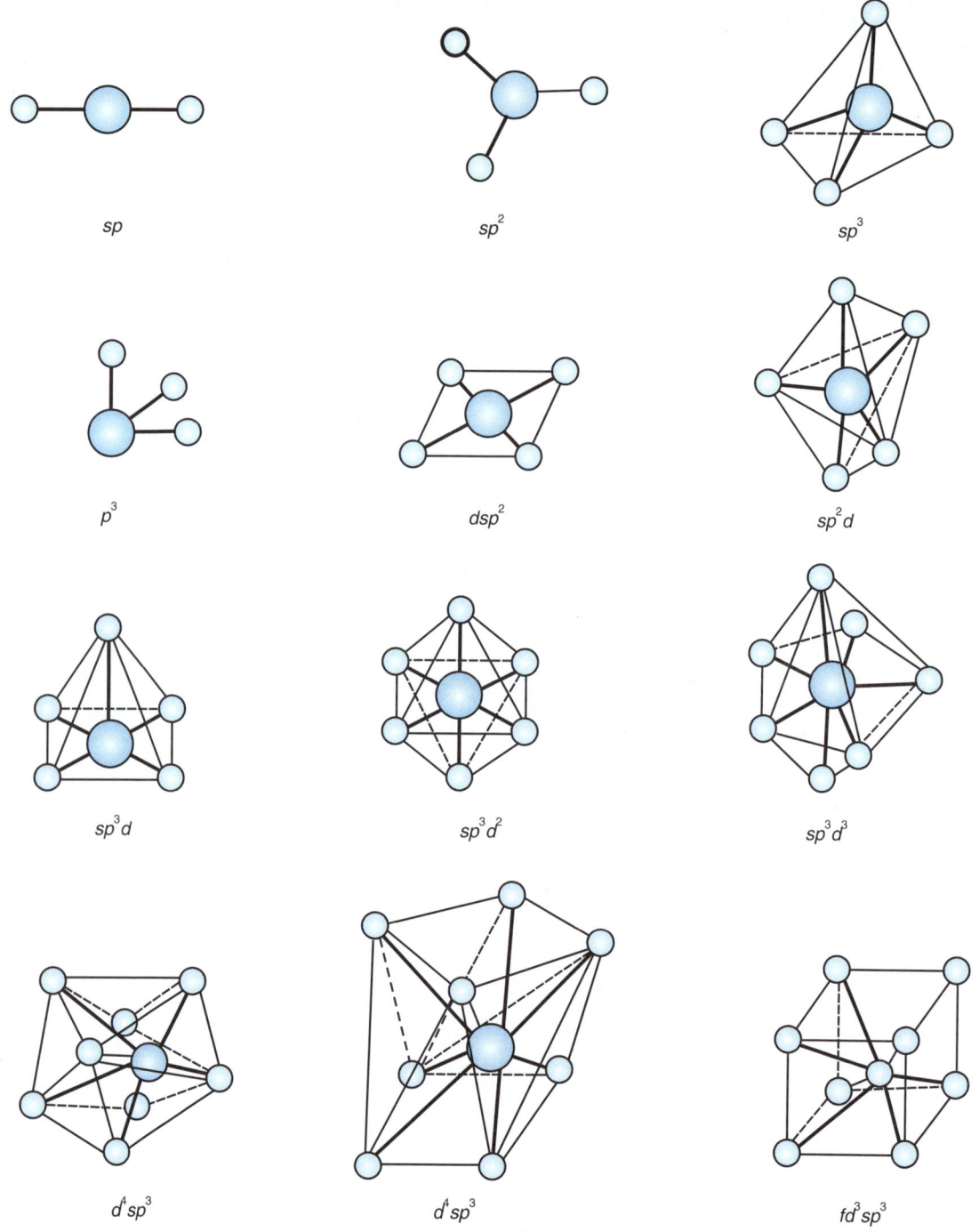

Fig. 2.12: The Various Shapes of Hybrid Orbitals that Serve as the Indicative of the Stereochemistry of the Central Atoms in Covalent Compounds Explicitly.

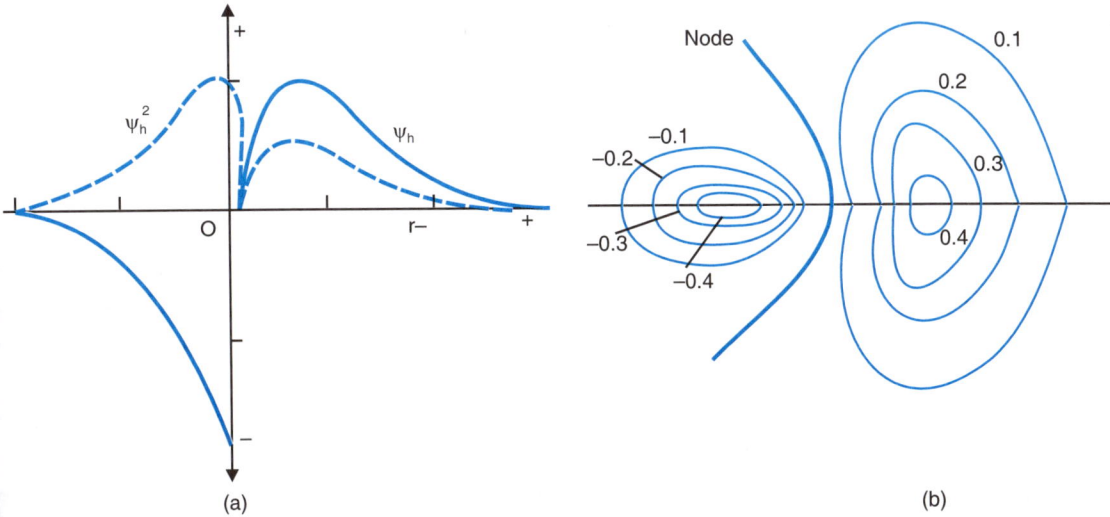

Fig. 2.13: Diagrammatic Illustration of the Explicit Electron Density (ED) contours: (a) an *sp*-Hybrid Orbital and (b) an *sp³* Hybrid Orbital.

Explanations– The various critical aspects involved in the above Fig: 2.14, may be duly explained as under:

1. Importantly, the **newly generated orbitals** do possess both specific and characteristic directional orientations; and hence, are found to be definitely much more appropriate for the so-called –**Covalent Bond Formation** perhaps on account of **better and superb overlap**.*

2. Since the particular–'**overlap integral**' gets duly enhanced in the following order:

$$p << sp^3 < sp^2 < sp$$

the resulting *two* **critial aspects**, namely:

- **Bond strengths**, and
- **Bond energies**,

$$p << sp^3 < sp^2 < sp$$

and invariably seen in the following order:

$$sp > sp^2 > sp^3 > p$$

Besides, the above observed criteria may also be depicted by:

- **Bond Energies** and
- **Bond Lengths**,

---

* That is, the *separate orbitals* is certainly *not symmetric* (in space).

which respect to the ensuing **C–H bonds** as could be seen in the following **spectrum of hydrocarbons:**

Radical	Ethyne $HC \equiv CH$		Ethene $H_2C = CH_2$		Methane CH_4		Hydrocarbon CH
k.J.mole^{-1}	198	>	444	>	410	>	336
pm	106	<	107	<	109	<	112

4.5.7. *Effect of the Precise Nature of Constituents*

It has been established beyond any reasonable doubt that the ensuing **electronegativity (EN)** of an atom rests exclusively upon the inherent and precise **nature of the substituents** duly attached to that **atom**.

Example: The **C-atom** present in **trifluoroiodo methane [CF₃I]** gives rise to the formation of an entity that essentially needs a **greater positive charge** *vis-a-vis* the **iodomethane (CH₃I)**.

Comments: Therefore, the **C-atom** duly present in **CF₃I** proves to be certainly *much more* **electronegative** (which means *more and better electron-pulling prevails*) in comparison to CH₃I. Besides, the observed difference in the so-called **electronegativities (ENs)** of an atom produced by the **substituents critically produces divergent chemical behavioural pattern of that atom perceptively**.

Example: **CF₃I** and **CH₃I** actually gives rise to the production of altogether *divergent chemical entities (compounds) being subjected to hydrolysis*, as shown below:

  (i) $CF_3I + OH^- \rightarrow CF_3H + IO^-$
  (ii) $CH_3I + OH^- \rightarrow CH_3OH + I^-$

4.6. Role of Ionization Energies (IEs) and Electron Affinity (EAs)

Mulliken proclaimed that **electronegativity (EN)** is intimately related to:

"**the average of the Ionization Energy (IE) and Electron Affinity (IA) of the atom.**"*

Since it is quite evident that the *higher* is the **Ionization Energy (IE)** as well as the Electron Affinity (EA),–the greater should be the observed **Electronegativity (EN) of the atom**.

Effective Nuclear charge ($Z_{effective}$) *vis-a-vis* **variation of Electronegativity (EN) in the Periodic Table–**

Alfred and Rochow (1958) argued vehemently that:

'**the inherent Electron Attracting Power of an** *Atom* **present in a** *molecule* **is directly proportional to the so-called** *Effective Nuclear charge* **[$Z_{effective}$]**'.

Therefore, in a broader perspective, any such specific factor also critically enhances the *electronegativity (EN)* of the **atom** perceptively.

Example: The actual value of $Z_{effective}$ gets reduced eventually as one moves down a particular *Group* in the **Periodic Table**. Perhaps it could be made effective based on the underlying fact that:

"**with the enhancement of the** *Atomic Number* **present very much within a** *Group*–**is responsible for the progressive increment in the respective** *Atomic Sizes* **perceptively**".

'**the Electronegativity (EN) profile decrease accordingly as one moves down a Group gradually**'.

---

* That is, as per the **Mulliken's Approach:** $X_m = 1/2 \,(IE + EA)$ [see section: 4.2]

Remarks: It may be inferred that the ensuing **electronegativity (EN)** of the so-called '*Halogens*' (**Group VII A**) gets decreased progressively in the following order:

$$F(4.0) > Cl\,(3.1) > Br\,(2.9) > I\,(2.6)$$

Cardinal Observations: Follow are the *two* cardinal observations, namely:

- ❑ **Increase in $Z_{effective}$** –with the critically observed *decrease* in the *size of the Atom*; whereas the corresponding *size of the Atom* gets decreased as one proceeds along a **Period from LHS to RHS** in the **Periodic Table**.

Comments: Thus, it may be concluded that:

'**electronegativity (EN) increases progressively with the respective increase in the *Atomic Number*,**'– as one proceeds along a *Period*.

Obviously, it leads to the ultimate dictum that the so-called **electronegativities (ENs)** of the various elements belonging to the *Second Period* (*viz.*, **Li, Be, B, C, N, O**, and **F**), do increase in the following order perceptively:

$$Li\,(1.0) < Be\,(1.5) < B\,(2.0) < C\,(2.5) < N\,(3.0) < O\,(3.5) < F\,(40)$$

- ❑ **Dependence of $Z_{effective}$ upon Number of Inner Electrons**–usually comes into being that critically *screen the last electron*. From this, one may infer that:

   '**bigger the number of *Inner Electrons*, the *higher* would be the screening profile, but *lower* would be the Effective Nuclear charge ($Z_{effective}$).**'

Comments: Based on the aforesaid dictum-the observed **electronegativity (EN)** would certainly get decreased with the respective increase in the precise and exact number of the ensuing number of **inner electrons** duly present in the *Atoms* of the elements belonging to the **same Group**.

Perhaps, it could serve as an '**additional valid and good reason**' why the electronegativity (EN) of the *Alkali Metals viz.*, **Li, Na, K, Rb**, and **Cs** (**Group IA**), –that gets decreased progressively in the order:

$$Li\,(1.0) > Na\,(0.9) > K\,(0.8) > Rb\,(0.8) > Cs\,(0.7)$$

4.7. Application of Electronegativities (ENs)

Based on the *scientific evidences*, *logical explanations*, and *broad-based utilities*– the applications of **Electronegativities (ENs)** are indeed of extremely *vital*, *gainful*, and *important* in the domain of **Inorganic Chemistry** in terms of knowledge, acumen, and research activities.

Following are the various perspective cardinal aspects that are intimately related to the–**applications of electronegativities (ENs)**–as stipulated under:

- (i) **Precise Calculations Leading to the Revelation of Partial Ionic ($Z_\pm$) Characteristic Feature of Covalent Bond,**
- (ii) **Enthalpies: Calculation for Generation of Chemical Entities (compounds),**
- (iii) **Calculation of Bond Length,**

(iv) **Bond Angles: Their Logical Explanations**,

(v) **Rationalization of Reaction Mechanisms**,

(vi) **Diagonal Relationship**, and

(vii) **Types of Bonds**,

which shall now be discussed individually in the sections that follows:

4.7.1. Precise Calculations Leading to the Revelation of Partial Ionic ($Z_{\pm}$) Characteristic Feature of Covalent Bond

Importantly, the useful development of the **ionic character (Z)** (or the *ionic charge*) **located strategically in a** *Covalent Bond* existing between *two* **distinct atoms** *viz.*, **A** and **B**, is solely due to the observed differences in the ensuing **electronegativities (ENs)** of both **A** and **B** perceptively. Furthermore, one may critically derive the following expression of interest that:

'**the higher being the prevailing difference in the ENs-values**,–the higher would be the crucial subsequent development of the Ionic [Z_{+}] character; and hence, the greater, should be the overall entire stability of the resulting *covalent Bond*.'

Pauling suggested strongly the undermentioned correlation existing prevalently between the **electronegativity (EN) difference**, X_A–X_B; and hence, the *observed* **per cent Ionic characteristic** feature of the **Covalent Bond** may be expressed as under:

$$\text{Per cent Ionic [Z] Character} = [1 - e^{-0.25} (X_A - X_B)] \times 100 \qquad ...(1)$$

Pauling proposed an alternative procedure that refers to an '**Emperical Equation**' that may be used for the calculation of the intended **per cent Ionic characteristic feautre**, that may be expressed as under:

$$\text{Per cent Ionic Characteristic Feature} = 18 (X_A - X_B)^{14} \qquad ...(m)$$

Remarks: As per **Eq. (m)**, the so-called observed **per cent Ionic characteristic feature of hydrofluoric acid (HF)** stands at **44.12**. Amazingly, this value appears to be extremely **closer to the one (*e.g.*, 44.8)** duly arrived at from the *Dipole Moment of hydro fluoric acid (HF)*.

4.7.2. Enthalpies: Calculation for Generation of Chemical Entities (Compounds)

It is now quite possible and feasible to precisely **calculate** the **enthalpies*** related to the formation of **chemical entities** (*compounds*) right from the **ENs** thereby using the following **Emperical Equation** elegantly proposed by **Pauling**:

$$\Delta H_f = 23\varepsilon(X_A - X_B)^2 - 55.4n_N - 26.0\, n_0 \qquad ...(n)$$

where,

ΔH_f = Enthalpy of formation (*i.e.*,) Enthalpy of formation (*i.e.*, *heat of formation*),

n_N = Number of nitrogen atoms in the molecule,

o_O = Number of oxygen atoms in the molecule.

ε = Values depending on the number of covalent bonds in the molecule.

---

* **Enthalpy:** Change in heat.

Example: The ε = 1 for hydrochloride acid (HCl); and ε = 2 for **beryllium dichloride (BeCl$_2$)**, and so on so forth.

Therefore, the respective values of **ΔH$_f$** duly accomplished as shown above do designate only the so-called–'**approximate values.**'

Examples: In order to calculate precisely the '**enthalpy for the formation of BeCl$_2$**', we may have to account for the *two* **Be–Cl X$_{cl}$ and X$_{Be}$** are found to be 3.1 and 1.5 respectively.

Thus, from **Eq. (*n*)** we have:

$$\Delta H_f = 23 \times 2\ [(3.1 - 1.5)^2] - 0 - 0 = 117.76\ \text{k.cal.mole}^{-1}$$

NOTE

(1) The values of n_s and n_o are '*zero*' since these atoms are completely devoid of in the BeCl$_2$ molecule itself.

(2) The experimental value for **ΔH$_f$** is 122 k.cal.mole^{-1} which is indeed very much close to the above calculated values.

4.7.3. *Calculation of Bond Length*

In a typical instance, when the *two* atoms **A and B being bonded together** *via* a **Covalent Bond** invariably differ in their ENs–values perceptively; and hence, it implements that the covalent Bond would certainly acquire certain critical **Ionic Characteristic Features,**–as depicted above.

Alternatiely, the **Covalent Bond** so formed does acquire an ***absolute polarity status***. Hence, it may be inferred that:

"**the higher being the polarity status, the lower would be the length of the *covalent bond* so formed between *A* and *B*.**"

Shoemaker and Stevenson proposed intelligently the following ***Emperical Equation*** for carrying out the calculation of the '**Bond length**' in all such typical instances accurately and precisely:

$$d_{AB} = \gamma_A + \gamma_B - 0.09\ (X_A - X_B) \qquad\qquad ...(o)$$

where,

d_{A-B} = Actual bond distance between **A** and B;
γ_A = Covalent radius of **A**,
γ_B = Covalent radius of B,

$X_A - X_B$ = Difference in Electronegativity (EN) value.

Thus, the *usual* and *normal* length of the **Covalent Bond** existing between A and B should have been actually equivalent to $\gamma_A + \gamma_B$. Therefore, one would visualize the actual and proven ***shortening of the bond existing between A and B*** on account of the ensuing **difference in the electronegativity (EN),**– that is equivalent to 0.09 $(X_A - X_B)$. Besides, by virtue of the ***reduction in length***, the accomplished **covalent Bond** will be certainly far **more stable**.

4.7.4. *Bond Angles: Their Logical Explanations*

It has been proven and ascertained that:

'**the fewer (lesser) being the inherent electronegativity (EN) of the so-called central atom present duly in a *Polyatomic Molecule*, the four would be the observed magnitude of the *Bond Angle*'.**

Interestingly, the above episode may be explained by the underlying fact, that in a situation when the **EN-of the central atom is observed to be less**, it will not be in a position to hold on the ensuing **bonding electron pairs** towards itself perceptively. Consequently, the resulting **bonding-electron pairs** would show a specific tendency to shift more towards the rest of the atoms which are duly attached to the **central atom**. Thus, it would ultimately give rise to an apparent decrease in the prevailing: *bond pair– bond pair* repulsions; and, therefore, a possible and feasible observed *decrease in the Bond Angle* is registered.

Remarks: The above analogy and dictum may be further expatiated by taking into consideration the observed **Bond Angles** and **Electronegativities (ENs)** of the so-called *trifluorides* of the elements belonging to Group 15 [*viz.*, **N, P, As**, and **Sb**],–as stated under:

S.No.	Trifluorides of Elements of Group 15	Electronegativity of Central Atom	Bond Angle
1	NF_3	3.0	102°
2	PF_3	2.2	97°
3	AsF_3	2.2	96°
4	SbF_3	2.0	88°

NOTE   As pre-empted, the observed lowering in the *Electronegativity (EN)* profile of the *central atom* gives rise to a lowering of the *Bond Angle* as well.

4.7.5. *Rationalization of Reaction Mechanism*

Importantly, when **methyl ioidide (CH$_3$I)** interacts with the **nucleophile [M$_n$(CO)$_5$]'**,–the following reaction occurs:

$$CH_3I + Na^{\oplus}[Mn\,(CO)_5]^{\ominus} \longrightarrow NaI + CH_3Mn\,(CO)_5$$

Methyl iodide   Nucleophile   Sodium iodide

Remarks: The critical formation of the aforesaid *reaction products* may be adequately explained as given under:

❑ **Iodine** present in **methyl iodide (CH$_3$I)** is certainly **more electronegative in nature** *vis-a-vis* the **C-atom**.

❑ The above phenomenon ensures a *partial* positive charge upon the prevailing **C-atom**.

❑ The *negatively* charged nucleophile, as shown above, attacks predominantly the *positive site* preferentially,–thereby giving rise to the formation of **CH$_3$Mn(CO)$_5$**

Interaction with trifluoroiodo methane [CF$_3$I]–In this specific case, under identical parameters, we may obtain absolutely **different reaction product**, –as expressed under:

$$2CF_3I + Na^{\oplus}[Mn\,(CO)_5^{\ominus}] \longrightarrow NaI + Mn\,(CO)_5I + C_2F_6$$

Trifluoroiodo   Nucleophile   Sodium iodide   Hexafluoro
methane   ethane

Explanation: These essentially comprise:

1. Since the crucial presence of the *highly* electronegative F atoms, the ensuing *electron charge cloud* present duly on the **C-atom** is virtually *attracted* very much towards the former.

2. Thus, it ensures a definite viable **electron charge deficiency on the C-atom**; and hence, is adequately balanced by attracting the *electron charge cloud present on the iodine atom* very much towards the **C-atom***.

3. In this manner, it specifically generates a definite *partial positive charge* upon the ensuing **iodine atom**; and hence, the **nucleophile [Mn(CO)$_5$]$^\ominus$**–now prominently attacks the **iodine site** thereby ultimately leading to the generation of **Mn(CO)$_5$I**.

4.7.6. Diagonal Relationship

It has been proven and established that the **EN-values** goes on increasing as one proceeds from: **Li (1.0) to Be (1.5)** *i.e.*, **the observed EN-values in a specfic Period**, nevertheless, it gets subsequently decreased as one moves from **Be (1.5) to Mg (1.3)** *i.e.*, the **recorded EN-value in a particular Group**.

Comments: Consequently, bearing in mind the aforesaid *two* **exactly opposite changes,** namely:

- **one along the Period**; and
- **second along the Group**.

Thus, as one gets along *diagonally* across these *two* above sited effects partially cancelling each other; and, therefore, there exists little marked alteration in the overall **EN-values** perceptively.

Perhaps it could be the *solid good reason* why **Li (1.0) and Mg (1.3)** do critically possess *extremely closer EN-values predominantly*.

4.7.7. *Type of Bonds*

Interestingly, the precise and exact **type of bonds** being duly formed between any *two* **distinct atoms** would obviously depend upon the prevailing difference(s) in their corresponding **electronegativities (ENs)**.

Nevertheless, in a situation as and when the difference in the **EN-values** becomes *less than 2.5*, but otherwise seems to be *quite significant*, the ultimate bond formed would turn out to be: '**Polar Covalent**'.

SUGGESTED READING REFERENCES	
Atkins PW	: **Physical Chemistry**, Oxford University Press, Oxford (UK), 1978.
Chanda M	: **Atomic Structure and Chemical Bond**, 2nd ed., Tata McGraw Hill, New Delhi, 1979.
Donohue J	: **The Structure of the Elements**, Wiley, New York, 1974.
Emsley J	: **The Elements**, 2nd ed., Clarendon Press, Oxford (UK), 1989.
Fergusson JE	: **Stereochemistry and Bonding in Inorganic Chemistry**, Prentice-Hall, Enlewood, Cliffs, NJ (USA), 1974.
Galasso PS	: **Structure and Properties of Inorganic Compounds**,
Karplus M and Porter RN	: **Atoms and Molecules**, Benjamin, New York, 1971.
Krebs H	: **Fundamentals of Inorganic Chemistry**, McGraw Hill, New York, 1968.

---

* That is, the **electron charge cloud of iodine (I$_2$)** is polarisable quite easily.

Parish RV   : **The Metallic Elements**, Longman, New York, 1977.

Schuster PG *et. al.*   : **The Hydrogen Bond, Vols. I to III**, North Holland, Amsterdam, 1976.

Smalley RE   : **In: Atomic and Molecular Chemistry**, ER Bermetein (Ed.), Elsevier, Amesterdam, 1990.

Troyer R   : **The Third Form in Carbon: A New Era in Chemistry**, *Interdiscip Sci Rev.*, 17: 161–170, 1992

Wells AF   : **Structural Inorganic Chemistry**, 4th ed., Clarendon Press, Oxford, (UK), 1975.

PROBLEMS WITH SOLUTIONS

Q.1. How would you calculate the Enthalpy of Formation of Ionic Compounds ΔH_f very much within a few per cent using the Born-Lande Equation and the Born-Haber Cycle. Given that the internuclear distance in NaCl crystal being 281.4 pm, Maoelung's constant for NaCl lattice being 1.74756, and the Born exponent, taking the average of the values for $Na^+(7)$ and $Cl^{-1}(9)$ is 8.

Solution: The Born-Lande equation provides the lattice energy U_C for the NaCl crystals as:

$$U_C = -\frac{1^2 (1.6\times10^{-19}C)^2\,(1.7456)\,(6.023\times10^{23})}{4(3.14)\,(8.85\times10^{-12}\,Fm^{-1})\,(281.4\times10^{-12}\,m)}\left(1-\frac{1}{8}\right)$$

$$= -7.552 \times 10^5 \text{ J.mole}^{-1}$$

$$= -755.2 \text{ kJ.mole}^{-1}$$

The actual heat capacity correction equals to 2.1 kJ.mole^{-1}

$\therefore \qquad\qquad U_C = -755.2 - 2.1 = -757.3 \text{ kJ.mole}^{-1}$

Hence, the **Born-Haber equation** gives:

$$\Delta H_f = S + I + 1/2D - X + U_C$$

$$= 108.4 + 495.4 + 120.9 - 348.5 - 757.3$$

$$= \mathbf{-381.1 \text{ kJ.mole}^{-1}}$$

NOTE The calculated value of -381.1 kJ.mole^{-1} (or -91.1 k cal.mole^{-1}) very much compares with the *observed ethalpy of formation of NaCl,*—that stands at -410.9 kJ.mole^{-1} (or -98 k cal mole^{-1}).

Q.2. When a mole of crystalline NaCl is prepared from 1 gramme atom of Na and 0.5 mole of Chlorine gas, 410 kJ of heat is produced. The heat of sublimation of Na-metal 108.8 kJ. The heat of dissociation of chlorine gas into atoms is 242.7 kJ, the ionization energy of Na is 493.7 kJ and the electron affinity of Cl is 368.2 kJ. Calculate the lattice energy of NaCl.

[Delhi, B.Sc., (Hons), 1972]

Solution: Thus, we have Q = 410 kJ. S = 108.8 kJ, I = 493.7 kJ, D = 242.7 kJ, E = 368.2 kJ, and U_o is to be calculated.

Substituting the above values with their appropriate signs in the following equation designating the heat of formation of NaCl(s) 'Q'–by **Hess's Law**–that being equal to the sum of all other energy terms; and hence, we have:

$$Q = +S + I + \frac{1}{2}D - E - U_o$$

∴ $\qquad -410 = 108.4 + 495.4 + 1/2 \times (241.8) - 348.5 + U_o$

or $\qquad -410 = 376.2 + U_o$

or $\qquad U_o = 410 + 376.2 = \textbf{786.2 kJ. mole}^{-1}$

(B) Problems Based on Pauling's Approach

Q.3. How would you calculate the Electronegativity (EN) of Fluorine (F) from the given data as stated under:

$E_{H-H} = 104.2$ k.cal.mole^{-1}; $E_{F-F} = 36.6$ k.cal.mole^{-1}; and $E_{H-F} = 134.6$ k.cal.mole^{-1}

Solution: According to the following emperical correlation related to **Pauling's Approach:**

$$E_{A-B} = (E_{A-A} \times E_{B-B})^{1/2} = \Delta^1$$

we have the following expression

$$\Delta = E_{H-F} - (E_{H-H} \times E_{F-F})^{1/2}$$
$$= 134.6 - (104.2 \times 36.6)^{1/2} = 72.85 \text{ k.cal.mole}^{-1}$$

As per the following relationship

$$0.182 \sqrt{\Delta^1} = X_A - X_B$$

we may have:

$$X_F - X_H = (0.182)(72.85)^{1/2} = 1.55$$
$$X_F = 1.55 + X_H = 1.55 + 2.1 = \textbf{3.65}$$

Q.4. How would you calculate the electronegativity (EN) of carbon based on the following given data: $E_{H-H} = 104.2$ k.cal.mole^{-1}; $E_{C-C} = 83.1$ k.cal.mole^{-1}, and $E_{C-H} = 98.8$ k.cal.mole^{-1}.

Solution: According to the following *empirical correlation:*

$$E_{A-B} - (E_{A-A} \times E_{B-B})^{1/2} = \Delta^1$$

∴ $\qquad \Delta^1 = E_{C-H} - (E_{H-H} - E_{C-C})^{1/2}$

or $\qquad \Delta^1 = 98.8 - (104.2 \times 83.1)^{1/2}$

or $\qquad \Delta^1 = \textbf{5.75 k.cal}$

As per the relationship given below:

$$0.182 \sqrt{\Delta^1} = X_A - X_B$$

we may have:

$$X_C - X_H = 0.182 \sqrt{5.75} = 0.436 \simeq \textbf{0.44}$$

∴ $$X_C = 0.44 + X_H$$
$$= 0.44 + 2.1$$
$$= 2.54 \text{ [or} \sim 2.55]$$

NOTE The electronegativities of $H(X_H)$ and that of $C(X_C)$ as given in Table 2.13 is given as: 2.1 and 2.5 respectively.

(C) Problem Based on Mulliken's Approach

Q.5. Calculate the electronegativity values of Fluorine (F) and Chlorine (Cl) on Mulliken's approach (scale) based upon following given data: $(IP)_F = 17.4$ eV. $atom^{-1}$; $(EA)_F = 3.62$ eV.$atom^{-1}$; $(IP)_{cl} = 13.0$ e.V. $atom^{-1}$; and $(EA)_{cl} = 4.0$ eV.$atom^{-1}$.

[*IP = Ionization Potential; and EA = Electron Affinity*]

Solution: Let us note the electronegativities of F and Cl by X_F and X_{cl}; and hence, they may be given as under:

$$X_F = \frac{(IP)_F + (EA)_F}{2 \times 2.8} = \frac{17.4 + 3.62}{5.6} = 3.75$$

and also,

$$X_{cl} = \frac{(IP)_{cl} + (EA)_{cl}}{2 \times 2.8} = \frac{13.0 + 4.0}{5.6} = \frac{17}{5.6} = 3.03$$

NOTE The *Mulliken's values* of electronegativity (EA) are found to be almost 2.8 times as large as the *Pauling's values*. Therefore, to render the former nearly equivalent to the latter, –the RHS of the following equation is divided by 2.8:

$$X_A = \frac{(IP)_A + (EA)_A}{2} \text{ i.e., } = \frac{(IP)_A + (EA)_A}{2 \times 2.8}$$

(D) Problems Based on Alfred and Rochow's Approach

Q.6. How would you calculate the electronegativity of Silicon (Si) by following Alfred-Rochow's approach? Given that covalent radius of Si = 1.175Å.

Solution: The electronic configuration of Si is: $1s^2\ 2s^2p^63x^2p^2$ According to Slater's Rules (*Chapter -1*), we have the following expression:

$$Z_{effective} = Z_{actual} - S$$

or
$$= 14 - (0.35 \times 4 + 0.85 \times 8 + 1.02 \times 2)$$
$$= 3.80$$

According to the following equation based on AR-approach:

$$X_{si} = 0.359 \times 3.80/(1.175)^2 + 0.744$$
$$= 2.50$$

Q.7. Calculate the electronegativity of Silicon (Si) using the Alfred-Rochow's Approach. Given that: Covalent radius of the Carbon (C) atom is 0.77Å.

Solution: Based upon the **Slater's Rules (Chapter 1)**–we have the expression:

$Z_{effective}$ at the periphary of Carbon (C) atom $= Z_{actual} - S$

$$= 6 - (0.35 \times 4 + 0.85 \times 2)$$
$$= \mathbf{2.90}$$

As per the following empirical relation, as proposed by Alfred-Rochow, the calculation for the electronegativity is given by the following expression:

$$X = 0.359 \times Z_{effective}/r^2 + 0.744$$

we have

$$X_c = 0.359 \times 2.90 \, (0.77)^2 + 0.744$$

or

$$X_C = \mathbf{2.50}$$

Q.8. How would you determine the electronegativity of Load (Pb) with the help of the given values: the screening constant (σ) of Pb = 76.70, atomic number of Pb(Z) = 82, and the covalent radius of Pb(γ) = 5.3Å

Solution: Substituting the values of σ, Z, and γ in the following equation:

$$(X_A)_{AR} = 0.359 \frac{Z_{eff}}{\gamma^2} + 0.744$$

we have:

$$(X_{Pb})_{AR} = 0.359 \frac{Z - \sigma}{r^2} + 0.744$$

we have:

$$(X_{Pb})_{AR} = 0.359 \frac{Z - \sigma}{r^2} + 0.744$$

or

$$(X_{Pb})_{AR} = 0.359 \frac{82 - 76.70}{(5.3)^2} + 0.744$$

or

$$(X_{Pb})_{AR} = \mathbf{1.55}$$

(E) Problem Based on Isoelectronic, Ionization Energies, Electron Affinity Values

Q.9. Arrange the following *Atoms* or *Ions* in the increasing order of their inherent individual size:
(a) F⁻, Al, Na⁺, Mg²⁺ and (b) Ca²⁺, Ar, K⁺, Cl⁻, S²⁻

Solution: (a) Since the ions: **F⁻, Na⁺, and Mg²⁺**–are *isoelectronic in nature*
[F⁻ = 9 + 1 = 10; Na⁺ = 11 − 1 = 10; Mg²⁺ = 12 − 2 = 10],–their respective **size** gets duly increased with the decrease of their **atomic number (Z)**. Thus, we may have the following decreasing order of preference:

$$Mg^{2+}(Z = 12) < Na^+ \, (Z = 11) < F(Z = 9)$$

Therefore, as F⁻ < Al, the *overall* increasing order becomes:

$$Mg^{2+} < Na^+ < F^- < Al$$

(b) In this particular instance,–as practically all the given species happen to be isoelectronic in character [$Ca^{2+} = 20 - 2 = 18$; $Ar = 18$; $K^+ = 19 - 1 = 18$; $Cl^- = 17 + 1 = 18$; $S^{2-} = 16 + 2 = 18$], –their corresponding **'size'** gets increases prgressively with the so-called respective **decrease** in their inherent **atomic number (Z)**.

Thus, we may have the following decreasing order:

$$Ca^{2+} < K^+ < Ar < Cl^- < S^{-2}$$

$$Z = 20 \quad 19 \quad 18 \quad 17 \quad 16$$

Q.10. How would you classify the under-mentioned species into *two* separate groups of species that must be isoelectronic in nature:

$$K^+, No^+, Ca^{2+}, C_2^{-2}, Sc^{3+}, CN^-, Cl^-, N_2$$

Solution: First and foremost let us determine the *number of electrons* actually present in the aforesaid (given) species:

$K^+ = 19 - 1 = 18$; $No^+ = 7 + 8 - 1 = 14$; $Ca^{2+} = 20 - 2 = 18$; $C_2^{2-} = 6 \times 2 + 2 = 14$; $S_c^{3+} = 21 - 3 = 18$; $CN^- = 6 + 7 + 1 = 14$; $Cl^- = 17 + 1 = 18$; $N_2 = 2 \times 7 = 14$;

Hence, the given species may be duly classified into the following *two* **distinct categories**:

(a) K^+, Ca^{2+}, Sc^{3+}, Cl^- : All possess **Eighteen electrons**; and

(b) No^+, C_2^{2-}, CN^-, N_2: All possess **Fourteen electrons**.

Q.11. How do we find out the exact number of the *Valence Electron* present in an *Atom* whose successive Ionization Energies (IEs) values (expressed in $kJ.mole^{-1}$) are as stated below:

$$IE_1 = 410, \ IE_2 = 820, \ IE_3 = 1100, \ IE_4 = 1500, \text{ and } IE_5 = 3200$$

Solution: The **Ionization Energy (IE) values** explicitly show that because the actual magnitude of energy required critically for knocking out the *5th electron* from the **atom** is indeed found to be more than **twice the quantum** of energy required for removing the *4th electron*; and, thus, the **Number of Valence Electrons in the atom is 4**.

Q.12. Arrange properly the given elements with their electronic configurations (ECs):

$$[He] \ 2s^1, \ [Ne] \ 3s^1, \text{ and } [Ar] \ 4s^1$$

in the decreasing order of their Ionization Energy [IE] values.

Solution: The **EC-values** show elegantly that the given elements critically belong to **Group 1B**. As we may observe that the **IE1 values** usually get decreased on *moving down the* **Group 1B**; and hence, the values of given elements decrease as shown under:

$$[He] \ 2s^1 > [Ne] \ 3s^1 > [Ar] \ 4s^1$$

Q.13. The observed Electronic Configuration (ECs) of four distinct elements *viz.*, A, B, C, and D are as given under:

$A = 1s^2, 2s^2, 2p^6, 3s^1$; $B = 1s^2, 2s^2, 2p^6, 3s^2$;

$C = 1s^2, 2s^2 \ 2p^1$; and $D = 1s^2, 2s^2 \ 2p^6, 3s^2 \ 3p^1$.

How would you predict precisely which of these elements does possess the so-called: Maximum Value of $(IE_2 - IE_1)$.

Solution: The respective **Ionization Energies IE$_1$ and IE$_2$ values** for the aforesaid *four* given elements do **correspond processes:**

A: $1s^2, 2s^2 2p^6, 3s^1 \xrightarrow[\text{IE}_1]{-3s^1(n=3)} 1s^2, 2s^2 2p^6 \xrightarrow[\text{IE}_2]{-2p^1(n=2)} 1s^2, 2s^2 2p^5$

B: $1s^2, 2s^2 2p^6, 3s^2 \xrightarrow[\text{IE}_1]{-3s^1(n=3)} 1s^2, 2s^2 2p^6, 3s^1 \xrightarrow[\text{IE}_2]{-2p^1(n=2)} 1s^2, 2s^2 2p^5$

C: $1s^2, 2s^2 2p^1 \xrightarrow[\text{IE}_1]{-2p^1(n=2)} 1s^2, 2s^2 \xrightarrow[]{-2s^1(n=2)} 1s^2, 2s^1$

D: $1s^2, 2s^2 2p^6, 3s^2 3p^1 \xrightarrow[\text{IE}_1]{-3p^1(n=3)} 1s^2, 2s^2, 2p^6, 3s^2 \xrightarrow[\text{IE}_2]{-3s^1(n=3)} 1s^2, 2s^2 2p^6, 3s^1$

For Element A – The very *first* electron is removed from the **3s-orbital** ($n = 3$) and the *second* electron is being removed from the **2p-orbital** ($n = 2$). Because the so-called 3s-orbital belongs to the *outer orbit* ($n = 3$) and the 2p-orbital critically belongs to the inner orbit ($n = 2$); and, therefore, IE$_2$ – IE$_1$, for the element 'A' is found to be the *maximum*.

NOTE	Amazingly, in the particular instance of other elements (*viz.*, B, C, and D),–the ensuing orbitals from which *electron(s)* being removed predominantly belong to the *same shell*.

Q.14. Arrange properly the elements: C, N, O, and F in the decreasing order of their IE$_2$ values.

Solution: following would be the **decreasing order of IE$_2$** values of the given elements:

$$O > F > N > C$$

Explanation: The observed **IE$_2$-value** of an *element 'A'* refers to the actual energy required to get rid of an **electron from A$^+$ cation** perceptively.

The various **electronic configurations (ECs) of C$^+$,N$^+$, O$^+$ and F$^+$ cations** are as given under:

$$C^+ = 2s^2 2p^1; \; N^+ = 2s^2 2p^2; \; O^+ = 2s^2 2p^3; \; \text{and} \; F^+ = 2s^2 2p^4.$$

Besides, since **O$^+$ cation** critically possesses the so-called **half-filled 2p-orbitals**, that eventually renders the $2s^2 2p^3$–electronic configuration reasonably stable.

Therefore, **IE$_2$** of the **O-atom** would be definitely higher *vis-a-vis* that of **F (O > F)**. Furthermore, since the **C-atom** happens to be *bigger in size* in comparison to the **N-atom**; hence, **IE$_2$ of C** would be certainly lower *vis-a-vis* that of **N (N > Cl)**.

NOTE	Therefore, the overall *decreasing order* the aforesaid elements would be:

$$O > F > N > C$$

Q.15. How would you arrange F, Cl, O, and S in the increasing order of their Electron Affinity (EA) values.

Solution: The **Electron Affinity (EA) values** of the above *four* elements, namely: F, Cl, O, and S–are found to be in the following *increasing order:*

$$O < S < F < Cl$$

Explanation: Based on the strategically located position of these four elements present in the so-called **Long-form of the Periodic Table** suggests that the observed variation in their respective **Electron Affinity (EA) values** may be depicted vividly as shown under:

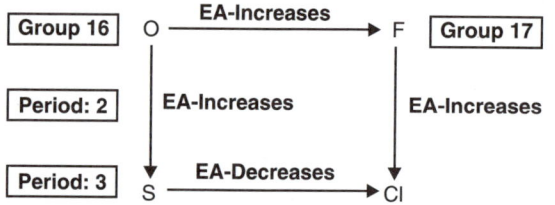

<table>
<tr><td>Group 16</td><td>O</td><td colspan="2">EA-Increases →</td><td>F</td><td>Group 17</td></tr>
<tr><td>Period: 2</td><td></td><td>EA-Increases</td><td></td><td>EA-Increases</td><td></td></tr>
<tr><td>Period: 3</td><td>S</td><td>EA-Decreases →</td><td></td><td>Cl</td><td></td></tr>
</table>

> **NOTE** Therefore, the net overall increasing order of the *Electron Affinity (EA) value* is found to be:

$$O < S < F < Cl$$

Q.16. **How would you justify which of the following species possesses the smallest value of Ionization Energy (IE):**

$$O, O^-, O^{2-}, \text{ and } O^+$$

Solution: It may be observed explicitly that the size of the given species increases progressively in the order:

$$O^+ < O < O^- < O^{2-}$$

Because, O^{2-} species being the largest in size in the present lot, it will be indeed the easiest way to get rid of *an electron right from this ion* perceptively.

> **NOTE** Therefore, the *IE-value* of the ion O^{2-} is found to be the small one.

Q.17. **Amongst the following four different processes–which one specifically absorbs the Ionization Energy (IE).**

(a) $Cl(g) + e^- \rightarrow Cl^-(g)$

(b) $O^-(g) + e^- \rightarrow O^{2-}(g)$

(c) $O(g) + e^- \rightarrow O^-(g)$

(d) $S(g) + e^- \rightarrow S^-(g)$

Solution: In a specific instance, when an *electron* is being added on to a **neutral gaseous atom (A)**– *energy gets released perceptively*. In this manner, the processes: **(a), (c), and (d)**– the **energy (IE)** gets released predominantly. However, when an *electron* is being added on to a given A⁻ion, **the energy gets absorbed vehemently**. Besides, in the process (b) the energy gets duly absorbed to overcome the so-called *repulsions* occurring between the **incoming electron** and the **negative charge** positioned strategically upon the **O⁻(g) ion**.

1. Explain why the metallic character increases as we move down the group.

    [Himachal Pradesh, 2000]

2. Write notes on: (i) **Ionization Potential**, (ii) **Electron Affinity**, and (iii) **Electronegativity**.

3. (a) Define **Ionization Potential** of an atom. Which factors influence the **Ionization Potential**? What is **Effective nuclear charge?** (b) IP of H is 13.6 eV. Calculate the energy required to produce H 0.5g of hydrogen.   [Kanpur, 2000]

4. (a) Given **Slater's Rules** for calculating **Shielding constant** or $Z_{eff.}$

    (b) Give **Pauling's Method** to determine the **electronegativity** of an atom.

    (c) On the basis of Slater's Rule, calculate the value of Z_{eff} for;

    (i) a **4s-electron**; and (ii) a **3d-electron** in **chromium**.

    [Himachal Pradesh, 2000]

5. What do you understand by the term–'Periodic Properties'? Distinguish between **Electron Affinity** and **Electronegativity**. **Fluorine** is more electronegative than **chlorine**, although its **Electron Affinity** is less than that of **Chlorine**. Explain.   [Lucknow, 2000]

6. Explain the following:

    (i) Electron Affinities of Be, Mg, and **Noble Gases** are 'zero', while those of **Nitrogen** (= 0.20 eV) and **Phosphorus** (= 0.80 eV) are very low.

    (ii) The formation of $F^-(g)$ **ion** from **F(g) atom** is exothermic; whereas; that of $O^2(g)$ ion from **O(g)** is *endothermic*.

    (iii) In a period **Ionization Energy** increases from **left to right**.

    (iv) In a group electronegativity decreases from **top to bottom**.

    (v) Group IA alkali metals from ionic compounds with **group VIIA Halogens**.

7. Explain **Ionization Potential** and **Electronegativity**.   [Lucknow, 2001]

8. (i) Arrange the following in the order of their increasing rod i.e., **F, Ne**.

    (ii) **First Ionisation Potential** of Nitrogen is more than that of oxygen but **second Ionization Potential** of Oxygen is more than that of Nitrogen. Explain.

    (iii) Which of the following has the *highest* **Electron Affinity?** F, Cl, Br.

    (iv) The **Electronic Configuration** of some elements are given as:
    (a) $1s^2, 2s^2 2p^6, 3s^2$ (b) $1s^2, 2s^2, 2p^6$ (c) $1s^2, 2s^2, 2p^4$ (d) $1s^2, 2p^6, 3s^2, 3p^6$. Which of these elements has the *lowest* **Ionization Potential**?

    (v) Arrange the **Bond Angle** in the decreasing order of the following **Hydrides**: NH_3, AsH_3, and PH_3.

    (vi) What is meant by Electronegativity? How does it depend upon **IP** and **EA** ?

    (vii) Arrange the following in the order of **decreasing covalent character**: C–Cl, B–Cl, Be–F, Cl–Cl.

    (viii) How is **Electronegativity** related with the **Polarity** of a Covalent Bond?

9   (a) What is Ionization Potential? Discuss the factors an which Ionization Potential depends. Arrange the following according to **increasing Ionization Potential:** Mg, Ra, Ba, Ca, Be.

(b) **Atomic radii** of the element in a series in the **Periodic Table** decrease with **increasing atomic number** upto VII Group, but **increase in inert gases**. Explain giving reasons.[Avadh, 2000]

10. What is Ionization Potential of an element? Explain why **IE of N** is higher than that of **O**? [Nagpur, 2002]

11. Write a note on **Electron Affinity**   [Lucknow, 2002]

12. Explain why **alkali metal cations** are generally smaller in size than their corresponding atoms? [Delhi, 2003]

13. (a) Define Electronegativity

    (b) Give reasons for the following

        (i) The **Second Ionization Energy** of sodium is very high as compared to its **First Ionization Energy**.

        (ii) **First Ionization Energy of Nitrogen** is higher than that of **Oxygen**.

    (c) Define the following and explain their trends across the **Period** and down the **Group** in **Periodic Table**:

        (i) **Atomic Radius**, and

        (ii) **Electron Affinity**   [Nagpur, 2003]

14. (a) Differentiate between **Covalent Reactions** and **Van derWaal's Radius**

    (b) What are **iso-electronic Ions**? Arrange the given **ions** in order of their decreasing size: F^-, O^{2-}, Mg^{2+} and Na^+.

    (c) What is **Screening Effect**? How does it govern the Ionization Energy of an atom?
                                                        [GNDU, Amritsar, 2004]

15. Explain why **EA-values** of N and P atoms are low.   [Meerut, 2005, 2008]

16. (a) Which of the following Ions has the smaller size? O^{2-}, F^-, Na^+ and Mg^{2+}.

    (b) Define EA and state how it changes in a period.   [HN Bahuguna, 2005]

17. Explain the following:

    (i) $(IP)_3$ of Mg is very high.

    (ii) **EA** values of N and P atoms are low.   [Meerut, 2006]

18. (a) Arrange N, O, Cl, and F in the increasing order of their **BA** values.

    (b) In a **Covalent Molecule AB**, **EN-values** of A and B are 2.1 and 3.5 respectively. Calculate v/o Ionic character in A–B bond.   [HN Bahuguna, 2006]

19. Explain why **Cl-atom** is smaller in size than **Cl⁻ ion**.   [Meerut, 2007]

20. (a) **First IP** of N is higher than that of O. Explain.

    (b) What is the trend in size of N^{3-}, O^{2-}, and F^-?

    (c) EA of Cl is higher than that of F. Explain.   [Purvanchal, 2007]

21. Define and discuss **Electronegativity** in detail along with the factors affecting it. [Meerut, 2008]

22. (a) Which of the following elements has the highest IP-value? Li, Na, K, and Rb.

    (b) Which of the following elements is the most electronegative? F, Cl, Br, and I. [Agra, 2008]

RADIOACTIVITY, ISOTOPES, ISOBARS AND ISOTONES

(A) RADIOACTIVITY

3.1. INTRODUCTION

Radioactivity refers to the property of *metastable atoms* that spontaneously emit *particles* and/or *photons* in order to assume a more *stable state*.

Examples: Following are *two* **typical examples:**

- ❑ **X-Rays** that are usually emitted as a result of changes in the inherent changes in the energy levels of the so-called **orbital electrons;** and
- ❑ **γ-Rays** which are mostly emitted as a result of changes occurring at the energy levels of the nucleolus.

Henri Baequerel (1895), observed critically soonafter the epoch making *discovery of the* **X-Rays** in an attempt to ensure logically a probable relationship prevailing between the following *two entities:*

- **fluorescence emitted by the glass walls of an X-Ray tube,** and
- **phosphorescence of some** *critical fluorescent salts* **duly caused by sunlight (i.e., UV-light),**

that ultimately left a little pile of a **Uranium (U) substance** (mineral) on a **photographic plate** carefully wrapped in a *thick black paper* lying in the **dark room.** Bacquerel observed, after due development of the said **photographic plate**–a distinct *dark spot* on it located strategically just below the position of the **pile of the mineral (Uranium, U).**

Bacquerel's Inference: It was vehemently observed that:

"**Uranium (U) did emit some critical rays which eventually passed through the paper wrapper; and subsequently, affected the photographic plate perceptively**".

Characteristics of Emitted Rays: They essentially include:

1. These rays are not visible to the naked eye; and hence, are duly emitted by **metallic Uranium (U)** along with the respective **uranium salts.**
2. Besides, these rays do have a tendency to pass *via* the so-called: **Glass sheets** and **Metal sheets.**
3. These rays critically give rise to the production of:

"**ionization in the gases *via* which they sail across**";

and discharge rapidly the so-called ***charged leaves of a properly insulated* 'Electroscope'.**

4. The aforesaid rays were termed as the **Radioactive Rays** and sometimes also known as:

- **Radioactive Emanations** or • **Radioactive Radiations**

5. The ensuing characteristic feature or the so-called phenomenon associated with the emission of such rays are invariably called as the **Radioactive Substances**.

6. **Marie Curie** was the first scientist who introduced the terminology–'**Radioactivity**'.

Remarks: In short, it may be added that the term '**Radioactivity**' predominantly relates to a *Nuclear Phenomenon*; and thus, entails a rather **drastic process** since the *element changes in kind* perceptively.

In addition, **Radioactivity** crucially represents both an ***irreversible and spontaneous disintegrating activity episode*** since the '*element*' undergoes complete cessation forever.

3.1.1. *Magnificent Discovery of Radioactive Elements [Polonium (Po) and Radium (Ra)]*

Marie Curie (1896) carried out an intensive and extensive systematic investigative studies pertaining to such critical entities (substances) emitting '**Radioactive Rays**' (or '*Radiations*') for the *Doctorate Thesis*. In the course of her detailed study she observed that

"***Pitchblende*–a mineral of Uranium (U) was indeed for more radioactive *vis-a-vis* on account of its inherent U-content**".

Observations: Following are a few **Important observations**, namely:

1. The greater activity invariably displaced by the element could be solely due to the critical presence of *certain* **new element present in it perceptively**.

2. **Marie Curie** and **Pierre Curie** (her husband) devotedly worked round the clock with the dogged determination to find a way out for the:

"**eventful and successful separation of the New hypothetical Element meticulously**".

Modus Operandi: A sufficient quantum of **Pitchblende** (*i.e.*, the *natural mineral deposit*) was subjected to the effective separation of its **various inherent constituents**; and thus, found the '**radioactivity profile**' of each and every constituent accurately and precisely.

- ❑ **Fraction Containing Element Bismuth (Bi)–Marie and Pierre** isolated successfully an altogether **New Radioactive Element** termed as: **Polonium (Po)***

- ❑ **Marier and Pierre (1985)**–After a prolonged and tedious process of *extraction* followed by *fractional crystallization* of the derived **Barium Fraction** obtained:

'**a few milligrammes (mg) of another *Radioactive Element*–that they named as–Radium (Ra)**'**

Remarks: In fact, **Radium (Ra)** was found to be almost **3-million folds more radioactive** *vis-a-vis* **Uranium (U)**. Later on, **Debierne (1900)**–reported the discovery of Actinium (Ac)–a radioactive element.

---

* That is, in honour of **Marie Curie's** native country–Poland.

** After the *Latin word* for '**Ray**'.

3.1.2. *Radioactive Ray Variants*

There are *three* commonly known and recognized variants of the so-called **Radioactive Rays**, such as:

- **Alpha (α) Rays**,
- **Beta (β) Rays**, and
- **Gamma (γ) Rays**,

which shall now be treated briefly in the sections that follows:

Rutherford *et al.* (1904) carefully passed the **radiations** duly emitted by the *radioactive elements via* a relatively **strong electric field between a cluster of parallel plates**; and hence, separated them into *three* **distinct types** *i.e.*, α-, β- and γ–Rays as illustrated in Fig: 3.1.

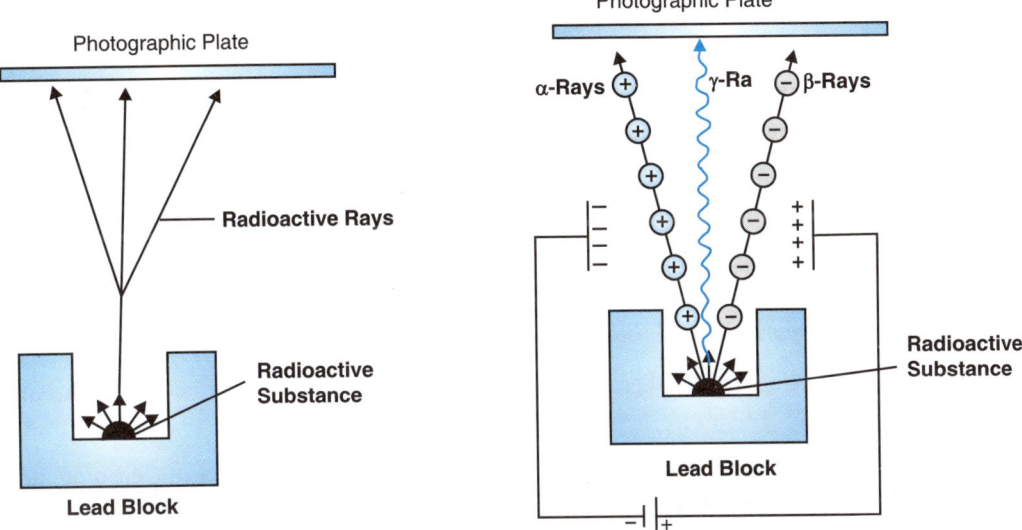

Fig. 3.1: Critical Emission and Subsequent Separation of the Radioactive Rays into: Alpha (α), Beta (β), and Gamma (γ) Rays.

(a) **Alpha (α) Rays**– They refer to the rays that are *deflected towards the negative plate*; and hence, are **positively charged** and are usually known as: • **Alpha (α) Rays** or • **Alpha (a) Particles**. In fact, these rays consist merely of the so-called **Helium (He) Nuclei**.

Alpha (α) Rays [Alpha (α) Particles] are invariably found to emit with energy about **(6–16) × 10^{-13}J**. They will penetrate *a few centimeters of air layer* thereby causing/affording the **ionization of certain molecules***, but they are mostly stopped by a few sheets of either 'Paper' or 'Very Thin Metal Foil'.

Cautions: These essentially comprise:

1. The particular dangers to health arising from α-particles do largely originate from the **risk of ingestion**.
2. Practically all the **α-particles** from a specific nucleus invariably do possess the **same level of energy**; and, therefore, the **same observed range**.

---

* That eventually forms the **basis of their determination**.

(b) Beta (β) Rays–These refer to the rays that are *deflected towards the positive plate* being **negatively charged**; and hence, are termed as **Beta (β) Rays** or **Beta (β) Particles**. In fact, they do merely comprise of **Electrons** exclusively.

Energies of β⁻ Particles– They usually range between **(0.03–5.0) X 10⁻¹³J**; and since they happen to be *much lighter in comparison to the α-particles* they do have a tendency to travel **much faster mode** and have a *range in matter a few hundred folds as great as the latter (α-Particles).*

Characteristic Features: These usually include:

1. The **total ionizing effect** of *β⁻ Particles* is found to be almost the same as that of the **α-Particles**, but it is certainly **effective over much longer distances**; and hence, the *degree of ionization per unit length of the traversed path is observed to be much lower.*

2. Besides, the **particulate matters having different energy levels** may be distinguished explicitly by their divergent inherent **penetrating powers**.

3. The incumbent energies of the **β⁻-Particles** pertaining to a specific '**nuclide**' do exhibit:

   "**a continuous distribution pattern up to a Maximum Value**".

Comment: However, the above critical observation, that was certainly amazing since the nuclei do possess absolutely discrete energy levels–that ultimately led to the *postulate* that particle with a *variable energy level* (*i.e.*, the antineutrino) was emitted simultaneously.

Beta Decay–It refers to a type of **radioactive decay** wherein a *radioisotope* does emit a *small*, *negatively charged*, and *fast moving particle* from the nucleus.

(c) Gamma (γ) Rays–Importantly, these rays are not deflected at all; and hence, are called the **Gamma (γ) Rays**.

Gamma Ray (or Gamma–Radiation)–It is of *very short wavelength**; and hence, the ensuing radiation that quite often accompanies both **α-or β-particle emissions** perceptively.

Important Observations: These essentially comprise:

1. Soonafter such critical emissions the resulting nucleus (**also known as '*daughter nucleus*'**) do often occur in an **excited state** (*i.e.*, *high-energy state*).

2. In a situation, when the prevailing **nucleus** in it usually rearrange themselves to provide the so-called *lowest energy level* of the **nucleus**–the ensuing '**energy**' gets emitted in the form a γ-Ray.

3. Besides, it hs been observed duly that the so-called: '**energies of γ-radiations are usually in the same range as those of the respective β⁻-particles.**'

Thus, the **γ-Rays** do even possess much greater and pronounced *penetrating power*–which could be duly stopped only by a **solid shield of several centimeters of Lead (Pb)**.

	Therefore, as a safety precautionary measure–the critical use of a *Heavy Shielding* is, therefore, an absolute necessity and must in all experimental work involving the *Gamma (γ) Radiation*.

---

*   Therefore, usually possess very high energy level.

3.1.3. *Characteristic Features of Radioactive Radiations*

There are several vital and important characteristic features of **Radioactive Radiations** that mostly provide an elaborated and effective sequential event with regard to scientific facts,–as expatiated under:

1. Amazingly, the emanated radiations are indeed extremely penetrating in nature; and hence, they critically affect:
   - **Photographic Plates** • **Ionize Gases** • **Cause Scintillations upon Fluorescent Screen**
   - **Develop Heat and** • **Cause Chemical Changes.**

2. The '**Emission of Radiations**' duly gives rise to the formation of altogether *Newer Elements* forward genuinely in an **irreversible phenomenon**; whereas, the so-called *Newer Elements* themselves being invariably **radioactive in nature**.

3. Besides, the '**Emission of Radiations**' is usually observed to be absolutely '*spontaneous*'; and thus, is not normally affected due to the *various* **external agents.**

4. In certain typical cases the ensuing '**emission**' fails to be instantaneous; however, it gets duly prolonged *i.e.*, it gets extended over a certain span;–otherwise, it would not have been probably discovered at all.

5. With the exception of *Radioactivity Profile*–one may not at all visualize any sort of an '**abnormality**' present/concerned about the so-called **Radioactive Elements** with respect to their **physical as well as chemical characteristic features**.

6. **Typical Effect of the *Electric* and *Magnetic* Fields–**

❏ Electrical Nature

> ➤ **Alpha (α) Rays**–In a particular instance, when a **narrow beam of radioactive radiations** emanating right from a so-called:

"**deep cavity located strategically in a *Block of Lead (Pb)* is being duly subjected to an 'electric field' existing between the parallel plates,–the alpha (α) particles are depleted particularly towards the negative plate.**"

Comments: It obviously suggests that the **Alpha (α) Rays** do comprise **positively charged particles –** that essentially possess *two* **units of positive charge plus a mass 4-folds the mass of a H-atom**. Alternatively, the so-called **Alpha (α) Particles** are indeed '**double charged Helium (He) ions**'–that eventually consist of *two* **neutrons and two protons (H^+), but possessing No Electrons (e^-) at all**.

Deflection of α-, β-, and γ- rays in a Magnetic Field.

Let us consider a specific instance when a **Magnetic Field** is being applied perpendicular (⊥) to:

   - **plane of the paper**, and
   - **directed from top to bottom**,

the emenated **Alpha (α) Rays** are duly deflected towards the left–as depicted in Fig: 3.2.

Applications of Fleming's Left-Hand Rule–It suggests vehemently that the **Alpha (α) Rays** do critically comprise the positively charged particles. In fact, the precise and accurate **value/ratio of 'e/m'** has been duly determined almost on the same lines as for the **Positive Rays**; and hence, the said determination clearly indicates that:

'Alpha (α) Particles are nothing but the Nuclei of Helium (He) atoms.'

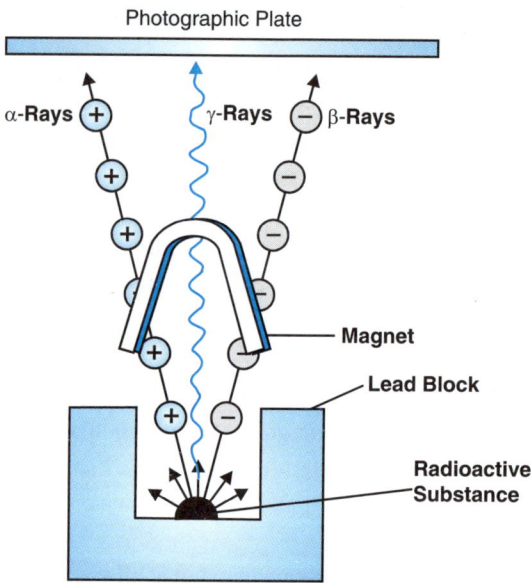

Fig. 3.2: Diagrammatic Representation of the Deflection of α-, β-, and γ- Rays in a Given Magnetic Field.

Explanations:

1. Rutherford and Royds (1909) proved and ascertained first and foremost that the **Alpha (α) particles** do designate specifically that:

   "they are the so-called doubly Helium (He) nuclei"

2. A specially designed **Discharge Tube**, usually made of glass, as shown in Fig: 3.3 was carefully employed for this purpose.

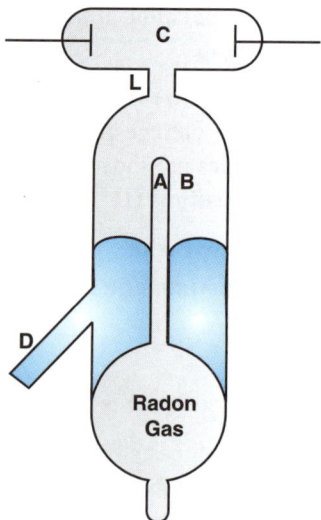

Fig. 3.3: The Experimental Evidence to Demonstrate that the Alpha (α) Particles Designate Doubly charged Helium Nuclei.

Modus Operandi

- A **thin-walled glass-tube (A)** containing the **Radon (Rn) gas** was duly sealed very much inside a **thick-walled tube (B)**.
- At its top-end is positioned strategically a **glass capillary tube (c) having** *two* **metallic electrodes** sealed *via* glass as depicted in Fig: 3.3.
- **Tubes (B) and (C)** were carefully evacuated *via* the **side tube (D)** after which the **Mercury (Hg)** was raised to a specific level as illustrated in the Fig: 3.3.
- After a little gap of time, a few **Alpha (α) particles** right from the **Randon (Rn) gas** do pass *via* the so-called **thin walls of (A)** and eventually enter the **tube (B)**; where they critically require the electrons (e^-) and subsequently turn out to be the so-called **neutral atoms of Helium (He)**.
- Having allowed it to stand for a couple of days, the level of **mercury (Hg)** is carefully raised upto to L;–thereby forcing duly the **Helium (He) gas** very much inside the **discharge tube (C)**.
- At this stage, a high-voltage is applied crucially to the respective electrodes of the **glass capillary tube (C)**; and thus, the **light originating from the discharge tube (C)** is observed with the aid of a **spectroscope**.
- Finally, the **line spectrum** so accomplished is critically observed to be absolutely identical with that duly obtained by carefully passing a **specific discharge** *via* **another tube filled with regular Helium (He) gas**.
  - ➢ **Beta (β) Rays**–It has been duly demonstrated that the b-rays are also deflected by the help of both:
    - **Electric Field** and
    - **Magnetic Field**

Importantly, their inherent emanated **direction of deflection** reveals explicitly that they are nothing but **Negatively Charged Particles**. Thus, the **e/m value** has been duly observed to be located strategically upon:

"**the same lines as seen for the Cathode Rays**";

and hence, this particular determination explicitly shows that the **Beta (β) Rays** are almost identical with the **Cathode Rays** *i.e.*, these do represent the so-called *fast-moving electrons*.

  - ➢ **Gamma (γ) Rays**–Interestingly, the **Gamma (γ) Rays** are found to be very much akin to the respective **X-Rays**; and hence, are more or less equivalent to:

"**the so-called short-wave electromagnetic radiations very much identical to light**".

Important Characteristics: These essentially comprise:

1. They (**γ-Rays**) do possess *extremely* **small wavelengths of** the order of 10^{-8} to 10^{-11} em.
2. The **Gamma (γ) Rays** possess *little mass*; and hence, they cannot be considered as the usual **material particles**.
3. Obviously, the **Gamma Rays** *cannot* be deflected even on being exposed to:

"**the strongest Electric or Magnetic field–thereby ascertaining the underlying fact that they certainly bear no charge at all**."

Properties of α-, β-, and γ- Rays–The various *typical, specific,* and *properties* of the **α-, β-, and γ- rays** shall now be discussed individually in the sections that follows:

[A] The Properties of Alpha (α) Rays

The array of **properties of Alpha (α) Rays** shall now be elaborated comprehensively as stated under:

1. **Velocity**– Mostly the **Alpha (α) rays** are emenated duly from the *radioactive material* having relatively large velocities ranging between **1.4×10^9 cms. sec^{-1}**. In fact, the *overall* velocity of **Alpha (α) rays** exclusively rests upon the respective **radioactive material** from which they are emenated perceptively; and hence, it invariably remains the same for a **given substance**.

> **NOTE**   Actually, the *Alpha (α) particles* do move with a velocity of **10,000 miles. sec^{-1}**.

2. **The Ionizing Power**–It has been proven and demonstrated that the **Alpha (α) rays** normally give rise to an '**intense ionization**' profile in the *gaseous medium* though which they happen to pass. Besides, their '**ionizing power**' is found to be *100 folds* greater in comparison to that of the corresponding **Beta (β) rays** and *10,000 folds* higher than that of the respective **Gamma (γ) rays**. Amazingly, an actual thickness of **0.000 5 cm of Aluminium** (Al) foil helps to reduce the *initial* '**ionizing power**' to almost **50%**.

> **NOTE**   Interestingly, the observed *tracks of Alpha (α) particles* by means of the *Wilson cloud chamber* are found to be **continuous, thick, and straight lines-having a small bend at the end in some typical/rare instances**.

3. **Penetrating Power of Alpha (α) Particles**–In general, the observed overall effect is indeed extremely **feeble in nature**; and hence, display the shortest penetrating power. Perhaps it could be due to the underlying fact that:

"**since the Alpha (α) particles do generate the so-called optimized number of ions in a given path; and thus, they usually penetrate the *shortest distance ultimately***".

> **NOTE**   In true sense, their (*Alpha particles*) so-called inherent *penetrating power* is found to be inversely proportional to their corresponding *ionizing power*.

4. **Fluorescense**–It has been duly observed that the **Alpha (α) rays** produced by a radioactive substance [*viz.*, **Radium (Ra) Salt**] do emit a *distinct fluorescence* (or *lumiscence*) as and when they happen to **strike a fluorescent screen** *e.g.* **Zinc Sulphate [$ZnSO_4$], Barium Platinous Cyanide [Ba Pt $(CN)_4$]**.

Alternatively, when an **Alpha (α) particle** happens to strike a **Fluorescent Screen** then a *small flash of light* termed as **scintillation** is generated. Thus, one may critically observe each and every *individual scintillation* with the aid of a **specially designed apparatus** known as **Spin Thariscope** (which, in fact, represents a **low-power microscope**).

Important Observations: These essentially include:

1. The observations of **fluorescence** *via* the **spin Thariscope** reveals the critical presence of *several* **successive scintillations** duly produced by the actual **impact of an individual Alpha (α) particle**.

2. In addition, the so-called impact of a **single Alpha (α) particle** may be observed so vividly; and hence, the actual number of such particles generated duly per second by means of a **radioactive material** may be observed perceptively.

3. In a broader perspective, these **scintillations** are critically given out exclusively due to the **Alpha (α) particles**; and certainly not due to either **Beta (β) or Gamma (γ) particles**. However, the aforesaid findings may be further proven and ascertained by:

"**interposing a thin-sheet of mica placed duly between the Radium (Ra) salt and the Fluorescent Screen.**"

Since the Alpha (α) rays do get duly absorbed by means of the *mica-sheet*, the so-called *scintillations disappear* completely.

5. **The observed Range of Alpha (α) Particles**–The scintillations given out from the **Alpha (α) particles** critically stop almost suddenly soonafter it has travelled a certain distance *via* the **radioactive matter**.

Range in Air: It relates to –'the distance that an Alpha (α) particle can travel in air, at the atmospheric pressure (760 mm of Hg).' In actual practice, it ranges from **2.70 cms for the Alpha (α) particles duly given by Uranium (U) to 8.62 cms for those of Polonium (Po)**.

Nevertheless, the ensuing range of an **Alpha (α) particle** usually depends upon the following *three* **cardinal factors**, namely:

(i) The **radioactive material** from which the *radiation rays* are emenated.

(ii) The precise and exact nature of the medium *via* which the **rays of radiation** actually traverse.

(iii) The inherent **velocity of emission of the radiation rays** *i.e.*, the range is found to be **directly proportional to V^3**.

Important Features– Following are the *two* **important features** of the **Alpha (α) Rays**:

❑ The precise number of **Alpha (α) particles in a beam** virtually remains constant predominantly up to the end of the observed range of **α-particles**; however, their inherent energy goes on reducing accordingly.

Finally, the so-called **Alpha (α) particles** do not possess enough energy to cause either:

• **Scintillations** or • **Ionizations**.

❑ The degree of **ionization** first and foremost enhances slightly approaches an **optimized level** and suddenly reduces to 'zero level'.

Measurement of the Range of an Alpha (α) Particle by Scintillation

Importantly, the movement of the *Radioactive Sample*, say **Polonium (Po)** farther away by carefully polling the **rod (R)**, one would reach a definite point–when the observed **scintillations** are no more observed via the **low-power microscope (M)**. In this manner, a **distance (d)** is duly attained where the so-called **Alpha (α) particles** *fail to reach the screen*; and hence, do not generate **scintillations** at all,– there by producing the range of the **α-particles perceptively**.

Fig: 3.4 shows explicitly the critical measurement of the ensuing range of an **Alpha (α) particles** due to the phenomenon of **scintillations** that subsequently provides the range of the **α-particles** particularly.

6. **Scattering of Alpha (α) Rays** on being Passed *via* Thin-sheets of Mica, **Gold (Au) Foil**.

It has been proven and ascertained that:

"**there occurs a definite divergence of the *Alpha (α) particle* from its straight-line passage is up to 2–3° (degrees).**"

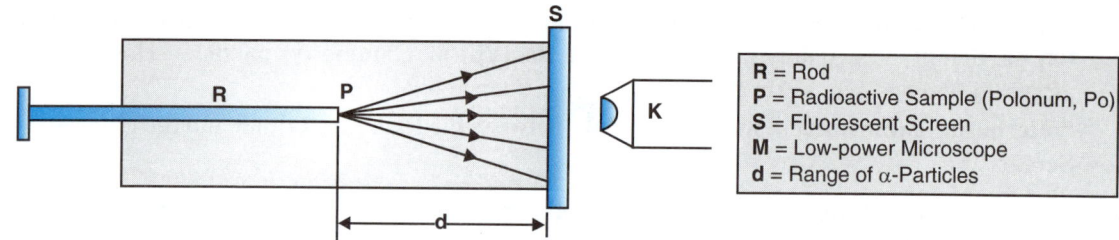

Fig. 3.4: Critical Measurement of the Range of an Alpha (α) Particle by Scintillations Produced Duly upon Fluorescent Screen.

Geiger and Marsden later on found that quite a few particles being specifically deflected a little. Rutherford subsequently expatiated the actual reason for such an observation is solely due to:

"**the prevailing repulsion occuring between the Alpha (α) particle and the nucleus of the atom causing its scattering perceptively**".

7. **Deflection of Alpha (α) Rays by Electric and Magnetic Fields**–Obviously, the **Alpha (α) rays** get duly deflected by both **electric and magnetic fields** there by suggesting vehemently that these are certainly the **charged particulate matters (particles)**.

8. **Heating Effect**–It has been established beyond any reasonable doubt that:

'**a certain quantum of Radium (Ra) always maintains itself at a temperature higher than that of its surrounding environments**'.

> **NOTE**   **The evolution of heat is perhaps due to the crucial blockade of the α-, β-, and γ-rays by the radioactive material**".

9. **Exposure Causes Incurable Burns**–The exposure of the **Alpha (α) rays** to the human body surface critically gives rise to incurable burns.

10. **Effect on the Photographic Plate**–The ensuing effect of the so-called **Alpha (α) rays** upon the **photographic plate** is generally found to be an *extremely* **feeble nature.**

[B] The Properties of Beta (β) Rays

There are *seven* cardinal properties of Beta (β) Rays–which are enumerated in a sequential manner as under:

1. **Velocity and Range of Beta (β) Rays**–In general, the β-particles do *not* emerge out of the nucleus having the *identical* **velocity and range**; but they rather exhibit a *wide spectrum of velocity* varying between **1% to 99%** of the **velocity of light**.

2. **Ionization in Air**–It has been duly observed that the **Beta (β) Rays** do produce **ionization in air**; however, the exact number of respective '*ions*' duly generated is hardly **1/1000th** of those duly produced by the corresponding **Alpha (α) Rays.** Even though their inherent **velocity seems to be pretty high**,–yet they do have a relatively **small mass**; and hence, inherit a **small kinetic energy** profile. Besides, since the **Beta (β) particles** invariably *change their path*, their tracks as observed in:

"the *Wilson cloud chamber*–are duly scattered and continuous as could be seen for the *Alpha (α) particles* perceptively".

3. **Penetration Profile**–Amazingly, the **Beta (β) Rays do have the ability to penetrate** *via.* a **broader thickness of material** *viz.* they may even pass across a **1 cm thickness of Aluminium (Al) sheet**.

4. **Deflection**–It has been established that the **Beta (β) Rays** do usually:

   "**get deflected by both Electric and Magnetic Fields effectively**".

   Besides, the *direction of deflection* vividly shows that these are nothing but the so-called–'**Negatively Charged Particles**.' The respective inherent **e/m value** has been duly found to be very much at par with the **Cathode Rays**; and hence, this critically determines that the **Beta (β) Rays** are obviously '*fast-moving electrons*',–which are certainly of the **Nuclear Origin**–although they fail to:

   'revolve in orbits very much similar to the extra–nuclear electrons'.

5. **Scattering of Beta (β) Rays**–The scattering of **Beta (β) Rays** invariably come into play on being passed *via* the matter; and, therefore, materializes due to their *extremely* **small mass**. In general, they usually do undergo the phenomenon of '*multiple scattering*'; whereas, they '*single scattering*' it also occurs by impact both with the '**Electrons**' and '**Nuclei**'.

6. **Beta (β) Rays** do affect the so-called **photographic plates** specifically rather **more strongly** *vis-a-vis* the respective **Alpha (α) Particles**.

7. Ultimately, the **Beta (β) Rays** do critically produce '*fluorescence*' in platinocyanide, calcium tungstate, Willmite (it occurs in nature as a **mineral** contains **Zinc Silicate, $(Zn_2 SiO_4)$** and the like.

[C] The Properties of Gamma (γ) Rays

Following are the *five* **important properties of Gamma (γ) Rays.**

1. **Velocity**–It has been observed that the **Gamma (γ) Rays** do have the *same* **velocity** as that of the **light** [*i.e.*, 3×10^{10} **cms. sec**$^{-1}$].

2. **Deflection Profile**–Importantly, the **Gamma (γ) Rays do not get deflected by electric and magnetic fields**. Furthermore, this property clearly ascertains that:

   '**these are basically the so-called–'Electromagnetic Radiations' that essentially comprise very energetic inherent photons'**.

   Besides, the **Gamma (γ) Rays** are duly emitted due to:
   - **an excited 'nucleus', and**
   - **hence do possess a very short wave length varying between *0.5 Å to 0.005 Å*.**

3. **Ionization in Gas and Producing Power**–The **Gamma (γ) Rays** do have a tendency to undergo **ionization** phenomenon duly generated by them,–which is found to be **relatively-small**. As a result, their inherent '**penetrating power**' is observed to be *extremely high* in comparison to those of the corresponding **Beta (β) Rays**. *i.e.*, in other words–the **Gamma (γ) Rays** may easily **sail across** *via* **a 30 cm thick iron sheet**.

4. **Diffraction**–It relates to the spreading of light waves behind a *grating*, leading to the production of *interference patterns*; and the bending/breaking of the light ray into its *component parts* (*i.e.*, the respective wavelengths).

   Thus, the **Gamma (γ) Rays** are duly diffracted by the **crystals** very much akin to the **X-Rays**.

5. **Produce Fluorescence**–The **γ-Rays** are usually capable of **producing fluorescence and affecting photographic plates** rather in a *more intensive manner* in comparison to **β-Rays**.

3.1.4. *Meritorious Uses of Radioactive Radiations*

Since the inception and discovery of the **Radioactive Radiations**–there are an avalanche of meritorious uses of the aforesaid classical and highly specific rays. In a broader perspective, these applications spread out almost overwhelmingly into a larger segment–that essentially comprise:

- **Medical and Health Sciences,**
- **Industrial Radiography,**
- **Electric Power by Thermo-electric Conversion,**
- **Space-related Applications,**
- **Preservation of Food Products,**
- **Controlling the Population of Harmful Insects,**
- **Evolution of Newer Varieties of Plants,**
- **Self-luminous Paints, and**
- **Elimination of Static Electricity to Control Fire-Incidents,**

which shall now be discussed briefly in the section that follows:

(a) **Medical and Health Sciences**–The **Gamma (γ) radiations** emenated from ^{60}Co [Cobalt 60] are being employed gainfully in the *hospitals* for the particular **sterilization of materials** *viz.,*
- **surgical dressings,**
- **hypodermic syringes/needles,** and
- **surgical sutures**.

(b) **Industrial Radiography**–Interestingly, an array of **radiation sources** *viz.,* ^{60}Co [Cobalt 60] *i.e.,* a *γ-Ray Emitter* mostly have been employed for **Industrial Radiography,**–which helps in carrying out the **investigating procedures/capabilities** for critically examining the so-called:
"interiors of the *Metallic Castings* for detecting any flaws/defects".

(c) **Electric Power of Thermo-electric Conversion**–The radioactive nuclides that specifically emit either **Alpha (α) or Beta (β) particles** have been used, both intensively and extensively, for the so-called **generation of electric power by means of the thermo-electric conversion perceptively**. Besides, in a situation when α- or β- particles do get duly absorbed in a matter, the **ensuing energy of the radiation gets converted** into the respective **heat energy,** –that could be eventually utilized for the production of **Electric Power**.

 NOTE In fact, such *Nuclide Power Generators* are suitable specifically for use exclusively in the so-called '*space vehicles*' since their light-weight, *long-life span*, and *absolute reliability.*

(d) **Space-related Applications**–Amazingly, the **radio nuclides** do find their abundant usage as: "compact sources of *heat energy* since they virtually find a plethora of *space related applications vehemently*".

Examples: The **radio nuclides** of:
- **Promethium–147,**
- **Polonium–210,** and
- **Plutonium–238,**

have been mostly employed to heat the so-called **Propellant Gas** (*viz.*, **Hydrogen**) specially in the **Low-thrust Rockets**.

(e) **Preservation of Food Products**–The so-called **Nuclear Radiations** [*viz.*, **Gamma (γ) Rays**] have been extensively employed for the particular effective, preservation of food. However, the critical–'**irradiation of food products**' specially such items:

- **Fish** • **Meat** • **Poultry** and • **Fruits**,

is duly accomplished by exposure to the **γ-rays** emenated from:

- ^{60}**Co [Cobalt 60]** or • ^{137}**Ca [Calcium 137]**.

Example: A specific dose of **2–5 million radiations**–has indeed proved to be just sufficient to destroy almost a large segment of the prevailing bacteria in the food products.

Comment: In fact, such a highly crucial/preventive measure of the **Radioactive Radiations** do go a long way in enhancing the *shelf-life* of these γ-ray exposed food items without any sort of refrigeration at all (either during *transportation* or storage in shelves).

NOTE	The Gamma (γ) Ray radiations do find their extended utilization particularly for the *disinfection of insects* in cereals viz., wheat, barely, rice, flour etc.

(f) **Controlling the Population of Harmful Insects**–In reality, the effective control of the ensuing population of the harmful insects that eventually causes a *substantial damage* to both:

- **Plant Crops** and • **Livestock**,

may be controlled and managed by irradiating:

"the so-called *male-members* of these insects so as to render them sterile germinely."

(g) **Evolution of Newer Varieties of Plants**–The intelligent and effective use of the **Radiation Mutations** in various plants have been duly practised for the evolution of a host of:

"an altogether newer breeds of the Plant Species".

A few typical examples may be seen amongst the so-called *Newer Breeds* of **Cereals** *viz.*, **Rice, Wheat, Pulses, Lentils**; fruits *viz.*, **Grapes, Oranges, Mangoes, Bananas, Apples, Kiwis, Pears, Dates, Apricots, Figs**, and **certain other citrus fruits**.

(h) **Self-Luminous Paints**–The wonderful and elegant discovery of the **self-luminous paints**– The wonderful and elegant discovery of the self-luminous paints for their specific use upon the so-called:

- **Dials of Watches** • **Instruments** • **Door-Handles** • **Car-Bumpers** and
- **Staircase Railings**,

that have been made by the critical incorporation of:

"a highly specific **Natural Alpha (α) Emitting Radioactive Material to Phosphors*"**

(i) **Elimination of Static Electricity to Control Fire Incidents**–It has been observed that the '**ionization produced by Beta (β) Rays**' has been bvroadly utilized for the particular intended *elimination of static electricity* which is considered to be a serious cause of fire and explosion hazards taking place mostly in the:

- **Paper** • **Plastic** • **Rubber** and • **Textile industries**.

---

* **Phosphor:** The nature of the **phosphor** used solely depends upon that of **radiation to be detected**.

3.2. SPECIFIC DETECTION AND PRECISE MEASUREMENT OF RADIOACTIVITY

The **specific detection and precise measurement of radioactivity** of a substance may be measured precisely by an array of duly time-tested and approved widely accepted methods, such as:

- **Electroscope Method,**
- **Wilson's Cloud Chamber Method,** and
- **Geiger-Muller (GM) Counter Method,**

which shall now be described individually in the sections that follows:

3.2.1. *Electroscope Method*

The **electroscope** being employed for the **measurement of radioactivity of a material** essentially comprises of the following *two* **components**, namely:

- **an Ionization Chamber,** and
- **an Electroscope Box,**

as illustrated in Fig: 3.5

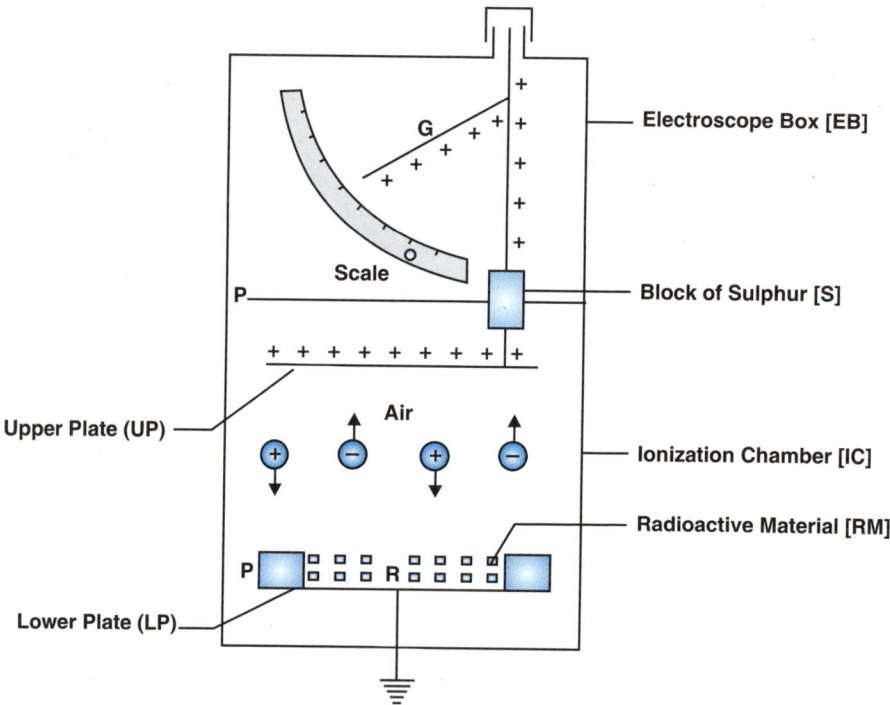

Fig.3.5: An Electroscope Being Employed for Measurement of the Radioactivity of a Material Under Investigations.

Modus Operandii

1. The **ionization chamber (IC)** is usually constructed out of a **thick sheet of lead (Pb)** and consists of *two* **metallic plates (P and P′)** duly separated by the '*gas*' that is being **exposed to the radioactive rays**.

2. The **lower plate (LP)** is carefully connected with the *wall* of the so-called **ionization chamber (IC)** and **earthed duly**.
3. The **upper plate (UP)** is duly insulated by a **block of sulphur (S)**; and subsequently, connected carefully with the **gold-leaf system (G)**,–which is charged initially to a reasonably **high-potential**.
4. The '**radioactive material (RM)**' is now placed very carefully in the cavity of the **plate (P′)**.
5. The so-called *entrapped air* between the *two* **metallic plates (P and P′)** gets adequately ionized; and the inherent charge slowly leaks between the *two* **metallic plates**.
6. At this material time, the **Gold Leaf (G′)** is found to fall slowly; and hence, its **rate of fall** is:

"**directly proportional to the actual rate of leakage borne by the inherent charge**"

and thus, remains absolutely *proportional to the intensity of the rays.*

 **NOTE** Amazingly, the observed '*rate of leakage in the inherent charge*' appears to be fairly comparable to that of a material of *known radioactivity* under the identical experimental parameters. Therefore, the observed–'**Total Radioactivity of the Material**' under the investigative study is determined finally.

3.2.2. *Wilson's Cloud Chamber Method*

The apparatus being used for the **measurement of radioactivity** was first and foremost devised by **Wilson (1912)**; and hence, is commonly termed as: **Wilson's Cloud Chamber**, –as depicted in Fig: 3.6.

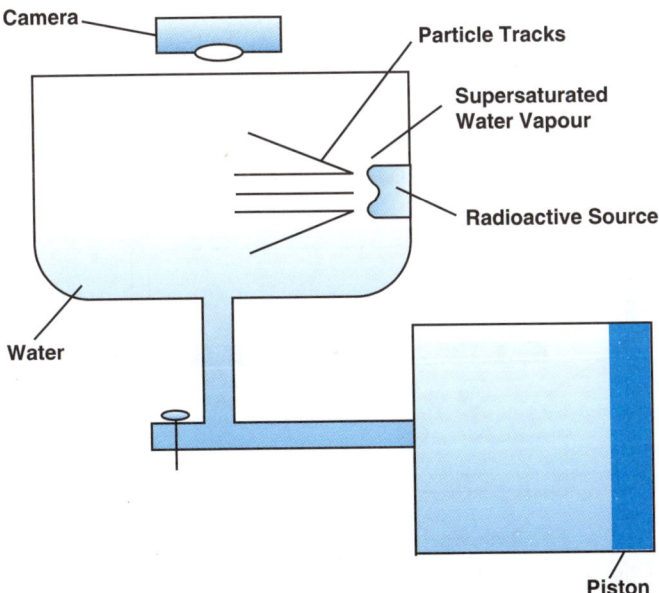

Fig. 3.6: Diagrammatic Representation of Wilson's Cloud Chamber Used for the Precise Measurement of a Radioactive Material.

In fact, it represents an altogether **New Technique** that is based exclusively upon the production of ions by the help of the so-called **radioactive emissions**. Besides, when such '*ions*' are duly generated along the path of a radioactive emission–the respective observed *track of fog* formed indicates explicitly the **actual path followed by particle** (which way be **photographed** simultaneously).

Modus Operandi: The various sequential steps involved include:

1. A **gas** or **dust-force air** duly saturated with either **alcohol** (or *water* vapour is carefully enclosed in the chamber.)

2. The **piston** is moved up quickly so as to afford the **desired cooling effect by** *sudden expansion in volume.*

3. In doing so, the **air contained duly in the chamber** gets adequately supersaturated with water vapours, which eventually paves the way:

   "**for enabling the passage of the** *Radioactive Emissions* **into the chamber**"

4. Thus, it immediately forms a so-called '**Fog Track**' very much along the path travelled by the ensuing **radioactive emission**.

5. The aforesaid '**Fog Track**' is photographed immediately by a pre-set camera.

> **NOTE** However, incertain more recently developed *sophisticated devices,* the so-called '*Cloud Chamber*' actually operates automatically.

3.2.3. *Geiger-Müller (GM) Counter Method*

The **GM-Counter** instrument is meant for detecting the **Beta (β) Particles**; and thus; it consists of a cylinder containing a **gas**, a **cathode**, and an **anode**, whereby the ions passing into the cylinder do cause a **flow-of-current** and each **β-particle** produces **a pulse of current** which gets **amplified and recorded duly**.

The **Geiger-Müller Counter** used in this method is illustrated in Fig: 3.7.

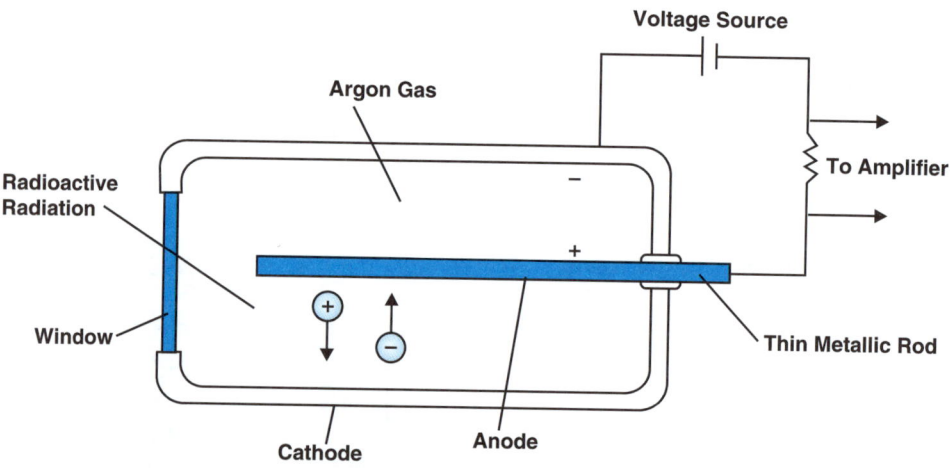

Fig. 3.7: Geiger-Müller Counter Used for the Detection and Measurement of Radioactivity of a Substance.

Modus Operandi:

1. The **GM-Counter** comprises a **hollow cylindrical tube duly fitted with a thin-metallic rod held along its axis,**–which being *insulated* perfectly from the said tube.

2. The **hollow cylindrical tube** is duly filled with a **gas** [*viz.,* **Oxygen (O_2) or Argon (Ar)**] at an extremely low pressure (*i.e.,* ~ **0.1 μm**).

3. Now, a reasonably **high-potential** (say ~ 1000 Volts) is being applied at the **thin-metallic rod** (acting duly as an 'Anode') and the outer tube (acting duly as a 'Cathode').

4. At this point in time, when the α- and β- particles of *high-energy* are subsequently permitted to enter the **thin-metallic tube** *via* the **quartz window** they do eventually **generate sufficient ions** that would certainly cause a pulse of a current to:

"flow between the *Rod* and the *Tube*"

5. Ultimately, this '*pulse*' may be duly amplified and recorded upon a pre-set **Loud-speaker–as a sharp click.**

 NOTE   **Nevertheless, in the so-called more sophisticated instruments, the age-old Loud-Speaker device has been duly replaced by an '*Automatic Recorder*'–that counts the number of pulses perceptively.**

3.3. THEORY OF RADIOACTIVE DISINTEGRATION

In order to understand fully the **theory of radioactive disintegration** one may have to get a proper in-depth knowledge with regard to:

- **Radioactive Disintegration**, and
- **Rate of Radioactive Disintegration**.

3.3.1. *Radioactive Disintegration**

It is an universal phenomenon that–"**the atoms of heavy elements** *viz.*, **Uranium (U), Thorium (Th), Polonium (Po), and Radium (Ra) etc.**"

are being subjected to cleavage (cessation) constantly into the corresponding '**fresh radioactive atoms**' wich the critical *emission of α-, β-, and γ- rays* emenating from **their nuclei**. Importantly, in accomplishing their phenomenon effectively, the so-called **original** (or *parent*) **atom** disappears completely; and hence, yields an altogether **New (daughter) Atom**.

Obviously, these '*Newer Atoms*' are indeed found to be having the **Radioactive Characteristics**; and, therefore, undergoes cleavage spontaneously in their turn,–thereby leading to a relatively *long-chain of divergent Radioactive Elements* usually in the form of a series of an '**inactive element**' ultimately (*i.e.*, invariably a '**lead**') is reached.

 NOTE   **The resulting spontaneous '*cleavage of the nucleus*' is usually termed as Radioactive Disintegration (or Radioactive Decay).**

3.3.2. *Rate of Radioactive Disintegration*

Let us assume that N designates the **exact number of atoms of a radioactive material** at any *material (or given) time*, and N represents the precise number of atoms of the substance after a time *t*, –the rate of disintegration of the *radioactive material* may be designated by the following expression:

$$\boxed{\dfrac{dN}{dt} = IN}$$

(i)

---

*   It is also known as the '**Radioactive Decay**'.

where, λ = Constant (usually known as the Disintegration Constant) of the element.

Simply carrying out the integration of Eq. (i) gives the expression:

$$\boxed{N = N_0 e^{-\lambda t}}$$

(ii)

Important Observations–These essentially include:

1. It may be observed that the **rate of radioactive disintegration** is independent of both *pressure and temperature* irrespectrive of any *external parametes*.

2. **Rutherford (1909)** discovered a rather '*simple law*' stating the– 'rate of radioactive disintegration **(or radioactive decay),**'–that may be stated as given under:

   "**A particular given constant fraction of an '*analyte*' (sample) of a radioactive element critically undergoes the disintegration process in a unit time.**"

Remarks: It further expatiates that the *actual number of atoms* that **disintegrate in a unit time** refers to a *constant fraction of the total number of atoms present perceptively*.

Since, the total quantum of the ensuing **Radioactive Element**, in fact, adopts a *decreasing trend with time i.e.,* the prevailing amount that **undergoes disintegration** in a *given unit time* also goes on decreasing accordingly.

Examples: Let us assume that the **50% of the actual quantum of a particular radioactive element disintegrates per day** (*i.e., in 24 Hrs.*). In case, one starts with **1g of the radioactive element**, then only *0.5 g shall be left at the end of the very 1st day*. Likewise, **at the end of the 2nd day, exactly 50% of 0.5 g i.e., 0.25 g shall be left, and so on so forth**, till it reaches on '*infinity*'.

 NOTE   **Importantly, the so-called *Radioactive Material* (element) never attains the *complete disintegration*, even though its actual quantum may virtually turn out to be too small to be measured precisely.**

3.4. HALF LIFE PERIOD [$t_{1/2}$ or T]

It relates to the amount of time a *radioactive material* (element) usually takes for half of its initial **amount to undergo disintegration**. Alternatively, in a *kinetic experiment, the time needed for the concentration of a specific reacting species to fall to almost 50% of its initial value.*

Let us suppose, after a time $t_{1/2}$,–almost half of the atoms of the radioactive substance have undergone disintegration *i.e.,* we may have the following expression:

$$\boxed{N = N_0/2}$$

substituting the above value of '**N**' in Eq. (ii), we have:

$$\boxed{1/2 = e^{-lt_{1/2}}}$$

(iii)

or $\ln 0.5 = \lambda t_{1/2}$

or

$$\boxed{t_{1/2} = 0.6931/l = \text{Constant}}$$

(iv)

Based on **Eq. (iv)**, we may safely infer that:

"the *time (t₁/₂)* **actually needed for the disintegration of one half of the original quantum of the Radioactive Material (element) is invariably termed as its—***Half-Life Period***".

Points to Ponder: These essentially include:

1. Obviously, the so-called **Half-Life Period** of a given *radioactive materials* **(RM)** is observed to be absolutely '**independent**' of the actual quantum of the **RM** duly present at the *initial stage*.

2. In true sense, the **Half-Life Period** solely rests upon the **disintegration constant (λ)** of the *radioactive element*; and hence, it could now be possible to:

   "**enable the precise and accurate characterization of each and every *radioactive element* by the ensuing value perceptively.**"

Example: The above findings and dictum may be further expatiated by considering the **Half-Life Period** of **Radium (Ra)**,–that is **1580 years**; whereas, the *same values* pertaining to a few of the **radioactive elements** stands at a **few seconds** only.

   Radioactivity Unit– In a broader perspective, the *inherent* **radioactivity** of a **Radioactive Material (RM)** is invariably measured:

   "**in terms of the observed *rate of disintegration* of the Radioactive Substance (RM)**"

   At this point in time, *two* situation arises, namely:

   First–when the so-called observed **rate of disintegrations** stands at: '**3.8 × 10¹⁰ disintegration sec⁻¹**'

the observed radioactivity of the *radioactive material (RM)* is said to be: '**one curie (ci)**'

From the above statement of facts, we may observe that:

$$\text{'1ci} = 3.8 \times 10^{10} \text{ disintegrations. sec}^{-1}\text{'}.$$

Remarks: The aforesaid relatively large number is exclusively based on the underlying fact that:

   "**1g of Radium (Ra) critically disintegrates at the rate of 3.8 × 10¹⁰ disintegrations sec⁻¹.**"

Second–when one encounters relatively smaller units *viz.*,

- **mci (*millicurie*)**; and
- **uci (micro curie)**,

are beingh employed commonly and extensively.

   Hence, we may have the following known equivalents:

1 mci = 3.8 × 10⁷ disintegrations . sec⁻¹

1 μci = 3.8 × 10⁴ disintegrations. sec⁻¹

3.5. RADIOACTIVE CONSTANT (λ)

Under section 3.3 (*Theorey of Radioactive Disintegration*) Eq. (i) related to the '*rate of radioactive disintegration*', we have:

$$\boxed{\frac{dN}{dt} = IN} \tag{v}$$

Now, we may re-write the above Eq. in terms of the **radioactive constant (λ)** as given under:

$$l = \frac{-dN/dt}{N}$$

(vi)

Based on the above relationship one may explicitly define the so-called **Radioactive constant (λ)** as:

"**the ratio of the amount of the substance that disintegrates in a unit time to the amount of substances present.**"

In another instance, when one assumes and puts $t = 1/\lambda$ in **Eq. (ii)** under *section 3.3 (theory of radioactive disintegration)*, we may have the following expressions:

$$N = N_0 e^{-1'1/l}$$
$$N = N_0 e^{-1}$$

or

or

$$N = \frac{N_0}{e} = \frac{N_0}{2.718}$$

or

$$N = 0.368\ N_0 = 0.37\ N_0$$

(vii)

Therefore, based upon the above relationship the **radioactive constant** may also be defined as:

"**the reciprocal of the time during which the exact number of atoms of radioactive material (RM) falls to 37% of its original value**"

3.6. AVERAGE LIFE-PERIOD (l)

It has been proven and established that the atoms of a **Radioactive Element** mostly *get disintegrated constantly* just one after the other; and hence, the '**life of each and every atom is altogether different**'. *Highlights:* These essentially include:

1. Obviously, the atoms that undergo disintegration at an early stage do have specifically **very short-life span**; whereas, those which happen to *disintegrate at the end* (or **latter stages**) do possess critically a **long-life span**.

2. Thus, one may definitely draw a *valid conclusion* that: "**the phenomenon of disintegration shall continue indefinitely.**"

3. Hence, the so-called '**Total Decay Period**' of any **Radioactive Element** is definitely moves towards an '**infinity**'.

> **NOTE**
>
> In short, it may be inferred that the utility of:
> • **Total Decay Period** and/or • **Total Life Periods**, for the *Radioactive Elements* are more or less absolutely meaningless.

Average Life Period (l) [Life Period or Life Expectancy or Mean Life Period]

Thus, one may consider instead of the **Total Decay Period**,– another terminology that is generally employed is **Average Life Period**, which being also sometimes known as:

• **Mean Life Period** or • **Life Expectancy**,

which may be defined as under:

"Average Life Period (l) of a radioactive element refers to the reciprocal of the Disintegration Constant (λ)"

Thus, we may have the following expression:

$$l = \frac{1}{\lambda}$$ (viii)

since we have already seen in **Section 3.4**, Eq. (iv) that:

$$t_{1/2} = 0.6931/\lambda$$

or

$$\frac{1}{\lambda} = \frac{t_{1/2}}{0.6931}$$ (ix)

From Eq. (viii) we have:

$$\frac{1}{\lambda} = l$$ (x)

Therefore, from Eqs. (ix) and (x), we get:

$$l = \frac{t_{1/2}}{0.6931}$$

or

$$l = 1.44 \times t_{1/2}$$ (xi)

Thus, we may have the following important expression:

$$\text{Average Life Period (l)} = 1.44 \times \text{Half - Life Period (t}_{1/2})$$

3.7. RADIOACTIVE EQUILIBRIUM [LAW OF SUCCESSIVE DISINTEGRATION]

In order to have an in-depth knowledge pertaining to the phenomenon of **Radioactive Equilibrium,–** one may take into consideration a so-called:

"long-lived Radioactive Element 'A'"

Furthermore, the said elment upon *prolonged standing* episode happens to disintegrate successively into:

'a respective Decay Series of Radioactive Atoms: B, C, D, … etc.',

as shown under:

$$A \rightarrow B \rightarrow C \rightarrow D \dots$$

Important Observations: These may comprise:

1. Amazingly, in the course of **successive disintegration** of **Radioactive Materials (RM)** *viz.*, **A, B, C, D** etc.,–any *two* **adjacent elements** may be regarded as:
   - **Parent** and • **Daughter,**

whereby the *former* being the one that by its own decay gives rise to the latter *viz.*, 'A'–which eventually produces 'B' thereby:

'designating the *Parent Element* with respect to *A* the Daughter Element'

 NOTE Obviously, the so-called '*Parent*' of the following element shall be '*Daughter*' of the preceeding element perceptively.

2. Interestingly, the **Law of Successive Disintegration** critically deals with the quantum of the "**Daughter Elements**" duly present at any specific instant; and hence, determines predominantly '**the prevailing parameter of '*Equilibrium*' existing between the *Daughter* and the *Parent* element respectively."**

3. Importantly, after a certain time gap-the rate at which the **element B** is duly produced from the **element A** almost becomes equal to the rate of its *subsequent decay into* **element C**.

In this manner, one may observe prevently that:

"**a stage will be accomplished when the exact number of atoms of B duly obtained from A. sec^{-1} shall become almost equal to the number of atoms that ultimately disintegrate into C. sec^{-1}."**

Comment: Based on the above revelations and observations one may take cognizance of the underlying fact that:

'**the element *B* is then said to be in a *Radioactive Equilibrium* with the given quantum of *A*.'**

Exactly, in the same manner the respective **element *C*** and **element *D*** would also attain, in due course, the state of **Radioactive Equilibrium** having the so-called *preceeding number of the decay series.*

Explanations: The **Radioactive Equilibrium** may be further explained as stated under in a sequential manner:

❑ In case, N_1, N_2, N_3 etc. designate the number of atoms of various numbers of the so-called **Decay Series** at the specific *steady state*; and λ_1, λ_2, λ_3 etc., represent their corresponding **Disintegration constants**, we have:

$$\text{Decay Series:} \qquad A \rightarrow B \rightarrow C \rightarrow D \rightarrow \dots$$
$$\text{Number of Atoms:} \quad N_1 \quad N_2 \quad N_3$$
$$\text{Disintegration Constants:} \quad \lambda_1 \quad \lambda_2 \quad \lambda_3$$

then $\lambda_1 N_1 = \lambda_2 N_2 = \lambda_3 N_3$

❑ Now, if we take into consideration the **elements A and B**, we may have:

$$\boxed{\frac{N_1}{N_2} = \frac{l_2}{l_1}} \tag{xii}$$

❑ Because the observed **Average Life Period (*l*)** is found to be the *reciprocal of disintegration constants* (λ), we may have the expression:

$$\boxed{\frac{\lambda_2}{\lambda_1} = \frac{l_1}{l_2}} \tag{xiii}$$

Now, with the help Eq. (xiii)–the Eq.. (xii) gets redue to:

$$\boxed{\frac{N_1}{N_2} = \frac{\lambda_2}{\lambda_1} = \frac{l_1}{l_2}} \tag{xiv}$$

whee, l_1 and l_2 = Average Life Periods of the **Radioactived Elements A** and **B**.

❑ **Eq. (xiv)** vividly depicts and ascertains that:

"at the steady state, the so-called *Radioactive Elements viz.*, **A, B, C etc., are duly present in the *direct ratio* of their corresponding Average Life Period (*l*).**"

Therefore, at the **steady-state**, the **Radioactive Element** possessing the *highest Average Life Periods (l)* is invariably present in the **largest quantum**.

NOTE — Importantly, the ensuing *Radioactive Equilibrium* **may be observed only when the** *Average Life Period (l)* **of the so-called** *Parent Radioactive Element* **is found to be greater than that of the ensuing** *daughter element*.

3.8. STABILITY OF NUCLEUS AND RATIO (N) AND PROTONS (P)

Based on the scientific evidences and subsequent authenttication of facts and figures the observed stability of a **nucleus** depends *solely and quantitatively* upon the *precise ratio of the* Neutrons (N) and Protons (P) duly present in it.

Amazingly, in the particular instance of **elements** having *atomic numbers* equal to **20** – the *observed ratio N/P* is almost equal to *unity* (*i.e.*, in all *such elements* the **exact number of Neutrons (N) and Protons (P)** is more or less the some).

Likewise, in the typical instance of *such elements* having the so-called **higher atomic number ratio,** *N/P gets enhanced progressively* to almost about **1.6** (*instead of 'I'*); and hence, the *observed incidence of repulsion between them gets increased accordingly*.

Fig: 3.8 illustrates the graphic representation between the number of **protons (P)** and **neutrons (N)** with the so-called **belt-of-stability** that essentially comprises a '**stable nuclei**'. Furthermore, the **nuclei** with **N/P ratio** lying *above or below the belt of stability* are found to be **fairly unstable**; and hence, do undergo the critical and specific **spontaneous disintegration**.

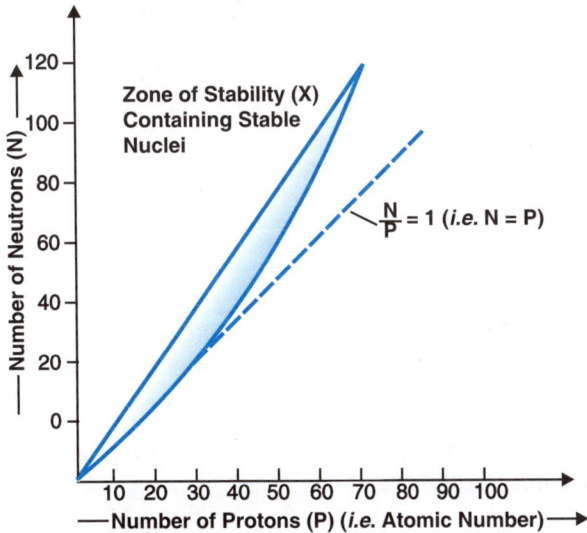

Fig. 3.8: The Graphic Representation Between Number of Protons (P) and Neutrons (N) Having the Zone of Stability (X) that Essentially Comprises Stable Nuclei.
[Nuclei whose N/P Ratio lies above or below the Belt-of Stability (X) are found to be fairly unstable; and thus, undergo the spontaneous Disintegration.]

Explanation: The various sequential steps involved in Fig: 3.8 may be duly explained as under:

1. By carefully plotting a graph between the **number of protons (P) and number of neutrons (N)** pertaining to the nuclei of different elements, it has been observed critically that a plethora of so-called **stable nuclei** (*i.e.*, the **ensuing non-radiactive nuclei**) could be seen located strategically in a **well-defined zone**, that is termed invariably as the '**zone or belt of stability**' since it particularly consists of the **stable nuclei**.

2. **Importantly, the specific nuclei whose N/P ratio falls either above or below this aforesaid *zone (or belt)* are found to be fairly stable in nature; whereas, those undergoing the spontaneous radioactive disintegration (or decay) from one atom to the other thereby yielding the respective *α-particle and β-particle*.**

3. Thus, the above **specified phenomenon of disintegration** usually continues unless and until a **fairly stable nucleus is accomplished**.

   Examples: The inherent **nucleus of Radium (Ra)** is observed to undergo changes almost successively into *nine* **other respective nuclei** thereby emitting in all: 5α- and 5β- particles; and hence, ultimately leads to the formation of: "**a fairly stable nucleus of Lead (Pb^{206}).**"

3.9. GROUP DISPLACEMENT LAW

Preamble–Fajan, Russel, and Soddy (1913) were pioneer in putting forward the **Group Displacement Law** which refers to the specific changes that eventually come into play when an **α-particle** or a **β-particle** is being duly emitted from a corresponding **Radioactive Element**.

Definition: Hence, the **Group Displacement Law** may be defined as stated under:

"**In a specific instance when an α-particle gets emitted from a *Radioactive Element* (*i.e.*, the *parent element*), –the newly formed element (*i.e.*, daughter element) formed has an atomic number two units less and the mass number *four* units less *vis-a-vis* the *parent element*. As a the ensuing elements gets duly displaced upto *two* places (*i.e.*, *two* groups) to the left-hand side of the *Periodic Table*. Likewise, as and when a β-particle gets duly emitted, the newly formed element possesses the atomic number *one unit more* but its respective mass number, almost remains the same as that of the parent element. Thus, ultimately, the *newly formed element* gets displaced by *one place* (*i.e.*, one group) to the right hand side of the Periodic Table.**"

Explanations: The various important aspects of the Group Displacement Law may be explained as under:

1. As the **α-particle** has a **Helium (He) nucleus** having a **mass number *four* (*i.e.*, 2-protons + 2-neutrons)**, the emission of the respective **α-particle** emitted from a **Radioactive Element** eventually causes the corresponding mass number to decrease by *four* units. Besides, since *two* protons have also been eliminated, the ensuing **atomic number** duly falls by *two* units.

NOTE	As a result, the newly formed element does move *two* places (*i.e.*, *two* groups) to the LHS of the *Periodic Table.*"

2. Since the **β-particle** is an **electron (e^-)** and also gets emitted from the **nucleus**. As there are no **electrons** present in the *nucleus*, the anticipated origin is quite uncertain. Thus, one may gainfully postulate that certainly.:

   'a Neutron gets changed into a Proton and an Electron perceptively'.

3. Consequently, the **electron** gets emitted as a **β-particle** thereby leaving a **Proton** behind. Thus, the actual number of **Protons** present duly in the *nucleus* gets enhanced by *one* unit (*i.e.*, the **atomic number** increases by *one* unit ultimately.

NOTE   In this manner, the so-called '*daughter element*' **critically moves one place (i.e., *one group*) to the RHS of the Periodic Table.**

Typical and Specific Examples to Expatiate the Group Displacement Law

   ❑ **The Radioactive Element Polonium** [$^{215}_{84}Po$], present in **Group VI A** of the **Periodic Table,** on losing particularly an α-particle [$^{4}_{2}He$]. gets duly transformed into the **radioactive lead** perceptively.

   ❑ $^{211}_{82}Pb$ belongs to **Group IV A** in the **Periodic Table.**

Observations: Based on the aforesaid *two* classical examples, one may observe particularly that the ensuing '**daughter element**' *viz.,* $^{211}_{82}Pb$ gets readily displaced upto *two* **places** to the **LHS** of the **Periodic Table,** Thus we may have the following expression:

$$\overset{VIA}{^{215}_{84}Po} \xrightarrow{-\alpha} \overset{IVA}{^{211}_{82}Pb} + ^{4}_{2}He \tag{a}$$

Again, when **radioactive** $^{211}_{82}Pb$ emits a β-particle [$^{0}_{1}e$] to yield the respective $^{211}_{83}Bi$ that eventually belongs to **Group VA of the Periodic Table** *i.e.,* located strategically **one places** to the **RHS of the parent element** *viz.,* $^{211}_{82}Pb$.

We may have the following expression:

$$\overset{VA}{^{211}_{82}Pb} \xrightarrow{-\beta} \overset{VIA}{^{211}_{83}Bi} + ^{0}_{1}e \tag{b}$$

It has been ascertained perceptrively that the **Radioactive Bismuth** $^{211}_{83}Bi$ critically sheds off a **β-particle** thereby giving rise to the formation of $^{211}_{84}Po$ that eventually belongs to **Group VIA** (in the *Periodic Table*) *i.e.,* one place to RHS of the so-called parent element $^{211}_{83}Bi$.

Thus , we may have the following expression: n:

$$\overset{IVA}{^{211}_{83}Bi} \xrightarrow{-\beta} \overset{VIA}{^{211}_{84}Po} + ^{0}_{-1}e \tag{c}$$

The various observed **transformations** in Eqs. (a), (b), and (c) may be elegantly summarized as shown under:

$$\overset{VIA}{^{215}_{84}Po} \xrightarrow{^{4}_{2}He} \overset{IVA}{^{211}_{82}Pb} \xrightarrow{^{0}_{-1}e} \overset{VA}{^{211}_{83}Bi} \xrightarrow{-^{0}_{-1}e} \overset{VIA}{^{211}_{84}Po}$$

Inferences: These essentially include:

1. Based on the above glaring expressions one may take cognizance of the underlying fact that both $^{215}_{84}\text{Po}$ and $^{211}_{84}\text{Po}$ are the respective **isotopes**; whereas, $^{211}_{82}\text{Pb}$, $^{211}_{83}\text{Bi}$, and $^{211}_{84}\text{Po}$ are termed as the 'isobars'*

2. Hence, it may be inferred that the so-called **β-particle emission** yields '*isobar*'; whereas, the observed:

"**combined emission of *one* α- and *two* β-particles results into the critical formation of an '*isotope*'**"

3.10. RADIOACTIVE DISINTEGRATION SERIES

The scientific *studies*, *revelations*, and *interpretations* do reveal the most critical and useful observations that:

"**the 'atoms of heavy elements' present duly in the Periodic Table *viz.*, Thorium (Th), Polonium (Po), and Radium (Ra) do undergo a continuous cleavage phenomenon thereby resulting into rather-*fresh radioactive atoms* having the crucial and enormous emission of α-, β-, and γ- rays originating from their nuclei in a very big way**".

Points to Ponder: These essentially include:

1. However, in the ensuing process the so-called **parent (original) atom** virtually disppears and provides to an altogether **new (*daughter*) atom**.

2. Subsequently, these newly formed **atoms** do retain **radioactivity**; and hence, undergo cleavage in their respective turn spontaneously thereby producing a relative long-chain of *various* **radioactive elements in the form of a series** unless and until it reaches an '**inactive element**' [**normally Lead (Pb)**].

3. Importantly, one may observe critically that:

"**the series of elements which are obtained duly by the so-called successive disintegration (decay) of the *newer atoms* is invariably termed as the *Radioactive Disintegration Series***".

4. Obviously, all the 'radioactive elements that essentially belong to any one of the following *three* series, namely:'
   - **Uranium Series: [(4*n* + 2) series],**
   - **Thorium Series: [4*n* series], and**
   - **Actinium Series: [(4*n* + 3) series].**

5. In general, these series have been named duly after the name of the element of at **or near the head of the corresponding series**. Thus, the aforesaid *three* **series** (*in 4 above*) are sometimes also known as the–'**Natural Radioactive Series**'; and of course, having a *stable isotope of Lead*.

6. Amazingly, not a single atom may ever go either-ways. Some **atoms** do go either ways; and hence, cause **definite/explicit branches in the series**.

    At this point in time, we may briefly discuss the following *four* **widely accepted and recognized** series, such as:

   (a) **Uranium (U) Series [(4*n* + 2) Series],**
   (b) **Thorium (Th) Series [4*n* Series]**
   (c) **Actinium (Ac) Series [(4*n* + 3) Seires], and**
   (d) **Neptunium (Np) Series [(4*n* + 1) Series]**

3.10.1. *Uranium (U) Series [(4n + 2) Series]*

Generally, the **Uranium (U) series** is also termed as the **[4n + 2] series** since the exact **Mass Number** of the specific elements belonging to these series are usually given by the *expression* **(4n + 2)**, where *n* designates an *integer the value of which gets increased by unity* as and when one moves from **one of the Radioactive elements to the next position just below it**. Interestingly, the **Mass Numbers** of the ensuing members of this series do usually provide *a remainder of 2 when divided by 4.*

Example: In the series, $^{238}_{92}U$ represents the *parent element*; and thus, *via* the inherent **successive disintegrations (decays)** it gets duly:

"**transformed into a 'stable isotope' of Lead (Pb).**"

Figure: 3.9 shows vividly the **Uranium (U) Series** or **(Un + 2) Series** for the *naturally occurring elements.*

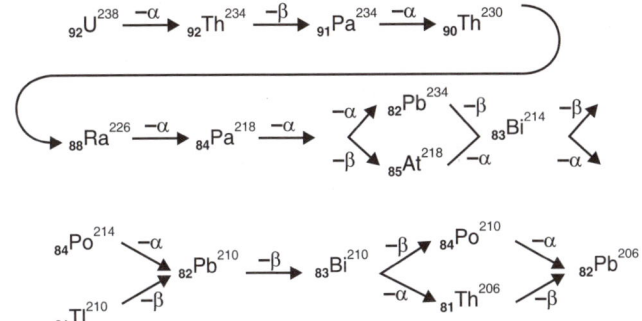

Fig.3.9: Sequence of Transformation of Uranium (U) or (4n + 2) Series for the Naturally Occurring Elements.

3.10.2. *Thorium (Th) Series [4n Series]*

It is also frequently known as **4n series** since the **Mass Numbers** of the ensuing members of this series are **divisible by 4**. Besides, this series usually commences with $^{232}_{90}$ **Th**.

Figure: 3.9 depicts explicitly a so-called **successive disintegration** or decay or **transmutation**–that eventually ends up in a **fairly stable** isotope of Lead: $^{208}_{82}$ **Pb**.

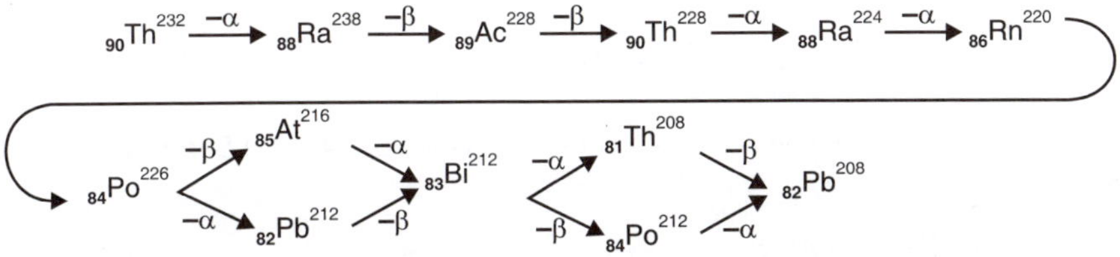

$$^{232}_{90}Th \xrightarrow{-\alpha} {}^{238}_{88}Ra \xrightarrow{-\beta} {}^{228}_{89}Ac \xrightarrow{-\beta} {}^{228}_{90}Th \xrightarrow{-\alpha} {}^{224}_{88}Ra \xrightarrow{-\alpha} {}^{220}_{86}Rn$$

Fig. 3.10: Sequence of Transformation of Thorium (Th) series or 4m series for the Naturally Occurring Elements.

3.10.3. *Actinium (AC) Series [(4n + 3) Series]*

It is also termed as the **(4n + 3) series** by virtue of the fact that the **Mass Numbers** of the variou members of this series do give a **remainder of 3'** on *division by 4*.

Remarks: At an earlier stage it was thought **Actinium (Ac)** to be the very starting element of the presen series; however, as on date it has been duly proven and established that:

"the *true starting element is* $^{235}_{92}$ U *–that eventually by successive transformations ends up in a stable*

isotope of Lead $^{207}_{82}$ Pb ."

Fig: 3.11 clearly depicts sequence of transformation of **Actinium (Ac) series or (4n + 3) series** o the so-called **naturally occurring elements**.

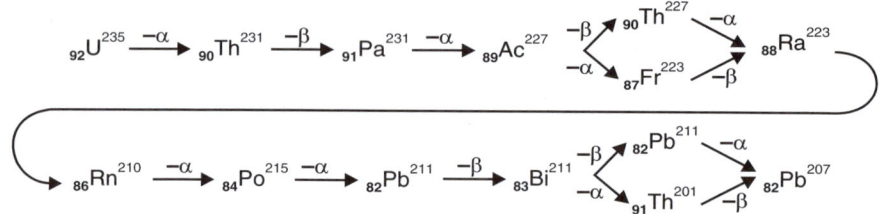

Fig. 3.11: Sequence of Transformation of Actinium (Ac) Series or [(4n + 3) series] for the Naturally Occurring Elements.

3.10.4. *Neptunium (Np) Series [(4n + 1) Series]*

The advent of meticulous research has recently introduced the *fourth* series that was duly obtained from:

"**an artificially generated** *Radioactive Material (RM)*".

so far we have already seen the very *first* element in this series to be $^{237}_{93}$ Np – after which it has been assigned a '**name**'. However, the *stable-end product* being the so-called:

"**ordinary Bismuth** $^{209}_{83}$ **Bi** rather than an *isotope* of Lead (Pb) as in the typical series: Uranium (U), Actinium (Ac), and Thorium (Th)," which has vividly illustrated in the Fig: 3.12.

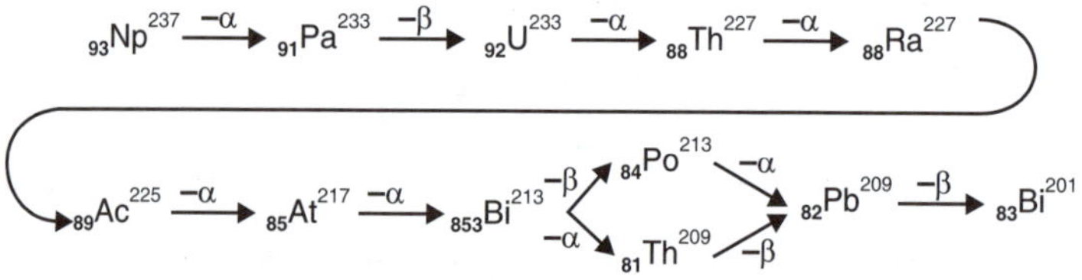

$$_{93}Np^{237} \xrightarrow{-\alpha} {}_{91}Pa^{233} \xrightarrow{-\beta} {}_{92}U^{233} \xrightarrow{-\alpha} {}_{88}Th^{227} \xrightarrow{-\alpha} {}_{88}Ra^{227}$$

$$_{89}Ac^{225} \xrightarrow{-\alpha} {}_{85}At^{217} \xrightarrow{-\alpha} {}_{853}Bi^{213} \begin{smallmatrix} -\beta \\ \\ -\alpha \end{smallmatrix} \begin{smallmatrix} {}_{84}Po^{213} \\ \\ {}_{81}Th^{209} \end{smallmatrix} \begin{smallmatrix} -\alpha \\ \\ -\beta \end{smallmatrix} {}_{82}Pb^{209} \xrightarrow{-\beta} {}_{83}Bi^{201}$$

Fig. 3.12: Sequence of Transformations of Neptunium (Np) Series or [(4n + 1) Series] for the Naturally Occurring Elements.

Remarks: These essentially comprise:

1. **The Neptunium (Np) Series** is also called as the **(4n + 1) series**. Besides, all the members belonging to this series are either:

'**unknown (so far) or extremely scarce in nature**'.

2. Amazingly, it is rather quite evident from the foregoing scientific revelations and observations that:

"**almost all the** *members of a specific series* **are duly characterized by having (4n + q) nucleous**",

where, q = Designates one of the numbers 0, 1, 2, 3, – characteristic of a given series; and

n = Represents any integer.

Obviously, **(4n + q)** provides the **Mass Number 'A'** of the respective members of the series.

Hence, we may have the *four* **distinct series** as under:

Thorium (Th) Series: $A = (4n + 0)$; $q = 0$;

Neptunium (Np) Series: $A = (4n + 1)$; $q = 1$;

Uranium (U) Series: $A = (4n + 2)$; $q = 2$; and

Actinium (Ac) Series: $A = (4n + 3)$; $q = 3$

4. RADIOACTIVITY–THE POTENTIAL CONTRIBUTOR

A splendid and meticulous researches conducted across the globe by various dedicated scientists have rightly concluded that:

"**Radioactivity is indeed the Potential Contributor**",

specifically in the following *five* **vital and important aspects**, namely:

(a) **Discovery Profile of Newer Fundamental Particles;**

(b) **Important Discovery of Isotopes, Isobars, and Isotones;**

(c) **Discovery of Newer Elements (in Periodic Table);**

(d) **Potential Release of Nuclear Energy;** and

(e) **Production of Radioisotopes,**

which shall now be treated individually in the sections that follows:

4.1. Discovery Profile of Newer Fundamental Particles

The intensive and extensive studies related to the critical **radioactivity** profile has finally led to the spectacular discovery of:

"**an array of** *Newer Fundamental Particles viz.*, **Neutrons (N), Positrons*** $\left(^{0}_{+1}e\right)$, **Deuterons** $\left(^{2}_{1}D\right)$, **and α-, β-, γ- particles.** "

In a broader perspective, these aforesaid particles exclusived as the so-called **bombarding particles** (or *projectiles*) usually in:

'**the artificial transmutation of elements; and hence, have been added onto our knowledge pertaining to the Atomic Structure episode.**'

Examples: Following are the *four* **distinct examples**, namely:

1. The phenomenon involved in the **scattering of α-particles** actually proved that the **nucleus is duly charged positive.**

2. The prevalent degree of **deflection of the α-particles** critically shows that the so-called *Atomic Number* gets numerically equivalent to the *Nuclear Charge.*

---

* **Positrons:** 'These are denoted by the **symbol β⁺, and so β⁻**, rather than β, and must be used for an **electron of nuclear origin only.**

3. The specific production of **Protons** $\left(^1_1H\right)$ due to the actual *bombardment* of the relatively **lighter** elements *viz.*, ^6_3Li by the **deutrons** $\left(^2_1D\right)$ vividly reveals the presence of **protons** $\left(^1_1H\right)$ in the *nucleus*. Thus, we may have the expression:

$$^6_3Li + ^2_1D \longrightarrow ^7_3Li + ^1_1H$$

4. Another glaring example may be cited by the typical bombardment of $^{24}_{12}Mg$ and ^4_2He to yield $^{27}_{14}Si$ –that proves to be certainly **radioactive** in nature; and, therefore, undergoes critical **disintegration (decay)** almost spontaneously to result into the formation of the so-called **stable isotopes**, such as: $^{27}_{13}Al$ and Positions $\left(^0_{+1}e\right)$.

Hence, we may ultimately have the following *two* **expressions**:

$$^{24}_{12}Mg + ^4_2He \longrightarrow ^{27}_{14}Si^+ + ^1_0n$$

$$^{27}_{14}Si \longrightarrow ^{27}_{17}Al^{++} + ^0_{+1}e \text{ Position}$$

$^+$**Radioactive in nature**

$^{++}$**Non-Radioactive Isotope (*i.e.*, Stable entity)**

Remarks: The particular generation of the **Positron** $\left(^0_{+1}e\right)$ in the course of the **disintegration (decay)** phenomenon of $^{27}_{14}Si$ explicitly indicates the presence of a positron $\left(^0_{+1}e\right)$.

4.2. Important Discovery of Isotopes, Isobars, and Isotones

The elaborated existence of **Isotopes, Isobars, and Isotones** was virtually **first ever recognized** in the course of the comprehensive study of the *Radioactve Series.*

4.3. Transmutation of Elements

It is, however, important to state here that the **critical investigations** and **commendable revelations** duly accomplished in the vast field of *Artificial Radioactivity*–which definitely paved the way in: **"the vital and important discovery of the so-called *Transuranium Elements* (*i.e.*, such elements that do possess '*Atomic Numbers*' more than 92); besides, they fail to exist in nature freely."** Interestingly, a good number of the **radioisotopes** with the *known elements* have been discovered elegantly by the so-called *Transmutation of Elements (or Artificial Transmutation of Elements)**.

Examples: Following are *two* **typical examples of transmutations in elements**, namely:

❑ **Transmutation by Deuterous** $\left(^2_1D\right)$

$$^6_3Li \quad + ^2_1H + \longrightarrow \quad ^7_3Li \quad + ^2_1H$$

Isotope of	**Isotope of**	**Proton**
Lithium	**Lithium**	
[Mass No. 6]	**[Mass No. 7]**	

---

* That is, the conversion of one element into another by artificial means.

❑ **Transmutation by Neutrons** $\left({}_{0}^{1}n\right)$

$$\overset{14}{\underset{7}{}} N + {}_{0}^{1}n \longrightarrow {}_{5}^{11}B + {}_{2}^{4}He$$

$$\overset{27}{\underset{13}{}} Al + {}_{0}^{1}n \longrightarrow {}_{12}^{27}Mg + {}_{1}^{1}H$$

$$\overset{23}{\underset{11}{}} Na + {}_{0}^{1}n \longrightarrow {}_{11}^{24}Na + \gamma\text{-Rays}$$

4.4. Potential Release of Nuclear Energy

The various known processes that are intimately involved in these cardinal aspects, as:

• **Nuclear Fission**, and   • **Nuclear Fusion**,

in fact, do release an **enormous quantum of energy** that may be used both judiciously and effectively in *Peace and War*.

4.5. Production of Radioisotopes

In a broader aspect, the so-called '**Artificial Radioactivity**' has legitimately gained a very impressive ground in their gainful utilization in:

"**the production of Radioisotopes of several elements that are used for various purposes**".

Examples: Following are the *four* glaring examples pertaining to the **production of radioisotopes**, for instance:

1. The intelligent usage of the '**Radioisotopes**' as *Radioactive Traces.*
2. The miraculous use of the '**Radioisotopes**' in the *relevant treatment of diseases in human beings, viz.,*
    • **post-operative Radiotherapy for the cancer patients**; and
    • **Radiotherapy with *Isotopes of Iodine* in the Goiter patients**.
3. Precisely determining the *actual age of the Earth* by the aid of **Rock-Dating Methodology**.
4. The critical and specific determination of the '*age of recent objects*' by means of the **Radiocarbon Dating Method**.

5. ISOTOPES, ISOBARS, ISOTONES, AND CYCLOTRON

The above mentioned '**terminologies**' shall now be expatiated with their appropriate examples in the sections that follows:

5.1. Isotopes

Preamble: A particular '**kind of atom**' (*i.e.*, a specific *nuclide*) is invariably characterized by its **mass Number (A)** and the **Atomic Number (Z)** or the respective **name of the element concerned** (*sometimes even by both*).

Example: **Carbon-12** is the name given to the *atoms of Carbon* ($Z = 6$) that have ($A = 12$) *i.e.*, the **Mass Number (Atomic Weight)**; and have, their nuclei do consist of *six* **Protons** and *six* **Neutrons**. Besides, the **Carbon-13 atoms** [${}^{13}C$] also possess $Z = 6$, but their *nuclei*, for which $A = 13$, do comprise *six* **Protons** and *seven* **Neutrons**.

Definitions: The **isotope** may be defined as:

"**Nuclides of the same element of different mass number (A) of that element**".

The *three* isotopes of C may be symbolized as given under truly:

- $^{12}_{6}C$ • $^{13}_{6}C$ • $^{14}_{6}C$.

since all the above *three* **species** having **Z = 6 are the C-atoms**–hence, the *subscript value of Z is* omitted quite often in actual practice.

The nomenclature '**Isotope**' was duly introduced by **Soddy**. The word **isotope** has been duly derived from the **Greek word**: meaning the *same position* (*isos* = same; *topes* = positions–since the **isotopes do** occupy the **same position** in the '**Periodic Table**'.

> **NOTE** In general, the '*isotopes*' usually possess similar chemical characteristic features and differ slightly with regard to such properties those depend exclusively upon their Mass Number.

The Isotopic Mass (m): in reality, the **isotopic mass (m)** of a '*nuclide*' is being expressed invariably in terms of the corresponding:

'**atomic mass unit, m_u**',

It is, therefore, may be defined as –'**one twelfth of the mass of an atom**' of carbon–12; and possesses the value **1.66057 × 10^{-27} kg**.

Points to Ponder

1. The most preferred choice of **Carbon-12** as the perspective '**standard**' is certainly based upon the fact that:

   "**all accurate and precise values of the *Isotopic Masses*–are now being determined by *Mass Spectrometry (MS)***".

2. The enormous quantum of the available number of the so-called **volatile C-compounds** belonging to divergent **Molecular Mass (m)** virtually facilitates the prevailing comparisons of **isotopes of unknown mass and C-containing species of a** *known mass*.

3. Thus, the ensuing **isotopic masses of the proton** $\left(^{1}_{1}H\right)$, **neutron** $\left(^{1}_{0}n\right)$, **and electron** $\left(^{0}_{1}e\right)$ – are given together along with the respective charges of these particulate matters (particles), –as recorded in Table: 3.1.

S.no.	Particle	Symbol	m/mo	m.kg^{-1}	Charge/C
1	**Proton**	$1p$	1.007277	1.67265 × 10^{-27}	+ 1.602 × 10^{-19}
2	**Neutron**	$1n$	1.008665	1.67495 × 10^{-27}	0
3	**Electron**	e	0.000548	9.1095 × 10^{-31}	−1.602 × 10^{-19}

Table: 3.1: The Properties of Proton $\left(^{1}_{1}H\right)$, Neutron $\left(^{1}_{0}n\right)$, and Electron $\left(^{0}_{1}e\right)$

4. Hence, for an element which essentially comprises an **admixture of the isotopes**, the observed relative *abundances of the isotopes* (may also be **determined by MS**) should be known as so to enable the **calculation of**:

   "**relative atomic mass, A, (or, as it is often loosely called the Atomic Weight) of the respective element**".

Example: The naturally occurring **chlorine (Cl)**, does contain **75.53%** ^{35}Cl and **24.47% of** ^{37}Cl; and hence, the precise **Isotopic Masses (*m*)** of which are **34.969** and **16.966 *m*$_u$** respectively.

Therefore, its *relative* atomic mass may be given by:

$$Ar = \frac{(34.969 \, m_u \times 75.53 + 36.966 \, m_u \times 24.47)}{100 \, m_u}$$

or   **Ar = 35.458**

5.1.1. *The Symbolic Representation of Isotopes*

In global practice, it is an accepted and recognized conviction to:

"**represent each '*Isotope*' of a given element by the help of a suitable symbol that is normally being adopted across the globe**".

In doing so, the ensuing '*isotope*' of a given element, the so-called **Mass Number (A)** is invariably depicted as the *superscript* at the head of the symbol of the specific element; whereas, the **Atomic Number (Z)** is shown as the *subscript* at its lower end.

Hence, both the '**superscripts**' as well as the '**subscripts**' may be shown actually at the: **Left or Right of the Symbol of Element.**

Superscript: It designates the **Mass Number (A)** and provides the sum of the number of **Protons and Neutrons duly present in the nucleus of an atom.**

Subscript: It represents the **Atomic Number (Z)** which gives the **number of Protons** or **Electrons**.

> **NOTE**   The Symbolic Representation of an 'Isotope' of an element is usually given by 'X'.

Fig: 3.12: illustrates the symbolic representation of an 'isotope of an element X'.

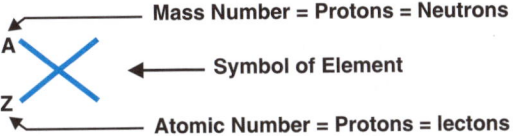

Fig. 3.12: The Diagrammatic Symbolic Representation of an Isotope of an Element 'X'.

Example: The classical example of *three* isotopes of Carbon (**Atomic Number = 6**) and having **Mass Number (A)** equivalent to: 12, 13, and 14 may be vividly depicted as given under:

- $^{12}_{6}$C or ^{12}C   or $^{13}_{6}$C or ^{13}C and   • $^{14}_{6}$C or ^{14}C.

In another way, the aforesaid *three* isotopes of **Carbon** may also be shown as: **C–12, C–13**, and **C–14,–** that usually may be read as:

- **Carbon Thirteen** and   • **Carbon Fourteen** – respectively.

5.1.2. *Typical Structure of Isotopes*

The typical structure of isotopes of an '*element*' may be duly expatiated based on the underlying fact that:

"**since the respective *isotopes* of an '*Element*' do possess the same *Atomic Number (Z)*, each of these isotopes does consist of equal numbers of protons present in its nucleus; and hence, a equal number of Electrons duly revolving in different orbits around the nucleus perceptively**".

Importantly, since they do critically inherit altogether different **Mass Number (A)**, hence they invariably possess different number of **Neutrons** in their corresponding **nuclei**. Amazingly, the precise number of **Protons (p)**, **Neutrons (n)**, and **Electrons (e)** present in an **Isotope**, with **Atomic Number (Z)** and **Mass Number (A)**.

We may have the following expression:

$$\frac{A}{Z} \text{ X Isotope}$$

that is given by the following relationships:

(i) **Number of Protons:** $p = Z$,

(ii) **Number of Neutrons:** $n = A Z$ [since, $A = n + p = n + Z$]

(iii) **Number of Electrons:** $e = Z$

Relevant Structures of Isotopes of Certain Elements

At this point in time, it would be worthwhile to take into consideration the so-called:

"**the structure of the 'Isotopes' of certain elements pertaining to their number of Electrons (e), Protons (p), and Neutrons (n) present in them.**"

(a) **The Isotopes of Hydrogen (H)**–In actual practice, one may duly encount *three* **divergent isotopes of Hydrogen (H)**, namely:

❑   **Ordinary Hydrogen or Proton or Protium** $\left[{}^{1}_{1}H\right]$,

❑   **Heavy Hydrogen or Deuterium** $\left({}^{2}_{1}H \text{ or } D\right)$, and

❑   **Tritium** $\left[{}^{3}_{1}H \text{ or } T\right]$.

Interestingly, each of these **isotopes** critically possesses **one electron only**. Nevertheless, the so-called **nucleus of Protium** $\left[{}^{1}_{1}H\right]$ **possesses one proton only, Deuterium has one proton + one neutron**, and **Tritium** has one **proton + two neutrons** respectively.

Obviously, the number of **electrons**, **protons**, and **neutrons** in each of *three* **classical isotopes** are as stated under:

S.no.	Isotope	Atomic Number (Z)	Mass Number (A)	Electrons (e) = Z	Protons (p) = Z	Neutrons (n) = (A − Z)
1	${}^{1}_{1}H$	1	1	1	1	$1 - 1 = 0$
2	${}^{1}_{1}H$ or D	1	2	1	1	$2 - 1 = 1$
3	${}^{3}_{1}H$ or T	1	3	1	1	$3 - 1 = 2$

The explicit structure of the *three* **isotopes of hydrogen** with specific reference of the ensuing numbers of: *Electrons*, *Protons*, and *Neutrons* may be depicted diagrammatically in Fig: 3.13.

(b) **The Isotopes of Oxygen (O)**–It has been duly proven and ascertained that **Oxygen (O)** essentially possesses *three* **isotopes** having **Atomic Number (Z = 8)**, and **Mass Number (A = 16, 17, or 18)**.

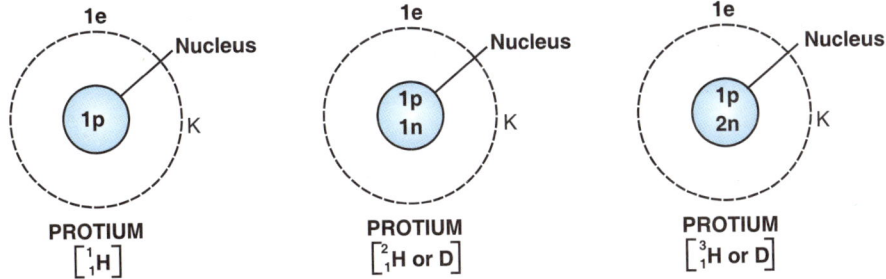

PROTIUM
$[{}^{1}_{1}H]$

PROTIUM
$[{}^{2}_{1}H \text{ or } D]$

PROTIUM
$[{}^{3}_{1}H \text{ or } D]$

Fig. 3.13: Diagrammatic Illustration of Three Isotopes of Hydrogen in Term of the Number of Electrons, Protons, and Neutrons.

Hence, the said *three* **isotopes of Oxygen** may be expressed as under:

- ${}^{16}_{8}O$, ${}^{17}_{8}O$, and ${}^{18}_{8}O$ respectively.

Besides, inherent precise and exact number of: **Electrons, Protons, and Neutrons**–*i.e.*, each of then is as given below:

S.No.	Isotope	Atomic Number (Z)	Mass Number (A)	Electrons (e) = Z	Protons (p) = Z	Neutrons (n) = (A – Z)
1	${}^{16}_{8}O$	8	16	8(2, 6)	8	16 – 8 = **8**
2	${}^{17}_{8}O$	8	17	8 (2, 6)	8	17 – 8 = **9**
3	${}^{18}_{8}O$	8	18	8 (2, 6)	8	18 – 8 = **10**

Thus, we may explicitly depict the prevailing structures of the *three* **aforesaid isotopes of Oxygen (O),**–as given in Fig: 3.14.

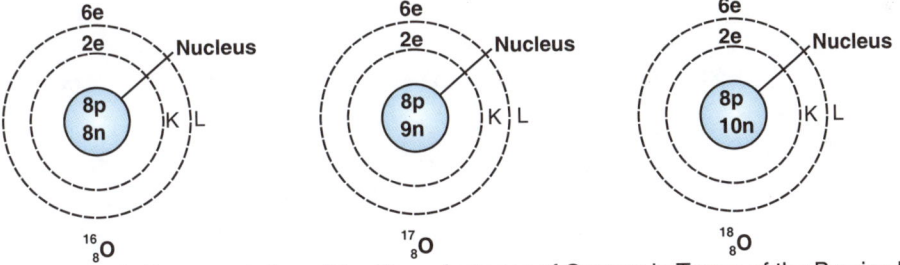

${}^{16}_{8}O$ ${}^{17}_{8}O$ ${}^{18}_{8}O$

Fig. 3.14: Diagrammatic Representation of the Three Isotopes of Oxygen in Terms of the Precise Number of: Electrons, Protons, and Neutrons.

(c) **The Isotopes of Neon (Ne)**–Neon (Ne) possesses *three* **Isotopes** having the Atomic Number (**Z = 10**), and the **Mass Number** equivalent to 20, 21, and 22. In fact, these are elegantly represented as:

- ${}^{20}_{10}Ne$ • ${}^{21}_{10}Ne$ and • ${}^{22}_{10}Ne$ respectively.

However, the exact and precise number of: **Electrons (e), Protons (p), and Neutrons (n)**–each of them is stated under:

The complete and detailed structure of all the *three* **aforesaid isotopes** may be adequately depicted in Fig: 3.15.

S.No.	Isotope	Atomic Number (Z)	Mass Number (A)	Electrons $(e) = Z$	Protons $(p) = Z$	Neutrons $(n) = (A - Z)$
1	$^{16}_{8}Ne$	10	20	10(2, 8)	10	$20 - 10 = 10$
2	$^{21}_{10}Ne$	10	21	10(2, 8)	10	$21 - 10 = 11$
3	$^{22}_{10}Ne$	10	22	10(2, 8)	10	$22 - 10 = 12$

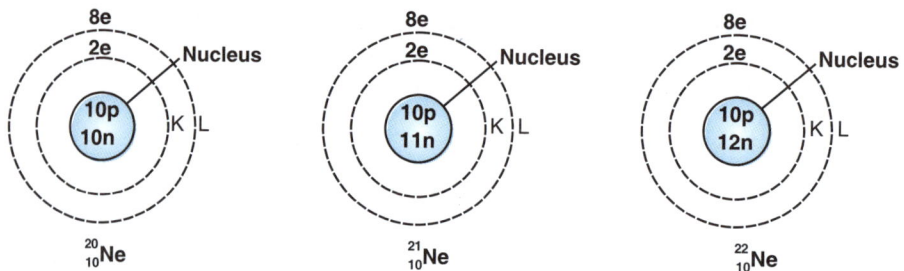

Fig. 3.15: The Elaborated Structure of the Three Isotopes of Neon (Ne) in Terms of the Precise Number of: Electrons, Protons, and Neutrons.

(d) **The Isotopes of Chlorine (Cl)**–Importantly **Chlorine (Cl)** happens to be an admixture of *two* **isotopes** having: **Atomic Number (Z = 17)**; and **Mass Number (A)** equivalent to **35 and 37**. Therefore, these *isotopes* may be duly represented as:

- $^{35}_{17}Cl$ and $^{35}_{17}Cl$ respectively.

Hence, the exact number of: **Electrons, Protons,** and **Neutrons** present in each of these the **isotopes** is as given under:

S. No.	Isotope	Atomic Number (Z)	Mass Number (A)	Electrons $(e) = Z$	Protons $(p) = Z$	Neutrons $(n) = (A - Z)$
1	$^{35}_{17}Cl$	17	35	17(2, 8, 7)	17	$35 - 17 = 18$
2	$^{35}_{17}Cl$	17	37	17(2, 8, 7)	21	$37 - 17 = 20$

However, we may explicitly illustrate the so-called complete structures of these *two* **aforesaid isotopes** is as shown in Fig: 3.16.

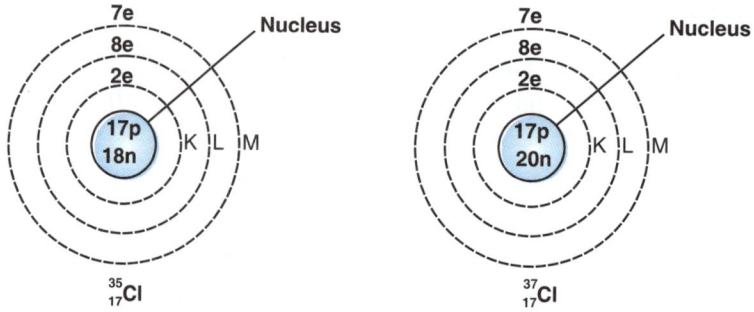

Fig. 3.16: The Explicit Structure of the Two Distinct Isotopes of Chlorine in Terms of the Precise Number: Electrons, Protons, and Neutrons.

5.1.3. *The Separation of Isotopes*

Since we all know that practically '**all the isotopes**' of a given element (from the ***Periodic Table***) do exhibit **absolutely identical characteristic features**, it becomes rather an extremely difficult and herculian task to:

'**separate them by means of the chemical methods of separation**'.

Therefore, the alternative feasible and possible methods for handling such a ***crucial task*** remains to depend on certain so-called **Physical Methods of Separation of the Isotopes**. Amazingly, in these proposed methodologies the methods are solely based upon:

'**the actual prevailing differences in the physical characteristics of the *Isotopes*–that exclusively rest on the *Mass of the Atom*' (being utilized perceptively).**

For this purpose, there are ***four*** **widely used and accepted methods** recognized for the '**Separation of Isotopes**', such as:

(a) **Fractional Distillation and Evaporation Method,**
(b) **Electromagnetic Method,**
(c) **Thermal Diffusion Method,** and
(d) **Gaseous Diffusion Method,**

which shall now be discussed individually in the sections that follows:

1. Fractional Distillation and Evaporation Method

Based on the **Graham's Law of Diffusion***–the observed **rate (r) of evaporation of a gas** is found to be:

'**inversely proportional to the square root of the density (d)**'.

$$ga\sqrt{\frac{1}{d}}$$

Observations: Thse essentially include:

1. Hence, the so-called **lighter isotope distills off** first and foremost, whereas, the **heavier isotope usually distills off at the end**.

Example: The ***liquid* oxygen (O_2)** on being subjected to distillation are may critically take cognizance of the fact that:

"**the lighter isotope (^{16}O) emerges first of all; whereas, the heavier isotope (^{18}O) escapes at the fag end**".

 NOTE   The *Isotpes of Neon* are being separated by the help of **Fractional Distillation and Evaporation Method.**

Classical Example of 2-Isotopes of Mercury (Hg)–The effective separation of the *two* **isotopes of Mercury (Hg) has been successfully accomplished by the fractional evaporation between 40 to 50°C under vacuum,**–as depicted in Fig: 3.17.

---

* **Graham's Law of Diffusion:** At constant conditions of temperature and pressure, the **rate of diffusion of a gas** is *inversely proportional* to the **square root of the density of the gas**.

Modus Operandi

(i) The emanated vapours of **Mercury (Hg)** are duly condensed upon the surface duly cooled in liquid air–that is being held a little above the **evaporating Hg**.

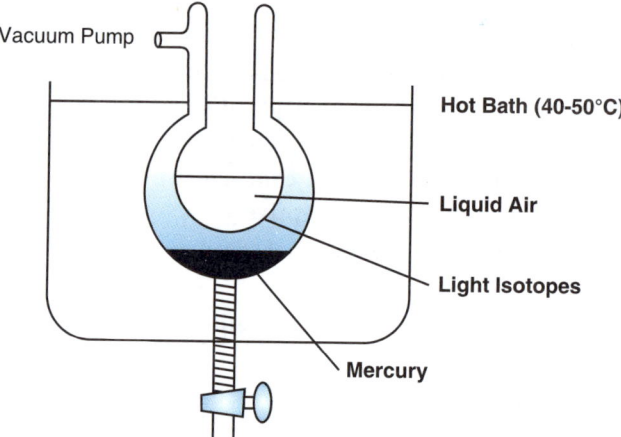

Fig. 3.17: Diagrammatic Separation of the Two Isotpes of Mercury (Hg) by Fractional Evaporation.

(ii) After a short-duration–the **Liquid Mercury (Hg) left behind** is critically observed to comprise the *Heavier Isotopes*–that may be collected separately by opening the stop-cock provided at the bottom.

(iii) However, the so-called **Frozen Mercury (Hg)** being collected at the **cooler surface** predominantly consists of the *Lighter Hg-Isotopes*.

(iv) Thus, the entire aforesaid process is normally repeated a good number of times.

2. Electromagnetic Method

In true sense, the **Electro magnetic Method** is solely based upon the underlying principle of:

"Dempster's Mass Spectrograph".

The aforesaid method has been used successfully for the gainful and effective separation of the *three* **distinct isotopes of Neon (Ne)**.

Fig: 3.18 shows the beam of ions produced by the following *three* **isotopes**, namely:

- ^{20}Ne • ^{21}Ne and • ^{22}Ne,

duly obtained by the *Dempeter's Mass Spectrograph* is actually made to travel between the *two* **poles of a magnet**.

Remarks: It may be observed that the *three* **divergent isotopes** duly get deflected at various angles, which are subsequently collected in *pre-colled chambers*; and thus placed meticulously in **different positions** perceptively.

| NOTE | Amazingly, the ultimate separation accomplished by the *Electromagnetic Method* is almost complete; however, the actual quantum of divergent isotopes duly obtained are too scanty. |

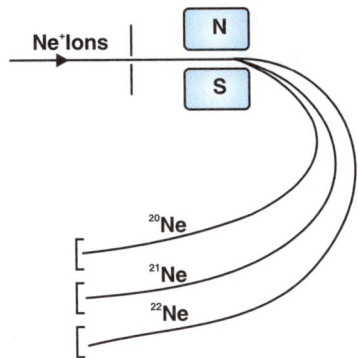

Ne⁺Ions

²⁰Ne
²¹Ne
²²Ne

Fig.3.18: Explicit Separation of the Three Isotopes of Neon (Ne): ^{20}Ne, ^{21}Ne, and ^{22}Ne by the Aid of Electromagnetic Method by using Dempeter's Mass Spectrograph.

3. Thermal Diffusion Method

The apparatus being employed for the **Thermal Diffusion Method** essentially comprises of: **"an elongated vertical cylinderical tube that is heated electrically upto 600°C or even above from within by the help of a *thin Pt-wise* almost passing *via* the axis of the tube (as depicted in Fig: 3.19)".**

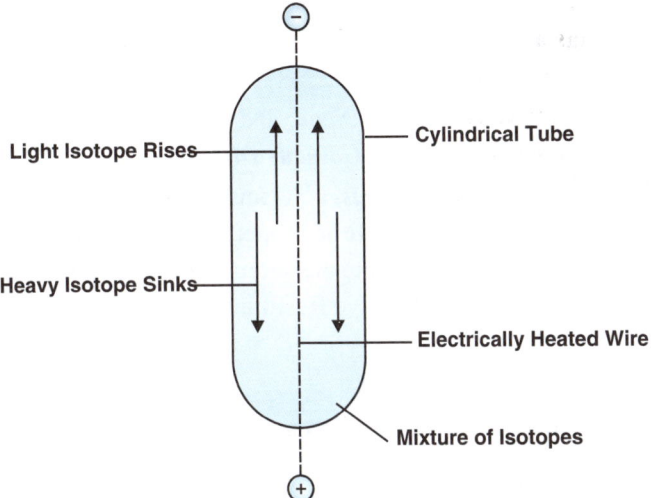

Light Isotope Rises — Cylindrical Tube

Heavy Isotope Sinks —

Electrically Heated Wire

Mixture of Isotopes

Fig. 3.19: The Illustration showing the Clear Separation of Isotopes by the Thermal Diffusion Method.

Remarks: These essentially include:

1. The above method is solely based upon the tendency of rather **specific heavier molecules of a gas concentrate** located critically in the cooler region of the **vertical tube.**

2. In this way, as and when an **admixture of the gaseous isotopes** is being introduced into the tube, the corresponding molecules of the rselectively **lighter isotope** get diffused more swiftly to the so-called:
   'hot central region of the vertical tube'

3. Interestingly, in this critical zone they are being carried upwards by the help of ensuing *convection currents of hot air.*

4. Thus, the molecules of the **Heavier Isotope** do travel to the respective **cooler inner surface** of the tube; and subsequently **sink to the bottom**.

5. Ultimately, the **lighter isotopes** gets duly collected at the top; whereas, the **heavier isotopes** collects at the bottom end.

6. The entire sequential process is repeated several times in routine studies as well.

 NOTE   The *Thermal Diffusion Method* **has been employed successfully by clausius for the respective critical separation of** ^{35}Cl **and** ^{37}Cl **isotopes of chlorine.**

4. Gaseous Diffusion Method

As per the **Grahm's Law of Diffusion** (see *section-1*), the *rate of diffusion* (**g**) of a gas is observed to be **inversely proportional to the squre root of the molecular weight (*m*)**.

Thus, we may have the following expression:

$$g = \sqrt{\frac{1}{M}}$$

Remarks: Hence, in a specific instance when an admixture comprising *two* **gaseous isotopes**,–is permitted to undergo '*diffusion*' *via* a **porous partition**, the called:

"**the Lighter Isotope has a tendency to soil across the porous partition rather more radily vis-a-vis** the *Heavier Isotope* **obviously.**"

Besides the gaseous ^{20}Ne and ^{22}Ne *isotopes of Neon*, and gaseous ^{16}O and ^{18}O *isotopes of Oxygen*– have been *Gaseous* efficiently separated by the *Gaseous Diffusion Method*.

Fig: 3.20 vividly depicts the *separation of gases* ^{16}O and ^{18}O isotopes Amazingly, the **admixture of gaseous isotopes** is made to pass *via* a **porous tube** properly sealed in an **outer jacket**. Consequently, the **Lighter Isotope** (*viz.*, ^{16}O -isotope) passes into the **outer jacket** easily; whereas, the **Heavier Isotope** (*viz.*, (^{18}O-isotope) passes right into the respective **Residual Gas**.

5.2. ISOBARS

The **nuclides** having the **identical Mass Number (A) but different Atomic Number (Z)**, for instance:

- $^{238}_{92}$Th   • $^{234}_{91}$Pa and   • $^{234}_{92}$U,

are invariably termed as–'**Isobars**'.

Since, the **isobars** do have the *same* **mass number (A)**, they may also be defined as–'atoms of different elements that have also the same sum of *Protons* and *Neutrons*' in the nucleus of each of these atoms but altogether *different* **atomic number (Z)**.

Besides, the term '**isobar**' has duly been derived from the *Greek* word, means **equally heavy** [*isos*=**equal**; *barys* = **heavy**].

Examples: $^{40}_{19}$K and $^{40}_{30}$Ca represent *two* classical examples of **isobars**, since each of them does possess the *same* **Mass Number (A)** *i.e.*, the sum of **Protons** and **Neutrons** remains the same.

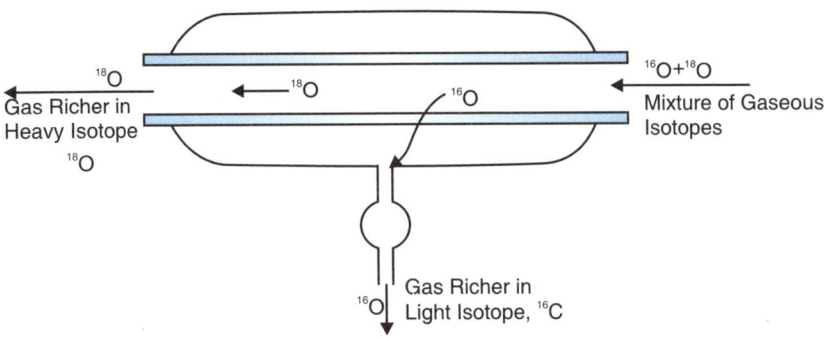

Fig. 3.20: The Diagrammatic Representation for the Separation of Gaseous ^{11}O and ^{18}O Isotopes of Oxygen by Gaseous Diffusion Method.

5.2.1. *The Structure of Isobars*

Because the **isobars** do have specifically the *same* **mass number (A)** i.e., the aggregate of **Protons and Neutrons** present in the nucleus of each of these **isobars** is more of less equal. Explicitly, the precise and exact number of **Protons (p)** and the number of **Electrons (e)** is virtually equal to the **Atomic Number (Z)**; whereas, the ensuing number of **Neutrons (n)** is found to be equal to **(A – Z)**.

Thus, we may have the following *two* expressions:

$$p = Z = e \tag{a}$$
$$n = (A - Z) \tag{b}$$

Based on the above *two* **Eqs. (a) and (b)**, the precise **number of Electrons Protons**, Neutrons duly present in each of the *three* isobars, such as:

- $^{40}_{18}\text{Ar}$

- $^{40}_{19}\text{K}$ and

- $^{40}_{20}\text{Ca,}$

may be duly shown as under:

S.no.	Isotope	Atomic Number (Z)	Mass Number (A)	Electrons (e) = Z	Protons (p) = Z	Neutrons (n) = (A – Z)
1	$^{40}_{18}\text{Ar}$	18	40	18	18	40 – 18 = **22**
2	$^{40}_{19}\text{K}$	19	40	19	19	40 – 19 = **21**
3	$^{40}_{20}\text{Ca}$	20	40	20	20	40 – 20 = **20**

Importantly, we may depict the structures of these '**isobars**' diagrammatically as given in Fig: 3.21.

NOTE	Obviously, in Fig: 3.21 are may observe that the isobars do have the same sum of the Protons and Neutrons in the Nucleus.

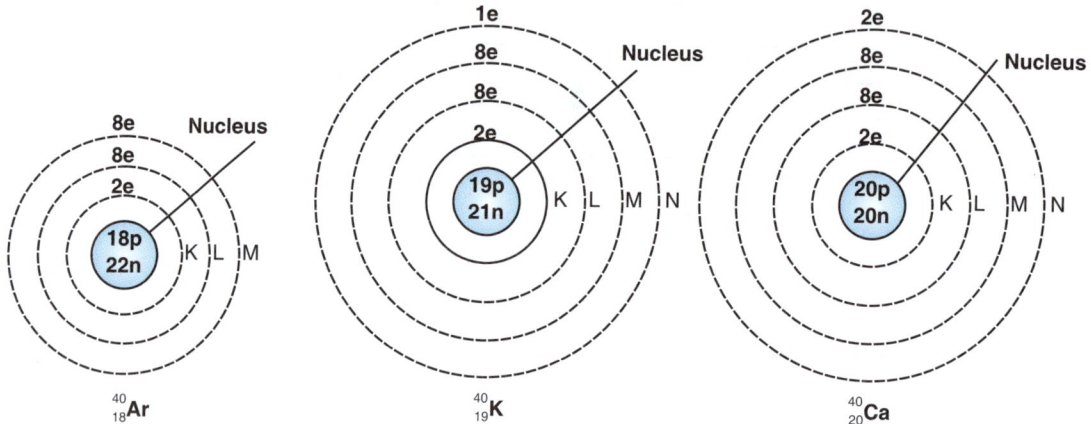

Fig. 3.21. The Structures of the Three Isobars: $^{40}_{18}Ar$, $^{40}_{19}K$, and $^{40}_{20}Ca$ in Terms of the Exact and Precise Number of the Inherent. Electrons, Protons, and Neutrons.

5.3. ISOTONES

The **isotones** may be defined as—"**the atoms of different elements that essentially do possess the *identical number of Neutrons* and *different atomic number (Z)*.**"

As the **isotones** do critically possess altogether divergent atomic number (Z), they may also be defined alternatively as:

"**the atoms of different elements that essentially inherit the same number of neutrons (n) but different number of *protons (p) and electrons (e)*.**"

Obviously, $^{14}_{6}C$, $^{15}_{7}N$, and $^{16}_{8}O$ are the 'isotones', since each of then does possess the identical number of **Neutrons (n)** as stated under:

S. No.	Isotope	Atomic Number (Z)	Mass Number (A)	Electrons $(e) = Z$	Protons $(p) = Z$	Neutrons $(n) = (A - Z)$
1	$^{14}_{6}C$	6	14	6	6	$14 - 6 = \mathbf{8}$
2	$^{15}_{7}N$	7	15	7	7	$15 - 7 = \mathbf{8}$
3	$^{16}_{8}O$	8	16	8	8	$16 - 8 = \mathbf{8}$

The explicit structures of the aforesaid *three* **isotones** may be illustrated as shown in Fig: 3.22.

5.4. CYCLOTRON

It has been established beyond any reasonable doubt that the **α-particles**, protons, and deutrons may be rendered much more effective and productive, in case, these are:

"**imparted with high velocity perceptively**".

In the recent past, quite a few specifically designed **sophisticated instruments** have been *designed, developed,* and used for:

'**accelerating critically these particles effectively**'.

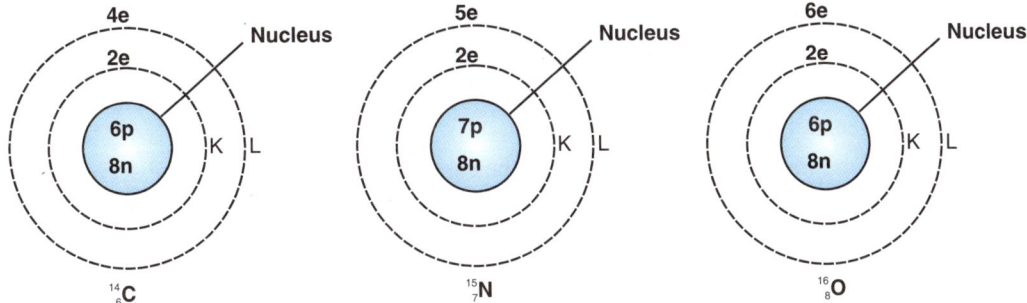

Fig. 3.22: The Structures of Three Isotones: $^{14}_{6}C$, $^{15}_{7}N$, and $^{16}_{8}O$ in Terms of Exact Number of Protons (*p*) and Neutrons (*n*).

Laurrence successfully developed an apparatus (device) that proved to be most effective for the above stated purpose and utility–which is called as–**Cyclotron.**

Fig: 3.23(a) shows the diagrammatic sketch of its commonest form that essentially comprises *two* **flat semicircular boxes**–known as the **'dees'***, duly labeled as **D₁ and D₂.**

The said *two dees* are adequately surrounded by a so-called **'vessel'**, containing: **Hydrogen or Helium** or **Hydrogen or Deuterium** in the form of **'gas' at a low pressure.** The said **'vessel'** is placed strategically between the *two* **poles of an 'Electromagnet'**– as depicted in Fig 3.23(b).

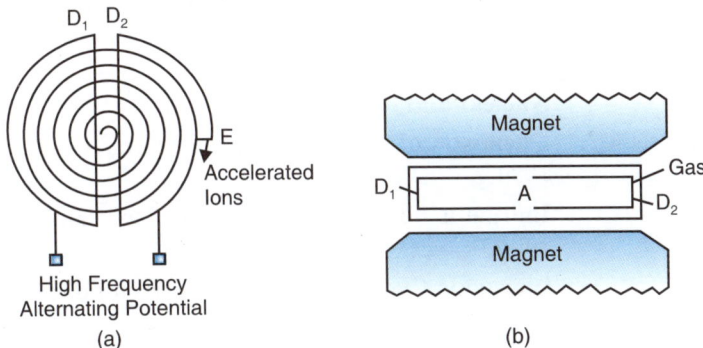

Fig. 3.23: (a) and (b): The Diagrammatic Sketch of Cylotron with Two 'Dees'.

Modus Operandi.

1. A **high-frequency alternating potential** measuring upto *several million cycles. sec⁻¹* is being duly applied **across D₁ and D₂.**

2. The production of the *Positive Ions* do commence at 'A' by the crucial bombardment of **Hydrogen, Helium or Deuterium by the electrons.**

3. At this point in time, let us consider that: *'applied potential'* is such that *D₁-is + ve, and D₂-is –ve;* then a +ve ion starting from 'A' would eventually move into 'D₂'–thereby taking a definite *semicircular path* perhaps on account of the certain deflection caused by the *Magnetic Field?*.

4. Subsequently, as and when the *ion gains entry* into the gap between **D₁ and D₂**, –it gets exposed promptly to the prevailing influence of the so-called **'Applied Potential'.**

* Since, they have a **D-like shape.**

5. Now, if the ensuing **Oscillation Frequency** is of scuh a magnitude that during the time the **+ve ion** sails through the **dee D_2**, –the observed sign of the **potential is reversed accordingly** *i.e.,*

"**D_1 becomes Negative; and D_2 becomes Positive**",

and hence, the ions gets duly accelerated towards **D_1**.

6. Now, the ion moves much more faster *vis-a-vis* the earlier stage; and hence, the observed **radius of the path increases** proportionately. Nevertheless, the actual time consumed by the said ion to travel the **dee** almost remains the same, since the **enhanced path** gets conveniently compensated by the **increased motion of the ion**.

7. In other words,–'if the oscillation frequency is properly adjusted, the ensuing ion should always remain perfectly in phase with the alternating potential prevailing between the *two* dees':

NOTE	Ultimately, the ion having an 'energy profile' as high as 'several hundred million electron volts and velocity as high as 40,000 km sec^{-1}',–gets duly emerged from the exit 'E' of the instrument perceptively.

SUGGESTED READING REFERENCES

Banwell CN : **Fundamentals of Molecular Spectroscopy**, 3rd. ed., McGraw Hill, New York, 1983.

Choppin GR and Rudberg J : **Nuclear Chemistry**, Pergamon Press Oxford, London (UK), 1980.

Edsworth EAV *et al.* : **Structural Methods in Inorganic Chemistry**, Blackwell, Oxford, London (UK), 1987.

Friedlander G *et al.* : **Nuclear and Radiochemistry**, 3rd ed., Wiley, New York, 1981.

Greenwood NN and Earnshow A : **Chemistry of the Elements**, Pergamon Press, Oxford, London (UK), 1984.

Smith D : **Inorganic Substances**, Cambridge University Press, London (UK), 1990.

Sharpe AG : **Inorganic Chemistry**, 3rd ed., Pearson Education Ltd., New Delhi, 1992.

PROBLEMS WITH SOLUTIONS

[A] Measurement of Radioactivity

Q.1. **One gramme of Radium (Ra) gets duly reduced by 2.35 mg over 5 years by a phenomenon of Radioactive Decay. How would you calculate the Half-life Period of Radium?**

Solution: Initial mass of **Radium (Ra)** (No) = 1 g

Mass of Radium (Ra) left at the end (N) = 1 – 0.00235 = 0.99765

Now, the Rate of Disintegration (Decay) is given by the expression:

$$N = N_0 e^{-\lambda t}$$

or

$$\frac{N}{No} = e^{-\lambda t}$$

Hence, we have:
$$\frac{N}{No} = \frac{0.99765}{1}$$

and
$$t = 6 \text{ years}$$

Therefore,
$$0.99765 = e^{-6\lambda}$$

or
$$e^{6\lambda} = \frac{1}{0.99765}$$

or
$$6\lambda = 2.3026 \log_{10}\left[\frac{1}{0.99765}\right]$$

∴
$$\lambda = \frac{2.3026 \times 0.0009}{6}$$

or
$$\lambda = 34.539 \times 10^{-6} \text{ per year}$$

$$T = \frac{0.693}{\lambda} = \frac{0.693}{34.539 \times 10^{-6}} = \textbf{2006 years}$$

Q.2. A Radioactive Material (RM) has a half-life period of 35 days. How would you calculate the following:

    (a) **Radioactive Disintegration Constant (λ),**

    (b) **Average Life Period (l),**

    (c) **Time taken by 3/4th of the original number of atoms to disintegrate (decay), and**

    (d) **Time taken by 1/8th of the original number of atoms to remain unchanged ultimately.**

Solution: We have, T = 35 days

(a) ∴
$$\lambda = \frac{0.693}{T}$$

or
$$l = \frac{0.693}{35} = \textbf{0.0198 day}^{-1}$$

(b) Average life-period: $l = \dfrac{1}{\lambda}$

or
$$l = \frac{1}{0.0198} = \textbf{50.50 days}$$

(c) Number of 'atoms' being disintegrated = 3/4 No

Number of 'atoms' left behind (N) = $No - \dfrac{3}{4} No = \dfrac{1}{4} No$

We have the decay equation expressed as:
$$N = Noe^{-\lambda t}$$

or
$$\frac{N}{No} = e^{-\lambda t}$$

Here,
$$\frac{N}{No} = \frac{No/4}{No} = \frac{1}{4}$$

or
$$\frac{1}{4} = e^{-\lambda t}$$

$\therefore$
$$\lambda t = \log e \, 4$$

or
$$t = \frac{2.3026 \log_{10} 4}{0.0198}$$

$$= \frac{2.3026 \times 0.6021}{0.0198}$$

$$= \textbf{70 days}$$

(d) We have,
$$\frac{N}{No} = \frac{1}{8}$$

and
$$\frac{N}{No} = e^{-\lambda t}$$

combining the two expressions, we have:

$$\frac{1}{8} = e^{-\lambda t}$$

or
$$\lambda t = \log e \, 8 = 2.3026 \log_{10} 8$$

$\therefore$
$$t = \frac{2.3026 \times 0.9031}{0.0198}$$

$$= \textbf{105 days}$$

Q.3. Calculate precisely the time wherein the Radioacting of an analyte sample of Thorium (Th) gets duly reduced to 90% of its original value.

Assumption being the Half-life period of Thorium as 1.4×10^{-10} years.

Solution: Since we know that
$$T = 1.4 \times 10^{10} \text{ years}$$

Now,
$$T = \frac{0.693}{\lambda}$$

or
$$\lambda = \frac{0.693}{T}$$

$\therefore$
$$\lambda = \frac{0.693}{1.4 \times 10^{10}} \text{ per year}$$

Based on the Decay, we have
$$N = No\, e^{-\lambda t}$$

or
$$\frac{N}{No} = e^{-\lambda t}$$

But
$$\lambda t = \log e \, 1.11$$

$$\therefore \qquad t = \frac{2.3026\,[\log_{10} 1.11]\times 1.4\times 10^{10}}{0.693}$$

or
$$t = \mathbf{2.108 \times 10^9 \ years}$$

Q.4. Given that the Half-life of Radium (Ra) (molar mass 226 g.mole^{-1} is 1580 year. How would you show that 1g of Radium gives 3.70×10^{10} disintegrations per second?

Solution: According to the following equation after time '$t_{1/2}$' *i.e.*, half of the radioactive substance that have disintegrated, we have:

$$t_{1/2} = 0.6911/\lambda = \text{constant}$$

or
$$\lambda = \frac{0.6931}{1580\times 365\times 24\times 60\times 60}$$

$$= 1.391 \times 10^{-11} \ sec^{-1}$$

However, according to the rate of disintegration of the radioactive substance would be expressed as:

$$\frac{dN}{dt} = \lambda N$$

$$= \frac{1.391\times 10^{-11}\ sec^{-1} \times 6.922\times 10^{23}\ mole^{-1}}{10^{26}\ g\cdot mole^{-1}}$$

$$= \mathbf{3.79 \times 10 \ disintegrations.\ sec^{-1}.g^{-1}}$$

Q.5. A *nuclide* has a half-life $^{220}R_n$ is 54.5 seconds. What would be the actual mass of the ensuing nucleus which is equivalent to 1 mci (millicurrie)?

Solution: As we have seen in **Q.4** above:

$$\lambda = \frac{0.6931}{t_{1/2}}$$

$$= \frac{0.6931}{54.5\ sec.} = \mathbf{1.27 \times 10^{-2}\ sec^{-1}}$$

Now,
$$1\ mci = 3.7 \times 10^7 \ disintegrations.\ second^{-1}$$

$$= -dN/dt$$

or
$$-dN/dt = \lambda N \qquad \text{[Rate of disintegration of the radioactive substance]}$$

$$\therefore \qquad N = \frac{dN/dt}{\lambda} = \frac{3.7\times 10^7}{1.27\times 10^{-2}\ sec^{-1}} = \mathbf{2.91 \times 10^9}$$

The precise value for $^{220}R_n$ may be duly converted into '**mass**' as follows:

Hence, the **mass of nucleus** equivalent to **1mci** is:

$$= \frac{2.91\times 10^9 \times 220\ g.mole^{-1}}{(6.022\times 10^{23})\ mole^{-1}}$$

$$= \mathbf{1.06 \times 10^{-12}g}$$

[B] Radioactivity, Isotopes, Isobars, and Isotones

Q.6. What do you mean by the terminology '*Magic Numbers*'? How does the stability of *Nuclei* related with '*Magic Number*'?

Solution: Obviously, the **nuclei** invariably do consist of: **2, 8, 20, 50, 82 or 126 protons** (p) or **neutrons** (n)–that are found to be perceptively:

• **Extra stable** • **Possess Huge Quantum of Isotopes** and • **Nuclear cells filled completely** (*i.e.*, **the closed shells**).

Besides, the prevailing numbers of Protons (p) or Neutrons (n) *viz.*, 2, 8, 20, 50, 82, and 126 are usually known as the **Magic Numbers** for the respective **Nuclear Shells**.

Typical Examples of Stable Nuclei–Following are the *three typical* **examples of the stable Nuclei**, namely:

1. These *three* **nuclei**, for instance:
   • $_{18}Ar^{38}$ ($n = 38 - 18 = 20$) • $_{40}Zr^{90}$ ($n = 90 - 40 = 50$) and • $_{82}Pb^{208}$ ($p = 82$; $n = 208 - 82 = 126$),

   three **nuclei** are indeed extra stable, because the actual number of **Neutrons** (n) present in these **nuclei** are, in fact, the **Magic Numbers**.

2. The following *form* **nuclei**, such as:
   • $_2He^4$ ($p = 2$; $n = 4 - 2 = 2$) • $_8O^{16}$ ($p = 8$; $n = 16 - 8 = 8$) • $_{20}Ca^{40}$ ($p = 20$; $n = 40 - 20 = 20$)j; and • **82 Pb^{208}** • $p = 82$; $n = 208 - 82 = 126$), these **nuclei** are observed to be indeed *Extra stable* since both the prevalent numbers of **Protons** (p) and **Neutrons** (n) duly present in these nuclei happen to be the **Magic Numbers**.

3. Amazingly, the **End-Products** duly present in the series **Thorium (Th) [4n]**, **Uranium (U) [4n + 2]**, and Actinium (Ac) [4n + 3] – are observed to be
   • $_{82}Pb^{208}$ • $_{82}Pb^{206}$ and • $_{82}Pb^{207}$ respectively.

Importantly, all these **isotopes** do represent the **stable nuclei of** ^{82}Pb, as the number of **protons ($p = 82$)** present strategically in each **isotope** refers to a Magic Number.

Q.7. How would you identity the *Nuclei*, out of the following *three* substances, that is expected to be Radioactive and why?

   • $_2^4He$ • $_{20}^{39}Ca$ and • $_{85}^{210}At$

Solution

(a) $_2^4He$ usually possesses **Protons** (p) = 2 and **Neutrons** (n) = 19 (odd number), but 20 designates one of the **Magic Numbers**. Therefore, it is expected to be **fairly stable** *i.e.*, **non-radioactive**.

(b) $_{20}^{39}Ca$ has **Protons** (p) = 20 (even number) and **Neutrons** (n) = 19 (odd number), but 20 designates one of the **Magic Numbers**.

However, the so-called n/p **ratio** is found to be **less than 1**. Therefore, it must be **far below the stability best** squarely. Hence, it would be certainly expected to be '**Radioactive**'?.

(c) $_{85}^{210}At$ is observed to be '**Radioactive**' since there exists practically **no stable nuclei having an Atomic Number (Z) more than 83**.

Q.8. Out of the following *four* species which are possesses the optimized value of the *n/p* ratio?
- Ne^{16} • O^{16} • F^{16} and • N^{16}.

Solution:

(a) $_{10}Ne^{16}$: $p = 10$; $n = 16 - 10 = 6$;

Hence, the *n/p* **Ratio** = $6/10 = $ **0.60**

(b) $_8O^{16}$: $p = 8$; $n = 16 - 8 = 8$;

Hence, the *n/p* **ratio** = $8/8 = $ **1**

(c) $_8F^{16}$: $p = 9$; $n = 16 - 9 = 7$;

Therefore, the *n/p* **ratio** = $7/9 = $ **0.77**

(d) $_7N^{16}$: $p = 7$; $n = 16 - 7 = 9$;

Hence, the *n/p* **ratio** = $9/7 = $ **1.22**

Q.9. What would be the ensuing relationship between the value of *n/p* ratio and *unstability* (or *radioactivity*) of the nuclei?

Solution: Obviously, the **nuclei** having a relatively **Higher Value** of the *n/p* **ratio** (or *lower value of p/n ratio*) are indeed **more stable**; and therefore, **more radioactive** in character.

Q.10. Importantly, in the *critical emission of β-particle* the *Atomic Number* of the daughter element gets usually enhanced by '1'. Explain the above statement of facts.

Solution: In order to expatiate the above statement of facts–let us consider the following **Nuclear Reaction** that involves the **β-elimination** perceptively.

$$_6C^{14} \longrightarrow {}_7N^{14} + \underset{\text{β-particle}}{1e^0} + \underset{\text{Neutrine}}{_0v^0}$$

$$\begin{array}{ll} p = 6 & p = 7 \\ n = A - p & n = A - p \\ = 14 - 6 & = 14 - 7 \\ = 8 & = 7 \end{array}$$

The above reaction explicityly shows that the precise and exact number of Protons (*p*) gets **enhanced duly by 1** ($6 \rightarrow 7$); whereas, the **Neutrons** (*n*) gets **reduced by 1** ($8 \rightarrow 7$). Therefore, the increase in the number of **Protons** (*p*) is exclusively on account of the underlying fact that–'one **Neutron** (*n*) gets converted into Proton (*p*) ultimately'.

$$_0n^1 \longrightarrow {}_1p^1 + _{-1}e^0$$

Therefore, by virtue of the increase in the actual number of Protons (*p*) the Atomic Number (Z) of the so-called **daughter element** gets increased by 1.

Q.11. Given that ^{131}I possesses a Half-life period of 13.3 hours. After a lapse of 88.7 hours what would be the residual fraction of ^{121}I in the body?

Solution: We have, $t_{1/2} = 13.3$ hours.

The actual number of **Half-lives** in 88.7 hours = $88.7/13.3 = $ **6.67 Half lives**.

The exact amount left after half lives (N) = $\dfrac{N_o}{2^n}$

or Residual Fraction left $\left[\dfrac{N}{No}\right] = \dfrac{1}{2^n} = \dfrac{1}{2^6} = \dfrac{1}{64} = $ **0.0156**

Q.12. Given that $^{234}_{90}Th \longrightarrow 7a + 6b + ^A_Z X$. **Identify** $^A_Z X$.

Solution: Frist of all let us balance the **Atomic Number (Z)** and **Mass Number (A)** on either sides of the given equation, and we have:

$$234 = 7 \times 4 + 6 \times O + A$$

or $$A = 234 - 28 = 206$$

Atomic Number (Z)

$$90 = 7 \times 2 + 6(-1) + Z$$

or $$Z = 90 - 14 + 6 = 82$$

Result –Hence, $^A_Z X$ is $^{206}_{82} Pb$

Q.13. ^{237}AC shows a **Half-life of 24 years** with regard to the ensuing **Radioactive Decay (Disintegration).** The observed *decay* explicitly adopted *two* distinct parallel (11) paths: 1st leading to ^{227}Th and 2nd to ^{223}Fr. However the net percentage yields of the aforesaid *two* daughter nuclides are 2.0 and 98.0 respectively. What would be the Decay constants (l) for each of the *two* separate paths?

Solution: The observed Radioactive Decay of ^{227}Ac and ^{223}Fr may be depicted as given below:

$$^{227}Ac \begin{cases} ^{227}TH\left[2\% = \dfrac{2}{100}\right] \\ \\ ^{227}TH\left[98\% = \dfrac{98}{100}\right] \end{cases}$$

In the above expression both λ_1 and λ_2 are the **disintegration constants (λ)** for the *two* **divergent paths** as well as the **overall disintegration constant** is given by λ.

Thus we have:

$$\lambda = \dfrac{0.693}{t_{1/2} \text{of} \ ^{227}Ac} = \lambda_1 + \lambda_2$$

or $$\dfrac{0.693}{24 \text{ years}} = \lambda_1 + \lambda_2$$

or $$\lambda_1 + \lambda_2 = 0.0288 \text{ years}^{-1}$$

Hence, the Fractional yield of $$^{227}Th = \dfrac{2}{100} = \dfrac{\lambda_1}{\lambda_1 + \lambda_2}$$

or $$\dfrac{2}{100} = \dfrac{\lambda_1}{0.0288} \qquad\qquad (a)$$

The Eq. (a) gives the value of λ_1 as:

$$\lambda_1 = 5.76 \times 10^{-3} \text{ years}^{-1}$$

Likewise, we have:

$$\text{Fractional yield of } ^{223}\text{Fr} = \frac{98}{100} = \frac{\lambda_2}{\lambda_1 + \lambda_2}$$

or
$$\frac{98}{100} = \frac{\lambda_2}{0.0288} \qquad \text{(b)}$$

Hence, Eq. (b) gives the value of λ_2 as:

$$\lambda_2 = 282.2 \times 10^{-4} \text{ years}^{-1}$$

Q.14. How would you calculate the exact amount of a *Radioactive Element* ($t_{1/2}$ = 10 days) to which 1 g of the said element shall reduce in 520 days?

Solution: From the above problem, we have:

$$\text{No} = 1\text{g}, \ t = 520 \text{ days, and } t_{1/2} = 130$$

Hence, the total number of Half-lives in 520 days $(n) = \dfrac{t}{t_{1/2}} = \dfrac{520}{130} = 4$

Amount of the Element left after 520 days or 4 Half-Periods

$$= \frac{\text{No}}{2^n} = \frac{1}{2^4}\,g = \frac{1}{16}\,g = 0.0625\,g$$

Q.15. The Radioactive Element $^{64}_{29}\text{Cu}$ [Half-life ($t_{1/2}$) = 12.8 hours] undergoes radioactive decay (degeneration) by means of the β^- emission (40%); β^+-emission (20%); and the Electron capture (45%). How would you express: Decay products and calculate the Half-lives pertaining to each of the *Radioactive Decay Phenomena?*

Solution: Based on the above given data we may have the following expression:

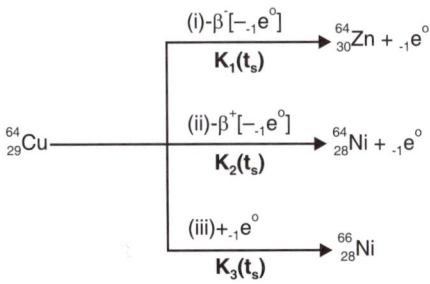

In a situation, when t_1, t_2, and t_3 are the half-life for the so-called partical ensuing phenomena in (i), (ii), and (iii) respectively,–then we may have the following expression for 't_1'

$$\therefore \qquad t_1 = \frac{0.693}{K_1} = \frac{0.693}{0.40K} \quad [K_1 = \text{Represents the disintegration}$$

constant for process (1); and

K = Rate constant for the entire overall process.]

$$= \frac{0.693}{0.40 \times 0.693 / t_{1/2} \text{of Cu}}$$

$$= \frac{0.693 \times t_{1/2}Cu}{0.40 \times 0.693}$$

$$= \frac{12.8}{0.40} = \textbf{32 hours (Answer)}$$

Likewise, we may have the following expression for 't_2':

$$t_2 = \frac{12.8}{0.20} = \textbf{64 hours (Answer)}$$

and,3

$$t_3 = \frac{12.8}{0.45} = \textbf{28.44 hours (Answer)}$$

UNIVERSITY QUESTIONS

1. Describe the nature and properties of the rays emitted by Radioactive Materials. **(Meerut, 1981)**
2. Describe the properties of α-, β-, and γ-Rays   **(North Bengal 81, Saugar, 1982, Kerala, 1980)**
3. Select any *two* member of the same Radioactive Series:
   $_{92}U^{238}$, $_{88}Ra^{224}$, $_{82}Pb^{206}$, $_{92}U^{235}$, $_{82}Pb^{208}$, $_{86}Rn^{22}$, $_{83}Bi^{224}$ **(Lucknow, 1985)**
4. Name the different Radioactive Series you know. How can you find that a particular isotope belongs to which **Radioactive Series**? To which Radioactive Series respectively the following belong?
   $_{90}Th^{231}$, $_{32}Pb^{266}$, and $_{89}Ac^{228}$.   **(Kanpur, 1986)**
5. Calculate the value of disintegration constant for $_{86}Rn^{222}$; if its Half-life period is 3.82 days.
   (Kanpur, 1985)
6. If the activity of a Radioactive Element drops to 1/6th of its initial value in one hour and 20 minutes, what are its disintegration constant and half-life period.   **(Rajasthan, 1984)**
7. Explain the theory of Radioactive Disintegration.   **(Meerut, 1981)**
8. Write short notes on '**Isotopes**'.   **(Banaras, 1980; Calcut, 1980; and Saugar, 1982)**
9. Explain the principle of **Aston's Mass Spectrograph**.   **(Bundelkhand, 1980; and Rajasthan, 1982)**
10. Why can the isotopes of an element not be separated by chemical methods?   **(Himachal, 1982; GND-University, 1982; and Kashmir, 1983)**
11. Discuss Dempeter's Mass Spectrographic Method for the identification of Isotopes.
    (GND-University, 1982)
12. Write note on '**Separation of Isotopes**'.   **(Jiwaji, 1984)**
13. Write a note on '**Isobars**'.   **(Banaras, 1980; Calicut, 1980; and Saugar, 1982).**
14. What is differences between **Isotope and Isotone**? Explain with examples.
    (Meerut 1989)
15. Discuss the Applications of Radioactive Isotopes.   **(Kanpur, 1999)**
16. Discuss the uses of **Isotopes** as tracers.   **(Gujarat, 2003)**
17. Name the isotopes of Hydrogen.   **(Nagpur, 2003)**

18. (i) What do you meen by Half-life period of a Radioactive Element? Deduce the formula for it

    (ii) 1 gramme of $79Au^{198}$ ($t_{1/2} = 65$ hrs) emite a β-particle and gives a stable isotape of Hg. Caleulate the amount of Hg after 260 hrs.

    (iii) Write the differences between **Nuclear Changes** and **Chemical Changes.**

    (iv) Write a note on **'Group Displacement Law'**. How many α-and β-particle should be emittend to change $_{90}Th^{232}$ to Pb^{208} (stable)   **(Meerut, 2009)**

PROBABLE QUESTIONS

1. Give a brief account on the -'**Characteristics of Emitted Rays'.**

2. Discuss briefly any two of the following.

    (i) **α-Rays,**

    (ii) **β-Rays,** and

    (iii) **γ-Rays**

3. How would you explain the various '**Characteristic Features of The Radiocactive Radiations'.**

4. Describe diagrammatically the- **'Deflection Profile of α,β, and γ-rays in a Magnetic Field.**

5. What are the properties of **α-, β-, and γ-Rays** ? Explain

6. Write short nites on the following:

    (i) **Measarement of the Range of an α-Particle by Scintillation.**

    (ii) **Properties of β-Rays**

7. Discuss brifly the- **'Meritorions Uses of Radioactive Radiations'.**

8. Explain the **Specific Detection and Precise Measurement of Radioactivity.**

9. Write a comprohensive account on the-**Theory of Radioactive Disintegration.**

10. Write brief notes on the following:

    (a) **Rate of Radioactive Disintegrations**

    (b) **Half-Life Period**

    (c) **Radioactivity Constant**

    (d) **Radioactivity Equilibrium**

11. How would you explain the '**Stability of Nucleus and Ratio of Neutrons and Protons (n/p)'**?

12. What do you mean by **Group Displacement Law? Explain.**

13. Explain the **Radioactive Disintegration Series** in an elaborated manner.

14. Explain any *two* of the following:

    (i) **Thorium (Th) Series [4n series]**

    (ii) **Actinium (Ac) series [(4n + 3) series]**

    (iii) **Uranium (U) series [(4n + 2) series]**

15. Write an essay on–'**Radioactivity the Potential Contributor'**

16. Give a detailed account on any *two* of the following:

    (a) **Isotopes**   (b) **Isobars**

    (c) **Isotones**   (d) **Cyclotron**

17. How would you carry out the '**Separation of Isotopes'**? Explain.

Chapter

4

CHEMICAL BONDING: LEWIS THEORETICAL CONCEPTS

1. INTRODUCTION

In a broader perspective, the '**atoms**' [except those of the *Nobel Gases viz.*, Helium (He), Neon (Ne), Argon (Ar), Krypton (Kr), Xenon (Xe), and Radon (Rn)], rarely have the '**free existence**' in nature. However, the **Nobel Gases** are found to be unreactive chemically; and hence, the '**atoms**' of rest of the elements do exhibit an unique tendency to get combined either *with each other* or *with the elements of other elements* to result into the formation of:

- **cluster of atoms**, or   • **aggregates of atoms**,

having a well-defined and definite composition perceptively.

Interestingly, the '**cluster of atoms**' so formed may be either:

➢ **a molecule** or

➢ **an ion**

Chemical Bond (or Valence Bond)–The prevailing observed attraction existing very much between the atoms thereby producing the respective cluster is invariably termed as:

"**a Chemical Bond or a Valence Bond**".

Hence, **a Chemical Bond** may be defined as–'**the attractive force that critically, holds two or more atoms together either as an ion or a** *molecule*.'

Another school of thought, relates the phenomenon of '**chemical bonding**' as:

'**the specific and critical union of two or more atoms** *via* **the so-called** *redistribution of electron* **specifically in their** *outermost shells* **thereby enabling the crucial involvement of either:**

- **transference of electrons amongst themselves,**
- **actual sharing of electrons amongst themselves,**

in order that all the atoms do need the so-called stable Noble Gas configurations of bare minimum energy profile'.

1.1. Causation of Chemical Bonding

An intensive and extensive study of **atoms** and **molecules** has revealed and proven that the '**atoms**' do combine chemically *i.e.*, the *causation of chemical bonding* is solely based upon the following *three* **cardinal reasons**, namely:

(a) **Overall Net Attractive Force Between Two Atoms,**

(b) **Octet Rule (or Rule of Eight),** and

(c) **Reducing the Energy of Atom configuration,**

which shall now be discussed separately in the sections that follows:

1.1.1. *Overall Net Attractive Force Between Two Atoms*

In general, the '**atoms**' do invariably comprise:

'**strongly positive *nucleus* and negatively charged *electrons***':

Mechanism: In a specific instance, when *two* **atoms** happen to come closer to each other in order to combine duly to form a bond between them, the resulting repulsive as well as attractive forces promptly commence to start operating between them effectively. Importantly, one may observe explicitly the so-called **attractive forces** actually prevail between the *electrons of one atoms* and the *nucleus of the other atom*; whereas, the **repulsive forces** do exist between the negatively charged *electrons* or the *nucleus* of the *two* **atoms**, as illustrated in Fig: 4.1.

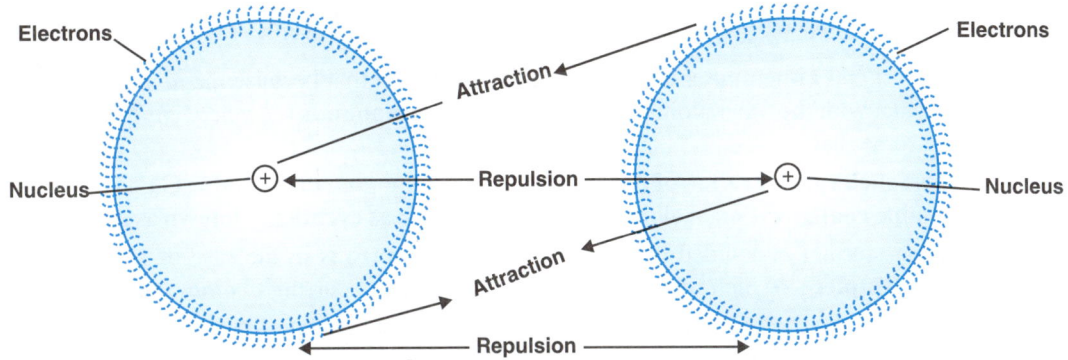

Fig. 4.1: Diagrammatic Representation of the Attractive and Repulsive Forces Operating Between Two Atoms when they Come Closer to Each Other.

Explanation: The various aspects of Fig: 4.1 may be explained as under:

1. As and when the *two* **atoms** come closer to each other, these forces do counteract each other obviously. Hence, the overall net result of these forces could be either:

   • **attraction**   or   • **repulsion,**

taking place between the *two* **atoms**.

2. Nevertheless, in a situation when the so-called '**attractive forces**' do tend to become absolutely dominant over the '**repulsive forces**', the net overall result is being apparently seen as the ensuing *attraction between the atoms*; and, therefore, they eventually combine together to produce a '**chemical bond**' between them.

3. Besides, in another instance when the **repulsive forces do become absolutely dominant over the attractive forces**,–the '**atoms**' fail to combine; and hence, practically **no chemical bond** gets duly established between then perceptively.

Example: Let us take the typical example of the **Hydrogen (H) Atoms:** it certainly shows the **net overall result** that happens to be: '*attraction*; and, hence, *two* **H-atoms do combine together to yield the He$_2$-molecule**'.

Helium (He) Atoms–Here, the **net overall result** is found to be: '**repulsion; and, therefore, the** *two* **He-atoms fail to combine together to yield the** *He₂ molecule*'.

1.1.2. *Octet Rule (or Rule of Eight)*

Kossel and Longmuir (1916) first and foremost introduced the so-called:

"**Electronic Theory of Valency or Octet Theory of Valency**",

so as to expatiate the underlying fact that *why the atoms actually get combined,*–based upon the inherent: '**electronic configuration of the Noble Gases**', having the following details:

S. No.	Noble Gas	Atomic Number (Z)	Electronic configuration
1	He	2	2
2	Ne	10	2, 8
3	Ar	18	2, 8, 8
4	Kr	36	2, 8, 18, 8
5	Xe	54	2, 8, 18, 18, 8
6	Rn	86	2, 8, 18, 32, 18, 8

Assumption: Kossel and Longmuir assumed the fundamental fact that because the *atoms* of the **Noble Gases** fail to interact with the other atoms to generate **newer compounds** (*chemical entities*),–it may be worthwhile to assume that:

"**the outermost shell configuration of the ensuing atoms belonging to the Noble Gases definitely has a fairly stable configuration of eight (8) electrons– that was eventually known as an '***Octet***'''**.

Conclusion: It was further concluded by them that the *two* electrons in the case of **Helium (He)** is usually termed as '**Duplet**', –which serves as a **Stable Octet** duly occurring in *other* **Noble Gases**.

It is, however pertinent to conclude the following important statement of facts:

"**As the inherent** *Octet of Electrons* **is found to be so stable in the** *Noble Gases*, **one may assume reasonably that as and when the so-called atoms of other elements do combine to result in the formation of a '***molecule*',–the electrons located strategically in their respective **outermost orbits are found to be arranged between themselves in such a fashion that they invariably accomplish an '***Octet*' **of electrons. It is indeed quite stable; and hence, the** *chemical bond* **gets formed between the atoms perceptively**".

Therefore, in a broader perspective there prevails an adorable tendency of the atoms to:

'**possess** *Eight (8) Electron* **in their outermost shall–which is called as the Octet Rule or Rule of Eight**'.

Points to Ponder: Because the **Helium (He) atom** possesses ruly *two* electrons–the ensuing '**rule**' is termed as the '**Doublet Rule**' or '**Rule of Two**' in this particular instance (of **He**).

Another school of thought designates the '**Octet Rule**' in terms of a '*theory*' being recognized profusely as:

'**Octet Theory of Valence or Electronic Theory of Valency**'.

Four Major Salient Features of Octet Theory–They may be summarized as under:

1. An **atom** with **eight electrons** present in the *outermost shell* [2 only in the case of **Helium (He)**] are duly observed to be **quite stable chemically**; and, therefore, are fairly capable of the **chemical combination phenomenon elegantly**.

2. However, an **atom** having *less than* **eight electrons** in its **outermost shell** appears to be fairly **active chemically**; and hence, does show an inherent tendency to:

   'specifically combine with other atoms progressively'.

3. In case, such **atoms** that eventually possess less than *four* **electron** duly present in their **respective ultimate shell** invariably shows a tendency to **gain** (*acquire*) **electrons** either:

   - **in the course of chemical combinations**, or
   - **bond formation so as to attain a** *stable configuration* **pertaining to the** *nearest* **Inert Gas**.

4. Amazingly, the **atoms** do *combine chemically* either due to the *actual transference of electrons right from the outermost shell of one atom to that of the other* or *by mutual sharing of 1, 2, or 3 electron pairs between the valence shell of both the so-called combining atoms*.

5. Lastly, there prevails a predominant tendency of:

   "an atom either for transference or sharing the electron pairs–that provides a reasonably accurate measure of its *chemical activity profile*".

1.1.3. *Reducing the Energy of Atom Combination*

It has been ascertained scientifically that when *two* **atoms precisely combined together so as to result into the formation of a** *chemical bond* **there exists:**

"an overall net decrease in the ensuing potential energy of the combined atoms perceptively".

Hence, it certainly indicates that a particular **system with bonded atoms has lower energy that with unbonded atoms**. Thus, it further implies that:

'the system of *bonded atoms having lower energy profile* is found to be more stable *vis-a-vis* the *unbonded atoms having higher energy profile*'.

Based on the above statement of facts and dictums we may infer that the **entire phenomenon of so-called '***Chemical Bonding***' existing between the atoms do lowers the energy of the** *combined atoms;* **and hence, results the formation of a system that eventually possessed the decreased energy profile; and thus, the** *higher stability* **profile.**

Following are the *three* **vital and important** aspects in the present context:

- **Potential Energy Curve**,
- **Variants in Bonds**, and
- **Modality of the Configuration of Atoms**,

which shall now be treated separately in the sections that follows:

1.1.3.1. *Potential Energy Curve*

The display of curve duly shown in Fig: 4.2 refers to the so-called:

"observed variation of the ensuing *Potential Energy** with the corresponding distance prevailing between the nuclei of *two* atoms A and B that are progressively approaching nearer to each other to result into the formation of a *bond* between A and B".

---

* That is, the energy an object has by virtue of its composition or position.

Remarks: The explicitly observed trend of the **curve** starting from *right to left* must be recorded duly. Thus, whenever the *two* **atoms A and B** happen to be far away *viz.*, *per se* **located strategically at an infinite distance** from each other,–one may observe that:

> "the attraction between A and B is found to be *zero*; and hence based on the accepted convention the ensuing energy of each of the atoms is arbitrarily taken to be *zero*; whereas, for the respective *stable system* it is considered to be *negative.*"

Therefore, based upon the above logical explanations and reasons one may safely infer that there exists absolutely **no possibility for the critical generation of any sort of a bond between them (A and B)**.

In Fig: 4.2, the aforesaid situation has been significantly represented by the crucial point 'X',–as depicted at the **far end toward the right hand side of the said curve**.

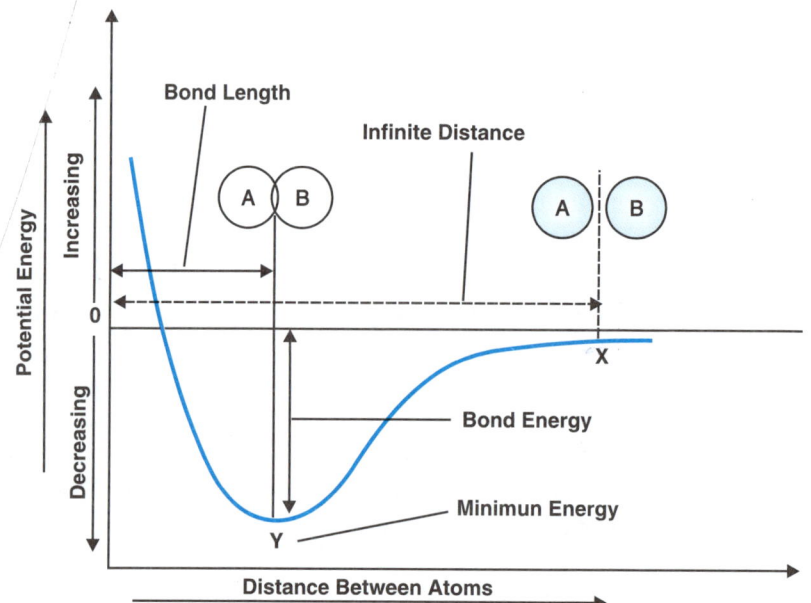

Fig. 4.2: Diagrammatic Representation of the Observed Variation of Potential Energy *vis-a-vis* the Distance Between the Nuclei of Two Atoms 'A' and 'B',–that are being Brought Closer to Each Other to Form a Bond Between them (*i.e.*, the Potential Energy Curve).

Explanation: The various vital aspects may be explained as under:

1. Since the *two* **atoms** are duly brought closer to each other (*i.e.*, decreasing the actual distance between the atoms),–the overall attractive forces prevailing between:
   - **Electrons (e^-)** • **Nucleus,**

   virtually render to be more dominant *vis-a-vis* the **repulsive forces** virtually functioning between the **electrons of the two atoms perceptively**.

2. Perhaps it could be the possible reason for the **observed energy of the system** to follow a *decreasing trend*–as depicted by the so-called downward trend of the curve (Fig: 4.2).

3. Amazingly, the said **decrease of energy** critically continues till such time a **certain minimum value** as illustrated by **point 'Y'** in the **said curve**.

4. At this point in time, if the **atoms (A and B)** are still duly brought nearer to each other, the so-called **repulsive forces** between the *two* **nucleus located strategically at such a small inter-nuclear distance** (*i.e., the internuclear distance prevailing between the nuclei of two atoms*) is found to be squarely dominant; and, therefore, the ensuing **energy of the system** commences to increase progressively, –as could be observed by the so-called '*upward trend of the curve*'.

The Characteristics of Point 'Y': It refers to the **lowest (minimum) segment of the curve** and possesses the following *three* characteristics, namely:

1. Importantly, at the ***critical point*** the *two* forces *viz.*, **attractive** and **repulsive** in attaining the **state-of-equilibrium** perceptively; and hence, the creation of '**bond**' occurs between the **atoms at this point 'Y'**.

2. Besides, the *phenomenon of overlapping* comes into play obviously between the so-called:
    '**orbitals of the *two* atoms that happens to be maximum at point 'Y'**'.

3. Amazingly, the prevailing *observed energy of the system at this point 'Y'*,–is indeed minimum; and, therefore, the '*AB Molecule*' does remain in the **most stable status**. However, the ensuing ***inter-nuclear distance*** is usually found to be at the **minimum ebb** at this **point 'Y'**. In addition, the observed **minimum energy valence** as well as the **inter-nuclear distance** with respect to this **point 'Y'**–are invariable termed as:
    • **Bond Energy** and • **Bond Length**,

pertaining to the '*AB Molecule*'.

Examples: Let us examine and explore the classical example of **molecular Hydrogen (H_2) Molecule**:

H_2 **Molecule** possesses the ***bond energy*** [**103.2 kcal. mole^{-1}**] and bond length [**0.74 Å**].

The aforesaid expression of interest renders it abundantly evident that:

"**a *chemical bond* is being duly formed between the *two* atoms (A and B)–as and when the ensuing *Potential Energy* belonging to the *Combined Atoms* is at the bare minimum ebb; and hence, they do remain at the minimum distance from each other**".

Thus, one may conclude that in the process of cleavage of the bond existing between the said *two* atoms (A and B) perhaps the same quantum of energy level has get to be provided–as was earlier consumed perceptively in the *creation of the bond*.

Strength of the Bond: It actually relates to the **quantum of energy** actually '**lost**' in either:
    • **bond-formation phenomenon**, or
    • **energy required to cause the cessation of the bond**, elegantly refers to the **measure of the strength of the bond**.

In short, it may be added that-'**more the energy decreased in the course of bond formation (or combination of atoms in a molecule),–the stronger would be the bond prevailing between the *two* atoms vehemently**'.

1.1.3.2. *Modality of Combination of Atoms*

The actual phenomenon whereby the so-called–'**atoms of the respective elements do rearrange their inherent outermost shell electrons so as to attain the 8-electron outermost shell configuration**'.

Importantly, it certainly leads to a **fairly stable configuration** that may predominantly occurs in the following *two* ways, such as:

❑ Due to the actual transference of one or more electrons right from the valence-shell of an atom to the corresponding valence-shell of another atom (i.e., the **formation of the Ionic Bond**); and

❑ Also, on account of the manner whereby *one, two* or *three electron* **pairs** of the **valence-shells** of both the ***combining-atoms*** are duly shared between them. In fact, the resulting **shared electron pairs** may be contributed almost equally by **both the atoms perceptively**–thereby leading to the ***formation of covalent bond*** or may be contributed by only **one of the combining atoms** thereby resulting into the formation of ***co-ordinate bond***.

1.1.3.3. *Variants in Bonds*

Thus, based on the aforesaid *two* explicit analysis whereby the ***two*** atoms do rearrange their *outermost shell electrons so as to obtain an **eight electron outermost shell configuration***, one may lay hands on the following *four* **variants in bonds**–that eventually *hold the atoms together in a molecule*, namely:

(a) *Ionic [or Electrovalence] Bond:*–The critical presence of such a '**bond**' is duly established due to the actual transference of **one or more valence electrons** right from one atom to the other.

(b) *Covalent Bond*–It is established perceptively by the *critical sharing of:* **1, 2, or 3 electron pairs** prevailing between the so-called **combining atoms**. Besides, each of the *two* '**Bonded Atoms**' eventually contributes at least ***one*** electron so as to produce the **shared electron pair** legitimately. Thus, it clearly ascertains the fact that:

"**the resulting electron pair does critically *fill up the outer-shell* of both the atoms; and hence, the said *two* atoms do attain the stable configuration of the closest *Inert Gas***".

(c) *Coordinate Bond*–It actually refers to a **covalent bond** wherein *both electrons* belonging to the so-called–'**shared electron pair**' really being provided by one of the *two* **atoms the nuclei.**

1. Amazingly, the *Metallic Bond* seems to be entirely different from the aforesaid *three* bonds (*viz.*, Ionic, Covalent, and Coordinate Bonds); and hence, certainly needs an elaborated discussion elsewhere (see section: 4.1.5).

2. Importantly, all the above *four* bonds are duly recognized to be the Strong Bonds.

Remarks: Nevertheless, the observed '**attractive interactions**' occurring between the *atoms* which are **relatively weaker** *vis-a-vis* the *four* bonds as stated above. Therefore, these **weaker interactions** are usually known as the **Weaker-Bonds.**

Weaker Bonds–In these typical instances one may critically take cognizance of the underlying fact that:

'**the bonding atoms fail to lose their original identity at all**'.

There are *two* variants seen in the domain of **Weaker Bonds:**

- **hydrogen Bond**, and
- **van der Walls Interaction (or Force)**,

which shall now be treated briefly as under:

Hydrogen Bond [H-Bond]: The **hydrogen bond** relates to a form of association between an **electronegative atom** and a **hydrogen atom**,–which is duly attached to a second relatively **electronegative atoms**.

Alternatively, the **Hydrogen Bond** critically involves the **bonding of a H-atom with *two* strongly electronegative atoms *viz.*, N, O, and F almost in a simultaneous episode.**

Van der Waal's Interaction (or Force): It refer to the attractive or repulsive forces occurring between the Molecular Entities (compounds), or between the groups (moieties) present duly within the same molecular entity, other than those due to bond formation or to the so-called electrostatic interaction of ions or of ionic groups with one another or with the neutral molecules.

In other words, the aforesaid bond critically involves the ensuing interaction between **atoms** or **molecules** with a typical **inert gas configuration**.

Following is the systematic classification of bonds:

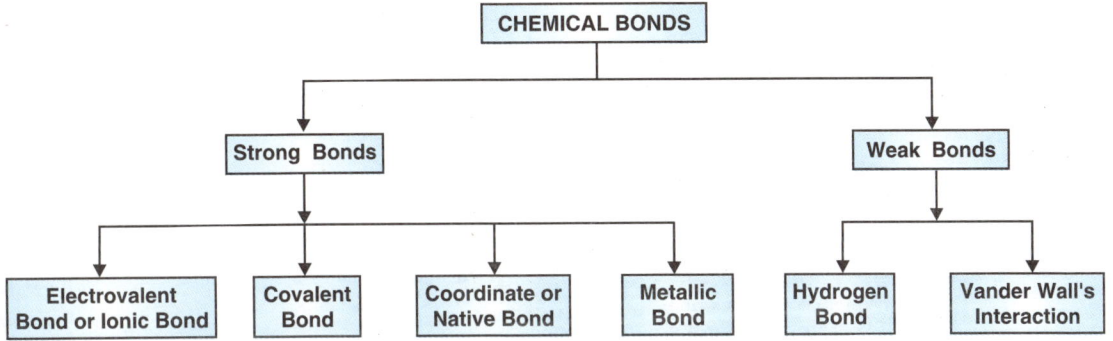

1.2. The Electrovalent (Ionic) Bonds

Preamble–The **electrovalent (ionic) bonds** do exist when *two* **oppositely** charged atoms actually share *not less than* one pair of electrons; however, the ensuing electrons spend definitely more time close to one of the atoms in comparison to the other atom.

Another school of thought defines the **electrovalent (ionic) bond as:**

"**the** *chemical bond* **duly formed between the two atoms due to the transference of one or more** *valence electrons* **from one atom to the other atom".**

NOTE   The *Ionic Bond* is also known as the *Polar Bond*.

1.2.1. The Ionic (or Electrovalent) Chemical Entities

Based on the scientific revelations and critical interpretations one may vehemently observe that: "one of the *combining atoms* possesses an excess of electrons *vis-a-vis* the stable number of electrons (2 or 8) duly present in its *valence shell*; whereas, the *other atoms* is found to be short of *electrons*; and hence, requires the electrons urgently to complete its octet".

Importantly, when the actual combination of atoms occur, the former surrenders the so-called inherent *surplus electrons* to the latter; and, therefore, as a concrete overall result of this *transference of electrons* each of the atoms certainly attains the anticipated 'stable configuration profile' pertaining to the respective *nearest* Inert Gas (*i.e., the desired* ns^2p^6 *configuration perceptively*).

Remarks: The resulting **chemical entities** (*compounds*) that essentially comprise the 'electrovalent bonds' are invariably termed as:

- **Electrovalent Compounds** or
- **Ionic Compounds.**

How does the Ionic Bond form?

The formation of an **Ionic Bond** may be expatiated vividly between *two* **atoms by taking into the consideration of a generalised instance** wherein an **atom 'A'** duly forms an **ionic bond** with another atom 'B'. However, the **atom 'A'** does possess only *one electron* in the so-called **Valence Shell**; whereas, the corresponding **atom 'B'** possesses *seven electrons*. In this manner, the **atom 'A'** has *one* **electron in excess** and atom **'B'** has *one* **electron less** *vis-a-vis* the so-called **stable octet**.

Comment: Therefore, under the aforesaid prevailing circumstances the **atom 'A'** critically **transfers one electron to atom 'B'**; and hence, due to this **amicable transaction** both the atoms definitely attain a so-called:

"fairly stable electron octet".

Fig: 4.3 illustrates the critical formation of an **Ionic Bond** being formed between **atom 'A' and atom 'B'**.

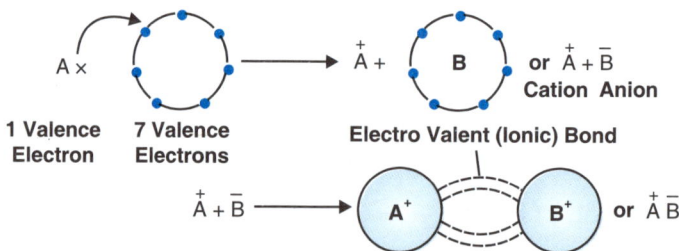

Fig. 4.3: Diagrammatic Illustration for the Formation of an Ionic Bond Between the Two Atoms 'A' and 'B'.

Explanation: The overall resulting **Positive Ion (Cation)** A^+ and the respective **Negative Ion (Anion)** B^- are being critically held together due to the prevailing *Electrostatic Force of Attraction*,–that is known as the **Electrovalent Bond** or **Covalent Bond**.

Based upon the above statement of facts and analogy one may also define the **Ionic Bond** (or **Electrovalent Bond**) as:

> "the **Electrostatic Force of attraction between** *Cation* **and** *Anion*–that are duly produced by the actual transference of electrons".

1.2.2. *Classical Examples of Ionic Compounds*

Following are the *three* **classical examples of Ionic Compounds,** namely.

- **Sodium Chloride (NaCl) Molecule,**
- **Calcium Fluoride (CaF$_2$) Molecule,** and
- **Aluminium Oxide (Al$_2$O$_3$) Molecule,**

which shall now be discussed individually in the sections that follows:

1.2.2.1. *Sodium Chloride (NaCl) Molecule*

In this particular instance, **Na atom (2, 8, 1)** eventually transfers its *one* **excess electron to the respective chlorine (Cl) atom (2, 8, 7)**, and in this manner the so-called **Na-atom** does acquire ultimately the *precise configuration of* **Ne atom (2, 8)**; whereas, **Cl atom** acquires the *exact configuration of* **Ar atom (2, 8, 8)**.

NOTE   Both *Ne* and *Ar* belong to the class of Noble Gases in the Periodic Table.

Explanation: The above sequence of events due to the observed phenomenon whereby the electron lost by Na-atom is being readily accepted by the respective **Cl-atom**. Consequently, the ensuing **Na-atoms** gets duly converted into:

"a *Positively charged Ion* (a cation) and Cl-atom gets converted into a *Negatively charged Ion* (an Anion). Besides, the *two* aforesaid ions so generated do have a tendency to attract each other by means of the inherent *electrostatic fores of attraction,*–that finally leads to the formation of an electrovalent (or ionic) bond between Na^+ and Cl^- ions."

The various intrigue steps being involved are explicitly depicted in Fig: 4.4.

(a) Na^X **Loss of Electron** ⟶ $\begin{bmatrix} {}^{XX}_{X}Na^{X}_{X} \\ {}_{XX} \end{bmatrix}$ or Na^+
   (2,8,1) $(-e^-)$ (2,8)

(b) $:\overset{\cdot\cdot}{\underset{\cdot\cdot}{Cl}}:$ **Addition of Electron Lost** ⟶ $\begin{bmatrix} :\overset{\cdot\cdot}{\underset{\cdot\cdot}{Cl}}: \end{bmatrix}$ or Cl^-
                   by Na-Ation $(+e^-)$ (2,8,8)

(c) Na^+ $+ Cl^-$ **Combination of Ions** ⟶ Na^+Cl^- (Obtained as 'Crystals')
   (2,8)   (2,8,8)

Fig. 4.4: A Simiple and Explicit Representation to Display the Formation of the NaCl Ionic Crystal.

Comments: The above *three equations:* (a), (b), and (c) leading to the ultimate formation of **sodium chloride (NaCl)** in Fig: 4.4. may also be depicted as stated below:

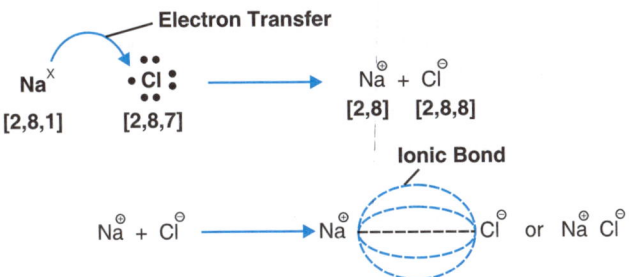

Inference: Following are the *two* **remarkable observations**, namely:

❑ *First*–in the Na^+Cl^- **crystal** the *two* respective ions do precisely acquires the so-called *Inert Gas Configuration,*–as depicted under:

$$Na^+ \longrightarrow 2,8 \quad [\text{Neon – Configuration}]$$

$$Cl^{-1} \longrightarrow 2,8,8 \quad [\text{Argon – Configuration}]$$

❑ *Second*–the **Dots** (·) and the **Crosses** (×) actually indicate the presence of **electrons** (e^-) in the outermost shells of the so-called **Cl and Na atoms**.

1.2.2.2. *Calcium Fluoride[CaF₂] Molecule*

Importantly, in the critical formation of **Calcium Fluoride (CaF_2) molecule**, one may vehemently observe that:

"each Ca-atom loses *two* electrons; and thus, gets converted to the respective Ca^{2+} ion; whereas, each F-atom duly accepts *one electron*; and thus gets converted into F^- ion".

Finally, it may be observed that *one* Ca^{2+} ion crucially accepts *two* F^- ions, –that eventually yields the **Calcium Fluoride [CaF$_2$] molecule.**

The following *four* sequential episodes will certainly expatiate the critical and explicit formation of **CaF$_2$**:

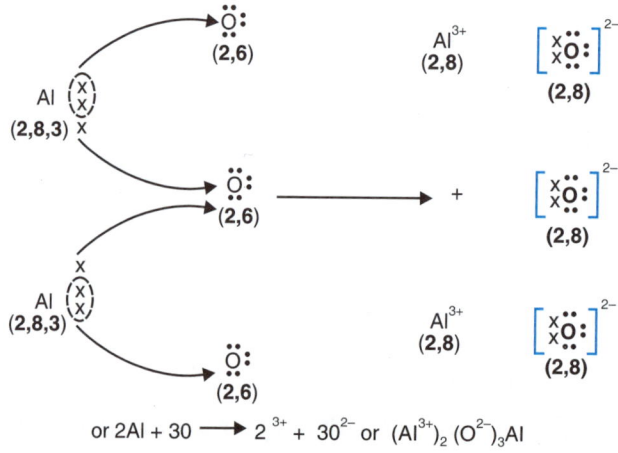

Hence, it may also be expressed as:

1.2.2.3. *Aluminium Oxide [Al$_2$O$_3$] Molecule*

In this particular case, the Al-atom possesses *three* electrons shell (2, 8, 3); whereas, the respective **Oxygen atom** has *six* electrons (2, 6). In reality, the *two* atoms of Al eventually transfer their *six* electrons to the ensuing *three* oxygen atoms perceptibly. In this manner, the *electron octets* pertaining to the *two* Al and *three* O atoms get accomplished duly.

Importantly, one may observe critically that:

'*two* **Al atoms are being meticulously deprived of their inherent *three* electrons each thereby yield *three* O^{2-} ions**'.

Interestingly, one may actually lay hands onto **Al^{2+}O$_3$$^{2-}$ or Al$_2$O$_3$ molecule**,–as illustrated vividly in Fig: 4.5.

Fig. 4.5: The Critical and Explicit Formation of the Aluminium Oxide [Al$_2$O$_3$] Molecule.

1.2.3. *The Prevalent Nature of Ionic Bond*

It is a universal phenomenon that as and when the combination of *two* **oppositely charged ions** take place,–they do exhibit an *obvious preferred mode of attraction* with each other by virtue of the so-called **electrostatic force of attraction**. Ultimately, these are held together by the prevailing–'*lines of force of attraction perceptively*'.

It has been amply proven and established that:

"**an ion may obviously attract other ions with the opposite charge upon them from any direction whatsoever; and hence, it may be concluded justifiably that the so-called *Ionic Bond* happens to be absolutely non-directional in nature that gets duly extended evenly in all possible directions**".

The aforesaid observed *concepts, ideas* and *dictum* may be further expatiated if the so-called '**Ions**' are duly considered as the **charged spheres.** Hence, we may show vividly the attraction between the ions with opposite charge on them,–as depicted under:

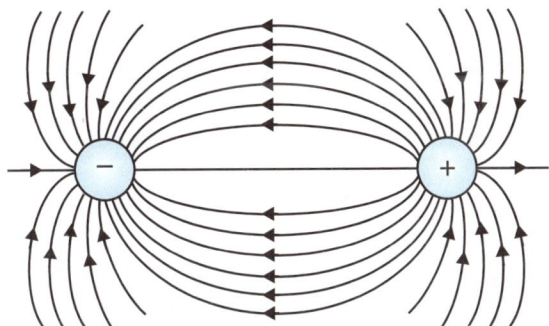

In another situation, when an **Ionic Compound** (or *chemical entity*) is made to dissolve in an aqueous medium (H_2O),–the corresponding **lines of force of attraction** critically anchoring the ions together are particularly broken by the *high* **Dielectric Constant Value of water**; and hence, the respective **ionic compound** get duly separated from each other as illustrated under for the typical–'**Na^+ Cl^- Ionic Crystal**' clearly.

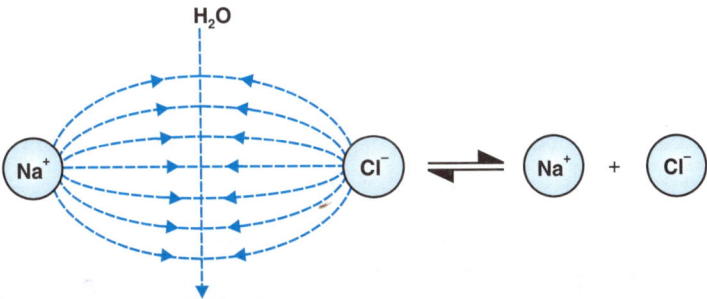

At this point in time, we may have an in-depth exposure to the following *four* **vital and important aspects** related to the nature of the **Ionic Bond**, namely:

(i) **Parameters for the Formation of Ionic Compounds,**

(ii) **Electron Affinity of Non-metal Must be High,**

    (iii) **Latent Energy of Ionic Compound Should be High**, and

    (iv) **Electronegativity Difference of A and B Must be Relatively High**,

which shall now be discussed separately in the sections that follows:

1.2.3.1. *Parameters for the Formation of Ionic Compounds*

In a broader sense, the *formation* as well as the *stability* of the '**Ionic Compounds**' rests exclusively upon:

"**the inherent ability and competence of the *ions* formation the so-called *Ionic Compound*".**

Some Important Factors: The *two* most vital and recognized **important factors** are as stated below:

  ❑ **Precise Number of Valence Electrons**–The atom '**A**'–that gets duly converted into the **cation A$^+$** must eventually, possess **1, 2, or 3 valence electrons**; whereas, the respective **atom 'B'** that gets converted into the **anion B$^-$** likewise must have **5, 6, or 7 valence electrons** perceptively. Thus, the elements duly present in the *Groups:* **IA, IIA, and IIIA** actually satisfy the parameter for **atom 'A'** intimately, and those of the *Groups:* **VA, VIA and VIIA** do closely statisfy the ensuing parameter of **atom 'B'**.

  ❑ **Observed Ionization Energy of the Metal Atom Must be Low**–It has been duly observed that the ensuing **Ionization Energy** happens to be at the *lowest ebb* with respect to the quantum of energy critically needed to get rid of the optimum number of the **loosely bound electron** right from a *neutral isolated gaseous atom*.

    Hence, we may have the following expression:

$$\underset{\substack{\text{Gaseous}\\ \text{Neutral}\\ \text{Atom}}}{A(g)} + \text{Ionization Energy} \longrightarrow \underset{\text{Cation}}{A^+ (g)} + e^-$$

Comment: Based on the above statement of facts it may be observed explicityly that:

"**if an atom does possess *low Ionization Energy*,–it shall be a lot easier for it to lose the electron; and, therefore, will certainly get converted into a '*cation*'. Thus, it would be absolutely clear that the prevalent presence of the *low Ionization Energy* of the metal with elegantly favour the formation of '*cation*'".**

1.2.3.2. *Electron Affinity of Non-Metal Must be High*

Amazingly, the ensuing **Electron Affinity** relates to the **exact quantum of energy being released**, as and when an electron is duly added on to be a **Neutral Isolated Gaseous Atom**.

    Hence, we may have the following expression:

$$\underset{\substack{\text{Gaseous}\\ \text{Neutral}\\ \text{Atom}}}{B(g)} + e^- \longrightarrow \underset{\text{Anion}}{B^- (g)} + \text{Electron Affinity*}$$

Comment: From the above explanations one may rightly infer that:

---

  * That is, the actual **Energy being Released**.

"since the ensuing *Release of Energy* shows the clear sign of stability of a system,–the atoms having *high* Electron Affinity shall definitely produce the respective 'anions' quite conveniently and easily".

1.2.3.3. *Lattice Energy of Ionic Compound Should be High*

It importantly refers to the **energy released duly:**

"when one gramme mole of a 'crystal' is formed right from the so-called *gaseous ions*; and hence, known as the Lattice Energy of the Crystal perceptively".

Thus, we may have the following expression:

$$A^+(g) + B(g) \longrightarrow A^+B^{-*} + \text{Lattice Energy}*$$

1 mole   1 mole   1 mole

Remarks:

1. Hence, one may critically take cognizance of the fact that–**higher being the observed value of Lattice Energy of a given *crystal*–the greater would be the ease of its actual formation *i.e.*, higher would be the observed stability of the *ionic crystal.***

2. Importantly, the emergence of the so-called **Energy Released** in the formation of the **Ionic Molecule** A^+B^- **will obviously lower the overall ensuing energy of the system.** In other words, for the critical formation of a **stable ionic compound there should be a net lowering of the energy**.

3. Therefore, for the **Lattice Energy** to retain a *high value*– the **electrostatic force of attraction** existing between the **constituent ions** of the ensuing **Ionic Compound must be high**.

4. Based on the **Coulomb's Law**–the observed **force of attraction (F)** prevailing between the *two* **oppositely charged ions** in air with the charges equal to q_1 and q_2 and separated by a distance equal to '*d*' is expressed as under:

$$F = \frac{q_1 q_2}{d^2} = \frac{q_1 q_2}{(g_{A^+} + g_{B^-})^2}$$

d = Sum of the radii of the +ve and –ve ions.

> **NOTE** From this relatioship it is quite clear that to inherit a large value of the force of attraction (or Lattice Energy) a few more conditions exist.

1.2.3.4. *Electronegativity Difference of A and B Must be Relatively High*

Based upon the scientific evidences and logical explanations right from the *1st* through *3rd* **parameters** (*as discussed above*), one may ascertain convincingly that:

"that *two* atoms 'A' and 'B' will definitely form an *Ionic Bond* provided they do possess extensively divergent Electronegativities".

---

 *   That is, a '**Crystal**'
 ** That is, the **Energy Released**.

Besides, it has been duly observed that a *definite* **difference** of **2 or even more** is an absolute must for the critical formation of an **Ionic Bond between the** *two* **atoms 'A' and 'B'.**

Example: Thus, we may vividly observe the so-called actual difference between Na and Cl stands at 2.1 [**since: Na = 0.9 and Cl = 3.0**]; and hence, Na and Cl shall obviously form an **Ionic Bond in the NaCl molecule** perceptively.

1.2.4. *Ionic Bond Formation Results in a Decrease in Energy Profile*

In a broad generalised rule one may vehmently observe the following critical revelations that:

"the formation of an *'Ionic Bond'* invariably involves a definite to decrease in the energy profile."

Interestingly, the above statement of facts may be further expatiated by taking into consideration the logistics involved in the critical formation of:

'1g **mole of solid sodium chloride (NaCl)s** *via* **the undermentioned** *five* **distinct sequential steps'**

Step-1: **Actual conversion of Na-atom from the** *Solid State* **to the respective** *Gaseous State* **[Na(S) ⟶ Na(g)].**

In scientific language, the conversion of the kind is usually known as the *sublimation* phenomenon *i.e.*,

'**the ensuing conversion of** *Solid Sodium (Na)* **into** *gaseous sodium (Na)*'.

Comment: **Obviously, in the solid state Na-atoms** are indeed fairly **close to each other**; whereas, those in the **Gaseous state** they do occur relatively **quite far apart.**

Therefore, the inherent conversion of **Na(S) into Na(g)** essentially needs the indulgence of *significant* **absorption of energy** *i.e.*,

'**the aforesaid conversion crucially involves an Endothermic Reaction**'.

Explanation–The total energy needed to **convert one mole of Na(S) to Na(g)** is invariably termed as the–'**Sublimation Energy**'. It has been observed to be equivalent to **108.81 kJ.mole⁻¹**.

Thus, we may have the following expression:

Na(S) + Sublimation Energy [108.81 kJ.mole⁻¹] ⟶ Na(g) (An Endothermic Reaction)

Hence, the energy absorbed in the reaction is duly shown at the **LHS** of the above expression.

Step-2: **Critical Dissociation of Chlorine [Cl₂(g)] Molecules into Chlorine [Cl(g)] Atoms [1/2 Cl₂(g) ⟶ Cl(g)]**

In this particular instance, the **requisite energy** needs to be adequately supplied so as to cleave the **bond** existing between *two* **chlorine atoms (in Cl₂ molecule) having the Cl–Cl bond.** Besides, in this *Step-2* one may take cognizance of the underlying fact that

'**half a mole of chlorine [Cl₂(g)] duly absorbs energy equivalent to half of the so-called** *Dissociation Energy* **of Cl₂(g)*;**'

and hence, gets converted into respective **Cl(g)** perceptively.

---

*   That is, the **Dissociation Energy of Cl₂(g)** has been found to be equal to **242.7 kJ. mole⁻¹**

NOTE Hence, one may safely conclude that the *absorption of energy* refers to the phenomenon of the so-called *Endothermic Reaction*.

Thus, we may have the following expression:

$$\frac{1}{2} Cl_2(g) + \text{Dissociation Energy } [242.7 \text{ kJ. mole}^{-1} \text{ of } Cl_2(g) = 121.35 \text{ kJ.mole}^{-1}]$$
$$\text{of } Cl(g) \longrightarrow Cl(g) \qquad \qquad \textbf{(An Endothermic Reaction)}$$

Step-3: Specific Removal of an Electron (e^-) from Gaseous Na-Atom [Na(g) $- e^- \longrightarrow$ Na$^+$(g)]

Here, the removal of an **electron (e^-)** from the **Gaseous Na atom** that critically needs energy equivalent to the so-called **Ionization Energy of Na***. Hence, the subsequent removal of an **electron (e^-) from Na(g)** is also observed to be an **Endothermic Reaction**.

$$Na(g) + \text{Ionization Energy } [495.8 \text{ kJ.mole}^{-1}] \longrightarrow Na^+(g) + e^{-1}$$
$$\textbf{(An Endothermic Reaction)}$$

Step-4: The Conversion of Cl(g) Atom into Cl$^-$(g) Ion [Cl(g) $+ e^- \longrightarrow$ Cl$^-$(g)]

Importantly, in this case the **Cl(g) atom** accepts the **electron (e^-)** provided critically by the **Na(g) atom** in the **previous** *Step-3* to result into the formation of **Cl^{-1}(g)**.

Amazingly, in this particular phenomenon one may vehemently observe that:

"**Cl(g) happens to release energy equivalent to its respective *Electron Affinity* (EA) [or sometimes referred to as *Affinity Energy*]–which stands equal to 348.7 kJ. mole^{-1}. The energy thus released is invariably depicted at the RHS of the equation**".

Hence, we may have the following expression:

$$Cl(g) + e^{-1} \longrightarrow cl^-(g) + \text{Affinity Energy } (= 348.7 \text{ kJ.mole}^{-1})$$
$$\textbf{(An Exothermic Reaction)}$$

Comment: The energy released in this manner implies that the inherent conversion of **Cl(g)** to **Cl$^-$(g)** designates an *Exothermic Reaction*.

Step-5: Ultimately Na$^+$(g) and Cl$^-$(g) Ions Combine to Produce NaCl(s) [Na$^+$(g) + Cl$^-$(g) $\longrightarrow$ NaCl(s)]

Importantly, it being the so-called **Final step** wherein the resulting **Na$^+$(g) and Cl$^-$(g) ions** so formed in *step-3* and *Step-4* respectively do eventually combine together to:

'**form one mole of *solid Sodium Chloride* [NaCl(s) crystal]**'.

Remarks: Thus, one may safely infer that the prevailing '*attractive forces*' existing critically between the entities **Na$^+$(g) and Cl$^-$(g) which do operate prevalently**; and thereby help in the **overal decrease in the ensuing Energy of the System.**

---

* **The Ionization Energy of Na. It is equal to 495.8 kJ.mole^{-1}.**

Based on the above revelations of scientific facts and logical explanations one may observe broadly that–

"**the critical formation of NaCl(s) right from the combination of Na$^+$(g) and Cl$^-$(g) is intimately associated with the Release of Energy**", which is widely recognized as the *Lattice Energy* of NaCl (*i.e.*, it is found to be equal to 769.2 kJ.mole^{-1}".

Hence, we may have the following expression:

$$Na^+(g)\ Cl^-(g) \longrightarrow NaCl(s) + \text{Lattice Energy } [= 769.2\ kJ.mole^{-1}]$$

$$\textbf{(An Exothermic Reaction)}$$

Finally, we may have the following *two* **vital derivations**, namely:

❑ **Total Energy** *Absorbed* = 108.21 + 121.35 + 495.80 = **725.96 kJ. mole**$^{-1}$

    [From Step-1]   [From Step-2]   [From Step-3]

❑ **Total Energy Released** = 348.7 + 769.2 = **1117.9 kJ. mole**$^{-1}$

    [From Step-4] [From Step-5]

Conclusive Remarks: Based on the above expressions of interest it is quite obvious that the **Total**

1. **Energy Released (1117.9 kJ. mole**$^{-1}$**)** is greater *vis-a-vis* **Total Energy Absorbed** (725.96 kJ.mole –1).

2. Besides, in the critical formation of **NaCl Ionic Crystal Energy** is being released; and hence, its **inherent magnitude** is found to be equal to:

$$1117.90 - 725.96 = 391.94\ kJ.mole^{-1}$$

Hence, one may observe critically that the so-called entire phenomenon for the formation of the NaCl crystal due to the actual transference of an electron right from the **Na-atom** to the **Cl-atom** essentially involves the **Evolution of Energy** (*i.e., Release of Energy*).

3. Ultimately, the formation of an '**Ionic Bond**' between the **Na$^+$ (cation) and Cl$^-$ (anion) due to the NaCl(s) ionic crystal explicitly involves a definte decrease in the energy level.**

Besides, the appreciable **release of energy** in the so-called formation of an *Ionic Bond* in **Sodium Chloride Solid [NaCl(s)]** also exhibits that:

'**the Ionic Bond in indeed an extremely *Strong Bond*'.**

1.3. Characteristics of the Ionic Compounds

The **Ionic Compounds** are the end-products duly obtained from the formation of the Ionic Bonds between a **Cation (Na$^+$)** and an **anion (Cl$^-$)**. In other words, when *two* oppositely charged atoms do share *not less than* one pair of electrons but the electrons virtually spend more time close to one of the atoms *vis-a-vis* the other atom.

Structure of Ionic Solids [*viz.*, NaCl(s)]

Let us examine critically a few common structures of certain chemical entities (*compounds*), such as:

- **MX or MX, and**
- **Double Oxide [Calcium Titanium Oxide (CaTiO$_3$)]***

---

* That is, **Perovskite**.

Highlights: These essentially include:

1. The explicit structures of these compounds are usually determined by **X-Ray Diffraction (XRD) methods***, because the *Scattering powers* of various incumbent ions solely rest upon the **total number of electrons** they actually contains.

Limitations of XRD-Methods: Nevertheless, the prevalent application of **XRD-methods** is usually subject to *two* **serious limitations**:

❏ *First*–These **XRD-methods** are not at all suitable for the precise and exact location of:

"**the presence of *light atoms* in the specific presence of relatively much heavier atoms; and hence, the highly useful *Neutron Diffraction method* may be an absolute must as the most preferred Complementary Technique**".

❏ *Second*–The **XRD-methods** are rarely capable of **identifying the *precise status of* ionization of the species** duly present:

"**only for certain specific substances that exhibit the *perfect electron density distribution* determined perceptively with significant accuracy for this purpose**".

 NOTE | **In a broader perspective, the fundamental *Ionic Profile* of a compound is being concluded from the actual possession of a *symmetrical 3D-structure*. Besides, it is also based upon the concrete evidences derived from a variety of other sources *viz.*,**

- *Conductivity* • *Hardness* • *Spectroscopic Data* and • *Thermochemical Data.*

2. *Three Basic Structures of the Ionic Solids*–Following are the *three* basic structures of the **Ionic Solids**, wherein the respective salts of **Formula Type MX are invariably crystalline in nature**, such as:

- **Caesium Chloride*** • **Zinc Blende Structure*** and • **Wurtzite Structure**.**

Fig: 4.6 illustrates the aforesaid *three* **structures** showing the **Ionic Solids** explicitly.

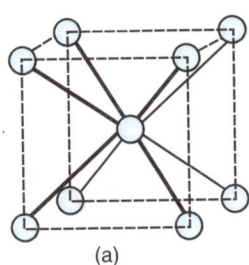 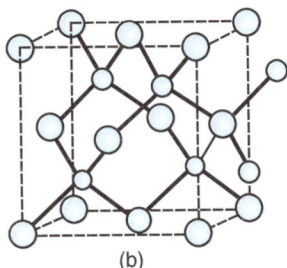 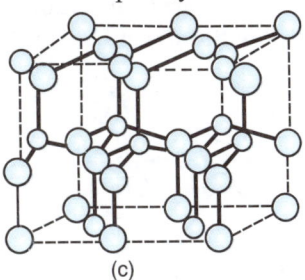

(a)   (b)   (c)

Figure: 4.6: Diagrammatic Structures of Ionic Solids: (a) Caesium Chloride; (b) Zinc Blende or Zinc Sulphide; and (c) Wurtzite Structure.

3. **Salts of Formula MX₂**–In this particular instance, the *three* **most important 3D-structure** are duly depicted in Fig: 4.7, wherein the *coordination number of 'X'* should obviously be **half that of M**.

❏ In the *cubic* **Calcium Fluoride (or Fluorite) structure**– one may critically observe that the *cation* has *eight* **distinct nearest neighbours** located strategically at the **corners of a cube.** Thus, the **anion four at the corners of a Regular Tetrahedron.**

---

* Kar A: **Pharmaceutical Drug Analysis**, 3rd ed., New Age International, New Delhi, 2016.

** That is, **Zinc Sulphide** (or *Zinc Blende*) structure. Both **Zinc Blende** and **Calcium Chloride**

*** **Wurtzite Structure:** It has a **Hexagonal Symmetry**. do have the **Cubic Symmetry.**

Examples (1): The various **chemical entities that** essentially possess this structure comprise are as stated under:

- **alkaline earth fluorides** [*viz.*, CaF_2, BaF_2, and SrF_2];
- **barium chloride** [$BaCl_2$], and
- **dioxides** of '*Lanthanide*' and '*Actinide*' **elements.**

In case, the compound possesses the **formula M_2X** *viz.*,

- **alkali metal monoxides** [*e.g.*, Na_2S, K_2S];

however, the precise positions of the '*Cations*' and '*Anions*' in the so-called '**Fluorite Structure**' do get usually interchanged; and therefore, the resulting substance is said to have:

"an 'antiflurite' structure".

2. Tetragonal Structure of *Rutile–In this case, the so-called **coordination** number of the *cation* and *anion* are *six (octahedral)* and *three* (**equilateral triangular**) respectively, such as:

- **Magnesium fluoride [MgF_2],**
- **Manganese difluoride [$Mn\ F_2$],**
- **Zinc Fluoride [ZnF_2], and**
- **Tin dioxide [SnO], Lead dioxide [PbO_2].**

3. Amazingly, in the **β-cristobalite*** each gilie atom possesses *four* **oxygen atoms** thereby surrounding it tetrahedrally. Besides, **each oxygen atom** possesses *two* **silicon atoms located** strategically as the *nearest neighbours.*

4. Since the prevailing *angle* [Si⌒O⌒Si] is somewhat **less than 180°–perhaps** provides a *solid reasoning* and *convincing fact* which suggests vehemently the ensuing structure happens to be:

"the inherent structure is certainly not a purely electrostatic structure".

Examples: **Berylium Fluoride [Be F_2]** and **Germanium Dioxide [GeO_2]** do possess the so-called **β-cristobalite structure** explicitly.

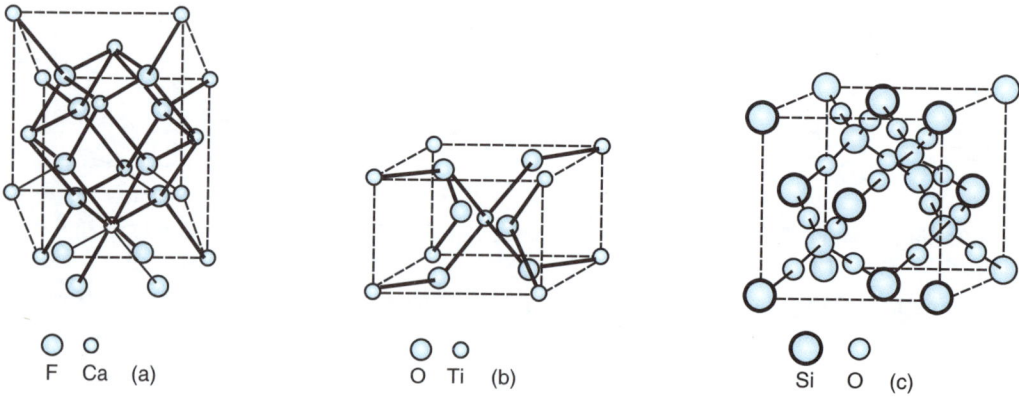

F   Ca   (a)   O   Ti   (b)   Si   O   (c)

Fig. 4.7: Diagrammatic Representation of Three Salts of Formula MX_2: (a) The Fluorite Structure: (b) The Rutile Structure; and (c) The β-Cristabalite Structure.

---

*   That is, one typical form of **Titanium Dioxide [TiO_2]**
**   That is, one typical form of '**Silica**'.

| **NOTE** | However, *Germanium Dioxide* also occurs in the *Rutile Structure; whereas, in rutile* itself one may critically observe the so-called *electron-density measurements* that categorically give the actual charge being held on the *cation as 13*. |

5. **Several Compounds of Formula 'MX$_2$' Crystallising in Layer Structures**-Importantly, the typical example belonging to this category of compounds is that of **Cadmium Iodide [CdI$_2$]**–that essentially possesses a *hexagonal symmetry*,–as illustrated in Fig: 4.8.

Explanation: The various aspects of **CdI$_2$ structure** may be expalined as under:

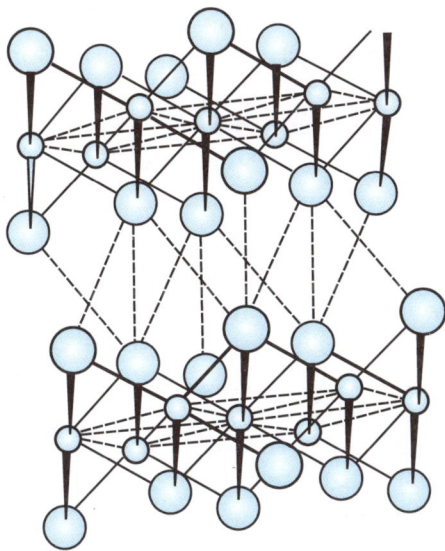

Fig. 4.8: Diagrammatic Representation of Formula MX$_2$ Crystallizing in Layer Structures: The Cadmium Iodide [CdI$_2$] Structure.

1. Obviously, each **Cadmium atom** is being surrounded by *six* **iodine atoms** located strategically at the **corners of a regular octahedron**. However, the *three* **Cadmium atoms** are critically located *nearest to each* **Iodine atom**; and duly positioned at the so-called:

   "Corners of the base of a *pyramid* of which the *Iodine Atom* is the apex (of the pyramid)."

2. Alternatively, the ensuing structure may explicitly be described as: "**comprising of a stacked sandwitches having the specific composition of** *Cadamium Iodide (CdI$_2$)*; **and hence, each sandwitch critically consists of a** *Layer of Iodine (I$_2$)* **atoms**–that eventually renders it almost *neutral electrically*".

3. The occurrence of only the **van der Waal's forces** operating meticulously between the sandwitches; and thus, the crystals having the requisite structure do exhibit predominantly a *marked and pronounced cleavage* almost parallel to the various layers.

4. Finally the so-called ensuing **unsymmetrical Environment of the Iodine Atoms** depicts beyond any reasonable doubt that:

   'the structure shown in *Fig: 4.8* may not a purely Ionic one'

Having been duly exposed to the intricacies of the structures of **Ionic Solids**, we may now venture into the so-called:

'Characteristics of the *Ionic Compounds* in a more elaborated manner under the following *four* sub-heads from 1.3.1. through 1.3.4.'.

1.3.1. *Physical State*

It has been proven and established that the **Ionic Compounds** do essentially comprise:

"**3D-solid aggregates of Cations and *Anions*– that are arranged meticulously in a well-defined geometrical profile perceptively**".

Hence, the **Ionic Compounds** (or *Ionic Chemical Entities*) are nothing but the so-called '*Crystalline Solid*' usually occurring at **room temperature (RT)**. Besides, they never could be observed either as **Liquids** or **Gases** under typical **ordinary parameters** *viz.*, *temperature* and *atmospheric pressure*,– since their respective '**Ions**' invariably do lack:

'**the absolute freedom with respect to their movement characteristic features both in the** *Liquid* **or** *Gaseous state*'.

1.3.2. *Electrical Conductivity Profile*

It is, however, pertinent to state at the very outset that:

"**the *Ionic Compounds* are unable to conduct electricity whenever these are critically present in the** *solid state*".

Perhaps the most plausible reason could be due to the fact that the *Cations* **and** *Anions* exert an *Electrical Force of Attraction* prevailing between than,–that eventually do remain together tightly very much in the **Ionic Compounds**; and hence, do *occupy their fixed positions in the* **Crystal Lattice**. The net overall effect may be seen vividly due to the fact that:

'**the *Ions* fail to move freely in a big-way whenever an *Electric Current* is made to pass** *via* **the Ionic Solids**.

Points to Ponder: They essentially include:

1. Obviously, the **Ionic Solids** [*viz.*, **NaCl(s)**] usually conduct electricity as and when they are observed either:

   • **in an aqueous medium**, or
   • **in the** *Fused Mother State*.

Comment: The most probable reason for this may be due to the fact that: "**as the temperature is being increased gradually, the ensuing *kinetic energy* of the *ions* also get enhanced accordingly**".

2. As soon as the **Ionic Compound** gets transformed into the respective *molten state*, the observed **Kinetic Energy** of the ions does become **so large** that:

   "**the ensuing *Attractive Forces* acting between the *Ions* are duly overcome; and hence, the corresponding** *well-defined arrangement of the ions* **found in the** *Ionic Crystal* **gets duly destroyed**".

Remark: As a result, the resulting '**Ions**' tend to become absolutely free so as to enable them to move about in the *liquid environment* under the influence of the so-called **Applied Electric Field** perceptively.

Hence, the **Ionic Compounds** prove to be the *superior* **conductors of Electricity** particularly in the **molten state** *i.e.*,

   • **the molten *Ionic Compounds*, or**   •   **their *Aqueous Solutions*,**

that eventually *conduct a flow of current* when placed in an **Electrolytic Cell**.

1.3.3. *Ionic Compounds Being Quite Hard Do Exhibit Low Volatility and Relatively Higher MP and BP–*

Based on the *stark reality* and *observed facts* that:

> "in the Ionic Solids–the cations and Anions are being duly held together in a reasonably tight manner very much within their *alotted positions* by the aid of their extremely *strong Electrostatic Forces of Attraction* perceptively".

Remarks: Amazingly, such an episode eventually needs a reasonably high quantum of energy* in order to separate the *cations* and *Anions* from one another very much against the so-called **Force of Attraction**; and hence, to render them absolutely free to move around (as could be seen in a *Liquid Environment*).

Therefore, as a result the **Ionic Solids**:

> "are invariably found to be fairly hard (though brittle in nature), do possess low volatility profile (having low-vapour pressure) and show higher range of MPs and BPs broadly".

1.3.4. *Observed Solubility Profile in Polar and Non-polar Solvents*

Importantly, the **Ionic Solids** are observed to be *freely soluble* in the **polar solvents** (*viz.*, H_2O, CH_3OH, NH_4OH) since the prevailing **electrostatic force** of **attraction** specifically holding the *Cations* and *Anions* together intimately in the so-called **Ionic Solids**:

> "gets reduced perceptively by the higher inherent value of *Dielectric Constant*** of the polar solvent."

Comment: It may be worthwhile to mention here that the ensuing reduction in the *Electrostatic Force of Attraction* due to the prevalent **high dielectric constant value** of the inherent *Polar Solvent*, fact, enables the respective ions to **move freely**; and subsequently, interact with the ensuing **Solvent Molecules** to form the so-called **Solvated Ions** perceptively.

Besides, one may also observe on the other hand that: 'the *Ionic Solids* are either *slightly soluble* or **even insoluble in the *non-polar Organic Solvents*'** (*viz.*, Benzene, *n*-Hexane, *Barbon tetrachloride*).'

 NOTE   In fact, such solvents by virtue of their relatively lower value of *Dielectric Constant* may not permit the *Ions* to move freely and also interest with them to result into the formation of the *Solvated Ions*.

1.4. The Crystalline Structure

Based on the survey of literature it may be revealed explicitly that the existence of **Single Ionic Molecules** (*viz.*, Na^+Cl^-) never takes place as such in the so-called:

> "solid form of the *Ionic Compounds i.e.*, the very existence of *Ionic Solids*, fail to occur in the form of Individual. Neutral Independent Molecules."

Hence, in a rather broader perspective a plethura of **cations** and **anions** do have an inherent tendency to attract each other on account of the '**Electrostatic Force of Attraction**' that.

- deem to be 'non-directional' in nature; and
- hence gets duly extended to all directions squarely.

---

* That is, **in the form of heat energy**.

** **Dielectric Constant:** A measure for the effect of a medium on the **potential energy of interaction between** *two* **changes** (*viz.*, Cations and Anions).

Therefore, to occupy the **Minimum Required Space**–the existing **ions** (*i.e.*, *cations and anions*) do have the most preferred and advantageous tendency to arrange themselves systematically by re-allocating themselves:

"in an alternating *cation-anion pattern* invariably termed as the crystal or Lattice".

In this manner, we may vividly visualise the appearance of the: "*Ionic Solids* that eventually comprise the remarkable formation of the 3D-Solid Aggregate perceptively".

Geometrical Structure: The appearance of the 'Geometrical Structure for an **Ionic Crystal** exclusively rests upon:

"the prevailing *Radius-Ratio* [$rc + r_a^-$ ratio] of the Cation and Anion"

Example: The *two* chemical entities *viz.*, **Caesium Fluoride (CSF)** and **Sodium Chloride (NaCl)** do have the *identical radius ratio*.

Ionic Crystals: In true sense, the **Ionic Crystals** do invariably possess a **typical structure of their own**. Thus, in each type of the structure of an **Ionic Crystal** one may critically take cognizance of the fact that:

"each Ion of *one* kind is being surrounded by a certain number of *Ions* bearing the opposite charge and located equidistant from it".

Thus, the explicit structure of **Sodium Chloride (NaCl) Ionic Crystal** may be considered as the most befitting example. It essentially possesses a *cubical shape*,–as illustrated in Fig: 4.9.

Remarks: The **X-Ray Diffraction (XRD)** investigative studies have duly revealed the explicit structure of the **Ionic Crystal of NaCl**,–as detailed under:

1. The actual distance prevailing between the *two* **strategically positioned adjacent ions** of the *same type* (*viz.*, Cl^- ions), based on the **actual length of the edge**, which is equal to **5.63Å**; whereas, the distance between the *two* **prevalent adjacent ions** of altogether *divergent type viz.*, Na^+ and Cl^- ions, is found to be equal to: **5.63/2 = 2.815Å**.

2. Interestingly, the Na^+ and Cl^- ions are virtually located in such a fashion that:

"each Na^+ is being duly surrounded by the *six* equally spaced Cl^- ions strategically positioned at the 'corners of a Regular Octahedron'; and likewise each Cl^- ion is also being surrounded by the so-called equally spaced *six* Na^+ ions duly located at the *corners of an Octahedron* (as depicted in Fig: 4.9)."

 NOTE   Hence, the observed *Coordination Number* of each is found to be *six*; and thus, the ensuing *NaCl-crystal* inherits an *Detahedral Structure* perceptively. That is to say– the NaCl structure possess at *6:6 coordination* predominantly.

Points to Ponder: These essentially comprise:

1. It is, however, pertinent to state at this point in time that the Na^+ and Cl^- ions are never connected to each other by '**pairs**', which is closely based on the underlying fact that:

'all the siz Cl^- ions are positioned almost the same distance away from any one Na^+ ion perceptively'.

2. Hence, for having the **Optimum Stability** of the NaCl-Crystal the so-called *oppositely charged ions* do remain quite close to one another; and thus, the *similarly* charged ions do occur *far away from one another as possible*.

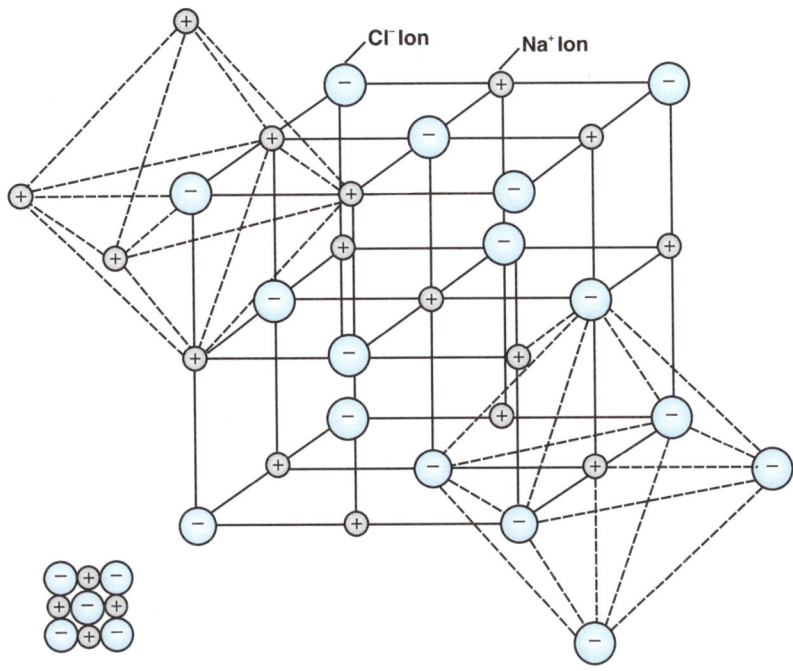

Fig. 4.9: The Cubical Structure of Sodium Chloride (NaCl) Ionic Crystal. [The Lower Segment Designates the Cross Section of the Sodium Chloride (NaCl) Crystal].

Comment: Besides, one may observe that although the so-called **Discrete Molecules [Na⁺ Cl⁻]** fail to exist in the **Solid Form** of the **Ionic Crystal**,–the ensuing **Independent Molecules** do largely occur in the *vapourised form* of the **Ionic Compounds** legitimately.

❑ The distinct **Crystalline Structures** usually exhibit the following *three* **characteristic features**, such as:

- **Extremely Brittle**,
- **Higher Inherent Density**, and
- **Display of Isomorphism**,

which shall now be treated separately in the sections that follows:

1.4.1. *Extremely Brittle*

It has been proven and established that the **Ionic Solids [NaCl(s)]** are **extremely brittle** in status *i.e.,* simply a small external force when applied upon the 'Ionic Crystals'–they immediately are **broken into small fragments**. Obviously, the said characteristic feature involving prompt cleavage is usually known as **Brittleness**; and hence, may be further expatiated as under:

Explanation–It is a known fact that the so-called **Ionic Solids** [*viz.,* **NaCl(s)**] are perceptively made up of **several parallel layers**–that predominantly consists of:

'**Cations and anions located strategically in the alternate positions in order that the** *opposite ions* **present in the different parallel layers do lie over each other**'.

Thus, whenever a *small external force* is being applied carefully upon an **Ionic Crystal**,–one may critically observe that:

'**one layer of ions meticulously slides over the other layer along a plane upto a small extent only**'.

Obviously, the phenomenon of sliding of a layer over the other duly suggests that the '**like-ions**' actually come in front of each other squarely; and thus, commence to repell each other-as shown in Fig: 4.10.

As a result, crucial application of a little **external force** brings about the *prevailing repulsion* existing between the *two* **layers**; and hence, the **Ionic Solid** [*viz.*, NaCl(*s*)] undergoes a cleavage promptly

NOTE	The above elaborated process strongly suggests that the so-called '*Ionic Solids*' are found to be *extremely brittle*.

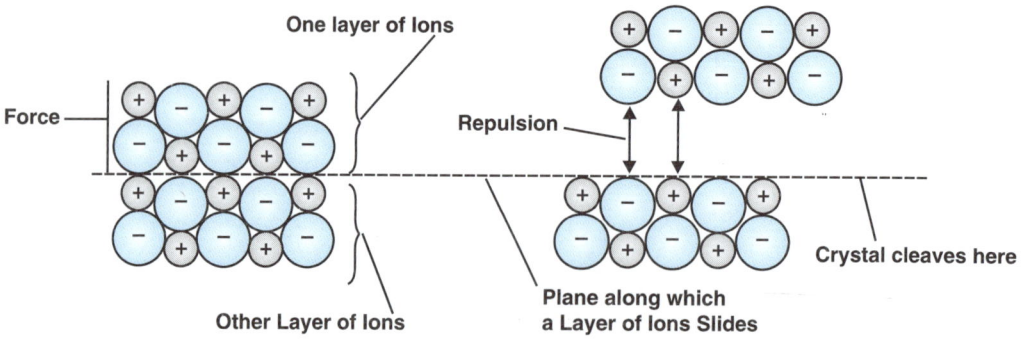

Fig. 4.10: An Elaborated Explanation Related to the Higher Brittleness Status of Ionic Solids.

1.4.2. *Higher Inherent Density*

It has been established beyond any reasonable doubt that the emanated **Electrostatic Force of Attraction** occurring between the *cations and anions* present in an **Ionic Crystal** helps to bring these ions very close to one another. It vehemently decrease the **Volume of the Crystal**; and hence, as a result the **Ionic Crystals** do exhibit high density profile.

1.4.3. *Display of Isomorphism*

The term **Isomorphism** refers to the *Ionic Solids* comprising the *ions* having an absolutely **identical electronic configurations** thereby explicitly displaying an **identity of the crystalline form**.

The *two* most commonly encountered *pairs of the* **Isomorphous Chemical Entities** are as given below:

❏ **Sodium Fluoride (NaF) and Magnesium Oxide (MgO)**; and
❏ **Calcium Chloride (CaCl₂) and Potassium Sulphide (K₂S)**.

The so-called **Electronic Configurations** of the above mentioned *two* **pairs** can be expatiated as under:

(a) Sodium	Fluoride	and	Magnesium	Oxide
Na^+	F^-		Mg^{2+}	O^{2-}
(2, 8)	(2, 8)		(2, 8)	(2, 8)

(b) Calcium	Chloride	and	Potassim	Sulphide
Ca^{2+}	Cl^-		K^+	S^{2-}
(2, 8, 8)	(2, 8, 8)		(2, 8, 8)	(2, 8, 8)

1.5. The Covalent Bond

Preamble–The **Covalent Bond** is also called as–'**the Bonding due to Mutual Sharing of Electrons**'.

Lewis Concept of Covalent Bond: It is also invariably termed as the **Octet Rule of Covalent Bond**.

Gilbert N Lewis (1916)*, an *American Chemist*, opined graciously that–'there are *atoms* that do **have the ability to combine with each other or even with other atoms by mutual sharing of the unpaired electrons in their outermost shells**'.

Obviously, in this aforesaid manner the so-called **Occupied Orbitals** belonging to the *outermost shell* pertaining to the **participating atoms** are duly filled with *two* **electrons** that critically inherits the *opposite spins* perceptively.

As a result, the typical prevailing–'**paired electrons are duly shared by both the atoms; and hence, circulate around the nuclei of the existing two atoms perceptively**'.

Thus, based solely on the above statement of facts one may infer conclusively that:

"**the attractive force of the *two* nuclei for the shared pair of electrons holds critically the *two* together firmly; and, therefore, results into the formation of a bond**".

It is usually known as the **Covalent Bond** or **Electron-Pair Bond**; and hence, the *chemical entities* comprising these **Covalent Bonds** are known as the **Covalent Compounds**.

Hence, we may now explicitly define the '**Covalent Bond**' as:

"**the chemical bonde existing prevalently between *two* atoms wherein the electrons (in pairs) are mostly shared by both the participating atoms**".

Formation of Shared Electron Pair: It is, worthwhile to state at this point in time that:

'**each of the *two* combining atoms does contribute one electron to the corresponding '*electron pair*'; and thus, ensures an *equal claim*' upon the resulting–shared–electron pair perceptively**'.

The Applicability of the Lewis Concept–According to the **Lewis Concept**–as and when *two* atoms do eventually form a **Covalent Bond** between them, each of the atoms certainly attains the so-called:

"**a *stable configuration* pertaining to the nearest *Inert Gas*, by completing its *Octet* (i.e., 8-electrons in the outermost shell) or *Duplet* i.e., 2-electrons in case of Hydrogen**)."

It is, however, pertinent to state here the **Covalent Bond** gets duly established between the atoms of either the **same or different elements**. Furthermore, because a **Covalent Bond** formed between the 2-atoms by the prevailing **interaction of their electrons** that eventually twin out to be common to **both the atoms**; and hence, it termed as '**Atomic Bond**'.

1.5.1. *Conditions for the Formation of Covalent Compounds*

Following are the various cardinal aspects of the *Covalent Bond*, namely:

(i) **Conditionalities for the formation of covalent Bonds**,

(ii) **Characteristics of Covalent Compounds**,

(iii) **Comparison between the Properties of Ionic and Covalent Compounds**,

(iv) **Deficiency of Octet Rule (Lewis Concept) in Covalent Compounds**, and

(v) **Percentage of Ionic characteristic feature in a Polar Covalent Bond**,

which shall now be treated individually in the sections that follows:

1.5.1.1. *Conditionalities for the Formation of Covalent Bonds*

The *five* most important characteristics for the critical formation of the **Covalent Bonds** are as stated under:

(a) Higher Level of Ionization Energy

It has been duly observed that the '**atoms**' which do possess higher level of **Ionization Energy (IE)** are indeed incapable of losing electrons to **yield the cations** ultimately. Perhaps this could be the most prevalent reason why these elements fail to form the **Ionic Bonds**; and instead do form the **Covalent Bonds** between the *two* **atoms** based on their **high inherent IE**:

(b) Equal Electron Affinities

In the particular instance of **Covalent Bonding** phenomenon one may take cognizance of the fact that:

> "**the *two* atoms should exhibit equal force of attraction for electrons; and hence, the so-called combining atoms should certainly inherit almost the equal electron affinities**".

(c) High Nuclear Charge vis-a-vis Small Internuclear Distance

Amazingly, in the typical case of **Covalent Bond** one may critically observe that the inherent *electron charge* gets eventually concentrated to a high level in the region between the **atom and the nuclei**. Thus, it predominantly helps in the **attraction of both the incumbent nuclei towards itself**.

Importantly, the ensuing force of **attraction** depends exclusively upon the following *two* **criteria**, namely:

- **the inherent charge on the nuclei**, and
- **the prevailing distance between them**.

Therefore, the presence of the **high charge upon the bonding nuclei** and also the relatively **smaller internuclear distance between** them do favour the ultimate formation of the **Covalent Bond**.

(d) Actual Number of Valence Electrons

Obviously, each of the *two* **atoms** must possess **5, 6 or 7 Valence electrons** (*viz.*, **the H-atom has only one electron**) in order that:

> "**both the atoms do accomplish the stable '*Octet*' by sharing critically 3, 2, or 1 electron pair in an articulated manner**".

 NOTE   **However, the non-metals of Groups: VA, VIA, and VIIA respectively satisfy the aforesaid condition befittingly**.

(e) Equal Electronegativity

In this case, both the atoms must have an *equal* **electronegativity profile** so as to afford the *transference of electron(s)* from one atom to the other may not occur at all *i.e.*, the **Ionic Bond** may not be generated. Thus, when the **electronegativity of both the atoms gets equalized**–the respective *sharing of the electron pairs* takes place vehemently; and hence, the so-called **Covalent Bond** is established duly.

1.5.2. *Characteristic Features of Covalent Bond*

There are eight recognized and prevalent characteristic features of **Covalent Bonds**,–which shall be discussed briefly in a sequential manner as under:

1.5.2.1. *The Physical Status*

Amazingly, the **Covalent Compounds** invariably comprises certain vital and important–'**Discrete Molecules**'; and, therefore, the prevailing force of attraction between the:

> "**so-called strategically positioned adjacent '***Covalent Molecules***' is found to be rather week in status**"

Based on the aforesaid underlying fact related to the '*weak forces*' that a majority of the **Covalent Compounds** do usually exist as:

> '***gases*** **or** ***liquids*** **of low boiling points under the influence of normal parameters** *viz.*, **Temperature and Pressure**'.

Nevertheless, they may also usually occur as: the '**Soft Solids**' exclusively,–whenever their **molecular weight (mw) remain high** *viz.*, **Chlorine [mw:71]** appears as a **gas**; **Bromine [mw:160]** occurs as a **liquid**; and **Iodine [mw:254]** appears as a **solid**.

1.5.2.2. *Crystal Structure*

In general, we may usually come across *two* **divergent types of crystals of Covalent Solids**, such as:

- ❑ *Giant Molecules:* It refers to those **crystal structures** that critically comprise: each and every atom which is duly bonded with other atoms by means of the **Covalent Bonds** thereby resulting into the formation of **Giant Molecules**; and

Examples: These include: • **Diamond** • **Silicon Carbide (SC)** • **Aluminium Nitride (AlN).**

- ❑ *Layer Lattice Structure*–It relates to those structures that essentially consist of altogether **separate layers**; and hence, are said to possess critically: **Layer Lattice Structure**.

Examples: These essentially include: • **Graphite** • **Cadmium Iodide (CdI₂)** • **Cadmium Chloride (CdCl₂)** • **Boron Nitride (BN).**

1.5.2.3. *MPs and BPs*

It has been established beyond any reasonable doubt that with the specific exception of '**Covalent Solids**' essentially comprising the **Giant Molecules [*viz.*, Diamond, Silicon Carbide (SiCN), Aluminium Nitride (AlN)** and the like,–several *other* **Covalent Solids** with relatively lower range of MPs and BPs *vis-a-vis* the **Ionic Solids** *viz.*, **BP of Silicon tetrachloride [SiCl₄]** (*i.e.*, the **Covalent Compounds**), and **Sodium Chloride [NaCl]** (*i.e.*, the **Ionic Compounds**)–are found to be: **58°C** and **144°C** respectively.]

NOTE — The crucial appearance of *low boiling points* are solely by virtue of the underlying fact that the inherent *attractive forces* prevailing between the so-called *Covalent Molecules* are nothing but–'*week van der Waal's forces*' perceptively.

1.5.2.4. *Electrical Conductivity*

Amazingly, the **Covalent Solids** essentially made up of *Giant Molecules* are observed to be:

"**bad conductors of electricity in a broader sense**",

because they fail to contain the so-called any sort of **charged particles** (*i.e.*, **ions or electrons**) in order to *carry the current* elegantly. Nevertheless, the ensuing **Covalent Bonds** essentially possessing the *layer lattices* (*viz.*, Naturally occurring **Graphite**) are obviously the well-known **good conductors of electricity** because in these types of 'solids'–the passage of electrons from one layer to the other may eventually brought into effect thereby enabling the flow of **current effectively**.

1.5.2.5. *Solubility in both Polar and Non-polar Solvents*

It has been proven and established that with the critical exception of **Covalent Solids** made up of the so-called **Giant Molecules**, the rest of the **Covalent Solids** are observed to be:

"*quite insoluble in the polar solvents viz.*, **H₂O, but are definitely** *soluble in the non-polar solvents viz.*, **Carbon tetrachloride (CCl₄), Benzene (C₆H₆), and n-Hexane (C₆H₁₄).**"

Solubility in Non-polar Solvents: Obviously, their actual *solubility profile* in the **non-polar solvents** is solely on account of the so-called:

'**observed similarity in the** *covalent nature* **of the molecules both in the solute and the solvent**'.

In other words, their solubility is entirely based upon the underlying principle:

"**Like dissolves like**".

Hence, the **Covalent Solids** with the *Giant Molecules* are certainly observed to be *insoluble* in almost all the solvents.

NOTE — The possible reason could be due to the fact that such '*Solids*' on account of their *big-size* fail to interact intimately with the respective solvent molecules.

1.5.2.6. *Covalent Compounds Being Neither Hard nor Brittle*

It may be observed in a critical manner that:

"whereas, the *Ionic compounds* **are usually** *hard* **in nature and also** *brittle*–the respective **Covalent Compounds are found to be neither hard nor brittle in character.**"

On the contrary, the **Ionic Compounds** are invariably *soft* and *waxy* in nature, perhaps because they quite often made up of separate molecules. Besides, there do exist rather *week forces* that firmly hold the molecules in the **solid crystal lattice.** Importantly, a so-called **molecular layer** occurring in the **crystal lattice** gets slipped conveniently *vis-a-vis* the other **observed adjacent layers**. Hence, there are complete absence of any kind of **repulsive forces between the said layers** *i.e.*, very much akin to those present commonly in the **Ionic Compounds**.

NOTE Therefore, the '*Covalent Crystals*' get cleaved very easily thereby ensuring that there exists practically no sharp cleavage between the prevailing layers upon the application of the *external forces*.

1.5.2.7. Molecular Reactions

As the **Covalent Bonds** invariably undergo the *molecular reactions in solutions*. Alternatively, in a '*solution*' – the **covalent compounds** often show such reactions where the *molecule as a whole* crucially undergoes a visible change.

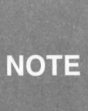

NOTE Because of the non-availability of '*electrical forces*' so as to speed up the prevailing reaction between the molecules, these reactions are observed to be rather *slow in nature;* and have, essentially require the definite control of both *pressure and temperature prevalently*.

1.5.2.8. Isomerism

Based on the underlying fact that the **Covalent Bonds** are found to be not only *rigid* but also *directional*,–which may eventually lead to:

'altogether divergent spatial arrangements of the *Atoms*'.

Hence, a **single molecular formula of a 'Covalent Compound'** may vehemently designate a good number of **divergent chemical entities (compounds) having various characteristic features**.

Thus, the '**Covalent Compounds**' may critically display '**Isomerism**'; and hence, the resulting *phenomenon of isomerism* is largely being observed amongst the '*Organic Compounds*' perceptively.

1.5.3. Comparing the Characteristics of Ionic and Covalent Compounds

The elaborated comparison of the characteristics of **Ionic and Covalent Compounds** has been duly provided in Table: 4.1.

Table: 4.1: A Detailed Comparison of the Characteristics of Ionic and Covalent Compounds.		
S.No. Characteristic Features	Ionic Compounds (Electrovalent Compounds)	Covalent Compounds
1. Physical State	These are crystalline solids at room temperature. They are never liquids or gases under the ordinary conditions of temperature and pressures.	Most of the covalent compounds exist as gases or liquids. However, the covalent compounds having high molecular weight can exist as solids.
2. Crystal Structure	They consist of three dimensional solid aggregates (ionic solids).	They crystals of covalent compounds are of two types: (a) those consisting of giant molecules (*e.g.* diamond etc.) (b) those consisting of separate layers (e.g. graphite, CdI_2 etc.)
3. Hardness and Brittleness	These are hard and brittle, since the cations and anions are held together very tigthtly by very strong electrostatic forces of attraction.	Solid covalent compounds are soft and waxy, since they usually consist of separate molecules. They are much readily broken.

4.	**Nature of Reactions**	They undergo ionic reactions (in solution) which are fast and instantaneous.	They undergo molecular reactions (in solution) which are slow.
5.	**Melting and Boiling points**	They have usually high melting and boiling points, since very high amount of energy (in the form of heat) is required to separate the cations and anions against the forces holding them together in the ionic solids.	With the exception of giant molecules, other covalent solids have low melting and boiling points. This is because of the reason that attractive forces between covalent molecules are weak van der Waal's forces.
6.	**Solubility in Polar and Non-polar Solvents**	They are freely soluble in polar solvents and insoluble or slightly soluble in nonpolar solvents. Their solubility in polar solvents is because of the reason that the solvation energy of the solvent is greater than the lattice energy of the ionic solid.	They are insoluble in polar solvents and readily soluble in non-polar solvents. Their solubility in non-polar solvents is based on the principle: "*Like dissolves like*".
7.	**Electrical Conductivity**	These are bad conductors when they are in the solid state. However, they conduct electricity in fused state or in solution in which their ions are free to migrate.	Covalent solids consisting of giant molecules are poor conductors because they do not contain charged particles or electrons to carry the current.
8.	**Nature of Bonds**	**(a)** Formed by the transfer of electrons from a metal to a non-metal. **(b)** Consist of electrostatic force between cations and anions. **(c)** Non-rigid and non-directional. **(d)** Cannot cause isomerism	**(a)** Formed by the sharing of electron pair(s) between non-metal atoms. **(b)** Consist of shared pair(s) or electrons between atoms **(c)** Rigid and directional **(d)** Cause stereoisomerism.

1.5.4. Deficiency of Octet Rule (Lewis Concept) in Covalent Compounds

Preamble–It has been amply proven and ascertained that in the usual formation of a **Covalent Bond**,–the '*atoms*' do acquire:

"**the so-called Inert-Gas Configuration having an Octet of Electrons (*i.e.*, ns^2p^6 configuration predominantly).**"

It is usually termed as the **Octet Rule** or **Rule of Eight** (see also *section 1.1.2*).

Incomplete Octet: The **covalent molecules** wherein the *central atom* is normally bonded covalently with other atoms having *less than* **eight electrons** are known as the **Incomplete Octet**.

Expansion of Octet: In the same vein, when the so-called **covalently bonded phenomenon** comes into play with other atoms with *more than* **eight electrons** are called as the **Expansion of Octet**.

Non-Octet Structure: Amazingly, one may also come across several such molecules that essentially exhibit the **non-octet structure**.

In the light of the aforesaid expression of thoughts pertaining to the '*Octet Rule*'–we may encounter *two* kinds of marked and pronounced deviations from the '**Octet Rules**' (or **Rule of Eight**) invariably in the **Covalent Compounds**, namely:

1.5.4.1. *The Incomplete Octet*

It refers to the molecules wherein the '**Octet**' **does remain specifically incomplete**.

In order to have a comprehensive grasp of the ensuing phenomenon involved in the **Incomplete Octet**, let us take into consideration the following *four* **classical molecules** separately:

- **Boron Trifluoride [BF₃]** • **Berylium Dichloride (BeCl₂)** • **Trimethyl Gallium [(CH₃)₃ Ga]** and • **Nitrous Oxide (NO)**

(a) **Boron Trifluoride [BF₃]**–In this instance, the **Central Boron (B) atom** possesses only *six* **electrons** in its outermost shell *i.e.,*

- *three* **electrons of its own**; and
- *three* **electrons derived from the 3-covalently bonded F-atoms**.

Thus, it may be expressed as under:

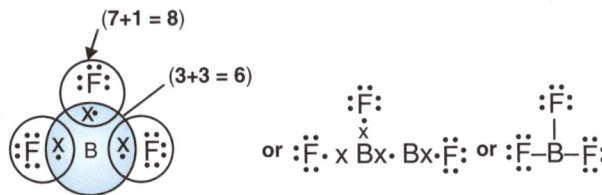

(b) **Beryllium Dichloride [BeCl₂]**–In this case, the molecule of **Berylium atom** (*i.e.*, the *central atom*) bears exclusively *four* **electron** in its *outermost shall* in the following manner:

- *two* **electrons of its own-self** (depicted by *crosses*); and
- *two* **electrons** (explicitly shown by *dots*),

duly derived from the *two* covalently bonded **Cl-atoms**,–as expressed under:

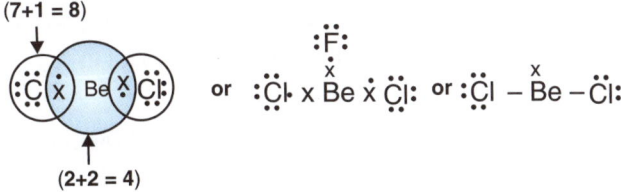

(c) **Trimethyl Gallium [(CH₃)₃ Ga] Molecule**–This typical example showing the Ga-atom bearing *six-electrons* in its *outermost shell*,–as given under:

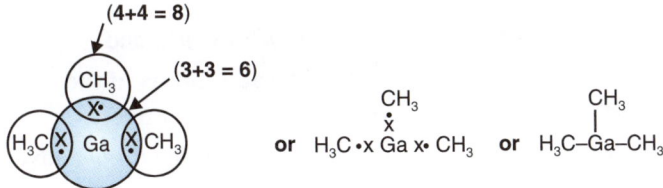

(d) Nitrous Oxide [NO] Molecule–In this classical example, the **N-atom** has only seven electrons (*septet*) *i.e.*, critically possesses:

- *five* of its own electrons in the outermost shell; and
- *two* remaining electrons from *O-atom* with which it gets hooked up by a *double bond* (*olefinic bond*).

> **NOTE**   The inherent O-atom critically bears *eight* electrons in its outermost shell perceptively.

Thus, it may be expressed as under:

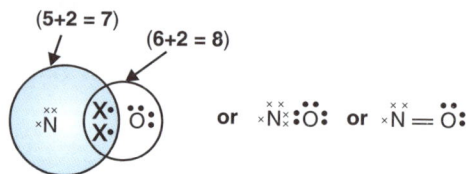

1.5.4.2. *The Expansion of Octet*

It also means that the **molecules** in which the **Octet gets duly expanded**. In order to expatiate the above phenomenon let us take into consideration the following **seven classical molecules:**

- **Phosphorus pentachloride [PCl$_5$]**
- **Carbon iodo-trifluoride [ICF$_3$]**
- **Carbon iodo-trichloride [ICl$_5$]**
- **Sulphur hexafluoride [SF$_6$]**
- **Iodo-heptafluoride [IF$_7$]**
- **Osmium octafluoride [O$_5$F$_8$]** and
- **Osmium tetroxide [O$_5$O$_4$]**.

(i) **PCl$_5$, ICF$_3$, and ICl$_3$ Molecules**–

❏ *PCl$_5$-Molecule:* In this instance, the *total number of electrons* that are duly present in the *outermost shell of the Phosphorus* (**P**) atom (*i.e.*, the **central atom**) totals *ten i.e.*,

- *five* electrons of its own [P → $3s^3p^3$]; and
- *five* electrons duly gained by it due to the critical formation of *five* **covalent bonds** with *five* **chlorine atoms squarely.**

❏ *ICF$_3$ and ICl$_3$-Molecules:* Likewise, in these *two* molecules the *total number of electrons* duly present in the **outermost shell** of the so-called **Cl and I atoms** (*i.e.*, the **central atoms**) is also *ten i.e.*,

- *Seven* electrons of their own [Cl → $3s^2p^5$; [$5s^2p^5$]; and
- *three* electrons duly gained by the Cl and I atoms in the actual formation of *three* **Covalent Bonds** with *three* **halogen atoms perceptively.**

The so-called electronic configuration structures of each of the *three above mentioned molecules* may be expressed as under in a sequential manner:

➢ **PCl₅ Molecule**

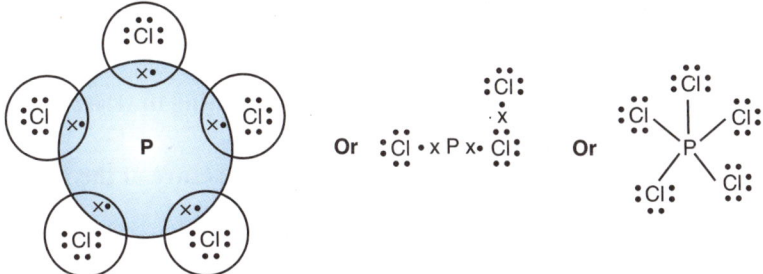

➢ **ICF₃ Molecule**

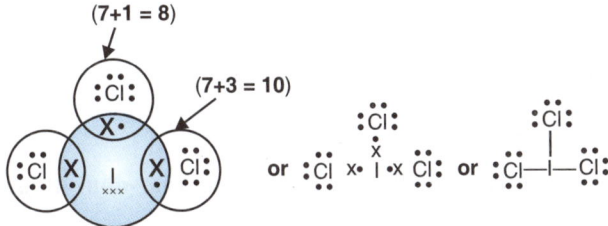

➢ **ICl₃ Molecule**

(7+1 = 8)

:Cl:

(7+3 = 10)

or :Cl x• I •x Cl: or :Cl—I—Cl:

(ii) **SF₆ Molecule**–Interestingly, in the **SF₆ Molecule** the precise and exact number of electrons that are duly present in the so-called-**Outermost shell of Sulphur (S) atom** (*i.e.*, the **central atom**) totals *twelve i.e.*,

- *six electrons of its own-self* $[S \rightarrow 3S^2p^4]$; and
- *six electrons acquired by S-atom* in the critical formation of *six covalent bonds* with *six F-atoms.*

The above electronic configurational structure may be expatiated as under:

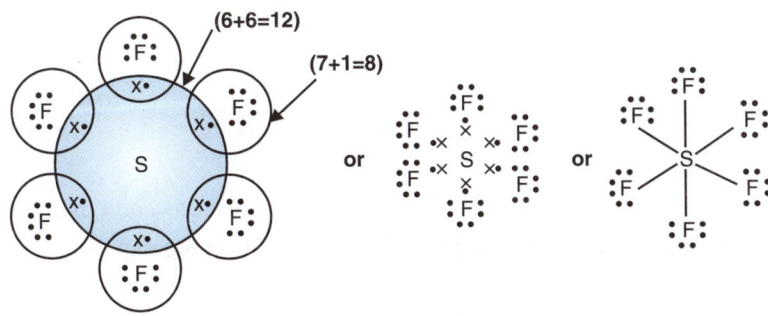

(iii) IF$_7$ Molecule–In the particular instance of the **IF$_7$ molecule** the precise and exact number of **electrons** duly present in the **outermost shell of Iodine (I) atom** (*i.e.*, the **central atom**) totals *fourteen* i.e.,

- *seven* electrons of its own [I → 5s²ps]; and
- *seven* **electrons gained by the *Iodine (I) atom* resulting in *seven* Covalent Bonds with seven Fluorine (F) atoms perceptively.**

Hence, we may have the following **electronic configurational structure of the IF$_7$ molecule:**

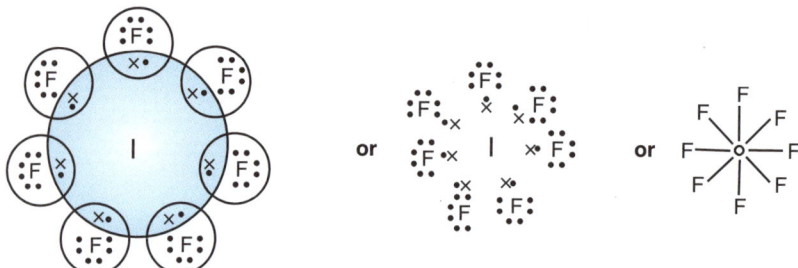

(iv) O$_S$F$_8$ and O$_S$O$_4$ Molecules–In this typical example the O$_S$F$_8$ molecule–the exact number of electrons that are present in the **outermost shell** equals to *sixteen* i.e.,:

- *eight* **electrons of its own [O$_S$ → 5d⁶6s²]; and**
- *eight* **electrons actually gained by O$_S$ in the formation of *eight covalent bonds* with eight F-atoms.**

Likewise, in **O$_S$O$_4$ molecule** the precise number of electrons duly present in the **outermost shell of O$_S$-atom** (*i.e.*, the **central atom**) totals *sixteen* i.e.,

- *eight* **electrons of its own-self [O$_S$ → 5d⁶6s²]; and**
- *eight* electrons duly gained by O$_S$ in forming *four* double covalent bonds with *four* oxygen (O) atoms.

➢ **O$_S$F$_8$-Molecule**

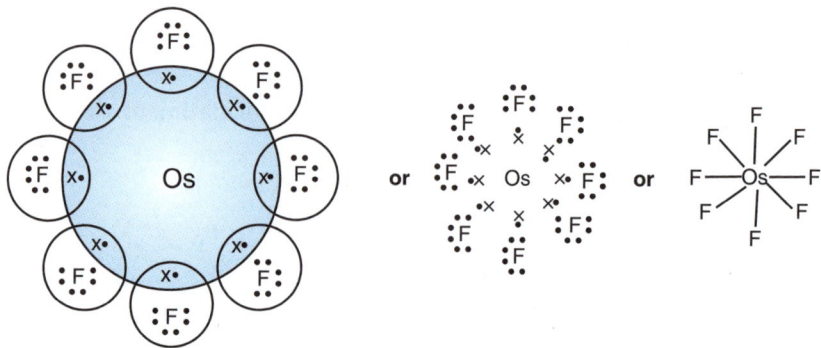

➢ **O$_S$O$_4$-MOlecule**

Super Octet Structures: It is, however, important to state here at this point in time that:

"**in all the above molecules the octet of the '*Central Atom*' is being expanded perceptively.**"

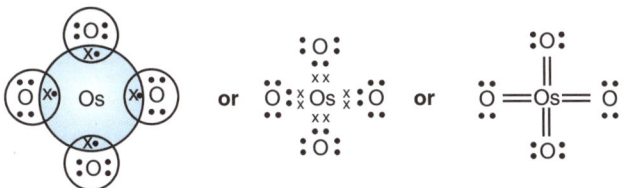

Based upon the above observations and dictum one may arrive at the underlying fact that the molecules containing the **Central Atoms** having **more than an Octet of electrons** are invariably known as the **Super Octet Structure**.

Remarks: The foregoing investigative studies render it quite evident and explicit that:

"**the Lewis concept fails to provide an adequate and necessary explanation the observed structure of the *Covalent Molecules* whose *central atom* possesses either less or more than eight electrons in its Outermost shell**".

Limitations of the Concept: Following are the *four* **glaring limitations** of the aforesaid concept, namely:

1. It fails to explain clearly the actual cause of **Covalent Bond Formation**.

2. It is unable to provide a logical explanation for the actual and precise nature of the attractive forces operating between the atoms genuinely.

3. It cannot provide a logical prediction pertaining to the quantum of energy lost in the course of bond formation precisely. Besides, it does not explain clearly as to why:
   "**divergent quantum of energy get released during the actual formation of altogether different molecules**".

4. It fails to predict the so-called '**geometry of a molecule**' *i.e.*, exact arrangement of atoms in space or the spatial arrangement of atoms.

Example: It cannot provide a logical and convincing explanations as it why **Boron trifluoride (BF$_3$) is planar** and Ammonia (NH$_3$) *pyramidal in shape*.

1.5.5. *Failure of the Octet Rule*

The so-called observed **failure of the Octet Rule**, as could be seen in the formation of the **Covalent Molecules** *viz.*, **BeCl$_2$, BF$_3$, (CH$_3$)$_3$ Ga, No, PCl$_5$, ICF$_3$, ICl$_3$, SF$_6$, IF$_7$, O$_5$F$_8$, O$_5$O$_4$** etc., has been expatiated with the aid of the following *two* **recognized concepts**, namely:

- **Sugden's Concept of Single Linkages**, and
- **Sidgwick's Concept of Maximum Valency**,

which shall now be discussed individually in the sections that follows:

1.5.5.1. *Sugden's Concept of Single Linkages*

Sugden advocated that the '*central atom*' of the molecules *viz.*, PCl$_5$, SF$_6$, IF$_7$, O$_5$F$_8$ and O$_5$O$_4$ essentially maintains its '**Octet**' squarely; and hence, in doing so the ensuing '*central atom*' gets duly hooked on to some of the **combining atoms** by means of the **single-electron bonds**, invariably termed as the '*Singlet Linkages*'. In the same vein, the *rest of the atoms* it gets duly linked by the aid of usual **normal two-electron covalent bonds** perceptively.

Singlet Linkage: Interestingly, the **Singlet Linkage** refers to:

"a *highly specific kind* of a bond that is critically produced by the so-called one-sided sharing of only one electron located strategically between the *Central atom* and its respective *combining atoms*".

NOTE   It is also called as the –'Single-Electron Linkage Half-bond' or simply a 'Singlet'.

Nevertheless, the **singlet-electron linkage** is invariably designated as–a **half-arrow** ← having its *arrow head* pointing explicitly from:

'**the Donor towards the Acceptor'.**

The **Singlet-Linkage** is observed to be a weaker *vis-vis* a **covalent Bond**; and hence, justifies significantly why **PCl$_5$** gets promptly dissociated to yield **PCl$_3$** and **Cl$_2$ respectively.**

Comment: We may once again vividly look at the **Singlet Linkages** in the structures of PCl$_5$ and SF$_6$ described earlier.

1.5.5.2. *Sidgwick's Concept of Maximum Covalency*

As per the basics of the **Sidgwick's Concept of maximum covalency**–one may critically observe that:

"**it may not an absolute must for an element to possess an optimized valency of *four***",

that further expatiates as well as suggests that it may not quite essential for an '**element**' to remain always surrounded by [**4 × 2 = 8 electrons**] for accomplishing its desired **stability profile**. Besides, ensuing **covalency of an element** may eventually **supercede *four***; and thus, the '**Octet**' may be expanded accordingly. Obviously, the **optimum valency of an element** depends perceptively upon the '*Period*' wherein the concerned element is actually present.

Examples: Hydrogen Atom (relates to the **1st Period having $n = 1$**) it is found to be **2**; and similarly the **elements related to 2nd Period having $n = 2$** *viz.*, (Li to F), it stands at 4.

Likewise, the **elements** related to the **3rd Period ($n = 3$)** and **4th Period ($n = 4$)** is found to be equal to **6**. Thus, for even **higher periods ($n > 4$) it is equal to 8.**

Remarks: As a result, the so-called *optimized capacity* of the **Valence Shell of an atom to:**

"**possess electron or the maximum number of electrons that are shared, for those elements stated above is found to be equal to: $2 \times 2 = 4$; $4 \times 2 = 8$; $6 \times 2 = 12$; and $8 \times 2 = 16$ respectively."**

NOTE   Therefore, one may not visualize any sort of discrepancy or anomaly in such *chemical entities* (compound) as: PCl$_5$, SF$_6$, O$_5$F$_8$, and the like,–wherein the *central atoms* making 5, 6, and 8 Covalent Bonds having the usual *Combining Atoms* respectively.

1.5.6. *Polar Covalent Bond Bearing Definite Percentage of Ionic Characteristics*

It has been observed vehemently that whenever *two* **distinct atoms A and B** are duly joined together by means of a Covalent Bond [$A^{d-} - B^{d+}$], the precise quantum of the **Ionic Character** in the aforesaid bond depends exclusively upon the so-called:

"**observed difference of the electronegativity values of A and B".**

It may be ascertained that–'**greater being the difference noticed ($x_A - x_B$), greater would be the percentage of the *Ionic Character*' inherited by the A–B bond perceptively.**

Importantly, it could be seen that the ensuing **electronegativity** of the **atom A** (x_A) has already been presumed to be at a *higher ebb* in comparison to that of the **atom B** (x_B). Consequently the above episode and dictum may be adequately expatiated by taking into consideration the following underlying fact that:

"**the inherent nature of the** *X–H Bond* **in the typical halogen acids (HX)** *viz.***, HCl, HBr, HI, and HF.**"

Remarks: These essentially include:

1. The observed *polarity of the* **H–X bond** in the particular instance of **halogen acids of HX type** that:

"**gets enhanced progressively along with the differences in the electronegativity profile** $[x_X–x_H]$**.**"

2. As the observed **electronegativity of the H-atom** (*i.e.*, x_H) virtually remains almost the same throughout the entire series; hence, **the polarity of the so-called** *H–X bond* **does increase with an increase of the ensuing electronegativity of X-atom (***i.e.***,** x_X**).**

Thus, the ensuing **polarity** or the so-called **Ionic Character** in the 'H–X bond' may be expressed in the following order:

$$H – F > H – Cl > H – Br > H – I$$

3. Consequently, from the above dictum: **the hydrofluoric acid (HF) happens to be the most polar compound; whereas, hydriodic acid (HI) is the least so.**

Pauling determined meticulously an *approximate percentage* of the **Ionic Characteristic profile** in an array of divergent **A–B polar Covalent Bonds** right from the **known** $(x_A–x_B)$ **values**; and thus, posted the observations in a Table: 4.2 showing the prevailing relation between $(x_A – x_B)$ values, the percentage ionic character in the A–B polar covalent bond; and also the precise nature of A–B bond $(x_A > x_B)$.

Table: 4.2: Showing the Relation Between $(x_A – x_B)$ Values, Percentage Ionic Character in A–B Polar Covalent Bond and Nature of A–B Bond $(x_A > x_B)$.

S.No.	$x_A – x_B$ Values	Percentage Ionic Character in A – B Bond	Nature of A–B Bond	Examples of Bonds
1.	0	0	Totally covalent (A – B)	H – H; Cl–Cl; O=O; N = N;
2.	0.1–0.8	0.5–15	Covalent (A – B)	C–S; C–I; N–Cl; P–H;
3.	0.9–1.6	19–47	Polar Covalent $(A^{d–} – B^{d+})$	The *two* O–H bonds in H_2O molecule are polar covalent bonds due represented as: $O^{\delta–} – H^{\delta+}$
4.	1.9	5.0	Ionic Bond (50%) and Covalent Bond (50%)	—
5.	1.9–2.0	55–63	Ionic Bond $(A^– – B^+)$	—
6.	2.0–2.3	67–74	Ionic Bond $(A^– – B^+)$	NaCl [Na^+ $Cl^–$]
7.	2.3 3.3	76–93	Ionic Bond $(A^– – B^+)$	CsF[Cs + F$^–$]

Important Points: Following are the *three* important points derived from Table: 4.2, namely:

1. In a specific situation when $(x_A – x_B) < 1.9$, one may clearly note that the **magnitude of the Ionic Character** observed in the so-called **A–B bond** is found to be more than 50%; whereas, that of the

covalent character happens to be less than 50%. As a result, the ultimate **A–B covalent bond i** **ionic predominantly**; and hence, is usually represented as: $A^- B^+$.

2. However, when $(x_A - x_B) = 1.9$, the ensuing magnitude of the inherent **Ionic Character profile** presents in the so-called **A–B bond** is found to be 50% and that of the corresponding **covalen** **character** also remains at **50%**. Consequently, the **A–B bond** is found to be **50% ionic and 50%** **covalent**.

3. In a typical instance, when $(x_A - x_B) > 1.9$, the actual observed magnitude of the **ionic character** **profile** in the **A–B bond** is found to be **more than 50%**, and that of the **covalent character profile** is certainly **less than 50%**. Therefore, the final **A–B bond** is predominantly of an **Ionic Character** and thus, may be represented as $A^- B^+$:

1.5.7. *Characteristic Features of the Polar Molecules*

Based on the concrete *scientific evidences* and *intelligent interpretations* have rightly put forward the following *six* **important characteristic features of the Polar Molecules**, namely:

(a) **An Intermediary between Purely Covalent and Ionic Compounds–** In fact, the prevailing characteristic features of the so-called **Polar Compounds** do serve as an important intermediary between the **covalent and ionic compounds** (*i.e.*, the *chemical entities* perceptively.

(b) **Dipole (*Two* Poles) character–**It has been duly proven and ascertaining that a **polar molecule** *viz.*, **HCl** $[H^{\delta+} - Cl^{\delta-}]$ that essentially possesses both **positive** and **negative** *charge centres* or *electrical poles*–that are invariably positioned strategically:

"**at the *two* terminals of the ensuing *Covalent Bond*– thereby rendering it '*dipolar in* *character*' genuinely**";

and hence, is usually termed as a **Dipole (*two* poles).**

Sidgwick (1960) vehemently suggested that the presence of a:

"***Dipole Bond* may be explicitly shown by an 'arrow' having a crossed tail**".

Besides, the 'arrow' is duly placed almost parallel to the line joining the so-called: **points of charge**; and hence, must be originating from the **positive** to the **negative end of the Dipole** perceptively.

Fig: 4.11 shows the **dipole of a HCl molecule** explicitly; and also the '**Dipole**' crucially behaves as a '*small magnet*'.

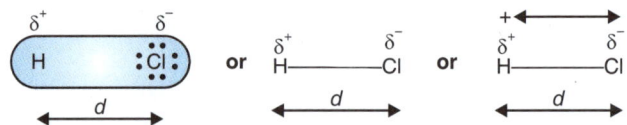

Fig. 4.11: The Representation of the Dipole of HCl Molecule.

NOTE — **In this case, '*d*' designates the distance between the +ve and –ve centres of the Dipole and is usually known as the Bond Length]**

(c) **Dissolution in Polar Solvents–**Amazingly, the dissolution of a **polar molecule in a polar** solvent (*i.e.*, *like dissolves like*) *viz.*, **Cl in H₂O**. In this particular situation the so-called **polar molecules of water** do surround the positive and negative terminals of the **polar covalent** **compound (NaCl) or solute**; and eventually split it into both:

"*cations* and *anions* that are surrounded by a certain definite but unknown number of water (polar) molecules".

Example: The **NaCl molecule** *i.e.*, a '**polar covalent molecule**' on being dissolved in **water (H_2O)** *i.e.*, a *polar solvent,*–the resulting polar molecules of water cause the cleavage of **NaCl [$Na^{\delta+}$ – $Cl^{\delta-}$]** molecule into the corresponding **$Na^{\delta+}$ and $Cl^{\delta-}$ ions** that are elegantly surrounded by '**water dipoles**', –as depicted in Fig: 4.12.

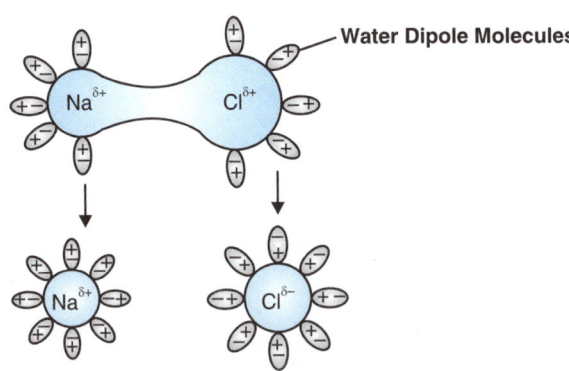

Fig. 4.12: Dissolution and Ionization of NaCl in Water (H_2O).

(d) **Dipole Moment**–One may critically observe that the **extent of polarity of a Polar Covalent Bond or of a Polar Molecule***–is being duly expressed in terms of its respective:

- *dipole moment* (**µ**) **that is equivalent to the product of the magnitude of inherent electric charge (*e*) in *esu*; and**
- *distance (d) in Å* **existing between the positive and negative centres (*i.e.*, the bond length).**

Hence, we may have the following expression:

$$\mu = e \times d$$

Based on the above expression, it may be inferred that since *l* has the order of $10^{-10} esu$, and *d* having the order of $10^{-8} cm$; and thus, the prevailing unit of µ is invariably is called as **Debys (D)**. Hence, we may have the following expression:

$$1D = 10^{-18} \text{ esu.cm}$$
or $\qquad 1D = 10^{-10} esu.\text{Å}$
or $\qquad 1D = 3.33 \times 10^{-28} \text{ coulomb.cm}$

NOTE Debye (D) designates duly a vector Quantity.

Remarks: From the aforesaid expression of scientific thoughts and revelation the precise and exact quantum of the **Dipole Moment** pertaining to a **Polar Molecule** depends solely upon the so-called observed differences in the *electronegativities of the bonded atoms*. Hence, it may be concluded safely that:

"**the greater being the difference in electronegativities,–the greater would be the observed** *Dipole Moment* **of a specific Polar Molecule (or chemical entity); and hence, greater shall be the** *degree of polarity* **of the** *Polar Covalent Bond* **existing between the joined atoms, –as given in Table : 4.3**".

---

* That is, a **molecule** with a **Polar Covalent Bond.**

S.No.	Molecule	Dipole Moment Value (Expressed in D)	% Ionic Character of H–X Bond [X = I, Br, Cl, F]
		Table 4.3: The Observed Relation Between Dipole Moment Value (μ) and Degree of Polarity (i.e., % of Ionic Character) of H–X Covalent Bond.		
1.	H–I	0.38	5
2.	H–Br	0.87	12
3.	H–Cl	1.03	17
4.	H–F	1.92	43

(e) **Association of Endless Chain of Polar Molecules**–Based on the underlying fact that the so-called **Polar Molecules** usually exhibit **great** attraction for each other; and thereby duly arrange themselves into an 'endless chain'. Hence, they get adequately associated with each other as depicted under in Fig: 4.13–for the $H^{\delta+} - F^{\delta-}$ **polar molecules** in an explicit manner.

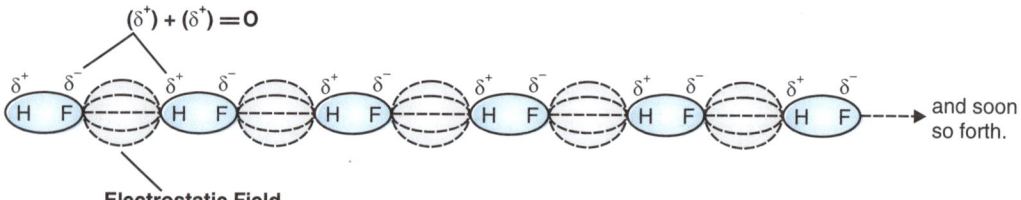

Fig. 4.13: Showing Explicitly the Association of Numberless $H^{\delta+}$–$F^{\delta-}$ Polar Molecules.

Remarks: The typical and vivid association of **Polar Molecules** to result into the formation of an 'endless chain' by virtue of the fact that:

"**the alternate δ^+ and δ^- charges positioned upon the so-called separate $H^{\delta+} - F^{\delta-}$ polar molecules do generate critically an *Electrostatic Field* between them perceptively**".

(f) **Orientation in an Electric Field**–By virtue of the so-called **Dipole Moment** – the *polar molecules* do show an inherent tendency to get:

"**duly oriented in an *Electric Field***".

Interestingly, the **Positive ends of dipoles** are critically directed towards the respective **Negative Electric Plate**; whereas, the corresponding **Negative ends of dipoles** towards the **Positive Electric Plate**,–as depicted in Fig: 4.14.

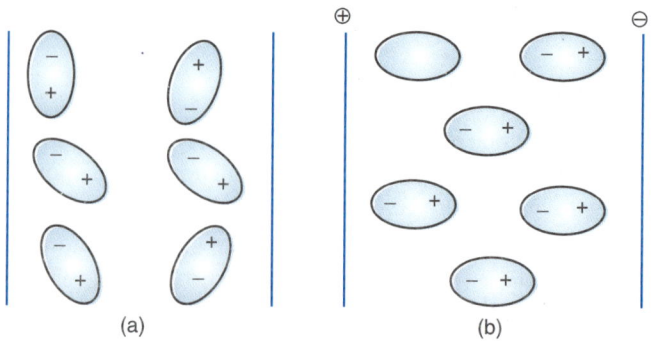

Fig. 4.14: The Orientation of Polar Molecules in the Electric Field (a) Field Off; and (b) Field On.

❑ **DIPOLE MOMENT AND ITS APPLICATIONS**–Following are the *two* **cardinal applications of the Dipole Moment**, namely:

- **Exact Quantum of Ionic character present in a Bond**; and
- **Spatial Configuration (Shape) of Molecules**,

which shall now be treated individually in the sections that follows:

(a) **Exact Quantum of Ionic Character Present in a Bond**–Importantly, the observed value of **Dipole Moment** may be employed for:

"**precise estimation in determining the so-called *ionic characteristics* present in a bond**".

Thus, we may note the following glaring facts that:

- The **Dipole Moment (μ)** of a *diatomic non-polar molecules* viz., H_2, Cl_2, F_2 etc., are found to be '*zero*'.
- Besides, *higher* being the *extent of inherent polarity* of a **diatomic molecule**–the *greater* would be its observed **Dipole Moment**.

Example: Determination of the **% of ionic characteristics** of *H–Cl bond* in *HCl molecule* by the aid of inherent **Dipole Moment of HCl–molecules**.

For this, let us assume that **HCl molecule is absolutely ionic in nature**. Hence, under the prevailing pre-assumed parameter **both H^+ and Cl^- ions** would certainly possess a **unit charge** equivalent to:

- **4.80×10^{-10} esu ($= e$)**; and
- **bond distance between H and Cl atoms $= 1.27$ Å ($= d$)**

Hence, we may have the following expression:

$$\mu = e \times d$$

or $\mu = (4.80 \times 10^{-10}$ esu$) \times (1.27$ Å$)$

or $\mu = 4.8 \times 1.27 \times 10^{-10}$ esu. Å

or $\mu = 6.19 \times 10^{-10}$ esu. Å

or $\mu = $ **6.09 D**

$$1D = 10^{-10} \text{esu.Å}$$

The above is the so-called **Theoretical Value of μ (Dipole Moment)**. Since, the **observed experimental value of μ for the HCl molecule is 1.03**; we have the following relationship:

$$\left.\begin{array}{l}\text{\% Ionic characteristics in H–Cl}\\ \text{Bond (in the HCl Molecule)}\end{array}\right\} = \frac{\text{Experimental Value of }\mu}{\text{Theoretical Value of }\mu} \times 100\%$$

$$= \frac{1.03}{6.09} \times 100\%$$

$$= 0.169 \times 100\%$$

$$= \textbf{16.9\% or 17\%}$$

From the above derivations one may vividly draw the conclusion that the '**H–Cl Bond**' duly present in the so-called **HCl Molecule** does possess **17% of the Ionic Character**; and ($100 - 17 = 83\%$) of the respective **covalent character** perceptively.

(b) Spatial Configuration (shape) of Molecules–The inherent **Dipole Moment (m)** of a given **molecule** (*chemical entity*) does help in enabling the precise and exact determination of its:

- **Geometry and** • **Shape.**

Explanation: It has been duly observed that the **Diatomic Molecules** explicitly composed of:

"*identical atoms* **(viz., H_2, Cl_2, F_2, Br, I_2 etc.) are certainly devoid of Dipole Moment (μ); whereas, those having** *different atoms* **(viz., HF, HCl, HBr, HI etc.) do possess the similar inherent value of Dipole Moment (μ).**"

Likewise, one may also take cognizance of the underlying fact that:

"**symmetrical linear polyatomic molecules** *viz.,* **Carbon tetrachloride (CCln), Carbon dioxide (CO_2), carbon disulphide (CS_2), Mercury dihalide (Hg)X_2–do exhibit** *zero* **resultant Dipole moment (μ); and hence, are non-planer in nature**".

Amazingly, since the ensuing **Dipole moments (m)** pertaining to all the bonds that are duly present in the molecule eventually cannel each other preferentially in the **opposite directions**.

Alternatively, the so-called **unsymmetrical nonlinear polyatomic molecules** *viz.,* **H_2O, SO_2, NO_2, H_2S etc.** do actually possess certain extent of value in **Dipole moment (μ)**; and, therefore, turn out to be the **Polar molecules**. In this way, the *observed inherent value of the* **Dipole Moment** of a molecule provides a solid evidence and concrete information pertaining to:

"**its spatial configuration (shape) of the molecules**".

Some Classical Examples: Following are a few **classical examples:**

1. **Carbon Dioxide (CO_2), Carbon Disulphide (CS_2), and Mercury Dihalide (HgX_2)**–First of all let us look into the **CO_2–molecule** [O = C = O], which being a **triatomic symmetrical molecule,** and **non-polar in character.** Even through the **Carbon ($>C = 0$) bonds** happen to be:

"*polar in status since the so-called* ***shared electron-pair* lies fairly closer to the respective O-atom** *vis-a-vis* **the C-atom, due to the relatively higher electronegativity profile of the O-atom.**"

Amazingly, the **CO_2-molecule** does possess **zero-Dipole Moment (μ).** Thus, we may explain both these typical characteristics *viz.,* **non-polar nature** and **zero Dipole Moment value** of the molecule may be expatiated clearly in a situation:

"**when it is assumed that the '***Molecule***' happens to a** *linear one* **so that the** *dipole moment of one* **$C^{3+} - C^{\delta-}$ bond located strategically on one side cancels that of the similar bond positioned on the other side**".

Hence, it may be duly expressed as:

$$O \overset{\delta^-}{=} \overset{\delta^+}{C} = O$$

(CO$_2$)

Remarks: Based on the similar arguments it would be possible to explain the status of the remaining *two* **molecules** *viz.,* CS$_2$ and HgX_2–that also specifically possess:

- **Non-polar status** and • **Zero Dipole Moment,**

and hence, are found to be *linear* **in configuration**–as shown under:

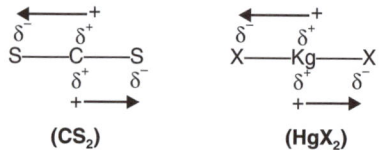

(CS$_2$)   (HgX$_2$)

2. **Water (H_2O), Hydrogen Sulphide (H_2S) Sulphur Dioxide (SO_2) and Nitrogen Dioxide (NO_2) Molecule** Interestingly, although the above mentioned *three* molecules, namely: H_2O, H_2S, and NO_2 are **triatomic in nature** that essentially bears a close resemblance to: CO_2, CS_2 and HgX_2 (*as in (1) above*), –yet these do exhibit and inherit certain value of **Dipole Moment** [*viz.*, H_2O = 1.85 D; SO_2 = 1.60 D etc.]

Based on the above findings and dictum it may be observed explicitly that:

"**there is absolutely little possibility for these molecules to possess vehemently a *Linear Shape* very much akin to CO_2, CS_2 anmd Hg X_2**".

On the contrary these molecules (H_2O, H_2S, SO_2, and NO_2) do invariably possess a *Bent Structure*,– thereby giving rise to the so-called *Resultant Dipole Moment*,–as depicted in Fig: 4.15.

Remarks: In short it may be added that in these molecules there exist *two* **critical Dipolar Bonds**; and hence, the so-called *Resultant Dipole Moment* of these separate **Dipolar Bonds** does provide the **Dipole Moment of the molecule**–usually known as the **Molecular Dipole Moment** (which has been depicted by a '**dotted line with an arrow**').

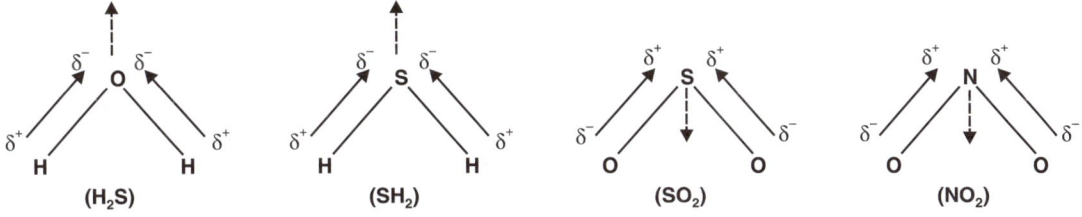

Fig. 4.15: Diagrammatic Representations of the Bent Structures of: H_2O, H_2S, SO_2 and NO_2 Molecules. [The Resultant Dipole Moment (or Molecular Dipolemoment) has been Explicitly Depicted by a Dotted Line with an Arrow.]

3. **Ammonia (NH_3) Molecule**–It certainly designates a **tetra-atomic molecule** and comprises *three* **distinct dipole bonds** [$N^{\delta-} - H^{\delta+}$] and their respective resultant does provide the so-called: **Molecular Dipole Moment** that has been determined experimentally equivalent to 1.46D, –as shown in Fig: 4.16(a) (b).

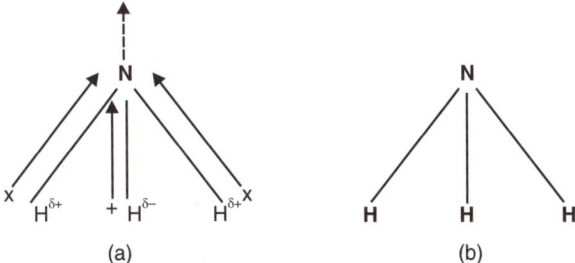

(a)   (b)

Fig. 4.16 (a) and (b) Display the Diagrammatic Representation of the Triangular Pyramidal Structure of Ammonia (NH_3) Molecule.

Remarks: Obviously, the crucial higher value of the **Dipole Moment** evidently reveals that *three* **H-atoms do not lie in the symmetrical configuration with regard to the N-atom**. In addition, the *high value* of the **Dipole Moment** may be expatiated further by providing:

"a typical *Triangular Pyramidal Structure* to the NH_3-Molecule, wherein the N-atom is located strategically at the *'apex'* wherein the *three* H-atom do lie at the other *three* corresponding corners perceptively (as depicted in Fig: 4.16(a) and (b))."

1.6. The Co-Ordinate Bonds

Preamble–We have already seen that the **Covalent Bond** usually results from the **critical sharing of a pair of electrons** existing between *two* atoms wherein:

"each atom contributes specifically *one electron* to the Bond".

However, it may also be feasible to have **an electron-pair bond in which both the electrons eventually originate from one atom and none from the other atom perceptively.** Hence, such bonds are termed as **Coordinate Bonds (or Dative Bonds).**

Because, in the so-called **Coordinate Bonds** we may observe the crucial involvement of:

"*two* electrons being shared intimately by *two* atoms; and hence, they do actually differ from the *Normal Covalent Bonds* only in the manner they are formed genuinely–thus, once formed these are very much similar to the aforesaid *Normal Covalent Bonds*".

Example: In the typical instance of the **Ammonia (NH_3)** molecule one may observe that:

'even through the NH_3–molecule possesses a *stable-electronic configuration,*'–it may vividly react with a H^+–ion by the donation of a share in the pair of electrons thereby resulting the ammonium ion (NH_4^+) ultimately'.

The above episode may be expressed as under:

$$H:\underset{\underset{H}{\cdot\cdot}}{\overset{H}{\underset{\cdot\cdot}{N}}}: \ + \ [H]^+ \longrightarrow \left[H:\underset{\underset{H}{\cdot\cdot}}{\overset{H}{\underset{\cdot\cdot}{N}}}:H \right]^+ \quad or \quad \left[H\!-\!\underset{\underset{H}{|}}{\overset{\overset{H}{|}}{N}}\!\rightarrow\!H \right]^+$$

❑ *Display of Covalent Bonds*–may be shown by means of *straight lines* linking the *two atoms* and the **Coordinate Bonds** (as *'arrows'*)–there by indicating which particular atom is actually donating the electrons.

Likewise, **ammonia (NH_3)** may perceptively donate the **lone pair of electrons** to **Boron Trifluoride (BF_3);** and thus, the so-called **Boron Attains** a *definte share in the 8-electrons-episode,* as shown under explicitly:

$$\underset{(NH_3)}{H:\underset{\underset{H}{\cdot\cdot}}{\overset{H}{\underset{\cdot\cdot}{N}}}:} \ + \ \underset{(BF_3)}{\underset{\underset{F}{\cdot\cdot}}{\overset{F}{\underset{\cdot\cdot}{B}}}:F} \longrightarrow H\!-\!\underset{\underset{H}{|}}{\overset{\overset{H}{|}}{N}}\!-\!\underset{\underset{F}{|}}{\overset{\overset{F}{|}}{B}}\!-\!F$$

In the same fashion a **molecule of BF_3** may duly give rise to the formation of:

'a *Coordinate Bond* by accepting duly a share in a Lone Ion Pair to form a Fluoride (F) ion.'

Thus, we may have the following expression:

$$\left[:\overset{\cdot\cdot}{\underset{\cdot\cdot}{F}}: \right]^- + \ \underset{\underset{F}{\cdot\cdot}}{\overset{F}{\underset{\cdot\cdot}{B}}}:F \longrightarrow \left[F\!-\!\underset{\underset{F}{|}}{\overset{\overset{F}{|}}{B}}\!-\!F \right]^-$$

Another school of thought, describes the **Co-ordinate Bond** as:
'**co-ordinate covalent bond (or Dative Bond)**".

❑ **Illustration of the Formation of a Co-ordinate Bond**–In reality, the formation of a **Co-ordinate Bond** between the *two* **atoms** (*viz.*, **A** and **B**) may be illustrated clearly as shown below:

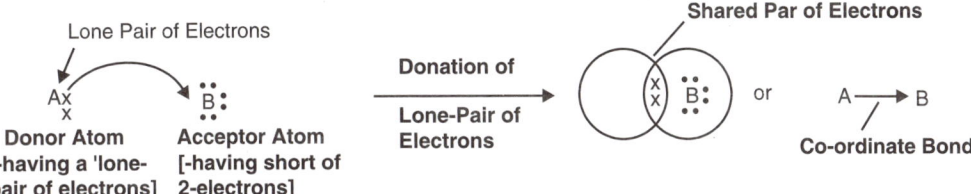

Lone Pair of Electrons

Donor Atom   **Acceptor Atom**
[-having a 'lone-   **[-having short of**
bair of electrons]   **2-electrons]**

Donation of
Lone-Pair of
Electrons

Shared Par of Electrons

or

Co-ordinate Bond

Explanation: The various steps involved may be explained as under:

1. The **donor atom 'A'** does possess a spare lone pair of electrons on it, whereas, the **acceptor atom 'B'** is rather short of *two* electrons *vis-a-vis* the **octet** in its *Valence Shell*.

2. Thus, **atom 'A'** eventually donates its so-called **lone-pair-of electrons** to **atom 'B'** that accepts the same gracefully.

3. In this way, the **2-electrons belonging to the lone pair** that originally belonged to **atom 'A'** are now being shared by **both the atoms**; and hence, the ensuing *mutual sharing phenomenon* of the so-called **electron pair** gives rise to the formation of a **Co-ordinate Bond** perceptively between:

> **'A' and 'B' (A → B)**

4. Even though the '**arrow head**' vividly indicates the actual origin of the electrons, once a **Co-ordinate Bond** has been established, it becomes rather **quite similar** with a *Normal Covalent Bond*.

NOTE   Since both the aforesaid bonds, namely: Co-ordinate Bond out *Normal Covalent Bond*–are duly established by the so-called: a shared electron pair; and thus, do exhibit almost identical characteristic features.

1.6.1. *Parameters Required for the Formation of a Co-ordinate Bond*

Based on the logical scientific evidences do reveal that the various important parameters that are found to be absolutely necessary and critical for the ultimate formation of a **Co-ordinate Bond** (®) are of *two* **kinds**, namely:

- **An** *atom* acting as a *Donor* must possess a *lone-pair of electrons*, **and**
- **An** *atom* acting as an *Acceptor* must have a *vacant orbital* so as to accept the electron-pair duly donated by the *Donor*.

The various compounds that essentially possess the **co-ordinate bonds** are usually known as the **Co-ordinate Compounds** [or **Lewis Formulae**, or **Electronic Formulae of Compounds**], such as:

➢ **Hydrogen Peroxide (H_2O_2) Molecule**–Amazingly, the critical formation of the **H_2O_2-molecule** and **Oxygen (O) atom**, it may be observed that:

'the O-atom of H_2O molecule possesses 2-lone pairs of electrons located on it perceptively.'

Hence, in the critical formation of **H_2O_2 molecule** due to the critical combination of **H_2O molecule** and **O-atom**, one may take cognizance of the fact that:

'one of the 2-lone pairs located on O-atom of H$_2$O-molecule is being donated to the *newly formed O-atom*';

and hence, the desired **Co-ordinate Bond** gets duly established between the **O-atom (in the H$_2$O molecule)** and the **Newer O-atom**, as shown under:

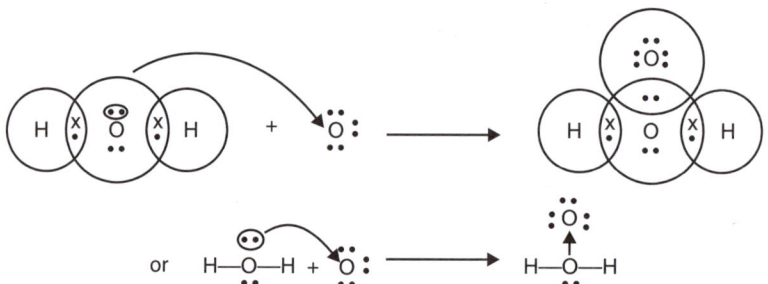

➤ **Fluoroborate Ion [BF$^-_4$]**–The formation of the **Fluoroborate Ion (BF$^-_4$)** comes into being when the **Borontrifluoride (BF$_3$) molecule** does share a pair of electrons duly provided by the fluoride (F$^-$) ion:

$$\text{F}-\overset{\overset{\text{F}}{|}}{\underset{\underset{\text{F}}{|}}{\text{B}}} \quad + \quad (\cdot)\overset{\cdot\cdot}{\underset{\cdot\cdot}{\text{F}}}: \quad \longrightarrow \quad \text{F}-\overset{\overset{\text{F}}{|}}{\underset{\underset{\text{F}}{|}}{\text{B}}}\leftarrow\text{F}^- \quad \text{or BF}_4$$

Boron Trifluoride   Fluoride Ion   Fluoroborate Ion
[B-Serves an Aceptor]

➤ **Carbon Monoxide [CO] Molecule**–Since, we know that the **C-atom** does possess *four* valence electrons **[2, 4]**; whereas, the **O-atom** has *six* valence electrons **[2, 6]**. Hence, due to the critical formation of **2-covalent bonds** between them [*i.e.*, **C** and **O**], the respective **O-atom** accomplishes an **Octet**,–but the corresponding **C-atom** bears only **6-electrons**. Hence, the **O-atom** donates crucially an **unshared electron pair to the C-atom**; and thus, a **Co-ordinate Covalent Bond** gets established predominately between the **2-atoms**.

Hence, the **Carbonmonoxide [CO] molecule** possesses **2-covalent bonds** plus **1-co-ordinate bond [O → C]**, as expressed under:

Distribution of Electrons in Orbitals $\overset{\cdot\cdot}{\text{C}}:$ **[2,4]** $+$ $\underset{\text{x}}{\overset{\text{xx}}{\text{x}}}\text{O}\underset{}{\text{x}}$ **[2,6]** $\longrightarrow$ ⊙ **[2,6]   [2,8]**   **or** $\overset{\cdot\cdot}{\text{C}}=\text{O}\underset{\text{x}}{\overset{\text{xx}}{}}$ **[2,6] [2,8]** $\longrightarrow$ $\overset{\cdot\cdot}{\text{C}}\equiv\text{O}\underset{\text{x}}{\overset{}{}}$ **[2,8]   [2,8]**

➤ **Sulphur Dioxide (SO$_2$) and Sulphur Trioxide (SO$_3$) Molecules**–It may be noticed that the **Sulphur (S) atom** accomplishes its *Octet* by specifically forming 2-Covalent Bonds with '**1**'– **O-Atom**, thereby yielding the **SO (Sulphur Oxide) species** perceptively. However, the S-atom present in the **SO** possesses **2-lone pairs of electrons** of which:

"one is being shared with a 2nd O-atom to form SO$_2$".

Thus, we may have the following explicit expression:

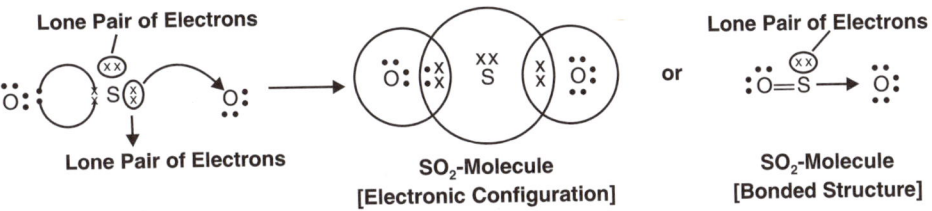

Lone Pair of Electrons

Lone Pair of Electrons

Lone Pair of Electrons

SO_2-Molecule
[Electronic Configuration]

Lone Pair of Electrons

SO_2-Molecule
[Bonded Structure]

Interestingly, the **S-atom** in the **SO_2-molecule** still possesses– 'one lone pair of electrons'–that it eventually donates critically to a *third O-atom thereby forming the SO_3-molecule ultimately, –as* shown below:

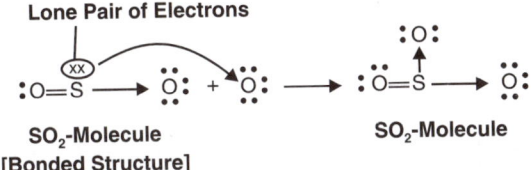

Lone Pair of Electrons

SO_2-Molecule
[Bonded Structure]

SO_2-Molecule

➤ **Sulphate [SO^{-2}_4] Ion**–Obviously, owing to the critical presence of:

"*two* **negative charges located strategically on the ion, the respective S-atom that possesses** *6-electrons* **in the** *Valence Shell (2, 8, 6)* **duly gains** *two* **additional electrons derived from the** *metal ion* **so as to complete its** *Octet* **perceptively**".

In this way, one may observe that in the **Sulphate [SO^{2-}_4] Ion**, the **S-atom** possesses *8-electrons* (*i.e., four* **pairs of electrons**):

- *six* **of its own, and**   • *two* **gained from** '*metal ion*'.

Thus, we may have the following expression:

$$\left[\begin{array}{c} \vdots O \vdots \\ \vdots O \quad S \quad O \vdots \\ \vdots O \vdots \end{array} \right]^{2-}$$

[SO^{2-}_4 Ion]

In this manner, the *four* **resultant electron pairs** located strategically around the **S-atom** present in **Sulphate Ion [SO^{2-}_4 Ion]** are being donated meticulously to the 4, **O-atoms** each of which thus **completes** its **Octet** perceptively.

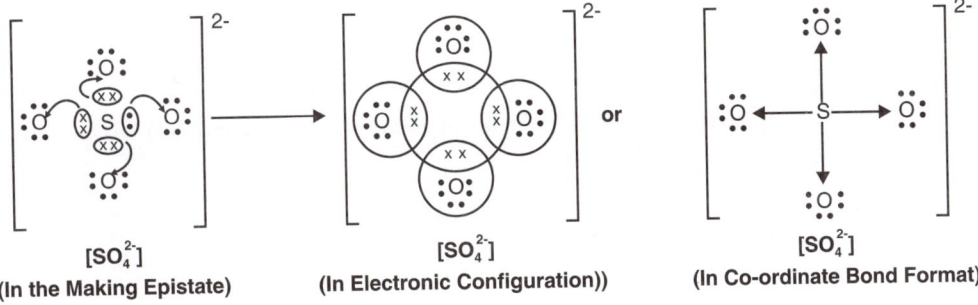

[SO^{2-}_4]
(In the Making Epistate)

[SO^{2-}_4]
(In Electronic Configuration)

or

[SO^{2-}_4]
(In Co-ordinate Bond Format)

Hence, we may vividly observe that the so-called **S-atom** is actually joined with **4, O-atom** by the aid of *four* **Co-ordinate Bonds**,–as illustrated below diagrammatically:

➤ **Ammonium [NH₄⁺] and Hydroxonium [H₃O⁺] Ions**–Importantly, the **ammonium [NH₄⁺] ion** may be duly considered to be obtained by the critical combination of **Ammonia (NH₃) molecule** and **Hydronium (H⁺) ion**. Thus, in the respective **ammonia (NH₃) molecule**– each of the **3, H-atoms** is duly linked to the **N-atom** by a **Covalent Bond**. Therefore, the **N-atom** present in this molecule (NH₃) is virtually left with a **pair of electrons** after completing its 'Octet' due to the critical **sharing of 3 of its 5 valence shell electrons with 3, H-atoms.**

Three Important Observations: These, essentially include:

1. Electrons present in the 'lone-pair' located on the **N-atom** are being donated to the respective **H⁺ ion**; and hence, **the formation of a $N \circledR H$ co-ordinate bond is established in NH₄⁺ ion** definitely, as indicated under:

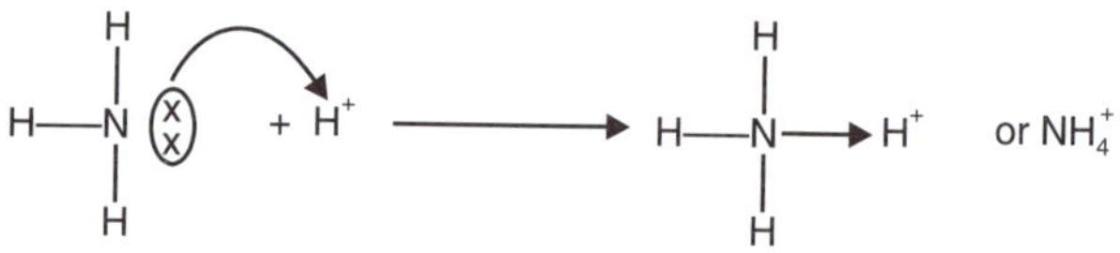

[NH₄⁺ Ion]
[In Coordinate Bond Format]

Soonafter the formation of the **ammonium (NH₄⁺) ion**–the *four* **N–H Bonds** do eventually become identical; and hence, the said **NH₄⁺ ion** can also be represented as 'shown in the margin' below:

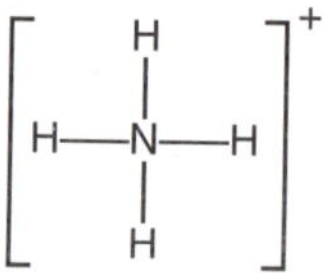

Hydroxonium Ion [H₃O⁺]–Amazingly, the **hydronium (H⁺) ion** in an *aqueous solution* is invariably found to be in absolute associated with **H₂O as H₃O⁺ ion** on account of the *inherent* **co-ordination phenomenon** of:

'the H⁺ion with the so-called Neutral Water Molecule'.

Hence, it may be expressed as under:

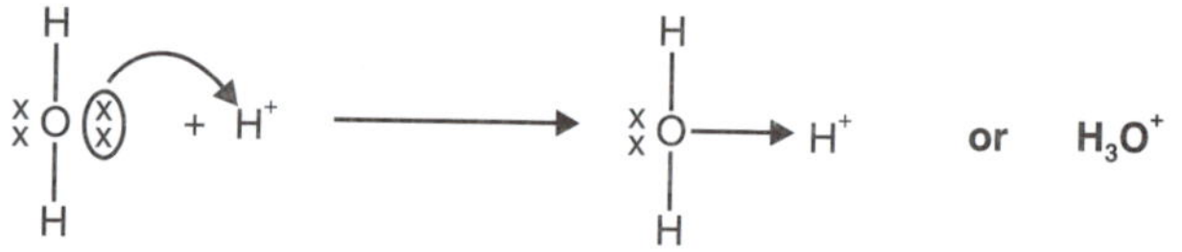

(Water)   (Hydronium Ion)   H₃O-molecule   Hydroxonium Ion

Copper (II) Tetra-Amine Ion [Cu(NH₃)₄]²⁺–In true sense, this ion, **cupric (II) tetra-amine ion**, is duly formed by:

"intricate linkage of four ammonia (NH₃) molecules together with the cupic (Cu²⁺) in by the aid of *four co-ordinate bonds* perceptively".

Thus, we may have the following expression:

4NH₃ + CU²⁺ → [H₃N → Cu ← NH₃ with NH₃ above and NH₃ below]²⁺ or [Cu(NH₃)₄]²⁺

Ammonium   Cupric
Molecule   Ion

[Coper (II)-tetra
Amine Ion]

[Coper (II)-tetra Amine Ion in
Co-ordinate Bond Format]

➤ **Aluminium Chloride [Al₂Cl₆]**–The **aluminium (Al) atom** essentially possesses **3-valence electrons**–which it shares intimately with **3Cl-atoms** to yield **3-covalent bonds** perceptively. Therefore, in the **Aluminium Chloride [AlCl₃]** molecule–the inherent **Al-atom** critically acquires **6-electrons** in its *outermost shell*. Thus, each of the **three prevalent Cl-atoms** present strategically in the **AlCl₃** molecule essentially possesses **3-lone pairs of electrons** *per sc*.

However, one of these **lone pairs of electrons is duly donated to the Al-atom of another AlCl₃-molecule**; and hence, suggests strongly that both the **Al-atoms** do achieve an 'Octet' plus a *fairly stable Al₂Cl₆* molecule,–as

Lone Pair of Electrons
Al₂Cl₆-Molecule
[Showing 2-separate lone pairs of Electrons]

Aluminium Chloride [Al₂Cl₆]

➤ **Addition Compounds**–It has also been established beyond any reasonable doubt that the so-called **Co-ordinate Bond** is being duly formed: "**critically between** *two* **atoms both of which virtually serves as the integral parts of a** *Stable Molecule*".

Example: The **Boron Fluoride Ammonia Complex [H₃N → BF₃]** serves as a befitting example of the **Addition Compound**, as given under:

Formation of the Addition Compound [H₃N →BF₃]

or   H₃N⟶ BF₃

Boron Fluoride Ammonia
Complex
[Addition Compound]

NOTE	In usual practice, the *Addition Compounds* are invariably represented by placing a dot (.) between the formulas of their respective constituents molecules *viz.,* NH₃.BF₃.

1.6.2. *Characteristics of the Coordinate Compounds*

There are *eight* different vital and important characteristics of the **Co-ordinate Compounds**, namely:

(a) **Melting Points/Boiling Points and Viscosity**–It has been already been proven that a **Co-ordinate Bond** is formed by the critical combination of:

➢ **an Ionic Bond**, and

➢ **a Covalent Bond**.

Perhaps based on the aforesaid glaring facts that the so-called **Co-ordinate Compounds** do exhibit **MPs and BPs** and **Viscosities**–that are observed to be relatively *higher vis-a-vis* those of purely *covalent compounds* but definitely *lower* than those of the corresponding *Ionic Compounds* perceptively.

(b) **Semi-Polar Character**–In a broader sense, these **chemical entities** (*compounds*) are indeed **semi-polar in character [A⁺–B⁻]** *i.e.*, these are certainly *more polar* in comparison to the **Covalent Compounds**; and thus, *less polar vis-a-vis* the **Ionic Compounds** perse.

(c) **Physical Status**–Mostly these compound could be seen as: **Solids, Gases, or Liquids**.

(d) **Solubility Profile**–In reality, these are invariably insoluble in the **polar solvents** *viz.*, water (H_2O), but are found to be soluble in the **non-polar organic solvents** (*viz.*, *n*-**Hexane, Benzene**).

(e) **Conductivity Status**–Very much akin to the **Covalent Compounds** the respective **Co-ordinate Compounds** are usually **non-ionic in nature** *i.e.*, these **fail to conduct electric current** *via* their so-called:

• **fused mass**, or

• **aqueous solutions**.

(f) **Molecular Reactions**–It is well-known that the **Co-ordinate Chemical Entities (Compounds)** are molecular in status; and, therefore, do undergo **molecular reactions**, but are found to be **rather slow**.

(g) **Stability Profile**–Amazingly, the **Co-ordinate Compounds** are indeed as stable as the corresponding **Covalent Compounds**. However, in a specific instance when these are actually made up of *two* **divergent stable molecules** (or **molecular compounds**) they **may not prove to be very stable at all**.

(h) **Isomerism**–The **Co-ordinate Compounds** do exhibit **isomerism** perceptively. Because, the ensuing *Coordinate Bond* is found to be reasonably **rigid** and **directional** *in terms of different space Models* (*i.e.*, the respective 'Stereoisomers') **of a single co-ordinate compounds** may be possible ultimately.

1.6.3. The Comparative Features of Ionic, Covalent, and Co-ordinate Bonds

The explicit comparison amongst the *three* types of bonds occurring in compounds *viz.*, **Ionic Bond, Covalent Bond**, and **Co-ordinate Bond** are provided in Table 4.4.

Table: 4.4: Explicit Comparison Amongst the Ionic, Covalent, and Co-ordinate Bonds in Various Compounds.		
S.No. **Ionic Bond**	**Covalent Bond**	**Co-ordinate Bond**
1. Ionic bond is formed by the transfer of electrons from a metal atom (A) which has 1,	(i) Covalent bond is formed by sharing two electrons between non-metal atoms having 1, 4, 5, 6	(i) It is formed by the sharing of two electrons between two atoms, both electrons coming from one atom.

(Contd....)

	2 or 3 valence-electrons to a non-metal (B) having 5, 6 or 7 valence electrons.	or 7 valence electrons.	

$$A\times + \ \overset{\bullet\bullet}{\underset{\bullet\bullet}{B}}\bullet \longrightarrow [A] + \left[\times \overset{\bullet\bullet}{\underset{\bullet\bullet}{B}}\bullet\right]^{-}$$

or A + B⁻

$$\overset{\times\times}{\underset{}{\times}}A + \ \overset{\bullet\bullet}{\underset{\bullet\bullet}{B}}\bullet \longrightarrow \overset{\times\times}{\underset{\times\times}{\times}}A + \ \overset{\bullet\bullet}{\underset{\bullet\bullet}{B}}\bullet$$

or A –B

$$\overset{\times\times}{\underset{}{\times}}A + \ \overset{\bullet\bullet}{\underset{\bullet\bullet}{B}}\bullet \longrightarrow \overset{\times\times}{\underset{\times\times}{\times}}A\times \longrightarrow \overset{\bullet\bullet}{\underset{\bullet\bullet}{B}}\bullet$$

or A⟶B

2.	Ionic bond consists of electrostatic force between cations and anions.	(ii) Covalent bond consists of two electrons that hold the atoms together.	(ii) It consists of an electron pair between the linked atoms.
3.	It is a weak bond, since the electro-static force can be broken easily.	(iii) It is a strong bond, since the paired electrons cannot be separated easily.	(iii) It is also a strong bond, since the paired electrons cannot be separated easily.
4.	It is a polar bond ($A^+ - B^-$)	(iv) It is non-polar bond (A–B)	(iv) It is a semi-polar bond ($A^+ - B^-$)

1.7. THE METALLIC BONDS

Preamble: In general, the **metals** are duly characterized by certain specific **physical characteristic features.** However, it is always important to mention at this material time that:

    "even though all these refer to the characteristics of the 'metals', –a few typical exceptions to practically all the properties do exist perceptively".

1.7.1. *Properties of Metals*

    Following are the various important properties of **Metals**, namely:

1. **Metals** do possess a *higher level of* **electric conductivity** that eventually gets decreased with the temperature apparently.
2. **Metals** do have high **MPs** and **BPs**.
3. **Metals** are found to be good conductors of **heat energy**.
4. **Metals** do possess a **classical and peculiar shine** upon their surface (usually called the *metallic lustre* and may retain **good polish elegantly**.
5. **Metals** usually exhibit **high density profile** and are **hard in state.**
6. **Metals** are invariably malleable and ductile *i.e.*, it may be **beaten into sheets** and may be drawn into thin wires.
7. **Metals** are able to withstand **high-level of stress** *i.e.*, they do exhibit **high elasticity profile**.
8. **Metals** are indeed opaque to light.
9. **Metals** do form '*solid solutions*' with each quite easily; and hence, these reaswulting *solid solutions* are usually called '*Alloys*' *viz.*, **Al-Alloy, Pb-Alloy**, etc. that eventually possess all **metallic characteristics.**
10. **Metals** do have a tendency to crystallize in such **systems** having *high-co-ordination numbers* of **8, 12, or 14**.
11. Metals also give rise to the formation of **non stoichiometric compounds** wherein the 'other atoms' (*viz.*, **H, B, C or N**) get strategically trapped into the so-called *interstices* **of the metallic structure perceptively**. In fact, these resultant substances do possess the **metallic characteristics features.**
12. **Metals** are invariably '**electropositive elements**'.

13. **Metal's** inherent reactivity profile gets enhanced along with the **state of subdivisions** specifically. Hence, the *finer particulates* **of metals** are found to be much more reactive chemically *vis-a-vis* the metal present in a **massive lump form**.

1.7.2. *Arrangement of Atoms in Metals*

The various physical characteristic features of 'metals', such as:

- **Density** • **Hardness** • **Melting Point** • **Heat of Fusion** • **Heat of Varporization and**
- **other Allied Properties,**

may be intimately correlated with the specific arrangement of the atoms in various inherent *Crystal Lattices*.

Obviously, the **X-ray diffraction (XRD) patterns do** evneutally indicate that: "**in** *solid state*–**the atoms are arrangement meticulously in the following** *three configurations profile*", such as:

❑ **The Hexagonal close Packed (hcp) format,**

❑ **The face centred Cubic (fcc) format,** and

❑ **The Body centered Cubic (bcc) geometrical arrangements**.

Remarks: It may be stated at the very outset that both hep (*first*) and fcc (**second**) arrangement of atoms do posses:

"**a minimum quantum of vacant space (~ 26%) provided the atoms are duly regarded to be occurring as** *Hard Spheres* **perceptively.**"

Besides, **bcc** (*third*) lattice is observed to be rathter more often having surely 32% vacant space.

Important Observations: These essentially comprise:

(1) Importantly, in a *Single Layer*–the only observed arrangement of **hard spheres** that certainly provides the *minimum vacant-space* is duly designated as and when **each sphere** is surrounded perceptively by and is in contanct with another *Six spheres* as illustrated in Fig: 4.17 (a) and (b).

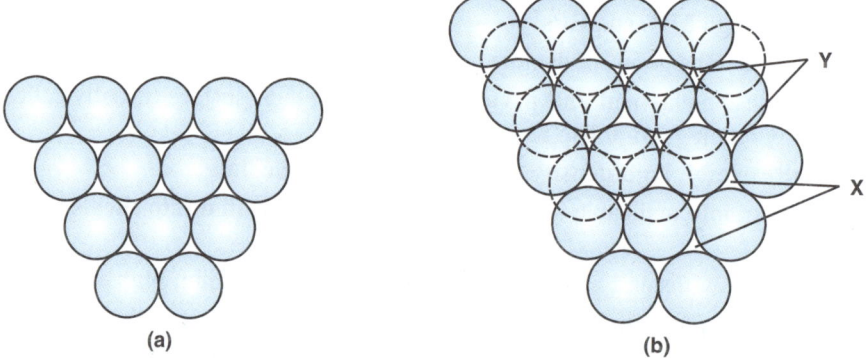

(a)   (b)

Fig. 4.17: (a) and (b): The Tetrahedral (X) and Octahedral (Y) (b) Holes Present in a Closed Packed Arrangement of Spheres.

Explanations: The various aspects of Fig: 4.17(a) and (b) may be explained as under:

1. Fig: 4.17(b) shows explicitly that a **second layer** of the so-called *closely packed spheres* may be placed strategically on top of the layer thereby providing bare minium void space (only in a situation when the spheres are duly placed strategically in the hollows of the layer,–as illustated by **(X) in Fig: 4.17(b)**).

2. Obviously, the so-called **hollow-space** is duly present in the centres of the **Equilateral Triangles** formed by joining the centres of any *three* **spheres** touching one another meticulously.

3. Hence, one may observe that the **2-layers** appear vividly as could be seen in **Fig: 4.17(b)**.

4. Besides, two equivalent but '*alternate sites*' are also available for the critical placement of the **3rd layer of the closely packed spheres**– both clearly provide the so-called:

"closest packing episode of the spheres"

5. Thus, one may place the spheres of the **3rd layer** in the so-called *hollow space* located almost right above the **spheres of the 1st layer**. Hence, it forms perceptively the so-called **Face-centered Cubic (fcc) lattice** in the spheres,–that are, in fact, so arranged that:

"the arrangement of the layer becomes: ABC ABC..."

6. In a typical instance, when the **3rd layer spheres** are positioned directly over the spheres of the **1st layer**–thereby given rise to the arrangement of the said layers as: ABA BAB..., the so-called:

"hexagonal close packed structure (hep) comes into being".

NOTE   **Amazingly, the vacant space in both these instances *viz.*, the '*fcc*' and the '*hep*' arrangements remain in the same (26%).**

2. **The Body-Centered Cubic (bcc) Lattice Arrangement**–Importantly, in the *bcc* **lattice**, one may observe that the spheres are arranged in such a fashion in a layer that **each sphere** is duly surrounded by and remains in contact with the **4-spheres of the adjacent layer**. Besides, the spheres of the **3rd layer** are placed strategically over the so-called **spheres** of the **1st layer** perceptively,–as depicted in Fig: 4.18.

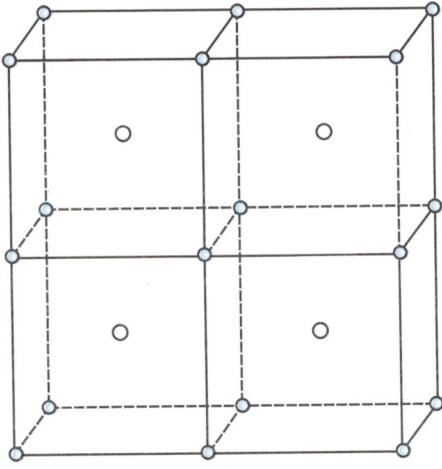

Fig. 4.18: The Body Centered Cubic (BCC) Arrangement.

Remarks: Hence, the arrangement of layers in the **bcc arrangement** is, therefore, observed to be: AB ABAB...and the ensuing **Vacant Space** is **32%**.

(3) **The *hcp* and *fcc* Arrangements**–In these two cases, the observed co-ordination number of the atom is 12, since the *central atom* of a layer is meticulously surrounded by:

- 6-*atoms* in its own layer; and
- 3-atoms each from layers at the top and the bottom,

as illustrated in Fig: 4.19.

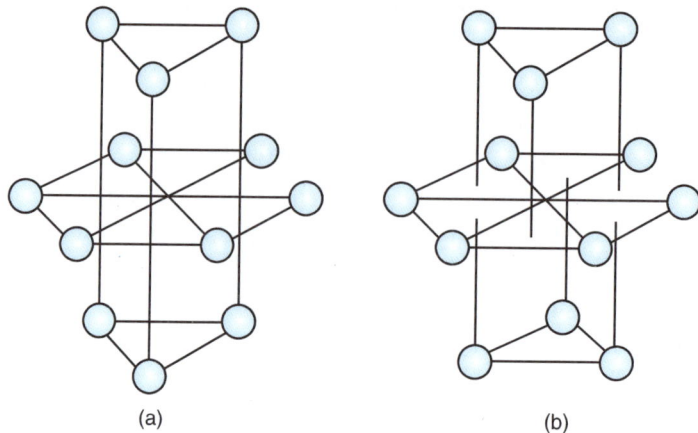

(a)   (b)

Fig. 4.19: The Coordination Numbers for (a) Hexagonal Close Packed (hep); and (b) Face Centered Cubic (fcc) Close Packing.

Conclusive Remarks: The structure (Fig: 4.19) essentially possesses a '*Single Atom*'–which is sandwitched between the **2-layers of** *four* **atoms** each, so that each atom bears a **coordination number of** *eight*–explicitly resulting in an **Open structure perceptively**. Nevertheless, as in **bee**, each atom critically possesses 6 more atoms at a slightly layer distance (~ **1.5 times**) the **coordination number** of the atom present in **bcc** may be taken as **14 also**.

1.8. THE HYDROGEN BOND

Preamble–In order to understand the basic concept of the **hydrogen bond** (or *H-Bond*), one may take into consideration a '**molecule**' *viz.*, AH, wherein the **H-atom** is duly linked with:

"**a reasonably strong electronegative but very small atom 'A' (where 'A' may be N, O, or F) by the help of a *Normal Covalent Bond*.**"

Another school of thought sometimes considers **Hydrogen Bonding as:** "**a 5th fundamental type of chemical interaction perceptively**".

Points to Ponder: These essentially include:

1. The actual inherent strengths of **H-bonds** do lie as an **intermediate** between those of:
   - **van der Waal's Bonds**, and
   - **Ordinary Covalent Bonds**.

2. The electron pair is being shared between the **H-atom** and the **strongly electronegative atom 'A'**. Ultimately, it will prevalently lie *far way* from the incumbent **H-atom**; and hence, the so-called: "**partial positive and negative charges will be developed upon the *H-atomI* and the *A-atoms* respectively.**"

3. As a result, the **AH-molecule** shall more or less behave a '*Dipole*',–that could be expressed as under:

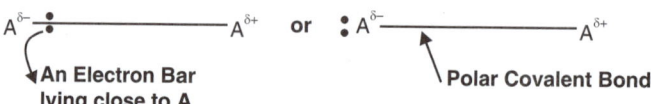

$A^{\delta-}$ ⋮ $A^{\delta+}$ or   ⋮ $A^{\delta-}$ $A^{\delta+}$

↓**An Electron Bar**   **Polar Covalent Bond**
lying close to A

4. Obviously, the '**Dipole**' critically shows the presence of:
   - '**A**'–as its *Negative* end, and
   - '**H**'–as its *Positive* end

5. Because, the inherent **electronegative atom 'A'** evidently attracts the *electron pair* constituting the respective **Covalent Bond** between the atoms '**A' and 'H**' almost completely towards itself. In doing so, the **atom 'A'** takes up practically the **complete possession of the ensuing 'electron pair'** *i.e.*, the **atom 'A' shall retain a love** pair of electrons perceptively.

6. The above episode shall render an event that eventually leaves the **H-atom having a large partial positive charge**; and hence, the **AH-molecule** turns into a **highly polar** and **its H-end** becomes **Proton (H)** or simply a bare H-nucleus *i.e.*, the **H-atom** gets reduced to a **Proton (H$^+$)** that is **absolutely devoid of electrons**.

7. Let us look into another typical situation where either AH or BH (**A and B being strongly electronegative atoms**)–that invariably gives rise to the formation of a **Dipole [A^{d-} –H^{d+} or B^{d-} – H^{d+}]** respectively. In this way, it is brought near **A^{d-} – H^{d+}** dipole, whereby these *two dipoles* shall be attracted towards each other by:

"**the ensuing** *Electrostatic Force of Attraction* **that may be duly represented either by a dotted line or dashed line; and is commonly termed as Hydrogen Bond or Hydrogen Bonding**".

The above sequence of events may be expressed as under:

Electro static Force
of Attraction

Covalent Bonds
Hydrogen Bond

Covalent Bonds

8. From the above expression with respect to the sequence of events we may observe that in **H-bond**, the **H-atom** duly forms a sort of '**Bridge**' existing between *two* **electronegative atoms**. Thus, based upon this logical explanation and dictum the **H-Bond** is also known as a **H-Bridge (or Hydrogen Bridge)**.

Hence, the **H-Bond** may be duly defined as:

"**the attractive electrostatic force between a H-atom which is being already attached covalently with a strongly electronegative atom of a molecule and another electronegative atom of certain other molecule is termed as Hydrogen Bond**".

The following are the *five* **different cardinal aspects of the Hydrogen Bond**, namely:

(a) **Nature of Hydrogen Bond,**

(b) **Characteristics of Hydrogen Bond,**

(c) **Variants in Hydrogen Bond,**

(d) **Consequences of Hydrogen Bond,** and

(e) **Significance of Hydrogen Bonding,**

which shall now be discussed individually and briefly in the sections that follows:

1.8.1. *Nature of Hydrogen Bond*

It is an universal fact that the so-called **Covalencey of H-atom** is confined solely to only one atom *i.e.*, the **H-atom** may be specifically linked to **only one atom (say 'A')**. The above crucial observation is solely based on the fact that:

"**1s atomic orbital of H-atom becomes completely filled soonafter it results into the formation of a Covalent Bond (A–H)**".

Explanations

1. Since H-atom has already bonded covalently to the **atom 'A'** in the **A–H bond**; and hence, it fails to form a *second* **Covalent Bond with another atom (say 'B')**.

2. The above expression of thought clearly reveals that **A–H–B**, wherein the **H-atom** depicts *bivalence* is quite unlikely to occur. If it is, so it would certainly require the use of either the **2s or 2p orbitals of H-atom**,–that are eventually of **much higher energy profile**; and hence, are of no use for bonding phenomenon at all.

4. Obviously, the only atoms *viz.*, **O, N**, and **F** do have **high electronegativity profile** as well as **small atomic size**; and hence, these atoms are capable of forming the H-Bonds squarely.

5. The so-called **H-bonding phenomeon** results predominantly either:
   - *critical formation* of long chains; or
   - **clusters of a relatively large number of closely associated molecules** *viz.*, *several small magnets*.

6. Just like the **Covalent Bond**, the **H-Bond** possesses a so-called:
   "**a preferred *Bonding Direction* characteristic feature**".

Importantly, the process of **H-bonding** comes into play *via* the **p-orbitals**–that essentially contain a **lone-pair of electrons** located on **atom 'A'**. Perhaps, it implies vehemently that:

"**all the *three* atoms present in A–H...A shall be aligned in a straight line by all means.**"

1.8.3. *Variants in Hydrogen Bonds*

Based on the *scientific findings* and *logical interpretations*, we may come across *two* **important variants in Hydrogen Bonds, namely:**

(a) Association [Inter-Molecule Hydrogen Bond]–Interestingly, this kind of H-bond invariably takes place between 2 or more molecules either:
   - **belonging to the same compound**; or
   - **belonging to the different chemical entity (compound)**.

Examples: The *three* glaring examples may be seen in the following *three* molecules, such as:
   - **Ammonia (NH_3)** • **Water (H_2O) and** • **Hydrofluric Acid (HF)**,

which are indeed *associated by the intermolecular H-bond*.

(b) Chelation [Intra-Molecular Hydrogen Bond]–Interestingly, this kind of of a **Hydrogen Bond [H-bond]** is usually formed between:
   - **a H-atom and** • **an Electronegative atom**,

present strategically in the *same molecule* (*Intra:* means **within**)

(c) Besides, in case **A–H–B** is presumed to be correct, one would certainly expert the location of the **H-atom** to be of **equidistant from the 'A' and 'B' atoms** (provided the electronegativities of 'A' and 'B' do remain the same.)

Comment: Based on the experimental evidences that *H-atom* **in H-bond** is found to be closer to that atom with which it does from the **Covalent Bond**. Hence, it may be seen vividly that:

> **"Hydrogen Bond is electrostatic in nature** *i.e.,* H-bond is simply an electrostatic force rather than an actual Chemical Bond".

1.8.2. *Characteristics of Hydrogen Bond*

Following are the *six* **important characteristics of Hydrogen Bond**, namely:

1. The **H-Bond** designates a *bond of hydrogen* prevailing between *two* **electronegative atoms** exclusively. However, it rever involves **more than** *two* **atoms (excluding the H-atoms).**
2. The **Bond Energy** of a *H-Bond* prevails within the **range of 3–10 k.cal.mole^{-1}**; whereas, that of a **Normal Covalent Bond** falls in the range of 50–100 k.cal.mole^{-1}.

Remarks: Thus, it may be observed explicitly that–

"with the enhancement of electronegativity profile of the ensuing atom to which the *H-atom* is linked covalently,–thereby the ultimate strength of *H-Bond* also gets increased accordingly".

In this way, the prevailing strength of **H-Bonds** present duly in:

- **N–H...N** • **O–H...O** and • **F–H...F**,

attains the following order perceptively:

Order of Strength: N–H...N < O–H...O < F–H...F

Order of Electronegativity Values: N(= 3.0) < 0 (= 3.5) < F (= 4.0)

NOTE   The 'numbers' shown in parenthesis do indicate the *electronegativity value* of the concerned elements.

3. Amazingly, the formation of a H-bond does not involve any start of **sharing of the electron pairs**; and hence, differs from the **Covalent Bond**.

Besides, one may take cognizance of the fact that in the typical instance of **Intramolecular H-Bonding [Chelation]** the so-called:

> **"H-atom gets duly bonded to *two* atoms of the same molecule".**

Therefore, this kind of **H-bonding** may eventually lead to the critical linkage of *two* **groups** so as to afford the ensuing formation of a **Ring Structure**; and hence, such an overall, effect may be considered as–'**one kind of** *Chelation* **phenomenon perceptively'**.

NOTE   Amazingly, the crucial occurrence of this kind of H-bond fails to distrub/dislodge the so-called *Normal Bond Angles*.

Examples: Following are a few **classical examples of molecules** that explicitly illustrate the **Intramolecular H-bonding**, which are duly observed in these *three* molecules, namely:

- *o*-**Nitrophenol** • *o*-**Hydroxy benzaldehyde** and • α-**Hydroxy benzoic acid**

| *O*-Nitro phenol | *O*-Hydroxy-benzaldehyde | α-Hydroxy benzoic acid [Salicylic Acid] |

Besides, we may have other examples also as: *o*-**Chlorophenol**, **Maleic Acid**, and *o*-**Nitrobenzoic Acid** etc.

Important Observations: These essentially comprise:

1. One may critically observe in the aforesaid molecules–that the **intra-molecular H-bonding** ultimately leads to:

   '**the critical formation of a 6-membered ring (phenyl) due to** *chelation*'.

2. Besides, the specific **occurrence of H-bond** of such a nature and kind may not be either possible or feasible in such compounds as:

| *m*-Nitro phenol | *p*-Nitro phenol |

which may be due to the ***prevalent size of the ring*** that would finally be accomplished.

3. Since the absolute non-existence of the '**H-bond**' in both *m*-**and** *p*-**isomers** which ultimately leads to the inheritence and display of various characteristic features *vis-a-vis* those of **other isomers** vividly.

Examples: These essentially comprise:

(a) The **MP of a-nitrophenol** stands at **214°C**; whereas, those of its respective *m*- and *p*- isomers are found to be: **290°C and 279°C respectively**.

(b) It has been observed that *p*-**Nitrophenol** certainly displays the so-called **inter-molecular H-bonding phenomenon** thereby critically ascertaining in the **perceptive association of two molecules as illustrated under:**

[2-Moles of *p*-Nitrophenol being enjoined *via* inter-molecular H-bending]

PROBLEMS WITH SOLUTIONS

[A] Problems based on Constructing Molecular Orbital Wave Functions and Valence Bond Wave Functions

Q.1. How will you explain the Molecular Orbital Wave Function for H_2^-–Molecule Anion?

Solution: Since it is known that H_2^-–**Molecule Ion** essentially comprises 3-electrons that may be labeled as 1, 2, and 3. It contains *two* **nuclei** which could be duly represented as **A** and **B**.

According to the following Equation:

$$\psi = \psi_1\psi_2\psi_3$$

where, $\psi_1\psi_2\psi_3$...etc. designate the **one-electron molecular orbital wave functions**.

$$t/mo = \psi_1\psi_2\psi_3$$
$$t/1 = c_1\psi_A(1) + c_2\psi_B(1)$$
$$t/2 = c_3\psi_A(2) + c_4\psi_B(2)$$
$$t/3 = c_5\psi_A(3) + c_6\psi_B(3)$$

Q.2. Based upon the underlying principle of LCAO for the Wave Function for the H_2^+ ion, how we may arrive at the so-called normalised *wave functions* for the BMO and ABMO.

Solution: Based upon the principle of LCAO, the ensuing Wave Function for H_2^+ ion may be expressed as under:

$$\psi = N(c_1\psi_A + c_2\psi_3) \tag{a}$$

where, N = Normalization constant.

According to the normalization parameter, for real ψ (*i.e.*, the amplitude of vibration for the wave in **wave Equation**), we have:

$$\int_{-\infty}^{+\infty} \psi^2 d \equiv <\psi \mid \psi>^{**} = 1 \tag{b}$$

Inserting the value of ψ from **Eq. (a)** in **Eq. (b)**, we have:

$$N^2 \left\langle (c_1\,\psi_A = c_2\psi_B) \mid (c_1\psi_A + c_2\psi_B) \right\rangle = 1$$

or

$$c_1^2 \left[\langle \psi_A \mid \psi_A \rangle\right] + c_1^2 \left[\langle \psi_B \rangle\right] + 2c_1c_2[\langle \psi_A \mid \psi_B \rangle] = 1/N^2 \tag{c}$$

However, it may also be shown that:

$$\left\langle \psi_A \mid \psi_B \right\rangle = \left\langle \psi_B \mid \psi_A \right\rangle$$

**   In fact, this kind of notation is known as 'bra' and 'ket' notation. Besides, in this notation $\int_{-\infty}^{+\infty} \psi_1\psi_2\, dr$ and $\int_{-\infty}^{+\infty} \psi_2\hat{H}\psi_2\, dr$ are expressed as $< \psi_1 \mid \psi_2>$ and $< \psi_1 \mid \hat{H} \mid \psi_2 >$ respectively.**

Let us now assume that both ψ_A and ψ_B are normalized:

i.e., $$\langle \psi_A | \psi_A \rangle = \int \psi_A^2 dr \quad \text{and} \quad \langle \psi_B | \psi_B \rangle = \int \psi_B^2 dr = 1,$$

we may have the following expression:

$$c_1^2 + c_2^2 + 2c_1c_2S = 1/N^2 \qquad \text{(d)}$$

where, S = Overlap integral; and may be defined as:

$$S = \langle \psi_A | \psi_B \rangle = \int \psi_1 \psi_2 dr \qquad \text{(e)}$$

From Eq. (d) we have:

$$N = (c_1^2 + c_2^2 + 2c_1c_2S)^{-1/2} \qquad \text{(f)}$$

Therefore, from Eq. (a) we may have:

$$\psi = (c_1^2 + c_2^2 + 2c_1c_2S)^{-1/2}(c_1\psi_A + c_1\psi_B) \qquad \text{(g)}$$

since, it is known that **BMO of H_2^+ ion**, $c_1 = c_2$ and for AMBO, $c_1 = -c_2$
Therefore from Eq. (g) we have:

$$\boxed{\psi_{BMO} = \left[\frac{1}{\sqrt{2(1+S)}} \right] (\psi_A + \psi_B)}$$

$$\boxed{\psi_{ABMO} = \left[\frac{1}{\sqrt{2(1-S)}} \right] (\psi_A - \psi_B)}$$

Q.3. Give the Wave Function for the BMO for a Heteronuclear Diatomic Molecule AB assuming that the *electron on an average* almost *consumes 90%* of its overall time upon *Nucleus A*; and remaining 10% of its time span upon the *Nucleus B* perceptively.

Solution: Amazingly in the **LCAO – MO schematic episode** we have:

$$1/MO = c_A\psi_A + c_B\psi_B$$

Hence, in the above **schematic episode**, the observed coefficients **c_A and c_B** are such that $[c_A]^2$ and $[c_B]^2$ do determine precisely the probability of locating the electron present in the **atomic orbitals (AOs)** viz., ψ_A and ψ_B respectively.

Thus, we may have explicitly:

$$[c_A]^2 = 90\% = 0.9$$
$$[c_B]^2 = 1\% = 0.1$$

Hence, $$c_A = \pm\sqrt{0.9} = \pm 0.95$$

and $$c_B = \pm\sqrt{0.1} = \pm 0.32$$

Therefore, we have:

$$\psi_{MO} = 0.95\psi_A + 0.32\,\psi_B$$

Q.4. How would you express the *Valence Bond* and *Wave Function* for the *hydrofluoric acid [HF] molecule* (assuming that it is duly formed from orbital of H atom and 2pz – orbital of F atoms) as could be seen in the following *three* specific instances:

   (a)   HF is purely covalent;

   (b)   HF is purely ionic, and

   (c)   HF is 80% covalent and 20% ionic in nature.

Solution: Let us assume that the **H-atom is being duly designated by A; and F atom by B.** Let the **electron numbered 1** be the *1s electron of H-atom*; whereas, the **electron numbered 2** be the *2pz electron* of the **F-atom**.

Thus, we may have the following *three situations:*

   (i) **For purely covalent structure of HF molecule**–for this the **Wave Function** may be expressed as under:

$$\psi_{covalent} = \psi_A(1)\,\psi_B(2) + \psi_A(2)\psi_A(1)$$

  (ii) **For purely Ionic Structure H^-F^-**–wherein both the electrons are duly positioned on the **F-atoms**; hence, we have

$$\psi_{ionic} = \psi_B(1)\,\psi_B(2)$$

  (iii) **HF – a Resonance hybrid** – and may be depicted as under:

$$H - F \longleftrightarrow H^+F^-$$

i.e., $\psi_B = c_1\psi_{covalent} + c_2\psi_{ionic}$ with $c_1^2 + c_2^2 = 1$

since $c_1^2 = 0.80$ and $c_2^2 = 0.20$ (given);

Hence, we have:

$$c_1 = \sqrt{0.80} = \mathbf{0.89}$$

and $c_2 = \sqrt{0.20} = \mathbf{0.45}$

Hence, $\psi_B = 0.89\,[\psi_A(1)\,\psi_A(2) + \psi_A(2)\,\psi_B(1)] + 0.45\,[\psi_B(1) + \psi_B(2)]$

[B] Problems Related to Percentage of Ionic Character

Q.5. How would you calculate the **Percentage of Ionic character** in C_s–F bond in C_sF molecule? Given the electronegativity values of C_s and F are 0.7 and 4.0 respectively. Predict precisely the nature of the C_sF molecule.

Solution: We have

$$x_A - x_B = x_f - xc_s = 40 - 0.7 = 3.3$$

∴   % Ionic character in C_s – F Bond $= [16 \times 3.3 + 3.5 \times (3.3)^2]\%$

$$= 90.9\%$$

since, the **Ionic Character is more than 50%,** the C_sF **may be regarded as a purely Ionic Molecule.**

Q.6. Calculate the exact Percentage of Ionic Character in the K–Cl bond in the KCl molecule. The electronegativity values of K and Cl are 0.8 and 3.0 respectively. Also predict the precise nature of KCl molecule.

Solution: As we know:

$$x_A - x_B = x_{Cl} - x_K = 3.0 - 0.8 = 2.2$$

∴   % Ionic Character in the K–Cl bond $= [16 \times 2.2 + 3.5 \, (2.2)^2]\% = \textbf{52.1}\%$

Because, the calculated Ionic Character is found to be more than 50%, KCl is an **Ionic Molecule** *per se.*

Q.7. How would you calculate the Percentage of Ionic Character in the Na–Cl bond in the NaCl molecule? Besides, predict precisely the nature of NaCl molecule [given: $x_{Na} = 0.9$; and $x_{Cl} = 3.0$]

Solution: Since we know:

$$x_A - x_B = x_{Cl} - x_{Na} = 3.0 - 0.9 = 2.1$$

Hence, % Ionic Character in Na – Cl Bond $= [16 \times 2.1 + 3.5 \times (2.1)^2]\% = \textbf{49}\%$

As the **Ionic Character is found to be less than 50%**; hence, NaCl is a **Covalent Molecule** [*i.e.*, according to the **Hanny and Smith's Equation**].

Q.8. Arrange the *four* molecules: HF, HCl, HBr and HI in the so-called *Decreasing Order* of the Percentage of Ionic Character in the H–X bond [*viz.* X = F, Cl, Br or I]. The Electronegativity values given are as: F = 4.0, Cl = 3.0, Br = 2.8, I = 2.2, and H = 2.1.

Solution: In a situation when one substitutes the **Electronegativity values of F, Cl, Br, I, and H atoms** duly present in the **Hanny and Smith's Equation as given under:**

$$\% \text{ Ionic character in A – B Bond} = [0.16(x_A - x_B) + 0.035(x_A - x_B)^2] \times 100\%$$
$$= [16(x_A - x_B) + 3.5(x_A - x_B)^2]\%$$

Hence, the **percentage (%)** of Ionic Character in **H–X bonds** in: **HF, HCl, HBr, and HI molecules** is found as stated under:

% Ionic Character in H – F Bond and HF molecule = **43%**

% Ionic Character in H–Cl Bond and HCl Molecule = **17%**

% Ionic Character in H–Br Bond and HBr Molecule = **13%**

% Ionic character in H–I Bond and HI Molecule = **7%**

Based on the above results we may infer that the percentage of Ionic character in the **H–X Bonds** in the aforesaid *four* molecules *viz.*, **HF, HCl, HBr, and HI** is found to be in the following **decreasing order:**

$$\boxed{H - F > H - Cl > H - Br > H - I}$$

Q.9. How would you arrange the following compounds in the order of their increasing Melting Points:

(i) **LiF₃, LiCl, LiBr, and LiI**

(ii) **LiCl, NaCl, KCl, RbCl, and CsCl**

Solution: The precise answer to the above question may be duly given on the basis of either **Fajan's Rules** or relatioship prevailing between the **electronegativity difference** occuring in the concerned *two* atoms.

Thus, we may observe:

(i) Greater the electronegativity difference between the *two* **concerned atoms**, greater could be the Ionic Character; and hence, higher would be the melting point:

$$\text{LiF} \rangle \text{LiCl} \rangle \text{LiBr} \rangle \text{LiI}$$

(ii) By virtue of the **Covalent Character LiCl** does posses **lower MP**.

$$\text{LiCl} \rangle \text{NaCl} \rangle \text{KCl} \rangle \text{RbCl} \rangle \text{CsCl}$$

Q.10. **The observed value of Dipole Moment of H_2O Molecule is found to be 1.84D. Calculate the H–O–H Bond angle in H_2O molecule–given that the Bond Moment of O–H bond is 1.5 D.**

Solution:

$$\text{Dipole Moment of Water: } \mu H_2O = 1.84 \text{ D (Given)}$$
$$\text{Dipole Moment of O–H Bond: } \mu OH = 1.5 \text{ D (Given)}$$

In case, the H–O–H Bond Angle is α, we may have: μOh

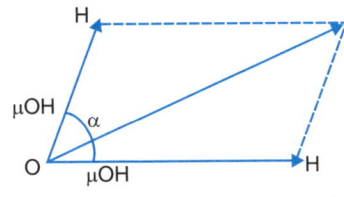

$$\mu^2 H_2O = \mu^2 OH + \mu^2 OH + 2 \times \mu OH \times \mu OH \times \text{Cos } \alpha$$
$$= (1.5)^2 + (1.5)^2 + 2 \times 1.5 \times 1.5 \times \text{Cos } \alpha$$
$$= 2.25 + 2.25 + 4.5 \times \text{Cos } \alpha$$

or
$$(1.84)^2 = 4.5 + 4.5 \cos \alpha$$

∴
$$\cos \alpha = \frac{(1.84)^2 - 4.5}{4.5} = \frac{3.3856 - 4.5}{4.5}$$

or
$$\cos \alpha = -0.2476$$

or
$$\alpha = 104°28' \text{ (Answer)}$$

Q.11. Explain which of the following compounds *viz.*, $SnCl_2$, $SnCl_4$, SnF_2, and SnF_4 happens to be the least covalent ?

Solution: Out of the *four* above stated halidas, **SnF_2** happens to be the **Least Covalent, which could perhaps be due to the underlying fact that.**

"According to *Fajan's Rule*–due to the so-called *smaller dimension of Sn^{4+} ion* in comparison to that of the corresponding S^{2+}_n ion, it may be critically observed that $SnCl_4$ is certainly for more covalent in nature *vis-a-vis* $SnCl_2$ (*i.e.*, $SnCl_4 > SnCl_2$)".

Likewise, SnF_4 is found to be more covalent in comparison to SnF_2 (*i.e.*, $SnF_4 > SnF_2$).

Besides, out of SnF_2 and $SnCl_2$, by virtue of the inherent large size of the Cl^- ion *vis-a-vis* F^-ion, the $SnCl_2$ is certainly more covalent than SnF_2.

Hence, it may be concluded that SnF_2 is the Least Covalent.

Q.12. Out of the following *four* pairs of compounds–which one is more Covalent and why?

(i) AgCl, Ag I ; (ii) Be Cl$_2$, MgCl$_2$; (iii) SnCl$_2$, SnCl$_4$; (iv) CuO, Cus.

Solution: Based on the **Fajan's Rules**, the result may be duly accomplished in each case, as given under:

(i) **Ag I is more *Covalent vis-a-vis* AgCl,** which is due to the fact that I^- ion is a lot bigger in size than Cl^- ion; and hence, gets more polarized than Cl^- ion.

(ii) **BeCl$_2$ is more *Covalent vis-a-vis* MgCl$_2$,** that is on account of the fact that the Be^{2+} ion is found to be smaller in size than Mg^{2+} ion; and therefore, possesses greater polarizing power.

(iii) **SnCl$_4$ is more *Covalent vis-a-vis* SnCl$_2$,** which could be due to the fact that Sn^{4+} ion has higher degree of charge and smaller in size than Sn^{2+} ion; and hence, possesses **much higher polarizing power**.

(iv) **CuS is certainly more *Covalent* than CuO,** which could be by virtue of the fact that S^{2-} ion has a *bigger* dimension than the respective O_2^- ion; and hence, is found to be more polarized than O^2 ion perceptively.

Q.13. BeCO$_3$ is certainly less stable than MgCO$_3$. Explain.

Solution: The observed **lower stability profile of BeCO$_3$** *vis-a-vis* that of $MgCO_3$ may be adequately expatiated based upon the **Fajan's Rules**. Because, Be^{2+} ion happens to be **small in dimension** in comparison to Mg^{2+} ion. Besides, **BeCO$_3$ is *Covalent* (*i.e.*, lesser ionic in nature)**; whereas, **MgCO$_3$ is more ionic.** In addition, being *Covalent,* $BeCO_3$ is *less stable than MgCO$_3$*–which is **ionic**. Besides, due to the **lesser stability of BeCO$_3$**–that eventually helps in to undergo decomposition to yield: **BeO** and **CO$_2$** upon heating:

$$BeCO_3 \xrightarrow{\Delta} BeO + CO_2^{\uparrow}$$

Q.14. How would you arrange the following molecules in the decreasing order of their Dipole Moment ?

CH$_3$Cl, CH$_2$Cl$_2$ [*cis–form*], and CHCl$_3$.

Solution: The various molecular structures of the aforesaid given molecules are given as under. The direction of **Bond Moment** has also been shown explicitly:

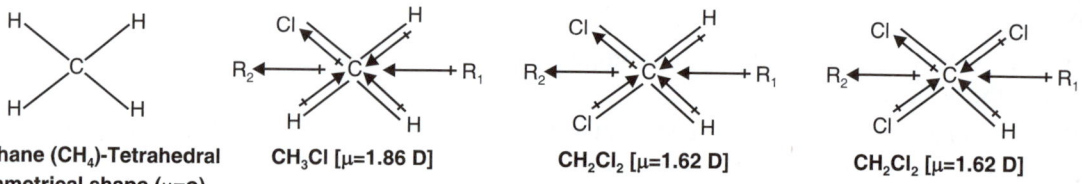

Methane (CH_4)-Tetrahedral   CH$_3$Cl [μ=1.86 D]   CH$_2$Cl$_2$ [μ=1.62 D]   CH$_2$Cl$_2$ [μ=1.62 D]
Symmetrical shape (μ=o)

- **CH$_4$ Molecule ($\mu = 0$)** – The molecule has a tetrahedral shape. (*i.e.*, symmetrical shape). Hence, μ for this molecule is **zero**.
- **CH$_3$Cl Molecule (m = 1.86D)**–In this molecule the **resultant (R$_1$)** of 2 C–H bond moments adds on to the **resultant (R$_2$)** and the **Cl–Cl bond** moments.

- **CH_2Cl_2 Molecule ($\mu = 1.62D$)** – Here, the **resultant (R_1)** of *two* **C–H bond moments adds** on to the **resultant (R_2)** of 2 **C–Cl bond moments** .
  - **$CHCl_3$ Molecule (m = 1.03D)**–In this case, the **resultant (R_1)** of **C–H bond and C–Cl bond moments** do oppose the **resultant (R_2)** of **2C–Cl bond moments**.

The above explanatios do reveal vividly that the **decreasing order of the Dipole Moments values** are as given under:

$$CH_3Cl \rangle CH_2Cl \rangle CHCl_3 \rangle CH_4$$
$$\quad 1.86D \qquad 1.62D \qquad 1.03D \qquad 0.0D$$

PROBABLE QUESTIONS

1. What do you mean by **Chemical Bond** (or **Valence Bond**)? Explain. How would you account for the **causation of Chemical Bonding?**
2. Write short notes on any **two** of the following
   (a) **Octet Rule (Rule of Eight)**
   (b) **Potential Energy Curve**
   (c) **Variants in Bonds**
   (d) **Modality of the Configuration of Atoms**
3. How does the Ionic Bond form? Discuss a few – '**Classical Examples of Ionic Compounds**'. *e.g.*, **NaCl or CaF$_2$ or Al$_2$O$_3$**.
4. (a) Explain diagrammatically the **Prevalent Nature of Ionic Bond.**
   (b) Discuss the parameters for the **Formation of Ionic Compounds.**
5. Elaborate the various **Characteristics of the Ionic Compounds.**
6. '**Several Compounds of Formula *MX$_2$* do crystallize in Layer Structures**'. Explain the above statement diagrammatically.
7. Give a brief account on the **Geometrical Structure for an Ionic Crystal. Explain.**
8. How would you explain the **Covalent Bond** in a comprehensive manner? Give details.
9. Explain the following
   (i) **Isomerism**
   (ii) **Extension of Octet**
   (iii) **Non-Octet Structure**
   (iv) **The Incomplete Octet**
10. How would you explain the **failure of the octet Rule** in terms of:
   (a) **Sugden's Concept of Single Linkages**
   (b) **Sidgwick's Concept of Maximum Valency?**
11. Discuss the following:
   "**Dipole Moment and Its Applications**".
12. How would account for the **Coordinate Bonds** in an elaborated manner? Explain.
13. Write short notes on:
   (a) **Metallic Bonds**
   (b) **Arrangements of Atoms in Metals**
14. Write an essay on **Hydrogen Bond**. Give examples in support to your answer.

UNIVERSITY QUESTIONS

1. (a) What is **Inert Pair Effect?** List the elements of **p-Block** for which **Inert Pair Effect** is predominant.
   (b) Explain the following:
   (i) Born forms no compounds in **unipositive Oxidation,** but **Thallium in Unipositive State** is quite stable.
   (ii) Al forms the ion $[AlF_6]^{3-}$ but Boron does not form the ion $[BF_6]^{3-}$,
   (iii) Bond angle in oxides decreases as $Br_2O > Cl_2O > OF_2$.   (Himachal Pradesh, 2000)
2. (a) Nitrogen forms only NCl_3 but phosphorus forms both PCl_3 as well as PCl_5. Explain Why ?
   (b) Write a note on (i) Sidgwick's Maximum Valencey Rule; and (ii) Variable Valence.
                                                                               (Kanpur, 2000)
3. (a) Write a note on "Free Electron Theory of Metallic Bonding".
   (b) Give an account of **Odd Electron Bond**.   (Kumaon, 2000)
4. (a) Explain the following:
   (i) NaCl is soluble in water but not in Chloroform.
   (ii) Though Ethanol is a **Covalent Compound** but it is soluble in water.
   (b) Write a note on '**Resonance**'.   [Gauhati (General), 2001]
5. Explain the following:
   (i) NCl_5 is not formed while PCl_5 is formed duly.
   (ii) Bond angle of PH_3 is less than NH_3.
   (iii) Al forms $[AlF_6]^{3-}$ ion but Boron (B) does not form $[BF_6]^{3-}$ (Lucknow, 2000)
6. (i) Explain the following:
   (a) AlF_3 is tonic while $AlCl_3$ is covalent
   (b) AgI is covalent whereas NaI is ionic.
   (c) Phosphorous forms PCl_3 and PCl_5 both while Nitrogen forms only NCl_3.
   (ii) Write short notes on:
   (a) **Diagonal Relationship**
   (b) **Inert Pair Effect**
   (c) **Hydrogen Bond**
   (d) **Electropositivity**
7. (i) **Explain why LiCl is insoluble in water.** (Avadh, 2000)
   (ii) **Write a note on "Inert Pair Effect"**
   (iii) **Explain σ and π Bonds.**
8. Which of the following compounds does not contain **Ionic Bond? NaOH, K₂S, HCl, and LiH.**
                                                                            (HN Bahuguna, 2006)
9. Which of the following is the most Covalent? $BeCl_2$, $MgCl_2$, $CaCl_2$, and $SrCl_2$.   (Agra, 2008)
10. (i) Write a not on **Fajan's Rule**
    (ii) Explain Intermolecular H-Bonding.   (Meerut, 2008)
11. Cu_2Cl_2 is more **Covalent** than NaCl. Why?   (Meerut, 2009)

Chapter
5

METALLURGY

1. INTRODUCTION

Metallurgy–It may be defined as the—'**critical laid-out, well-defined process of extraction of '***Metals*'
from their respective ores *viz.,*'

- **Sulphide Ores:** An array of metals *e.g.,* **Iron Pyrites (FeS), Galena (PbS), Cinnabar (HgS),**
  Copper Pyrites (CuFeS$_2$), and Zinc Blende (ZnS) and the like.
- **Oxide Ores:** A good number of metals *e.g.* **Iron (Fe), Aluminium (Al), Manganese (Mn), Tin**
  (Sn), Zinc (Zn) etc., usually occur as their **oxides** *viz.,* **Haematate (Fe$_2$O$_3$), Bauxite**
  (Al$_2$O$_3$.2H$_2$O), Pyrolusite (MnO$_2$), Zincite (ZnS), Copper Pyrites (CuFeS$_2$) etc.
- **Carbonate Ores:** The important **Carbonate Ores** are invariably of **Mg, Fe, Cu, Zn and Pb**
  viz., **Magnesite (MgCO$_3$), Dolomite (CaCO$_3$, MgCO$_3$), Siderite (FeCO$_3$), Malachite**
  (Cu(OH)$_2$, CuCO$_3$)], Cerussile (PbCO$_3$), Calamine (ZnCO$_3$) etc.
- **Halide Ores:** The most common and abundantly available **Halide Ores** are designated by the
  so-called **Chloride Ores** *e.g.,* **Chlorides of Na, K, and Mg** are usually found in *Salt Beds* either
  in *sea water* or upon the **Earth's surface**. A host of *typical* **chloride ores** are: **Carnallite (KCl,**
  MgCl$_2$. 6H$_2$O), Rock Salt (NaCl) and **Horn Silver (AgCl)**. However, the respective **Bromide**
  and Iodide ores of K and Mg are duly seen in small quantum only in **Sea water**; whereas, the
  corresponding **Fluoride Ores** do comprise: **Fluorspar (CaF$_2$) and Cryolite (AlF$_3$, 3NaF)**.
- **Sulphate Ores:** A plethora of **Sulphide Ores** do get usually converted to: *Sulphates*–due to
  atmospheric oxidation. The most prevalent and common **Sulphate Ores** are: **Barytes (BaSO$_4$),**
  Anglesite (PbSO$_4$), Epsonite (MgSO$_4$.7H$_2$O) etc.

Metallurgy: Based upon the aforesaid reasonably sufficient exposure to the related subject that
predominantly deals with both the **intrigue science** and **superb technology** intelligently applied for:

"**the critical *extraction of 'Metals'* economically on a large scale from their respective natural ores**".

Nevertheless, another equally important and vital glaring aspect of the **Metallurgy** critically centres
and deals vehemently with:

"the making of '*Alloys*'–that are regarded to be the so-called '*Metallic Solutions*' duly composed of *two*
or more elements *e.g.,* **Aluminium Alloy, Stainless Steel, Silver Alloy etc.**

Therefore, in order to understand the in-depth of **Metallurgy**, we may have to know the precise and exact differences that largely exist between the 'Mineral' and 'Ore' perceptively.

❑ *Mineral:* Importantly, the divergent compounds of a metal that are naturally available in the **Earth's Layers (*Crust*)** may be obtained by '*mining*' are invariably called as **Minerals**.

Besides, a specific **Mineral** may eventually comprise of one or even more **metallic chemical entities** (*compounds*) with a pre-determined (*fixed*) **chemical composition** largely.

❑ *Ore:* The naturally occurring deposits of '**minerals**' from which a metal may be extracted *viably*, *economically*, and *conveniently* are invariably termed as **Ores**.

At this material time, one may have to clearly take cognizance of the underlying glaring and interesting fact that:

"All ores are Minerals, but All Minerals are Not Ores".

Example: **Iron Pyrite (FeS$_2$)** is found abundantly in the **Earth's layers** but may not be employed as an '*Ore of Iron*'. The above statement can be argued justifiably upon the following chemical reactions that are duly involved in the **actual conversion of iron pyrite to iron** are:

$$2FeS_2 + 5\frac{1}{2}O_2 \longrightarrow Fe_2O_3 + 4SO_2 \uparrow$$

$$Fe_2O_3 + 3C \longrightarrow 2Fe + 3CO \uparrow$$

NOTE

1. The meticulous removal of *Sulphur(S) from FeS* do involve crucially a relatively high cost of production as well as the critical presence of higher percentage of S in iron renders it *quite brittle*; and hence, of little usage.

2. *Impurities* present in the *ore* are usually called as '*Gangue*' e.g., Silica (SiO$_2$) happens to a common impurity in a majority of the available *Iron Ores*.

Different Types of Ores

Fig: 5.1. shows the different types of Ores that are known.

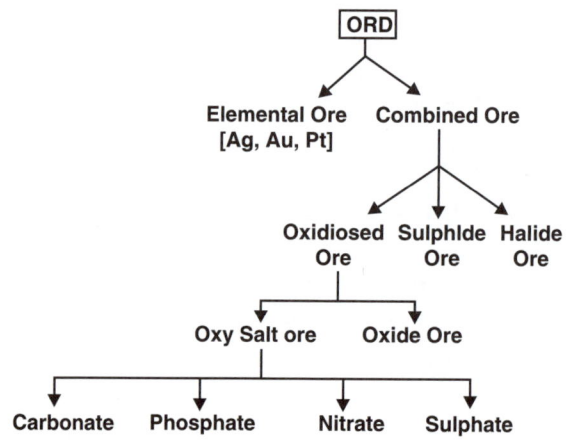

Fig. 5.1: Different Types of Ores.

Following are certain **Important Ores** and their **respective Formulae**:

S. No.	Type of Ore	Name of Ore	Chemical Formula
1.	**Oxide Ore**	Zincite	ZnO
		Haemalite	Fe_2O_3
		Magnetite	Fe_3O_4
		Bauxite	$Al_2O_3\ 2H_2O$
		Limonite	$Fe_2O_3\ 3H_2O$
		Cuprite or Ruby Copper	Cu_2O
		Pyrolusite	M_nO_2
		Tinstone or Cassaterite	S_nO_2
		Rutile	TiO_2
		Chromite Ore	$FeCr_2O_4(FeO + Cr_2O_3)$
		Borax or Tincal	$Na_2B_4O_7\ 10H_2O$
		Pitch Blende	U_3O_8
		Hemenite	$FeOTiO_2$
2.	**Sulphide Ores**	Galena	PbS
		Cinnabar	HgS
		Zinc Blende or Sphalerite	ZnS
		Copper Glance or Chalcocite	Cu_2S
		Copper Pyrite or Chalcopyrite	$CuFeS_2$
		Iron Pyrite or Fools Gold	FeS_2
		Silver Glance or Argantite	Ag_2S
3	**Halide Ores**	Rock Salt	$NaCl$
		Sylvine	KCl
		Fluorspar	CaF_2
		Cryolite	Na_3AlF_6
		Horn Silver	$AgCl$
		Carnalite	$KCl\ MgCl\ 6H_2O$
		Atacamite	$Cu_2Cl(OH)_3$
4	**Oxy Salt Ores**		
	(a) *Carbonate Ores*	Limestone	$CaCO_3$
		Magnesite	$MgCO_3$
		Dolomite	$CaCO_3\ MgCO_3$
		Siderits	$FeCO_3$
		Cerrusite	$PbCO_3$
	(b) *Sulphate Ores*	Gypsum	$CaSO_4\ 2H_2O$
		Epsum Salt	$MgSO_4\ 7H_2O$
		Anglesite	$PbSO_4$
		Baryte	$BaSO_4$

(Contd...)

	Glauber Salt	Na_2SO_4 $10H_2O$
	Chalcanthyte	$CuSO_4$ $5H_2O$
(c) *Nitrate Ores*		Na_2SO_4 $10H_2O$
	Indian Saltpeter	KNO_3
	Chile Saltpeter	$NaNO_3$

[**Adapted From:** Lee JD: **Concise Inorganic Chemistry,** 2nd ed. Wiley, New Delhi, 2013]

2. MAJOR SEQUENTIAL STEPS INVOLVED IN OPTIMIZED RECOVERY OF METAL FROM ITS ORE

The various cardinal steps that are involved in a sequential order in obtaining the so-called *Optimized Recovery* of a **metal** from its respective '**Ore**' are enumerated as under:

   (a)   **Crushing (Grinding) of Ore,**
   (b)   **Calcination and Roasting (Conversion of Concentrated ore into its Respective Metallic Oxides)** ,
   (c)   **Various Practiced Reduction Processes,**
   (d)   **Purification (Refining of Metal),**
   (e)   **Thermodynamics of Reduction Process,**
   (f)   **Alloys and Amalgams**
   (g)   **Furnace variants in Metallurgy,**
   (h)   **Extraction of Silver,**
   (i)   **Extraction of Gold (Au) By Cyanide Method,**
   (j)   **Extraction of Magnesium (Mg),**
   (k)   **Extraction of Tin (Sn),**
   (l)   **Extraction of Aluminium (Al),**
   (m)   **Extraction of Lead (Pb),**
   (n)   **Extraction of Copper (Cu),**
   (o)   **Extraction of Zinc (Zn), and**
   (p)   **Extraction of Iron (Fe),**

which shall now be treated individually in the sections that follows:

2.1. Crushing (Grinding) of Ore

The **Rock Ores** after completion of the mining episode are usually obtained as **Big Lumps**–that are eventually **Crushed** (*ground*) to their corresponding *smaller pieces* by the aid of:

   • **Jaw Crusher**
   • **Hammering** or   • **Grinders**.

2.1.1. *Jaw Crusher*

It essentially comprises *two* **plates** (having the cut grooves in opposite manner) that predominantly serves as a kind of '*jaw*'. Here, one of the '**plates**' **is fixed**; whereas, the **other is movable**. The so-called '**lumps of the ore**' are being fed duly in between the plates; and thus, the **movable plate** is made to move with the aid of a *motorized* **Mechanical Device**. The **crushed** (*ground*) **ore** are now collected in **dry polyester empty bags** (and *sealed*), as depicted in Fig: 5.2.

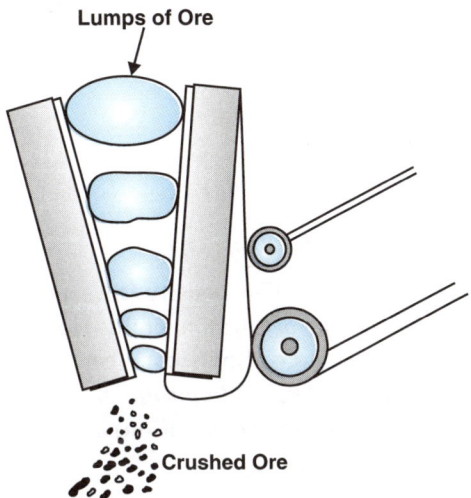

Fig. 5.2: Jaw Crusher Being Used to Crush the Lumps of Ore into Smaller Particles.

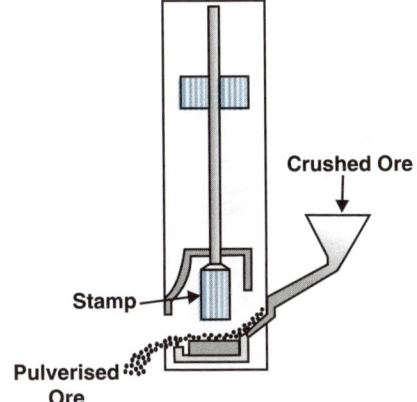

Fig. 5.3: Stamp Mill Being Employed to Pulverize the Crushed Pieces of Ore.

2.1.2. *Pulverizing of Ore*

The **crushed pieces** of the '*ore*' are now fed into a **Stamp Mill** (*Pulveriser*) whereby the former are finely powdered and collected from the bottom, as illustrated in Fig: 5.3. Obviously, the **coarse ore particles** is carefully fed into the **Stamp Mill** from *one side* (as shown in Fig: 5.3) and the fine powder is discharged by the help of a continuous flow of water.

Occasionally, the *Ball Mills* are also used for accomplishing the pulverization process effectively.

2.1.3. *Ore Dressing (Concentration of Ore)*

In actual practice, the **Ore Dressing** (or **Concentration of Ore**) may be duly accomplished by *two* **distinct procedures**, namely:

1. **Physical Method of Separation**
   (a) **Levigation (or Gravity Separation),**
   (b) **Magnetic Separation**, and
   (c) **Froth Floatation (Oil Floatation)**.
2. **Chemical Method of Separation (or Leaching)**

2.1.3.1 (a) *Levigation (or Gravity Separation)*

In fact, this method makes use of the prevailing **Density Differences** existing between the '*ore*' and '*impurity present*' in order to *concentrate the ore* meaning fully. However, it is mostly applicable to the **Oxide Ores** specifically *viz.*, **Fe_2O_3, $FeCr_2O_4$, SnO_2** etc.–that are adequately concentrated by the **Levigation Method** right from the so-called *Silicons Impurities* perceptively. Fig: 5.4 gives an elaborated representation of the typical **Gravity Separation (Levigation)** by means of the **Wilfley Table:**

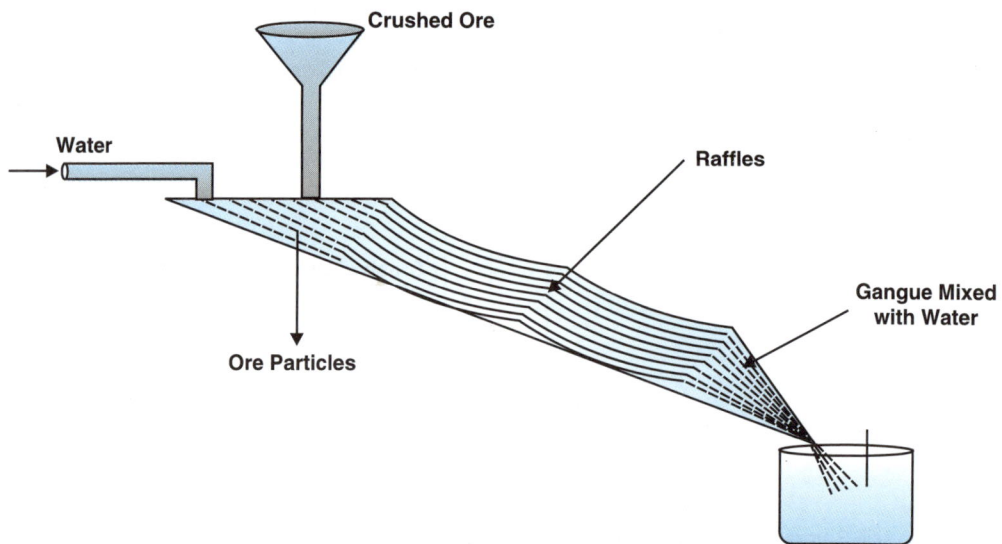

Fig. 5.4: An Exclusive Representation of Levigation (or Gravity Separation) by Means of the Wilfley Table.

Modus Operandi

1. The crushed ore is first and foremost washed thoroughly by a stream of water upon a strategically placed **Sloping Table** duly fitted with a series of *corrugated boards* invariably termed as **Wilfley Table** (see Fig: 5.4).
2. **Wilfley Table** is vibrated in a continuous mode mechanically by virtue of which the so-called **lighter particulate matters** do move in the *downward direction*; and hence, the relatively **heavier particulate matters** are duly left behind the **corrugations** (*on the boards*) thereby serving as a **Barrier**.

3. Finally, an appropriate arrangement is made to get rid of the rather **here in particulate matters** almost continuously rest the accomplished separation procedure will be rendered **practically ineffective** after a certain span of time since the *heavier* and *lighter* **particulate matters** will commence to pass through the '**Barrier**' collectively.

2.1.3.1. (b) *Magnetic Separation*

In a broader sense, the **Magnetic Separation** is invariably being employed to cause the separation of the highly specific **Magnetic Impurities** perceptively right from the so-called:

"**Particular Non-Magnetic Ore**".

Examples: **Tin Stone (SnO_2)** is being separated efficaciously in the presence of the **magnetic impurities**– '*Wolframite*' [**$FeWO_4 + MnWO_4$ (min)**] by making use of **Magnetic Separations** perceptively.

Fig: 5.5 illustrates the diagrammatic representation of **Magnetic Separation**.

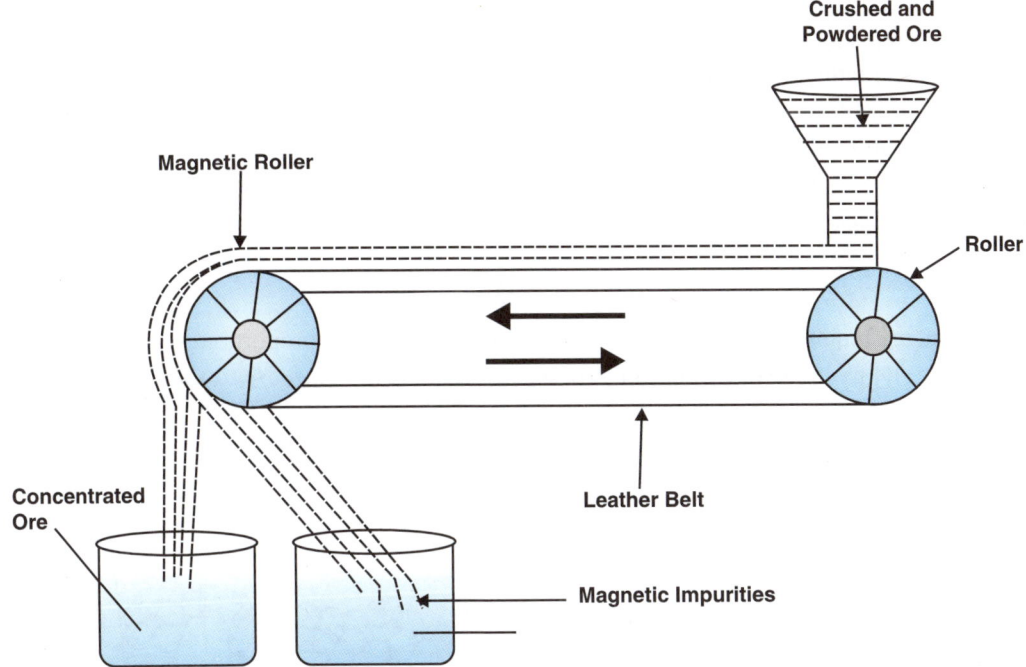

Fig. 5.5: The Diagrammatic Representation of Magnetic Separation of Collected Ore.

Modus Operandi: In the **Magnetic Separation** procedure the so-called **Powdered Ore** is allowed to move in the direction of the *Magnetic Roller*; and subsequently, fall **downwards (gravitationally)**.

 NOTE   Interestingly, the *Magnetic Material* does form an altogether *New Heap* since it is being held to the *Magnetic Roller* certainly for a longer duration perceptively.

2.1.3.1. (c) *Froth Floatation (Oil Floatation)*

In true sense, the **Froth Floatation (Oil Floatation)** is solely confined to cause the effective separation of the **sulphide ore** as in impurity present in the 'ore'. However, it is based entirely upon:

"the crucial wetting of the *'ore'* by an *'oil'* and the *'impurity'* by H_2O."

Modus Operandi

1. The 'ore' is first ground to *five powder* and then mixed with water so to obtain a **slurry (with operational consistency)**.

2. Thus, any one of the so-called **'Oily Components'** *viz.*, **Pine Oil, Eucalyptus Oil, Cresols,** and **Crude Coal Tar**–is now incorporated to it along with a *'collector' viz.*, **Sodium Ethyl Xanthate**.

3. A low-pressure **'Air'** is slowly bubbled *via* the mixture that: **'eventually serves' as a sort of 'Agitator' and creates bubbles effectively'**.

4. Ultimately, the **'ore'** is carefully floated to the **froth**; and thus, the ensuing **silicons (Sandy) impurities** present are duly settled at the bottom of the tank.

5. Now, the **Froth** is collected into an altogether cleaned separate **container**; and subsequently, washed thoroughly and dried.

Remarks: The 'oil' added duly serves as the so-called **'Frothing Agent'**; and hence, predominantly in causing the critical reduction in the ensuing **Surface Tension of Water**–that, in fact helps to produce a **Stable Froth**.

Besides, the causation of forming the *'Bubbles'* do involve an enhancement in the so-called *'air-water surface'* which eventually means doing a work totally against the *Surface Tension*.

Therefore, the incurred energy to produce a **bubble of radius 'r'** is expressed by:

$$r = 2 \times 4pr^2 \times E$$

where, E = Energy required to create unit surface area and is directly proportional to surface tension.
Fig: 5.6 shows vividly the **Froth Floation Process** for the so-called *Sulphide Ore*.

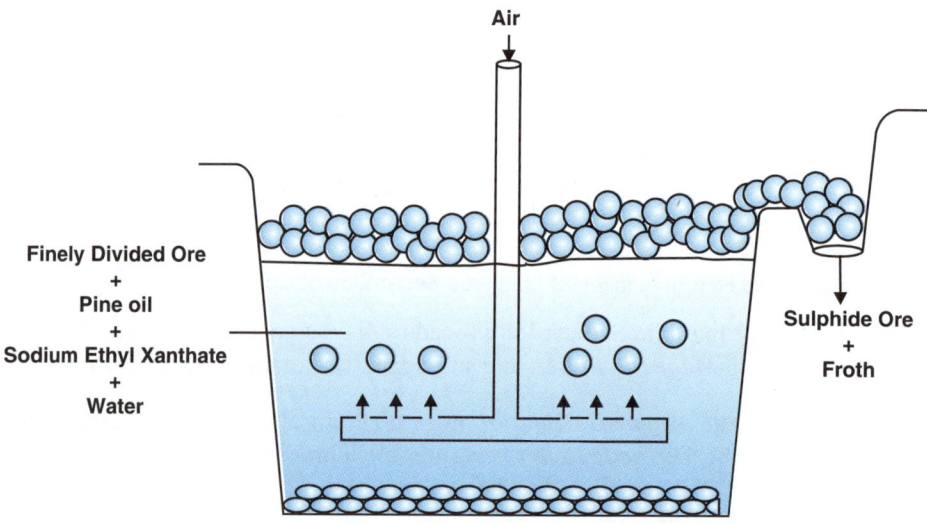

Fig. 5.6: Diagrammatic Representation of the Froth Floatation process for the Sulphide Ore.

Following structural description of the **Sodium Ethyl Xanthate**–critically serves as a **Collector of the Sulphide Ore:**

Fig: 5.7 clearly depicts the presence of Sodium Ethyl Xanthate $\left[C_2H_5-O-C\underset{S^-}{\overset{S}{<}} \bullet N_a^+ \right]$ located strategically at the so-called **Air-Water Interface.**

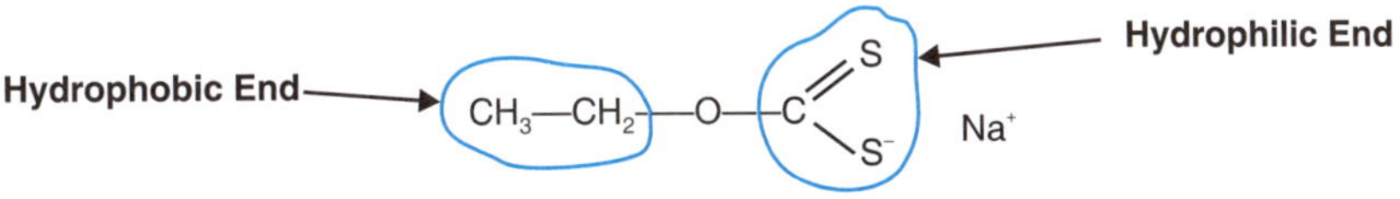

Fig. 5.7: Representation of Sodium Ethyl Xanthate at the Air-Water Interface Showing Both Hydrophobic and Hydrophilic Ends.

Remarks: In fact, the *Hydrophobic End* critically prefers to stay preferentially in '*Air*'; whereas, the respective *Hydrophilic End* crucially prefers to remain in '*Water*'. Because, the ensuing **molecule** essentially inherits both **Hydrophobic** and **Hydrophilic segment**; hence; the most preferred location for it to stay is:

"the so-called Air-Water-Interface";

and, therefore, it vehemently shows the **Floating Characteristic Features**. Thus, in Fig: 5.7 we may critically observe that the **Hydrophilic End attracts the sulphide (S^-) particle** and thereby produces the *Surface-coated Particles*.

Observations: Ultimately, the so-called accumulated '**Sulphide Particulate Matters**' do get floated up to the '**Froth**'; and hence, subsequently being transferred to an altogether **cleaned separate vessel**, washed thoroughly with water, dried, and passed into the **next step of extraction of the 'ore'.**

Besides, we may also take into consideration the following *two* **major and cardinal aspects**, namely:

❑ Nevertheless, the other '*Alkyl Moities*' present in **Xanthate** may be:

- **Amyl [C_5H_{11}]** or • **Octyl [C_8H_{17}],**

instead of the other respective **Ethyl Moiety.**

❑ A few other important '*Collectors*' that may also be used effectively are:

1. **Sodium Layryl Sulphate** $\left[C_{12}H_{25}-O-S\underset{O}{\overset{O}{\updownarrow}}-Na \right]$

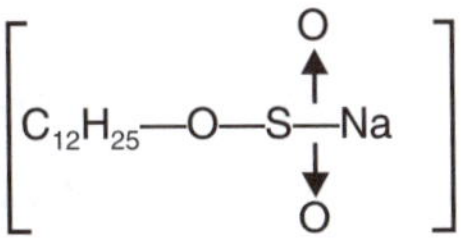

2. **Dicresyl Dithiophosphate** $\left[(CH_3-\bigcirc-O)_2 \cdot P\underset{S-H}{\overset{S}{<}} \right]^*$

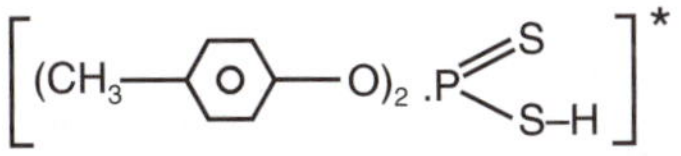

3. **Trimethyl Cetyl Ammonium Bromide** $[C_{16}H_{13}N^+(Me)_3]Br^-$

4. **Mercaptobenzthiazole** $\left[\bigcirc\underset{S}{\overset{N}{<}}C \right]^{**}$

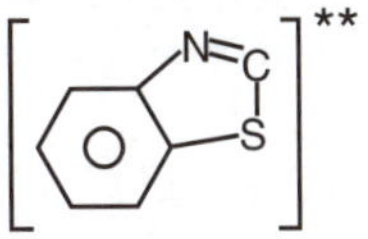

* Also known as the '**Aerofloat**'

** Also called as '**Floatogen**'

NOTE   Amazingly, the overall *'Ore-Collecting Phenomenon'* occurs through adsorption; and hence, depends solely upon the so-called *'Unbalanced forces'* located strategically upon the so-called Solid-Surface.

2.1.3.2. Chemical Method of Separation (or Leaching)

This process refers to the treatment of the **required substance** (either *Metal Component* or *Metal impurities*) being dissolved out from the *body of the ore* by utilizing an appropriate reagent.

Example: A few classical examples are stated as under:

1. Aluminium Oxide (Al_2O_3) is first dissolved out from the **Red Bauxite ore** by using **10% (*w/v*) NaOH**,–the following reaction takes place:

$$\underbrace{Al_2O_3 + Fe_2O_3 + SiO_2 + TiO_2}_{\textbf{Impurities Present}} \rightarrow NaOH\, H_2O \rightarrow Impurity + \underset{\textbf{A Soluble complex}}{Na^+\,[Al(OH)_4]^-}$$

2. Calcination and Roasting (Concentration of Concentrated Ore into its Respective Metallic Oxides)–In a broader perspective, the actual conversion of the **concentrated ore** into its corresponding **metallic oxide** is carried out means of the following *two* **methods**, namely:

- **Calcination**, and
- **Roasting**.

3. Calcination–**Calcination** may be defined as:

"**as process wherein the 'ore' is heated strongly in the absence of air**"

In this way, we may eventually achieve *three* **important changes**:

(a)   It helps to get rid of the so-called **Volatile Impurities** *viz.*, **SO_2, CO_2, organic matter**, and **moisture (H_2O)** present in the **ore**.

(b)   It also eliminates **water (H_2O)** from the *hydrated* **Oxide ore** *viz.*, **Bauxite [$Al_2O_3 \times 2H_2O$]**, and **Limonite [$2Fe_2O_3.3H_2O$]**–as given under:

$$\underset{\textbf{Bauxite}}{Al_2O_3 \times 2H_2O} \xrightarrow{\text{Calcination}} \underset{\textbf{Aluminium Oxide (Anhydrons)}}{Al_2O_3 + 2H_2O}$$

$$\underset{}{2Fe_2O_3 \cdot 3H_2O} \xrightarrow{\text{Calcination}} \underset{\textbf{Iron Oxide (Anhydrons)}}{2Fe_2O_3 + 3H_2O}$$

(c)   It usually removes **Carbon dioxide (CO_2)** from the *Carbonate Ore viz.*,

$$\underset{\substack{\textbf{Lime stone}\\ \textbf{(A Natural Deposit)}}}{CaCO_3} \xrightarrow{\text{Calcination}} \underset{\textbf{Calcined Oxide}}{CaO_3} + CO_2 -$$

$$\underset{\textbf{Malachite}}{Cu(CO_3) \times Cu(OH)_2} \xrightarrow{\text{Calcination}} \underset{\textbf{Copper Oxide}}{CaO_3} + CO_2 - + H_2O$$

$$\underset{\substack{\textbf{Lime stone}\\ \textbf{(A Natural Deposit)}}}{CaCO_3 \cdot MgCO_3} \xrightarrow{\text{Calcination}} CaO + MgO + 2CO_2 \uparrow$$

4. Roasting–Roasting relates to a specific process–

"whereby the '*Ore*' (invariably the *Sulphide Ore*) is subjected to strong heating in the presence of a excess of air".

Importantly, the '**heating**' must be carried out at a temperature a little below the *melting point* of the '*ore*'.

Amazingly, '**Roasting**' process helps to achieve the following aspects squarely:

- It aids to dry the '**ore**' as far as possible.
- An array of **Volatile Impurities** *viz.*, **CO₂**, **SO₂**, **Organic substances**, and moisture are all eliminated to a great extent.
- Also the '*ore*' gets duly converted into the **Metallic Oxide**–that may promptly be reduced to the respective '**metal**'.

Examples:

(i) $2PbS + 3O_2 \xrightarrow{\text{Heat}} 2PbO + 2SO_2\uparrow$
   Gelana

(ii) $2HgS + 3O_2 \xrightarrow{\text{Heat}} 2HgO + 2SO_2\uparrow$
   Cinnabar

(iii) $2ZnS + 3O_2 \xrightarrow{\text{Heat}} 2ZnO + 2SO_2\uparrow$
   Zinc Blende

(iv) $2Cu_eS + 3O_2 \xrightarrow{\text{Heat}} 2Cu_2O + 2SO_2\uparrow$
   Copper Glance

Important Observation: It has been observed that quite often the so-called **Sulphide Ores**,–that are usually *insoluble* in nature are being duly converted into the respective '**sulphates**' perceptively–that are mostly *water soluble* in character.

Examples:

(i) $\underset{\text{(Insoluble)}}{PbS} + 2O_2 \xrightarrow{\text{Heat}} \underset{\text{(Soluble)}}{PbSO_4}$

(ii) $\underset{\text{(Insoluble)}}{CuS} + 2O_2 \xrightarrow{\text{Heat}} \underset{\text{(Soluble)}}{CuSO_4}$

(iii) $\underset{\text{(Insoluble)}}{ZnS} + 2O_2 \xrightarrow{\text{Heat}} \underset{\text{(Soluble)}}{ZnSO_4}$

2.2.3. *Various Practiced Reduction Processes*

The '**metal**' may now be obtained from the so-called *pre-roasted* or *pre-calcined* '**ores**' by any of these following recognized procedures:

2.2.3.1. *Smelting [or Carbon Reduction Method]*

Importantly this method is being used primarily for the extraction of:

- **Pb** • **Zn** • **Fe** • **Cu**   and • **Sn**

Smelting: It may be defined as:

"**a process wherein the '*Oxide Ore*' in the *fused state* gets duly reduced by C to the ultimate *Free Metal*".**

Another school of thought, refers **smelting**–'**as the extraction of a, metal from the *oxide ore* by a typical phenomenon involving *melting*'.**

Modus Operandi: Various steps involved include:

1. The properly **roasted oxide ore** is adequately mixed with C (*viz.*, **coke, charcoal** or **coal**) along with a flux.

2. The resulting mass is now heated to a **high temperature** in an appropriate **Furnace**. Thus, the **C** helps to cause reduction of the ensuing **Oxide Ore** to the corresponding **Free Metal**.

3. The reduction of **Cassiterite Ore** (SnO_2 *i.e.*, **an oxide ore of Sn**) is mixed with *powdered* **Anthracite** (a *reducing agent*).

4. Hence, the **metal** duly obtained initially in the form of **vapours** (*viz.*, **Zn**) which are subsequently condensed or in the typical state of **Molten Metal** (*viz.*, **Fe, Sn**)

Examples: Following are *five* **clasical examples**:

(a) $\underset{\text{Zincite}}{ZnO} + e \xrightarrow{\text{Heat}} \underset{\substack{\text{Molter} \\ \text{Metal}}}{Zn} + \underset{\text{(Vapours)}}{CO}$

(b) $\underset{\text{Cassiterite}}{SnO_2} + 2C \xrightarrow{\text{Heat}} \underset{\substack{\text{Molter} \\ \text{Metal}}}{Sn} + \underset{\text{(Vapours)}}{2CO}$

(c) $\underset{\text{Massicot}}{PbO} + C \xrightarrow{\text{Heat}} Pb + CO\uparrow$

(d) $\underset{\text{Haematite}}{Fe_2O_3} + 3C \xrightarrow{\text{Heat}} 2Fe + 3CO\uparrow$

(e) $CuO + C \xrightarrow{\text{Heat}} Cu + CO\uparrow$

Points to Ponder

1. In usual practice, the **Smelting Process** is performed elegantly either in a '*Blast Furnace*' or a '*Reverberatory Furnace*', but in a **perfect controlled supply of air.**

2. For **Volatile Metals** like **Zinc (Zn)**–In the typical case, the **Smelting Process** is not at all feasible/ possible (in an *Open Furnace*); and hence, may be carried out in a–'**Fire Clay Vertical. Retorts**'– wherein heating is done by a **Producer Gas.**

3. **Flux**–It relates to a–"**substance that is incorporated to the respective '*Furnace Charge*' (calcined or roasted oxide ore plus coke) in the course of *Smelting Process* so as to get rid of the so-called non fusible impurities of the earthy matter, such as:**"

- **Metallic Oxides** • **Silicates and** • **Silica**–that are present duly in the caleined or roasted Oxide Ores."

NOTE During *Smelting* process the 'Flux' has a tendency to combine intimately with the available *Non-fusible Impurities* to help in the conversion into a respective *Fusible Material* invariably termed as '*Slag*'.

Besides, the 'Slag' designates a substance having relatively low MP; and hence, at a *higher temperature* the 'Slag' is in a **Liquid State**–which being quite distinctly **insoluble in the molten metal**.

4. *Function of Flux:* The major **function of Flux** is concerned with the perceptive removal of the so-called **non-fusible impurity** in the form of a **Fusible Slag**.

The **Flux** are generally of *two* types, namely:

- **Acidic Fluxes** and • **Basic Fluxes**.

(a) *Acidic Fluxes*–The so-called **acidic fluxes** *viz.*, **Borax, Silica** etc., are usually being added in order to discard the **inherent basic impurities** *viz.*, **metallic oxides: CaO, FeO**–as the ultimate *Fusible Slag*. The above episode may be expressed as under:

Basic Impurity		Acidic Flux		Fusible Slag
CaO	+	SiO_2	$\longrightarrow$	$CaSiO_3$
				Calcium Silicate
FeO	+	SiO_2	$\longrightarrow$	$FeSiO_3$
				Iron Silicate

(b) *Basic Fluxes*–A good number of **Basic Fluxes** usually occur *viz.*, **Limestone [$CaCO_3$], Magnesite [$MgCO_3$]**, and **Haematite [Fe_2O_3]**–which are added right into the *Furnace Charge* so as to remove the so-called **acid impurities** (*viz.*, , **Silicates**) as the **fusible slag**.

$$SiO_2 \quad + \quad MgCO_3 \quad \longrightarrow \quad MgSiO_3 \quad + CO_2\uparrow$$

Silicon dioxide   **Magnesium**   **Magnesium**
(An Acidic Impurity)   **Carbonate**   **Silicate**
     (Basic Flux)   **(Fusible stage)**

2.2.3.2. *Carbon Monoxide Reduction Method*

It has been duly observed that in certain typical cases, the **carbon monoxide (CO)** gets generated in the '*furnace*' itself which is being duly employed as a **Reducing Agent** *viz.*,

(i) $Fe_3O_4 + 4CO \xrightarrow{\text{Heat}} 3Fe + 4CO_2\uparrow$
   Magnetite   **Iron Metal**

(ii) $Fe_3O_3 + 3CO \xrightarrow{\text{Heat}} 2Fe + 3CO_2\uparrow$
   Magnetite   **Molten Iron**

(iii) $PbO + CO \xrightarrow{\text{Heat}} Pb + CO_2$
   Lead Oxide   **Molten Lead**

(iv) $PbO + CO \xrightarrow{\text{Heat}} Cu + CO_2\uparrow$
   Copper Oxide   **Molten Copper**

2.2.3.3. Hydrogen Reduction Method

In certain cases, the **H₂-gas** has also been employed as a **reducing agent**,–which may be expatiated by the following *three* reactions:

(i) $\underset{\substack{\text{Tungsten} \\ \text{Trioxide}}}{WO_3} + 3H_2 \xrightarrow{\text{Heat}} \underset{\substack{\text{Tungsten} \\ \text{Metal}}}{W} + 3H_2O$

(ii) $\underset{\substack{\text{Tungsten} \\ \text{Trioxide}}}{MoO_3} + 3H_2 \xrightarrow{\text{Heat}} \underset{\substack{\text{Molybdenum} \\ \text{Oxide}}}{Mo} + 3H_2O$

(iii) $\underset{\substack{\text{Germanium} \\ \text{Dioxide}}}{GeO_2} + 2H_2 \xrightarrow{\text{Heat}} \underset{\substack{\text{Germanium} \\ \text{Metal}}}{Ge} + 2H_2O$

2.2.3.4. Magnesium Reduction Method

It has been proven and ascertained that the **Oxides** of a few metals are reduced adequately by **Magnesium (Mg)**,–as could be seen in the following *two* specific examples:

(i) $\underset{\substack{\text{Rubidium} \\ \text{Oxide}}}{Rb_2O_3} + 3Mg \xrightarrow{\text{Heat}} 3MgO + \underset{\substack{\text{Rubidium} \\ \text{Metal}}}{2Rb}$

(ii) $\underset{\substack{\text{Titanium} \\ \text{Chloride}}}{TiCl_4} + 2Mg \xrightarrow{\text{Heat}} \underset{\substack{\text{Titanium} \\ \text{Metal}}}{Ti} + 2MgCl_2$

Thus, **Titanium (Ti)** is obtained by the reduction of **TiCl₄** by **Mg** in an inert atmosphere of *Argon Gas*.

2.2.3.5. Aluminium Reduction Method

It is also sometimes referred to as the–**Alumino-Thermic Method**. In actual practice, one may encounter *certain* **metallic oxides** that cannot be subjected to **reduction** by *carbon* since the *prevailing* **affinity of Oxygen (O₂)** for the ensuing 'metal' is found to be for greater *vis-a-vis* its actual affinity for **carbon**.

Amazingly, such classical *Metallic Oxides* may be reduced effectively by the aid of **aluminium powder**. Based upon the aforesaid analogy and dictum,–**metallic oxides** into the respective 'Metal' by making use of the *Al-powder* is termed as:

"alumino-termic process";

and hence, this process has been mostly employed to reduce:

'the so-called TiO₂, Cr₂O₃ and Mn₃O₄ Oxides into Ti, Cr, and Mn metals respectively'.

| | Both Stroutium (Sr) and Barium (Ba) are duly obtained by the reduction of their respective oxides into the corresponding metals *viz.*, Ti, Cr, and Mn. |

2.2.3.6. *Self-Reduction Method [Auto-Reduction Method; Air-Reduction Method]*

In a situation when the **Sulphide Ores** pertaining to some of the so-called known *less-electropositive metals viz.*, **Hg, Cu, Pb, and Sb**–are duly subjected to **heating in the air**–then one may critically observe that:

"**a certain portion pertaining to these 'Sulphide Ores' gets duly altered either into the respective** *oxide* or *Sulphide*–**that may eventually reacts vehemently with the remaining portion of the** *Sulphide Ore* **so as to yield the corresponding** *metal* **and** *sulphur dioxide (SO₂)*."

The various reactions that explicitly illustrate the extraction episode pertaining to: **Hg, Cu,** and **Pb** right from their **Sulphide Ores** may be stated as under:

(i) $2HgS + 3O_2 \longrightarrow 2HgO + 2SO_2\uparrow$

  Mercury   Mercury
  Sulphide   Oxide
  (Cinnabar)

  $2HgO + HgS \longrightarrow 3Hg + SO_2\uparrow$

                    Mercury
                    Metal

(ii) $Cu_2S + 3O_2 \longrightarrow 3Cu_2O + 2SO_2\uparrow$

  Copper   Copper
  Sulphide   Oxide

  $2Cu_2O + Cu_2S \longrightarrow 6Cu + SO_2$

                    Copper
                    Metal

(iii) $2PbS + 3O_2 \longrightarrow PbO + 2SO_2\uparrow$

  Lead Sulphide   Lead
  (Galens)   Oxide

  $2PbO + PbS \longrightarrow 3Pb + SO_2\uparrow$

                    Lead
                    Metal

Comment: Obviously, the alternate method of extraction of **Pb** by heating its **sulphide ore [Galena (PbS)]** in air may also be expressed as under:

$$PbS + 2SO_2 \longrightarrow PbSO_4$$

$$PbSO_4 + PbS \longrightarrow 2Pb + 2SO_2\uparrow$$

                    Lead
                    Metal

2.2.3.7. *Electrolytic Reduction Method*

Interestingly, the **oxide of the active metal** *viz.*, **Alkali Metals** (Na, K, Li), **Alkaline Earth Metals** (**Ca, Sr, Ba**), Al, Zn, etc. are indeed **extremely stable**; and hence, to *cause reduction of their oxides* to the respective **Free Metal**,–the **oxides** have get to be heated at a **high temperature along with C**. Besides, at **high temperature** *these metals* do combine together with C to result into the formation of Metallic Carbides.

Therefore, the aforesaid '**metals**' may not be extracted by the so-called **reduction of their oxides with C.** Alternatively, such metals are duly extracted by the *phenomenon* **of Electrolysis** of their corresponding:

- **Oxides** • **Chlorides** or • **Hydroxides,**

in the **fused (molten) state** perceptively; and thereby the '**metal**' gets liberated strategically at the '**Cathode**' specifically.

Example: **Manufacture of Na-Metal**–It can be accomplished by the careful electrolysis of the *fused admixture of NaCl and CaCl$_2$*, commonly known as **Down's Process.**

Following are the different reactions that do occur at the **Electrolytic Cell** are as given under:

Electrolysis:

❑ **At Cathode (Negative Electrode)**–

$$Na^+ + e^- \longrightarrow Na \text{ (\textbf{Due to Reduction})}$$

❑ **At Anode (Positive Electrode)**–

$$Cl^- \longrightarrow Cl + e^- \text{ (\textbf{Due to Oxidation})}$$

$$Cl + Cl \longrightarrow Cl_2^{\uparrow} \text{ (\textbf{As 'gas' escapes into Air})}$$

2.2.3.8. *Amalgamation Phenomenon*

Importantly, the method is being employed mostly for the extraction of **Noble Metals** *viz.*, **Ag, Au** from their respective '**Ores**'. However, the entire process is solely based upon the underlying analogy and dictum that:

"**Noble Metals do eventually get dissolved in liquid Mercury (Hg) metal; and thus, forms the** *Amalgams*"

Therefore, it retains the procedural nomenclature as the '**Amalgamation Phenomenon**' perceptively.

Modus Operandi

1. Here, the finely powder **ore** and **water** are duly mixed to obtain a '**slurry**', which is then made to flow over either **Brass** or **Copper plates** duly coated with mercury (Hg) and held in a *slanted position*.
2. Thus, the **metal particulate matters** do form an '**amalgam with Hg**'; and hence, are retained on these plates nicely.
3. The '**amalgam**' is scrapped off from these plates mechanically and later on subjected to distillation in '**Iron Retorts**'–when Hg distills over thereby bearing the **Free Metal** in the Retort.

2.2.3.9. *The Wet Process [Hydro-Metallurgy Process]*

In the procedure the *concentrated ore* is leached meticulously using an aqueous solution of an appropriate **chemical reagent**–that helps to release the '**Metal**' as a Soluble Salt. Subsequently, the resultant '**Metal**' is recovered from the ensuing **Salt Solution** either:

- **Using the electrolysis process,** or
- **Adding certain electropositive metal to it,**

whereby the 'Metal' need to be extracted is retained as a **precipitate**.

Examples: Following are the *two* classical examples:

1. **Extraction of Au and Ag**–It is usually acomplished by the conversion of Argenite (**Ag_2S**) or Horn Silver (**AgCl**), both being the *Ag-ores*; and **Gold (Au)** into the corresponding **soluble complex chemical entities** (compounds):

- **$Na[Ag(CN)_2]$** and • **$Na[Au(CN)_2]$**

by making use of **aqueous solution of NaCN**.

2. **Extraction of Cu**–A sufficiently large quantum of the **Copper Glance Ore [Cu_2S]** is adequately exposed to **air** and **water**. After a span of **one full** year the treated ore gets duly oxidised to $CuSO_4$ (copper sulphate). Now, the solution of $CuSO_4$ is either:

- **electrolysed**; or
- addition of *Scrap Iron* is made to render **Cu-as a precipitate**.

The reactions may be expressed as under:

(i) $\underset{\substack{\text{Copper Sulphide} \\ \text{Copper Glance Ore}}}{2Cu_2S} + SO_2 \longrightarrow \underset{\text{(Solution)}}{2CuSO_4} + 2CuO$

(ii) $\underset{\text{(Solution)}}{CuSo_4} + \underset{\text{(Scrap Iron)}}{Fe} \longrightarrow \underset{\text{(Solution)}}{2CuSO_4} + 2CuO$

2.2.4. *Purification (Refining) or Metal*

In a broader perspective, the '**metals**' so obtained by any of the aforesaid **Reduction Process** (see *Section: 2.2.3*) are not absolutely pure (*100%*); and hence, do essentially need further purification (refining). Following are the *two* **cardinal procedures** for the specific refining of the **Crude Metals**, namely:

(a) **Thermal Refining**, and

(b) **Electrorefining**,

which shall now be described in the sections that follows:

2.2.4.1. *Thermal Refining*

In actual practice, one may come across *five* **variants in Thermal Refining**, such as:

- **Oxidation by Air and Cooling**,
- **Fractional Distillation**,
- **Liquation Phenomenon**,
- **Zone Refining**, and
- **Vapour-Phase Refining**.

(a) **Oxidation by Air and Cooling**–Amazingly, this method is indeed most suitable for the refining of **Copper (Cu) and Tin (Sn) metals** perceptively. Nevertheless, this process critically involves the melting of the so-called '*Crude Metal*' with a continuous blasting of presurized air through the molten mass of metal. After a considerable volume of air is made to pass through the melt, it is subsequently– stirred using a **raw wooden pole**. Thus, its *unburnt condition* yields a significant quantum of **Carbon (C) and Carbon monoxide (CO)** in order to help in the crucial **reduction of the metallic oxide formed**– thereby producing finally the '**Refined Metal**'.

Fig: 5.8 shows the diagrammatic representation of the oxidation crude metals *viz.*, Cu and Sn by air and poling (*i.e.*, stirring with a raw wooden pole).

Remarks: A thin-uniform layer of '**coke powder**' is duly maintained at the top of the surface so as to prevent the reoxidation of the **metal** so formed (however, a small quantum of metal to be refined may get oxidised in the above process).

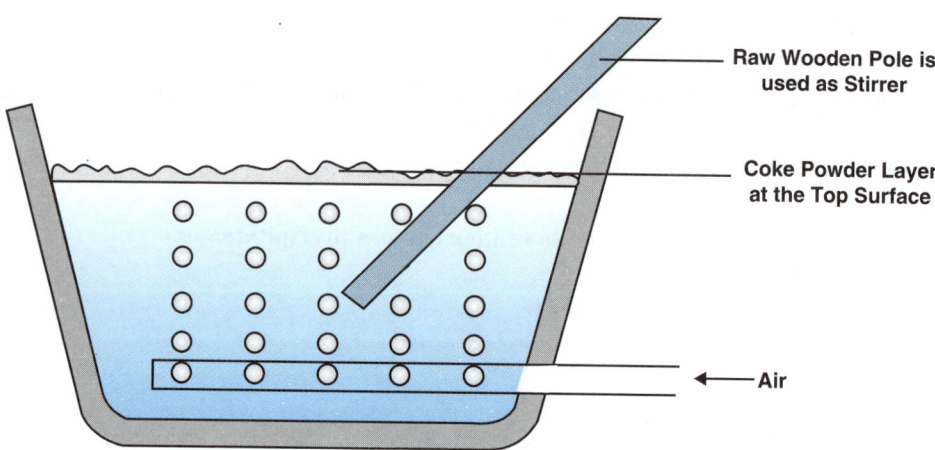

Raw Wooden Pole is used as Stirrer

Coke Powder Layer at the Top Surface

← Air

Fig. 5.8: The Diagrammatic Representation of Oxidation by Air and Poling for metals like: Cu and Sn.

Interestingly, the so-called *more* **basic metallic impurities** are being subjected to the preferred oxidation by *oxygen (O₂) of air* thereby resulting into the formation of either:

- **Volatile Oxides** or • **Non-Volatile Oxides (or serum)**

(b) Fractional Distillation–Importantly, the so-called **Fractional Distillation** phenomenon does essentially utilize the particular prevailing difference between the metal and that of the impurity perceptively.

Example: **Crude Zinc (Zn) containing: Cd, Fe, and Pb** as the inherent *impurities* may be purified by **fractional distillation** as stated under:

$$\text{CRUDE ZN} \begin{cases} \text{Cd : bp = 767°C} \\ \text{Fe :} \\ \text{Pd :} \end{cases}\begin{array}{l} \\ \text{bp> 1500°C} \end{array} \xrightarrow[-Cd\ (vap)]{T>767°C} \text{Zn}\begin{Bmatrix} \text{Pb} \\ \text{Fe} \end{Bmatrix} \xrightarrow{T/920°C} \begin{array}{l} \longrightarrow \text{Zn (vapour)} \\ \\ \longrightarrow \text{Fe and Pb} \end{array}$$

(bp=920°C)

Comment: Amazingly, at a specific **temperature > 767°C**– *Cd* does separate as its vapours; whereas, at **temperature > 920°C – *Zn* (pure)** separates as its **exclusive vapours** thereby leaving behind: Fe and Pb–as the so-called *impurities* retained in the '**malt**'.

(c) Liquation Phenomenon–In fact, theis method is exclusively applicable for such typical metals, for instance: Sn, Pb, and Bi–that do have rather **low melting points** *vis-a-vis* their **corresponding impurities**. However, in the aforesaid method–the '**Block of Crude Metal**' is held strategically at the top-end of the **sloping furnace**; and thus, heated just above the **melting point of the metal to be refined**.

In this manner, the **pure metal** melts and flows down the **floor of the fireplace**; and hence, duly gets collected in a **reservoir** kept at the bottom of the slope,–as shown in Fig: 5.9.

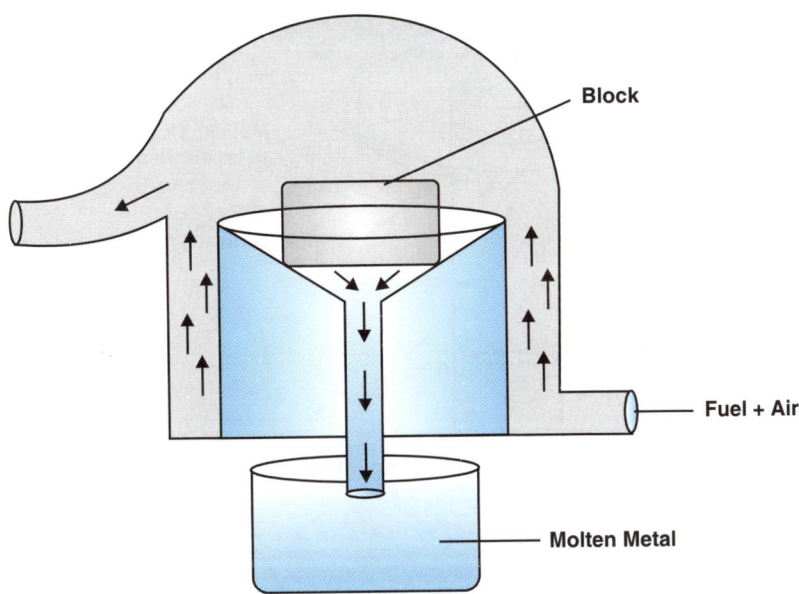

Figure: 5.9: A Explicit Diagrammatic Representation of the Liquation Phenomenon for the Purification of: Sn, Pb, and Bi.

Remarks: The impurities so collected is thrown (rejected) as the **perforated block**.

NOTE   **The actual content of the impurities present in the crude metal must be high enough so as to enable the proper purification, otherwise the *impurities* will tend to flow-down along with the molten metal itself.**

(d) **Zone Refining**–Certain metals *viz.*, **Silicon (Si), Germanium (Ge), and Gallium (Ga)** of high purity profile[*] are being adequately purified by means of **Zone Refining**. It is invariably termed as **Ultrapurification** since it results in the *ultimate* impurity level pegged down to only **ppm level**.

Zone-Refining–is exclusively based upon the so-called:

"**fractional crystallization because the inherent impurity tends and prefers to remain in the '*melt*' itself; and, therefore, upon solidification only the *Pure Metal* gets duly notified at the *top layer* of the 'melt.'**"

Modus Operandi: The **Zone refining** is carried out **in a 'Ring Furnace'**–which is being heated to a *desired temperature* for enabling to melt the so-called **metal rods,**–as depicted in Fig: 5.10. Thus, it eventually produces a **distinct thin-Zone** throughout the **entire available cross sectional area,**–as illustrated vividly in Fig: 5.11.

Remarks: It is always desirable and advisable that the actual diameter of the **Metal Rod (*d*)** (to be *purified*) must be small enough so as to yield a *uniform melt*.

Besides, in a situation when the '*Melted Zone*' in the *metal rod* becomes ready, the furnace is allowed to *move downwards* very gradually along with the **Melted Zone**–as shown in Fig: 5.12. At the end the **furnace is termed off, allowed to cool down, and again taken to the top for the entire repitation of the process.**

---

*   Mostly used in **Semiconductors**.

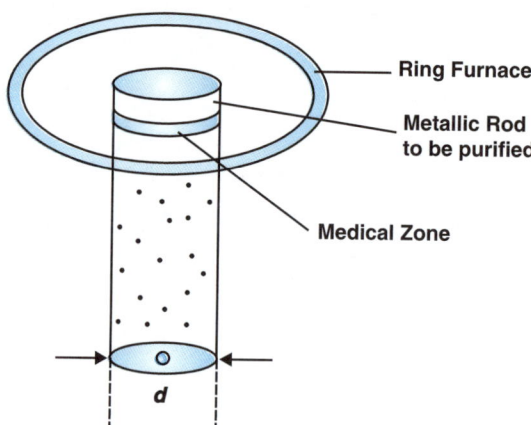

Fig. 5.10: Description of Metal Rod being Heated by Ring Furnace.

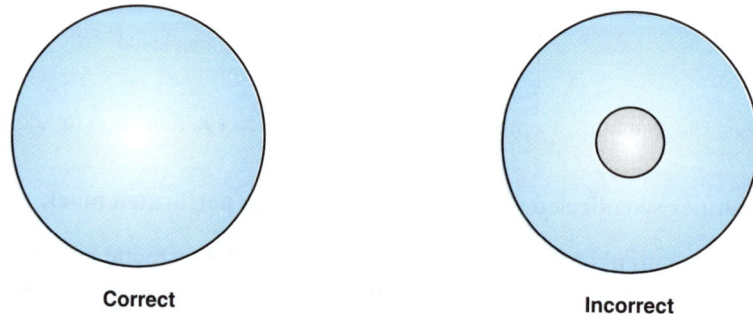

Correct   Incorrect

Fig. 5.11: The Cross-Sectional View of the Metal Rod.

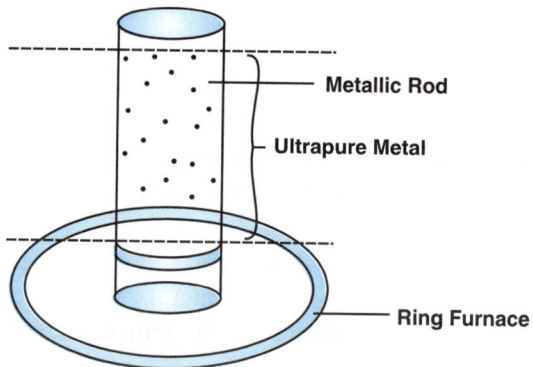

Fig. 5.12: Illustration of the Ring Furnace Moving Down the Metal Rod.

| NOTE | Practically a large segment of the '*Impurities*' present is being swept out to the bottom of the furnace after several successive operations of the process. |

(e) **Vapour-Phase Refining**–The *two* most vital and important criteria for the so-called **vapour-phase refining** phenomenon are as given under:

❑ The **intermediate chemcial entity** so formed has got to be *volatile* in nature, and

❑ In addition, the **intermediate compound** so formed must be quite **unstable relatively** *i.e.*, it should decompose readily on being subjected to heating at almost **achievable temperature** perceptively.

Interestingly, the aforesaid **refining methodology** is vehemently employed in the following **purification methods**:

➤ **Moud's Process [for the specific purification of Nickel (Ni) Metal]**–Let us look into the following expression:

$$\underset{\textbf{Impure Form}}{Ni(Solid)} + CO\,(gas) \xrightarrow{50-60°} Ni(CO)_4\,(gas) \xrightarrow{80-00C°} \underset{\textbf{Pure Form}}{Ni(Solid)} + \underset{\textbf{Recycled}}{CO(gas)}$$

Remarks: In case, $Ni(CO)_4$ is not volatile in character it cannot be separated from the *bulk of the impurities*. Thus, its **inherent volatile nature** gas as long way to aid in the escape it from the **original bulk of impurities** legitmately. Besides, $Ni(CO)_4$ does require to undergo the so-called; **thermal decomposition process** readily (otherwise it would certainly fail to yield the **pure Ni-Metal**.

➤ **Van-Arkel De Boer Process [for the specific Purification of Zirconium (Zr), Boron (B) and Titanium (Ti)]**–

We may examine the following expression:

$$\underset{\textbf{(Impure Form)}}{Zr\ or\ Ti} \xrightarrow[\text{(250°C)}]{I_2\,(Vapour)} \underset{\textbf{(Volatile Entities)}}{Zr\,I_4\ or\ Ti\,I_4} \xrightarrow[\text{(1400°C)}]{On\,Tungsten\,Filament} \underset{\textbf{(Pure Form)}}{Zr(s)\ or\ Ti(s)} + \underset{\textbf{(Recycled)}}{2I_2\,(gas)}$$

$$\underset{\textbf{(Impure Form)}}{B} + 3/2 I_2\,(vapour) \longrightarrow \underset{\textbf{(Volatile Form)}}{BI_3} \xrightarrow{Heat} \underset{\textbf{(Pure Form)}}{B} + \underset{\textbf{(Recycled)}}{3/2\ I_2\,(gas)}$$

2.2.4.2. *Electrorefining*

In general, the **Electrorefining** process is exclusively applicable for the crucial **purification** phenomenon pertaining to such metals as:

Cu, Z, Sn, Ag, Au, Ni, Pb, and Al

In order to attain an **optimized refining target** the *cathode* is made up on a ***thin-strip of pure metal**[*]*; whereas, the *anode* is duly made of a **'large-slab of Impure Metal'** (*i.e.*, *intended to be refined*). The *electrolyte* being utilized is nothing but the aqueous solution of an *appropriate* **salt of the metal (to be refined)** or the respective '**Melt**' of the oxide/salt occasionally,–as shown in Fig: 5.13.

Comments: The 'metal' gets slowly corroded from the 'anodes' – *i.e.*, the slab of **impure metal**; and thereby, the **Pure Metal** gets adequately deposited at the '*Cathode*'. Finally, it gets duly purified in this manner.

Reactions Occurring at the two Electrodes:

➤ **At Cathode:** $M^{n+} + ne \longrightarrow M$

➤ **At Anode:** (i) $M \longrightarrow M^{n+} + ne$ **(Desired)**

(ii) $X^{n-} \longrightarrow {}^1X_2 + ne$ **(Undesired)**

---

[*] That is, very much akin to the '**Metal**' intended to be refined.

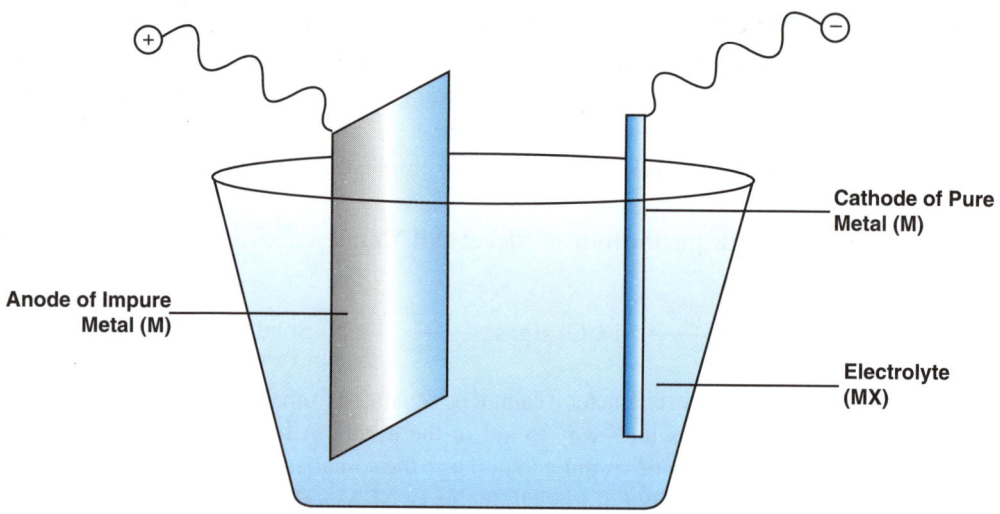

Fig .5.13: An Elaborated Diagrammatic Representation of the Electrorefining Process.

Based on the aforesaid expressions (*two reactions*) competing strictly at the 'Anode',–one may observe that the so-called 'anionic segment of the electrolyte' needs to be selected in such a manner that the 'Reaction (ii)' fails to occur at the 'Anode'.

NOTE	Nevertheless, the *metallic impurities* with lower oxidation potential *vis-a-vis* that of the metal to be refined are being duly separated in the form of the so-called 'Anode Mud' at the bottom (see Fig: 5.13).

2.2.5. *Thermodynamics of Reduction Process*

Evidence from the literature and scientific revelations do ascertain that:

"extraction of '*metal*' right from their respective oxides by making use of either *Carbon (C)* or *other 'metals'* and also by the *Thermal Decomposition phenomenon* invariably involves a plethora of reasonably valid and important aspects that certainly merit a comprehensive discussion perceptively".

Table: 5.2 records the so called **Reduction Potentials** as well as **Extraction Procedures** of an array of elements along with their respective: element, **standard electrode potential difference [E°(V)], materials**, and **extraction method**.

S.No.	Element	E°(V)		Materials	Extraction Procedure
1.	Lithium	Li$^+$ | Li	−3.05	LiCl	
2.	Potassium	K$^+$ | K	−2.93	KCl, [KCl, MgCl$_2$, 6H$_2$O]	Electrolysis of fused
3.	Calcium	Ca^{2+} | Ca	−2.84	CaCl$_2$	salts, usually chlorides
4.	Sodium	Na$^+$ | Na	−2.71	NaCl	
5.	Magnesium	Mg^{2+} | Mg	−2.37	MgCl$_2$, MgO	Electrolysis of MgCl$_2$ High temperature reduction with C

Table: 5.2: The Reduction Potentials and Extraction Procedues of Various Elements.

(Contd...)

6.	Aluminium Al^{3+}		Al	−1.66	Al_2O_3	Electrolysis of Al_2O_3 dissolved in molten $Na_3[AlF_6]$
7.	Manganese	Mn^{2+} | Mn		−1.08	Mn_3OI_4, MnO_2	Reduction with Al
8.	Chromium	Cr^{3+} | Cr		−0.74	$FeCr_2O_4$	Thermite process
9.	Zinc	Zn^{2+} | Zn		−0.44	Fe_2O_3, Fe_3O_4	Chemical reduction
10.	Iron	Fe^{2+} | Fe		−0.44	Fe_2O_3, Fe_3O_4	Chemical reduction
11.	Cobalt	Co^{2+} | Co		−0.27	CoS	of oxides by C
12.	Nickel	Ni^{2+} | Ni		−0.23	$NiS, NiAS_2$	Sulphides are
13.	Tin	Sn^{2+} | Sn		−0.14	SnO_2	converted to oxides
14.	Lead	Pb^{2+} | Pb		−0.13	PbS	then reduced by C, or
15.	Copper	Cu^{2+} | Cu		+0.35	Cu (metal), Cus	Found as native
16.	Silver	Ag^+ | Ag		+0.80	Ag(metal), Ag_2S, AgCl	metal, or
17.	Mercury	Hg^{2+} | Hg		+0.85	HgS	compounds easily
18.	Gold	Au^{3+} | Au		+1.38	Au(metal)	decomposed by heat (also cyanide extraction)

Remarks: These essentially include:

1. For a **spontaneous reaction, the observed Free Energy Change (ΔG) should be negative:**

$$\boxed{\Delta G = \Delta H - T\Delta S}$$ ....(a)

Thus, in the above expression Δ**H**, vehemently desigantes the actual *enthalpy change* taking place in the **course of a reaction, T** represents the **absolute temperature**, and ΔS relates to the **change in entropy** during the reaction.

Now, let us take into consideration a typical reaction *viz.*, the critical formation of an '**oxide**': We my have the following expression:

$$\boxed{M + O_2 \longrightarrow MO}$$ ....(b)

The **dioxygen (O_2)** is critically being consumed during the reaction. The gases do have a rather **higher entropy profile** *vis-a-vis* the respective **solids or liquids**. However, in this reaction, the S designates the **entropy** or **randonness** that gets decreased progressively; and hence, ΔS is found to be **negative**. Thus, when the temperature is raised, the **TDS** eventually renders more **negative**.

In **Eq. (a)**, because **TDS is being subtracted**, –then ΔG turns out to be **less negative perceptively**. Thus, ultimately one may observe most prevalently that:

"the *Free Energy Change* gets enhanced with an increase of temperature profile".

Therefore, when the **Free Energy Changes** that rightly take place as and when 1g, molecule of a **common reactant (***viz.***, dioxygen)** is employed may be plotted graphically *Vs* the temperature pertaining to a good number of:

'**reactions of metals to their oxides respectively**'.

Ellingham Diagram (Fig: 5.14) is the actual graph plotted for **Oxides** that particularly depicts the critical change in the **Free Energy (ΔG)** with temperature for oxides (*i.e.,* **actually based upon 1g mole of dioxygen in each case**).

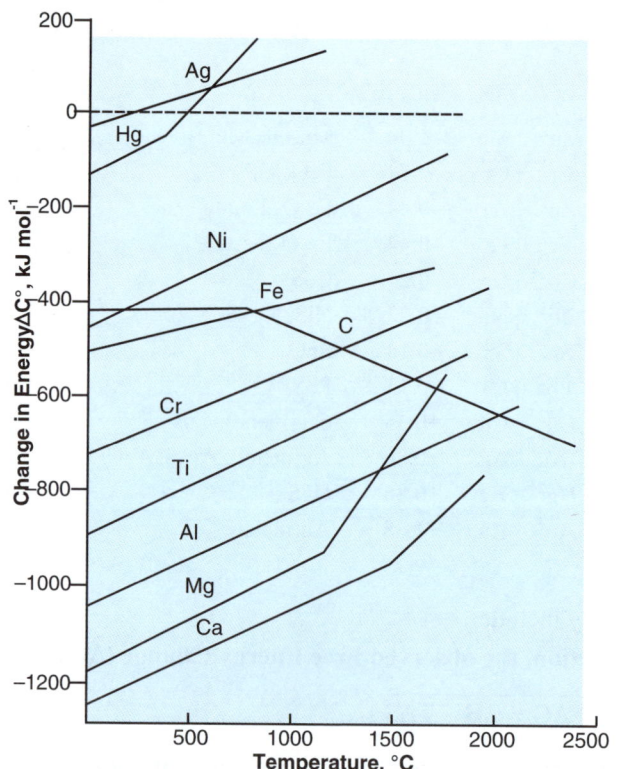

Fig. 5.14: Diagammatic Representation of Ellingham Diagram Depicting the Perceptive Change in Free Energy ($\Delta G°$ kJ mole⁻¹) *Vs* Tempeature for oxides [based upon 1 g mol of Dioxygen (O_2) in each case].

Certain Important Characteric Features of Ellingham Diagram for Oxides

There are *four* **most important characteristic features of Ellingham Diagram for '*oxides*', namely:**

1. Generally, the '**graphs**' for the *metal to metal oxide* mostly **slope** in the **upward direction**, since the observed **Free Energy Change** critically enhances with an increase of temperature perceptively.

2. Amazingly, almost all the **Free Energy Changes** do actually follow a **straight-line** unless and until the **given material(s)** either *vaporizes* or *melts*–as and when there prevails an enormous visible change in the '**entropy-associated with the change-of-state**'–that appears to be solely responsible in:

"changing the slope of the line".

Example: It has been duly observed that the **Hg–HgO line** helps to alter the **slope at 365°C**–when **Hg-boils**; and likewise, **Mg-MgO** changes at 1120°C when **Mg boils** vigorously.

3. Obviously, with the steady rise in the termperature, one would reach at a specific point when the ensuing '**Graph**':

"Crosses the so-called ΔG = O line; whereas, just below this temperature the observed Free Energy of formation of the 'Oxide' is found to be negative–that renders the ensuing 'Oxide' fairly stable in nature."

Therefore, above this specific temperature the so-called:

"**free-energy of formation of the corresponding oxide is *positive*; and hence, the oxide turns out to be unstable, and must decompose into the respective *Metal* and Dioxygen (O_2).**"

 NOTE Importantly, the oxides of: Au, Ag, and Hg are the *three* critical oxides– that may be easily decomposed at a temperature which could be attained conveniently; and hence, these '*metals*' may be extracted by means of the Thermal Decomposition of three Oxides.

4. Interestingly, we may come across a good number of processes whereby **one metal is being employed to cause the reduction of the oxide of another metals**–that:

"**predominantly lie just above it in the so-called Ellingham Diagram since the *Free Energy* may tend to be more negative by an amount almost equivalent to the actual difference existing between the *two* graphs at that specific temperature**".

Remarks: From the above expression of **factual statements** and **critical observations**, we may conclusively infer that:

'**Al helps to cause the reduction of oxides of Fe, Cr, Ti–in the recognized *Thermite Reaction* perceptively; whereas, Al shall fail to reduce MgO at a temperature below 1500°C germinely**'.

2.2.6. Alloys and Amalgams

In a broader sense, an **Alloy** represents –"**a homogenous metallic substance that essentially comprises *two* or more metals as a 'solid solution'.**"

However, in case one of the integral components of the **Alloy** happens to be *Mercurry (Hg)*,–it is usually known as '**Amalgam**' perceptively.

2.2.6.1. Classification of Alloys

In general, the Alloys may be classified as:

(a) *Ferous Alloys*–In a situation when the '**Alloy**' contains Iron (Fe)–as *one of the integral constituents*,–it may be termed as the **ferrous Alloy,** such as:

- **Stainless Steel [Cr + Fe + Ni]**; and
- **Ferrosilicon [Fe + Si]**

(b) Non-Ferrous Alloys–In case, an **Alloy** is completely devoid of **Iron (Fe)**,–it is known as a **Non-ferrous Alloy**, namely:

- **Brass [Cu + Zn]**,
- **Bell metal [Cu + Sn]**, and
- **Solder [Sn + Pb]**.

2.2.6.2. Properties of Alloys

In usual practice, the **Alloys** are prepared in such a way so as to develop critically.

"**certain highly specific characteristic features that are not found usually in the so-called constituent elements at all**".

Besides, both the inherent *characteristic of metals* whidh may be grossly improved upon by preparing **Alloys**–are as stated under:

➢ **Superior Casting**–Amazingly, the crucial formation of the **Alloys** predominantly enhances the so-called–'*Casting Characteristics*' of the inherent **Metal selected**.

Example: Following is a classical example:

- **Type Metal [Pb (80%) + Sb(16%) + Sn(4%)]** essentially comprises **Antimony (Sb)**–that prevalently:

  ➢ expands upon '*solidification phenomenon*', and

  ➢ adapts a prompt impression of the mould.

NOTE   **Hence, it certainly improves upon the casting Property of Metal.**

- **Hardness**–The **Alloys** are found to be definitely harder and stronger *vis-a-vis* the **metals** and **non-metals** from which these are prepared:

Examples–**Steel** is perceptively **harder** than **Cast Iron.**

- **Corrosion Resistant**–Generally, the **Alloys** are found to be more resistant toward corrosion.

Examples–**Stainless Steel (SS)** is definitely more resistant towards corrosion in comparison to **iron**.

- *MP*–Mostly the **Alloys** do have **MPs** fairly comparable to their **constituent elements**.

- *Tenacity**–The tenacity of **Copper (Cu)** almost gets doubled upon the addition of 5% silicon into it.

2.2.6.3. *Preparation of Alloys*

Following are the *four* **different methods** that are being used largely in the preparation of alloys, such as:

1. **Fusion**–The *two* most prevalent **Alloys** move by **fusion** are:
   - **Brass [Cu (90%) + Zn (10%)]   and   • Bronze [Cu(90%) + Sn(10%)].**

2. **Reduction**–The **Alloy:** *Aluminium Bronze* is usually prepared by heating the following *two* **substances**:
   - **Aluminium Oxide (Al$_2$O$_3$) and   • Carbon,**

meticulously in the presence of the **requisite quantum of Copper (Cu)**. Hence, we may have the following expression:

$$Al_2O_3 + 3C + Cu \longrightarrow \underset{\textbf{Aluminium Bronze}}{(Al + Cu)} + CO\uparrow$$

3. Compression–The *selected metals* are first and foremost beaten into **thin metallic sheets; and** subsequently **rolled together and hammered under very high pressure** to produce the desired **Alloy**.

Example: **Solder** (an **Alloy**) is duly prepared by compression method using **Pb(50%) and Sn (50%).**

4. Simultaneous Electrodeposition Technique–In this typical instance, the respective '**Aqueous Solution of the salts of the component Metals**' is loaded into a large **Electrolytic Cell**, which is then subjected to the passage of **Electric Current**. In this manner, the so-called **Selected metals** do get uniformly deposited simultaneously upon the '**Cathode**' to yield the intended **Alloy**.

---

*   **Tenacity**–Determination in holding a position (or status).

Example: **Brass** (an Alloy) is duly prepared by this method due to the critical electrolysis of a solution comprising:

"Copper (Cu) and Zinc (Zn) cyanides in a KCN solution".

2.2.7. *Types of Furnaces Employed in Metallurgy*

There are *four* different types of **Furnaces** that are frequently being employed in **Metallurgy**, such as:

(a) *Blast Furnace*–It is mostly used in the *Smelting of Roasted ore* [already described under **Extraction of Iron (Fe)**].

(b) *Reverberatory Furnace*–This kind of a furnace is chiefly being employed in the **Roasting** and **Calcination**, as illustrated in Fig: 5.15.

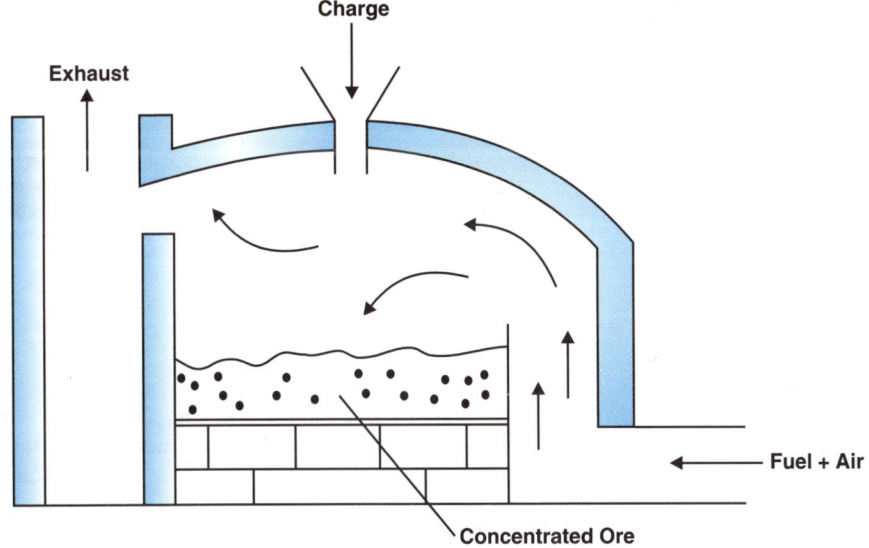

Fig. 5.15: Diagrammatic Representation of a Reverberatory Furnace.

(c) **Muffle Furnace**–In fact, it relates to a '**Closed Chamber**' that is being heated via an external heating arrangement in order that:

'**the substance to be heated never comes in contact with the fuel directly**'.

Hence, it is invariably being employed the so-called **small-scale metallurgical processes** exclusively.

(d) *Electric Furnace*–It is solely used in such instances where an extremely high range of temperature is an absolute must. Obviously, the **high temperature.**

The high temperature is usually accomplished due to:

"an *Electric Arc* struck between *two* thick *Graphite Electrodes*"

2.2.8. *Extraction of Silver (Ag)*

The *four* most abundantly available and important **ores of Silver (Ag) are** as given below:

- **Horn Silver: AgCl;**
- **Argentite [or Silver Glance]: Ag$_2$S;**

- **Ruby Silver: 3Ag₂S.Sb₂S₃,**and
- **Stromeyrite [or Silver-Copper Glance]: Ag₂S.Cu₂S.**

Nevertheless, the actual **Extraction of Silver (Ag)** is invariably carried out by the help of the following *three* **established processes**, namely:

➢ **Cyanide Process,**

➢ **Parke's Process*,** and

➢ **Parkinson's Process*.**

Cyanide Process [or McArthur Forest Process]–A detailed and labeled description of the **Cyanide Process** is illustrated duly in Fig. 5.16.

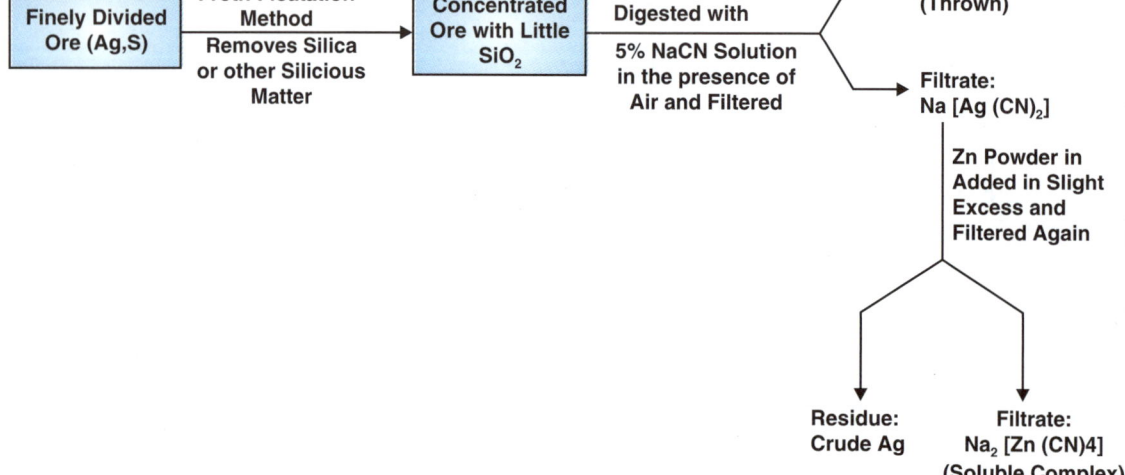

Fig. 5.16: A Detailed Flow Chart for McArthur Forest Process for the Extraction of Silver (Ag) Metal.

Following are the *two* **different** *stages of reactions* that eventually come into being in the **Cyanide Process** for the extraction of silver (Ag):

Stage-1: We may have the following *two* expression:

$$\underset{\substack{\text{Argenite}\\ \text{[or Silver Glance]}}}{Ag_2S} + \underset{\substack{\text{Sodium Cyanide}\\ \text{(Excess)}}}{4NaCN} \rightleftharpoons \underset{\text{Silver Cyanide}}{2[Ag(CN)_2]} + Na_2S + 2Na^+$$

$$\underset{\text{(Air)}}{4Na_2S + 5O_2} + 2H_2O \longrightarrow 2Na_2SO_4 + 4NaOH + 2S$$

Comments: Because the aforesaid reaction is **reversible** in nature, the observed **conversion ratio** is not fairly good. Therefore, preferentially the entire process is performed in the **presence of air** [*i.e., Oxygen (O₂)*]–that eventually helps in the proper conversion of the **sodium sulphide (Na₂S)** so produced into:

- **Sodium Sulphate (Na₂So₄)** and • **Sulphur (S);**

---

*   That is, these processes are actually beyond the scope of discussion here.

and thus, the overall reaction turns out to be unidirectional genuinely.

Stage–2: In the particular step involving the **precipitation of Silver (Ag)** a small quantum of **Zn-powder** in excess is added–that renders the **sodium silver cyanide**, $Na[Ag(CN)_2]$, to attain the status of a '**limiting reagent**'–thereby preventing the so-called excessive **under loss of silver (Ag)** to a significant extent.

Comments: In this instance, **Zinc (Zn) metal** is chosen since it is certainly **more electro positive** *vis-a-vis* to **Silver (Ag)**; and, therefore, the replacement takes place quite easily and conveniently.

We may have the following expression:

$$2Na[Ag(CN)_2] + Zn \longrightarrow Na_2[Zn(CN)_4] + 2Ag \downarrow$$

Sodium Silver Cyanide   Sodium Zinc Cyanide   Silver Metal

*Refining of Silver (Ag) Metal–*The effective refining of **Silver (Ag)** metal is accomplished by the **Electrolytic Process** mostly:

$$\text{Crude (Ag)} \xrightarrow[\text{(Mobius Process)}]{\text{Electrolytic Refining}} \text{Pure (Ag)}$$

$$\begin{bmatrix}\text{Comprising Zn, Cu,} \\ \text{Au as Impurity}\end{bmatrix} \qquad \begin{bmatrix}\text{Obtained Pure Ag} \\ \text{upto99.9\% Purity}\end{bmatrix}$$

Electrolyte Used   :   Silver Nitrate $(AgNO_3)$ Solution + HNO_3 (10%)

Anode   :   Impure Silver (Ag) Slab

Cathode   :   Pure Silver (Ag) Strip

Reactions at Anode   : $Ag \longrightarrow Ag^+ + e^-$

Reactions at Cathode : $Ag^+ + e^- \longrightarrow Ag \downarrow$

2.2.9. *Extraction of Gold (Au) By Cyanide Method*

The **extraction of gold (Au) by cyanide process involves several sequential stages**, as depicted in Fig: 5.17.

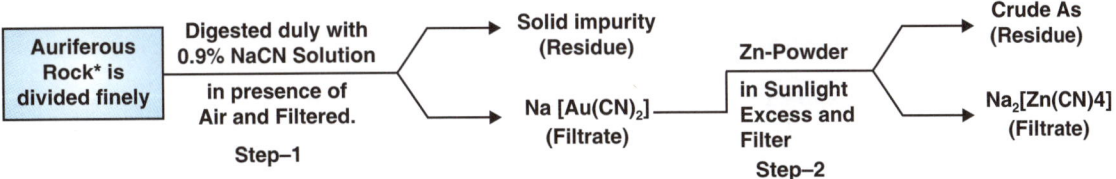

Fig. 5.17: An Explicite Flow Diagram for Cyanide Process for Extraction of Gold (Au).

The chemical reactions that occur in **different steps are:**

❑ **Step-1–**

$$4\,Au + 8NaCN + 2H_2O + O_2 \longrightarrow 4Na[Au(CN)_2] + 4NaOH$$

In this case, the **oxidation of gold (Au)** is not all possible and feasible without the presence of **air**; and **NaCN** does act as the so-called *Complexing agent* to yield **sodium-gold cyanide complex.**

---

*   **Auriferous Rock–**It is a rock of **Quartz** duly contaminated with **Gold (Au) linings**.

❑ **Step-2**–The following reaction takes place:

$$2Na[Au(CN)_2] + Zn \longrightarrow Na_2[Zn(CN)_4] + 2Au \downarrow$$

Refining of Gold (Au) Metal–Fig: 5.18 provides the various steps that are involved intimately in the **refining of Gold (Au) metal**. Importantly, in the last but one step (*i.e.*, **Step-3**)–on being subjected to heating with Borax [$Na_2B_4O_7$]–the resulting **soluble metaborate of Cu [$Cu(BO_2)_2$]** so formed is now washed thoroughly with water. Likewise, in **step-4** the **silver (Ag)** usually dissolves out as **soluble silver sulphate (Ag_2So_4)**; and thus, leaving **pure Gold (Au)** as a residue.

NOTE	1. Removal of *silver metal (Ag)* may also be accomplished by making use of chlorine (Cl_2) or by electrolysis process.
	2. *Cupellation*–It refers to a process whereby *Crude Gold (Au)* is first taken in a *Cupell (i.e.,* a small bowl) and melted in the presence of air (O_2). Perhaps due to the optimized high O_2-affinity of Lead (Pb)–it gets readily converted into *Lead oxide* (**PbO**)–a volatile substance that escapes from the system rapidly.

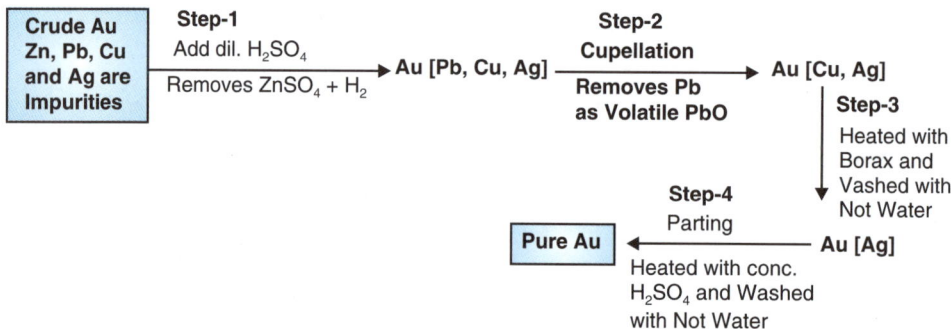

Fig. 5.18: An Elaborated Flow Chart for Refining of Gold (Au) Involving Four Sequential Steps (1, 2, 3, and 4).

2.2.10. *Extraction of Magnesium (Mg)*

Following are the *Seven* **important ores of Magnesium (Mg), namely:**

1. **Magnesite** : $MgCO_3$
2. **Dolmite** : $MgCO_3$, $CaCO_3$
3. **Carnalite** : $kcl.Mgcl_2.6H_2O$
4. **Kainite** : $K_2SO_4.MgSO_4.Mgcl_2.6H_2O$.
5. **Kieserite** : $MgSO_4.H_2O$
6. **Asbestos** : $CaMg (SiO_3)_4$
7. **Spinel** : $MgO.Al_2O_3$

The effective **extraction of Magnesium (Mg)** may be accomplished by means of the following *two* widely used methods, namely:

➢ **Electrolytic Reduction**, and

➢ **Carbon Reduction**.

(a) Electrolytic Reduction–In general, the **electrolytic reduction process being used for the extraction of Magnesium (Mg)** comprises predominantly of *three* distinct and sequential steps, such as:

(i) Preparation of Hydrated Magnesium Chloride [MgCl$_2$.6H$_2$O],

(ii) Conversion of *Hydrated MgCl$_2$* to *Anhydrous MgCl$_2$*, and

(iii) Electrolytic Reduction of Anhydrous MgCl$_2$,

which shall now be discussed individually in the sections that follows:

(a) **Preparation of Hydrated Magnesium Chloride [MgCl$_2$.6H$_2$O]**–It may be duly prepared by either of these *two* common procedures:

(1) **From *Carnalite* (aMg-ore) [KCl.MgCl$_2$.6H$_2$O]**–The various sequential steps being involved in the preparation of **Hydrated Magnesium Chloride (MgCl$_2$.6H$_2$O)** are illustrated in Fig: 5.19.

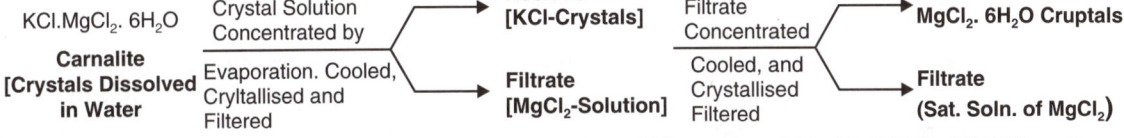

Fig. 5.19: Detailed Flow Chart for the Preparation of Hydrated Magnesium Chloride [MgCl$_2$.6H$_2$O] from the Magnesium Ore 'Carnalite'

Comment: The underlying principle of the above process being that **Potassium Chloride (KCl)** is less soluble *vis-a-vis* **Magnesium Chloride (MgCl$_2$)**; and hence, crystallizes out as the *First Crop*.

(2) **From Sea Water**–Amazingly, the **sea water** duly comprises a good quantum of **magnesium chloride (MgCl$_2$)**.

Dow Sea Water Method–The critical process engaged for the extraction of **Magnesium Chloride (MgCl$_2$)** right from the *sea water* is usually called as **Dow Sea Water Method**.

Fig: 5.20 illustrates the various sequential steps that are crucially involved in the **extraction of MgCl$_2$ from sea water**. However, the **principle** based on the process being that **calcium hydroxide [Ca(OH)$_2$]** is found to be **water-soluble**; whereas, **magnesium hydrocide [Mg(OH)$_2$]** being **sparingly soluble in water**.

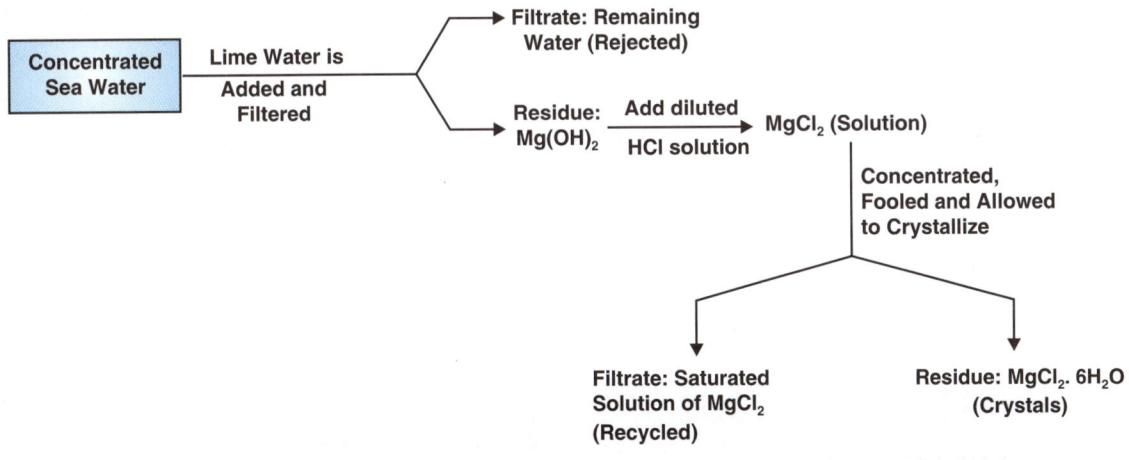

Fig. 5.20: Description of a Flow Chart for Preparation of Hydrated Magnesium Chloride (MgCl$_2$) from Concentrated Sea Water.

(b) **Conversion of Hydrated MgCl$_2$ to Anhydrous MgCl$_2$**–It has been duly ascertained that direct heating aof the so-called **hydrated magnesium chloride [MgCl$_2$.6H$_2$O]** may not yield the corresponding **anhydrous magnesium chloride [MgCl$_2$]** perhaps due to its **hydrolysis phenomenon**,–as stated below:

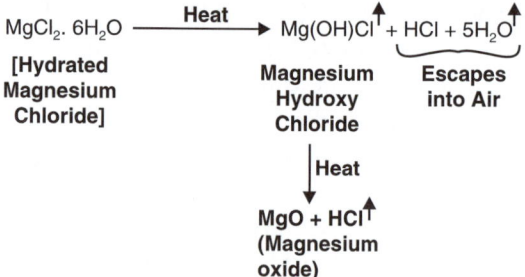

Comment: Thus, MgO is obtained as an **unwanted material** in the aforesaid **Electrolytic Reduction** step based upon its extremely elevated **MP(~ 2850°C)**.

Plausible Way Out: Hence, the so-called **hydrated magnesium chloride [MgCl$_2$.6H$_2$O]** is heated to **175°C** under reduced pressure (vacuum) in a **steady current of dry HCl-gas**–thereby resulting into the formation of **anhydrous MgCl$_2$**, as stated under:

$$MgCl_2 \cdot 6H_2O \xrightarrow{\text{Heat}} MgCl_2 + 6H_2O \uparrow$$
Hydrated Magnesium Chloride

$$MgCl_2 + H_2O \longrightarrow Mg(OH)Cl + HCl$$
Hydrated Magnesium   Magnesium Hydroxy
Chloride   Chloride

Comment: It may be observed that as and when '*dry HCl*' is duly present in the ensuing '**system**'–the so-called:

"**hydrolysis equilibrium duly shifts towards LHS, but on account of the enhanced thermal energy of the system, one may notice that the decomposition reaction proceds continuously and gives ultimately an anhydrous MgCl$_2$ perceptively**".

(c) **Electrolytic Reduction of Anhydrous MgCl$_2$**–The explicit and schematic representation of the electrolytic reduction of anhydrous **MgCl$_2$** is depicted in Fig: 5.21.

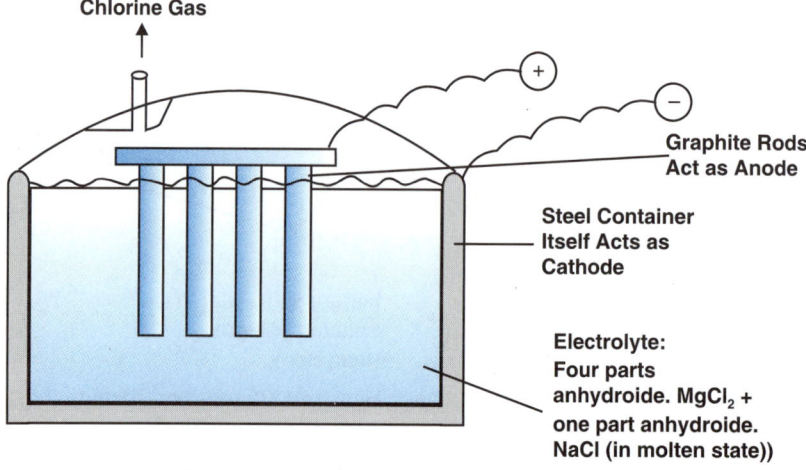

Fig. 5.21: The Schematic Representation of the Electrolytic Reduction of Anhydrous Magnesium Chloride (MgCl$_2$).

Reactions Occuring at the Electrodes

$$MgCl_2 \longrightarrow Mg^{2+} + 2Cl^-; NaCl \longrightarrow Na^+ + Cl^-$$

- **At Cathode:** $Mg^{2+} + 2e^- \longrightarrow Mg$

- **At Anode:** $2Cl^- - 2e^- \longrightarrow Cl_2^\uparrow$

Remarks: These essentially include:

1. Container meant for the **Electrolytic Cell** is required to be covered and the air present inside is flushed cut by a stream of coal gas–so as to check and prevent the so-called **oxidation of Mg So** formed and *floating upon the top surface of the molten electrolyte*.

2. **Electrolyte (molten form)** comprises: **4 parts of anhydrous MgCl$_2$ plus 1 part of anhydrous NaCl**. At this point in time, **one part of additional NaCl** is being incorporated so as to **lower the MP of the electrolyte** from **1200°C down to 700°C**. Thus, it also enhance the **electrical conductivity of the melt** significantly.

3. Utilization of the **fused carnalite [KCl.MgCl$_2$.6H$_2$O]** as an *electrolyte* may be done since both **Na and K** are more **electropositive** *vis-a-vis* Mg; and hence, Mg^{2+} gets discharged preferably at the **cathode**.

(b) **Carbon Reduction**–In this particular process utilized for the extraction of **Magnesium (Mg)**– the following *two* **steps** are being adopted one after the other as given under:

- **production of magnesium Oxide (MgO) from the Calcination of magnesium carbonate (MgCO$_3$)**; and

- **subsequent direct heating of MgCO$_3$ with coke powder at ~ 2000°C preferably in a *closed electric furnace.***

Thus, these chemical reactions usually occur:

$$MgCO_3 \longrightarrow MgO + CO_2^\uparrow$$
$$MgO + C \longrightarrow \underset{\text{(Vapour Form)}}{Mg^\uparrow} + CO^\uparrow$$

Comment: The **Mg** so obtained evolves primarily in the '**Vapour form**'–which on being subjected to **sudden cooling** to ~ **200°C** due to immediate dilution with a relatively *huge volume of H$_2$ gas* so as to check and prevent the so-called **reoxidation of Mg** sequarely **(BP of Mg = ~ 100°C)**

2.2.11. *Extraction of Tin (Sn)*

Cassiterite designates an important **ore of Tin(Sn)** having nearly 1 to 5% tin dioxide (SnO$_2$) present in it. However, the major impurities that is present in **Cassiterite happens to be sand [*i.e.*, Silicon dioxide (SiO$_2$)], pyrite of Copper (Cu) and Iron (Fe)**, and **Wolframite** (an *ore of Tungsten*) [FeWO$_4$ + MuWO$_4$]. Fig: 5.22 shows the elaborated flow chart for the extraction of Tin(Sn) from its finely divided ore.

The various reactions taking place in the extraction of Tin(Sn) are expressed duly as under:

---

*   It is also knonws as **Tin Stone**.

Step-1:

$$SnO_2 \quad + 2C \longrightarrow Sn + 2CO\uparrow$$

Cassiterite
[1–5% SnO_2]

Step-2

$$SnSiO_3 + CaO + C \longrightarrow Sn + CaSiO_3 + CO\uparrow$$

Tin Silicate   Calcium
                              Silicate

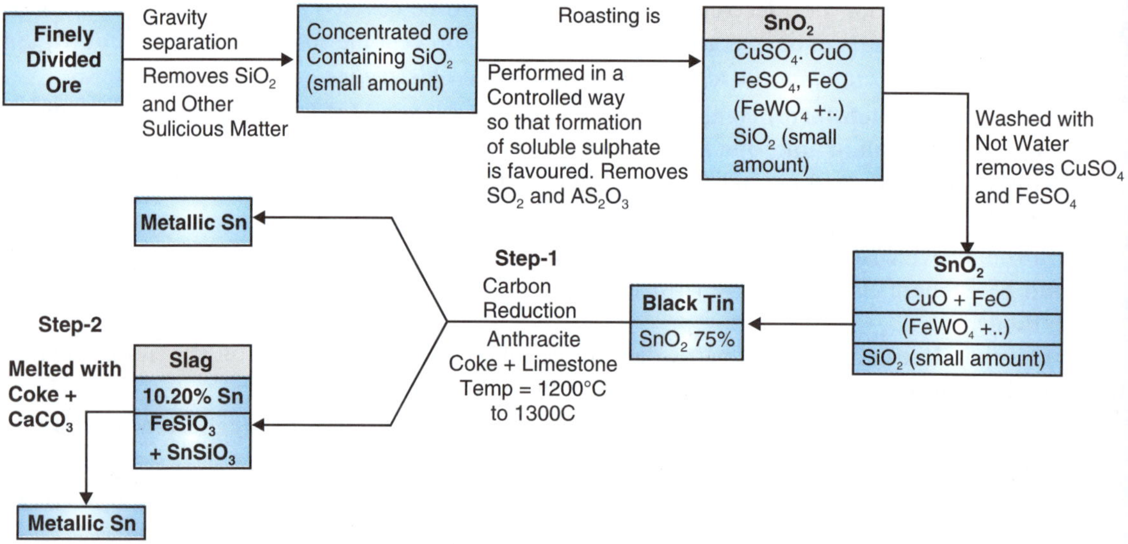

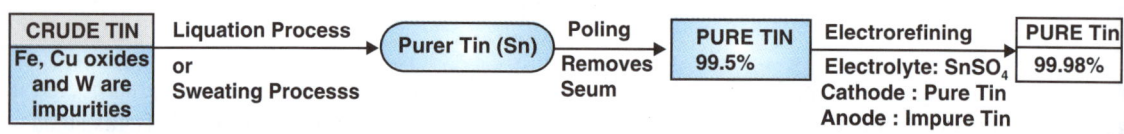

Fig. 5.22: Explicit Representation of Flow Chart Pertaining to the Extraction of Tin(Sn) Metal.

Purification of Tin(Sn) Metal: Importantly, the various steps being involved in the refining of Tin(Sn) metal are duly depicted as under:

CRUDE TIN	Liquation Process		Poling	PURE TIN	Electrorefining	PURE Tin
Fe, Cu oxides and W are impurities	or Sweating Processs	Purer Tin (Sn)	Removes Seum	99.5%	Electrolyte: $SnSO_4$ Cathode : Pure Tin Anode : Impure Tin	99.98%

2.2.12. *Extraction of Aluminium [Al]*

The *six* most prevalent and common minerals of Aluminium (Al) are as stated under:

- **Oxide** : **Corundum [Al_2O_3]**
- **Hydrated Oxides** : **Bauxite [$Al_2O_3.2H_2O$]; Diaspore [$Al_2O_3.H_2O$]; and Gibbsite [$Al_2O_3.3H_2O$]**
- **Sulphate** : **Alunite [$K_2SO_4.Al_2(SO_4)_3.4Al(OH)_3$]**
- **Fluoride** : **Cryolite [$3NaF.AlF_3$]**
- **Aluminate** : **Spinel [$MgO.Al_2O_3$]**
- **Silicate** : **Feldspar [$K_2O.Al_2O_3.6SiO_2$]; Kaolin or China Clay [$Al_2O_3.SiO_2.2H_2O$]**

Benevolence of Bauxite Ore—Several well-defined and accepted procedures being largely adopted for the benevolence of both:

- **Red Bauxite Ore of Mg** and
- **White Bauxite Ore of Mg**, are briefly treated as under:

(a) Bayer's Process—It is solely being recommended for the benevolence of the so-called **Red Baxite Ore of Magnesium (Mg)**.

Fig: 5.23 illustrates the different sequential steps involved in the flow chart explicitly:

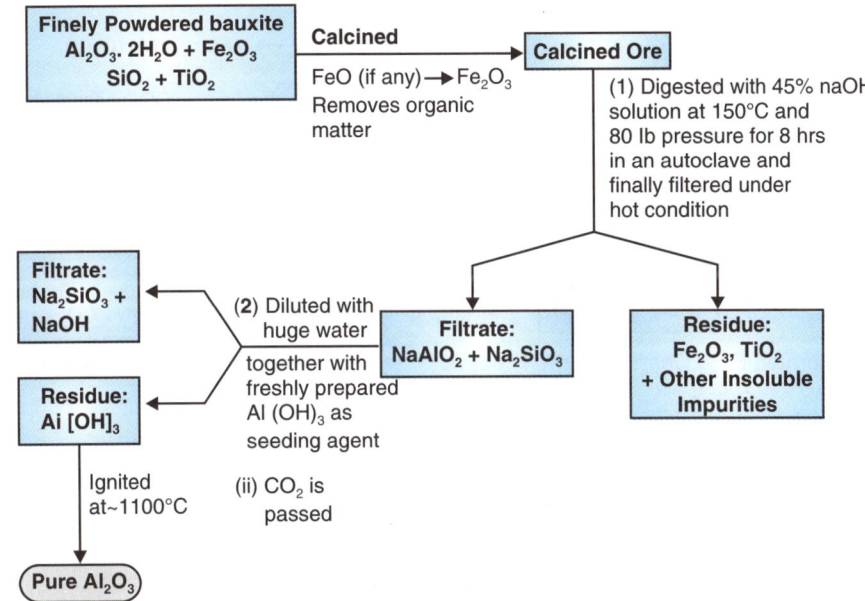

Fig. 5.23: Diagrammatic Representation of the Flow Chart for Bayer's Process for Benevolence of Bauxite Ore.

The *two* distinct reactions that are actually involved in the the aforesaid process are:

$$Al_2O_3 + 2OH^- + 3H_2O \longrightarrow 2[Al(OH)_4]$$

Bauxite

Step-1

$$\underset{\textbf{Sand}}{SiO_2} + 2NaOH \longrightarrow \underset{\textbf{Sodium Silicate}}{NaSiO_3} + H_2O$$

$$[Al(OH)_4] \rightleftharpoons \underset{\substack{\textbf{Aluminium}\\\textbf{Hydroxide}\\\textbf{(White precipitate)}}}{Al(OH)_3} \rightleftharpoons Al^{3+} + 3H_2O$$

Step-2

> **NOTE**
> Importantly, the *Bayer's Process* may not be adopted for the *White Bauxite Ore of Magnesium* since the *chief impurity SiO$_2$* (sand) gets duly departed together with Al_2O_3; and hence, ultimately-the resulting Al_2O_3 is cetainly of an *Inferior Quality*.

(b) Hall's Process–In true sense, the **Hall's Process** is exclusively adopted for the benevolence of the so-called **lower grade of magnesium (Mg) ore** *viz.*, **Red Bauxite**. Fig: 5.24 explicitly deliberates the different steps engaged in the said process.

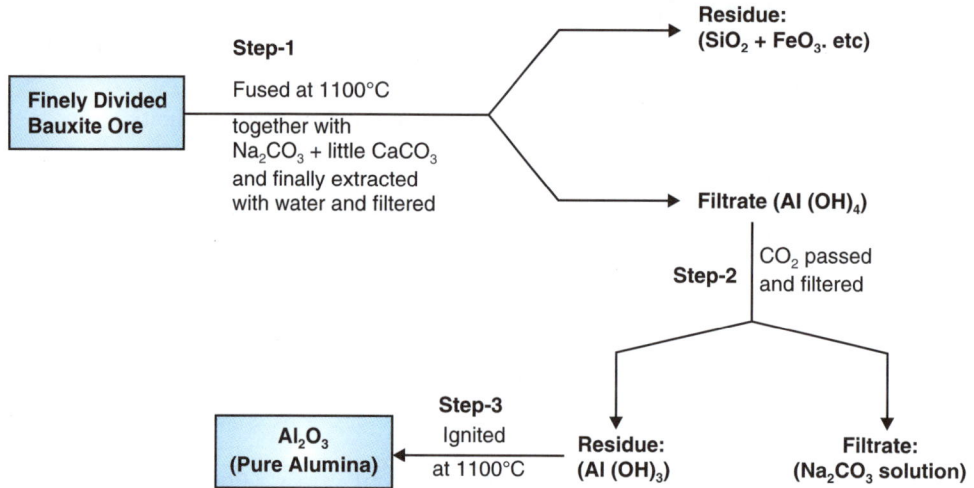

Fig. 5.24: An Elaborated Flow Chart for Hall's Process for the Benevolence of Low Grade Magnesium Ore (Red Bauxite).

The various reactions that are involved in the aforesaid process are in *three* stages:

Step-1

$$Al_2O_3 + Na_2CO_3 \longrightarrow 2Na\,AlO_2 + CO_2 \uparrow$$

$$SiO_2 + Na_2CO_3 \longrightarrow Na_2SiO_3 + CO_2 \uparrow$$

$$Fe_2O_3 + Na_2CO_3 \longrightarrow 2Na\,FeO_2 + CO_2 \uparrow$$

$$CaO + SiO_2 \longrightarrow CaSiO_3$$

Step-2

$$2NaAlO_2 + CO_2 + 3H_2O \longrightarrow 2Al(OH_3)\downarrow + Na_2CO_3$$
$$\text{White ppt.}$$

Step-3

$$2Al(OH)_3 \xrightarrow[-3H_2O]{\text{Heat}} \underset{\text{(Pre Aluminium)}}{Al_2O_3} + 3H_2O$$

(c) Serpeck's Process–In fact, this process is solely employed for the benevolence of the **White Bauxite ore of Magnesium (Mg)**. Fig: 5.25 describes the various steps involved in the **Serpeck's Process** by means of a flow diagram using **White Bauxite**.

There are *three* **distinct steps**–that are duly involved in the various steps reactions in the above episode:

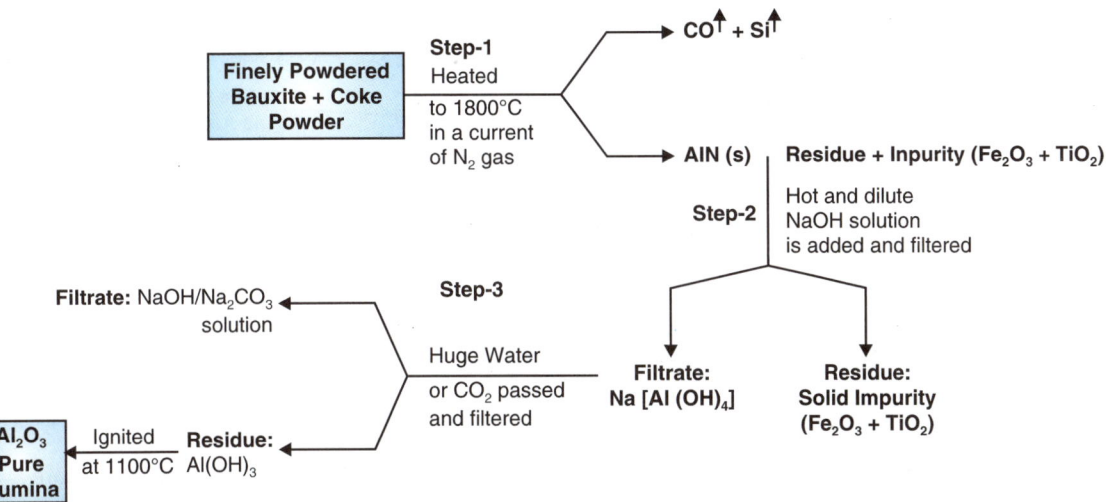

Fig. 5.25: Representation of the Flow Chart for Serpeck's Process for the Benevolence of White Bauxite Magnesium (Mg) Ore.

Step-1

$$Al_2 + 3C + N_2 \rightarrow 2AlN(s) + 3CO \uparrow$$

$$SiO_2 + 2C \rightarrow Si \uparrow + 2CO \uparrow$$

Step-2

$$AlN + NaOH + 3H_2O \rightarrow Na[Al(OH)4] + NH_3 \uparrow$$

Step-3

$$Na[Al(OH)_4] \rightleftharpoons Al(OH)_3 \downarrow NaOH$$

or

$$[Al(OH)_4] \rightleftharpoons Al(OH)_3 \downarrow OH^-$$

$$CO_2 + 2OH^- \rightarrow CO_3^{2-} + H_2O$$

THE ELECTROLYTIC REDUCTION OF PURE ALUMINA (AL$_2$O$_3$)

In this particular instance of the electrolytic reduction of **alumina (Al$_2$O$_3$)** [*i.e.*, molten alumina (20%) mixed duly with cryolite (3NaF.AlF$_3$) 60%, and Fluorspan (20%)] is taken up in an **Iron Tank** having a smooth inner **C-lining** that serves as the *Cathode*. Besides, a thick **Graphite Rod** hanging duly from the **top of the Iron-Tank** acts as the **Anode**. However, a layer of powdered coke at the top of the Fe-Tank is maintained always.

Fig: 5.26 shows the **Electrolytic Reduction of Alumina (Al$_2$O$_3$)**

- **At Cathode:** After electrolysis the **Molten Aluminium** gets duly deposited at the Cathode.
- **At Anode:** The liberated Oxygen (O$_2$) gas emerges at the **Anode**.

NOTE   **Since aluminium metal (Al) is definitely heavier *vis-a-vis* the electrolyte, it readily gets deposited at the bottom (see Fig: 5.25).**

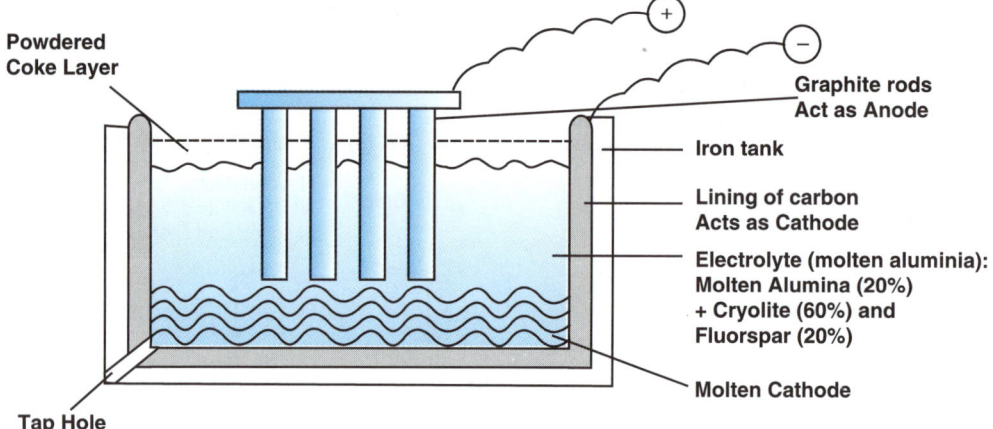

Powdered
Coke Layer

Graphite rods
Act as Anode

Iron tank

Lining of carbon
Acts as Cathode

Electrolyte (molten aluminia):
Molten Alumina (20%)
+ Cryolite (60%) and
Fluorspar (20%)

Molten Cathode

Tap Hole

Fig. 5.26: Diagrammatic Representation of the Electrolytic Reduction of Alumina (Al_2O_3) to Obtain Pure Molten Aluminium.

Following are the *four* cardinal aspects related to the **Electrolytic Reduction of Alumina (Al_2O_3)**, namely:

(c) **Reactions–** $Al_2O_3 \longrightarrow 2Al^{3+} + 3O^{2-}$

At *Cathode:* $Al^{3+} + 3e^- \longrightarrow Al$

> **NOTE**   Since Na and Ca are found to be more electropositive vis-a-vis Al, hence only Al^{3+} gets deposited promptly at the *Cathode*.

At *Anode:* $2O^{2-} \longrightarrow O_2 + 4e^-$

(b) **Typical Functions of Flurospar [CaF_2]**–It has *two* functionalities:
  - It critically helps to lower the MP of the mixture to **900°C from 2050°C** (*i.e.*, MP of pure Al_2O_3); and
  - It definitely improves upon the inherent **electrical conductivity** of the ensuing melt in comparison to **molten Al_2O_3**.

(c) **Functionality of Cryolite [$3NaF.AlF_3$]**–These essentially include:
  - It aids in the reduction of the so-called MP of the mixture.
  - It also serves as a **solvent** and thereby aids to **dissolve alumina (Al_2O_3)**.

(d) **Functionality of the Coke-Powder Layer at the Top of Iron Tank–**
  - The liberated **Oxygen (O_2) at the Anode** helps to corrode the Anode surface due to the interaction with **Graphite** to yield CO and CO_2. Thus, ultimately the Anode crucially *cuts down at the bottom*; and hence, the **electrical conductivity** is virtually finished, –as shown in the following Fig: 5.27.

Explanation–Amazingly, at the critical junction of so-called **Liquid-Solid-Air interface**, the available energy level seems to be **optimum***; and, therefore, the extent of **corrosion** is also maximum at this point particularly. Hence, to check and prevent this corrosion, the **coke powder**** layer is maintained duly at the top that eventually interacts with the **liberated oxygen (O_2)** perceptively.

---

*   which may be proven **thermodynamically**.
**   That is, having large surface are for reaction.

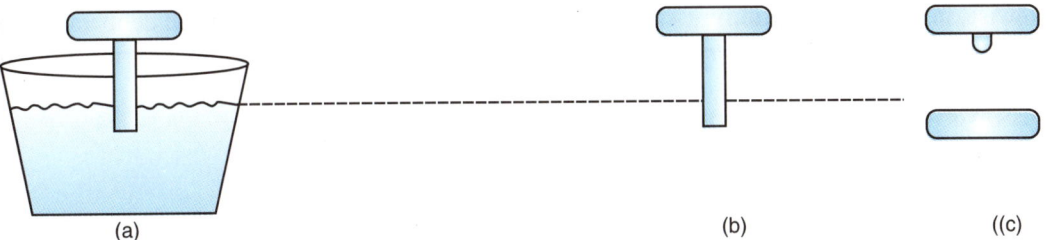

Fig. 5.27: The Corrosion of Graphite Anods.

- It may, however, be observed that the *surface* eventually turns out to be rough unlike the **shiny mirror** *viz.*, the **molten electrolyte**; and, therefore, the usual **radiation loss of heat** gets adequately prevented.

THE ELECTROREFINING OF ALUMINIUM (AL)

The electrorefining of Al may be carried out efficaciously by mixing *impure aluminium* duly with the **copper Melt** in an **Iron Tank** using a **Graphite Lining**. However, the ensuing layer of **pure-Al** does act as the **Cathode**. The critical position of the thick **Graphite Rods** at the top segment of the **Iron Tank** are an absolute must so as to facilitate the **electrical connections** a lot easier.–as depicted in Fig: 5.28.

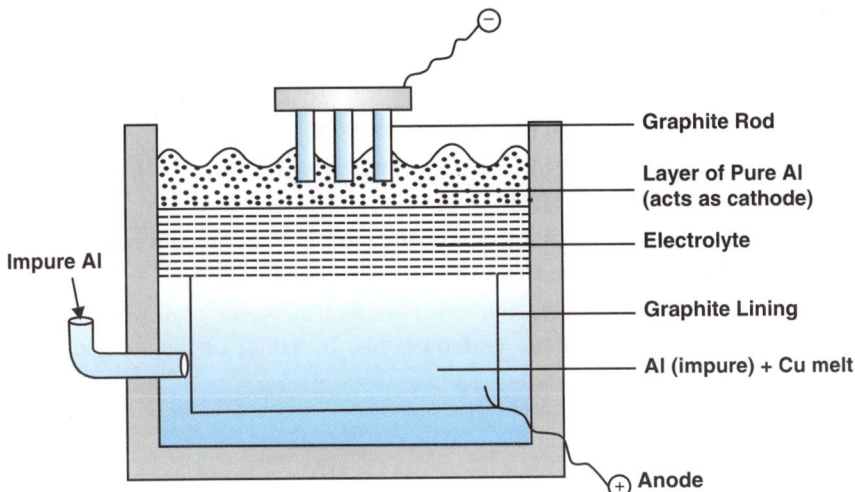

Fig. 5.28: Diagrammatic Representation of the Electrorefining of Aluminium (Al).

Modus Operandi

1. **Electrolyte** refers to the molten admixture of:
   - **Cryolite [3NaF.AlF$_3$]** and • **Barium Fluoride [BaF$_2$]**,
   adequately saturated with **alumina (Al$_2$O$_3$)**.

2. Importantly, **Barium fluoride (BaF$_2$)** is duly incoporated instead of **Calcium fluoride (CaF$_2$)** so as to adjust precisely the so-called:
   "**density in such a manner that it does exist as the separate Middle Layer**".

3. Likewise, in **impure aluminium (Al)**, the addition of the **Cu-melt** is being made deliberately in order to enhance precisely the *density* in such a manner that it really prevails as the **separate bottom layer perceptively**.

Reactions Involved in the Entire Process of Electrorefining of Al–

- **At Anode:** $Al \longrightarrow Al^{3+} + 3e^-$

In this case, the *top-surface of bottom layer* serves as the **Anode**. Besides, **Al** only gains entry right into the electrolyte as Al^{3+} since we have:

$$\boxed{E^o_{Al/Al^{3+}} > E^e_{cu}/Cu^{2+}}$$

- **At Cathode:** $Al^{3+} + 3e^- \longrightarrow Al$

In this instance, the **bottom surface of the top layer (relates to the pure Aluminium (Al) melt)** and eventually acts as the **Cathode**; and hence, Al^{3+} **gains entry into as Al** from the electrolyte itself perceptively.

NOTE

1. **Various impurities** *viz.*, **Fe, Si, and Cu do invariably remain very much intact in the bottom layer.**
2. **Amazingly, when the thickness of the top layer gets enhanced to a certain extent, it immediately gets drained into an altogether separate container (or vessel).**

2.2.13. *Extraction of Lead (Pb)*

Following are the *five* most importnt **ores of lead,** namely:

- **GalenaPbS;** - **Cerrusite: $PbCO_3$;** - **Anglesite: $PbSO_4$;** - **Crocosite: $PbCrO_4$;** and - **Lanarkite: [PbO, $PbSO_4$].**

Points to Ponder: **Galena (PbS)** is considered to be the most commonly and commercially used ore of **Lead (Pb)** for the extraction of the **pure Lead (Pb)**.

Nevertheless, based upon the precise and exact content of *impurity present in the* **ores of Lead (Pb)**,–**Lead** may be extracted from its ore **Galena (PbS)** by either of the following *two* recognized methods:

➢ **Carbon Reduction***, and
➢ **Self Reduction****

1. **Carbon Reduction Process**–The various steps that are being involved sequentially in the so-called **Carbon Reduction Process** for the extraction of **Lead (Pb)** from **Galena (PbS)** are illustrated in the Fig: 5.29 based on its flow-diagram:

The *three* **sequential steps** involved in the aforesaid sequential procedure are being expiated as under:

Step-1: **Froth Floatation Step: PbS,** present in **Galena**, fails to react with **NaCN solution**; whereas, **ZnS** gets duly **dissolved in NaCN solution**; and hence; its **floating characteristcs** are practically eliminated:

---

* That is, when the **content of impurity is high enough**.
** That is, when the **content of impurity is fairly low**.

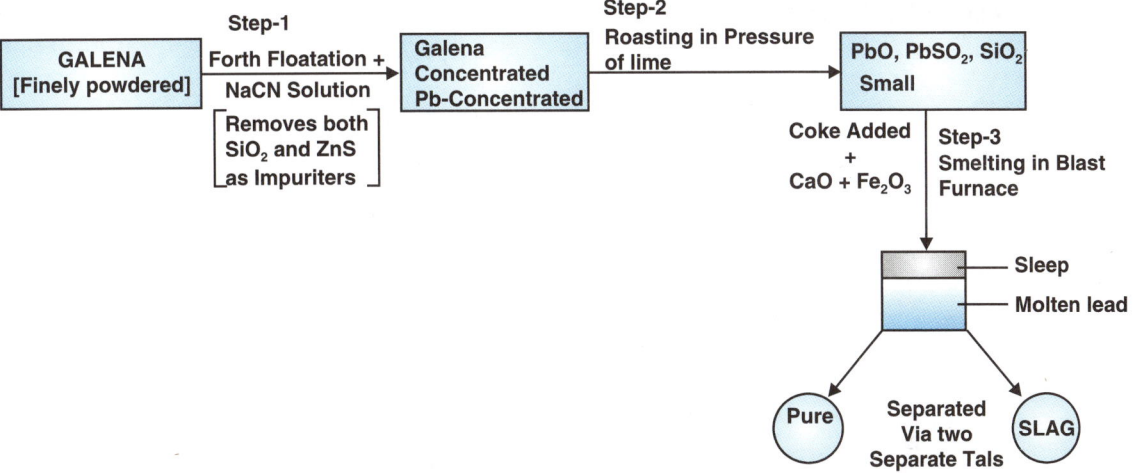

Fig. 5.29: The Flow Diagram Depicting the Carbon Reduction Process for Extraction of Pure Lead (Pb) from Galena (PbS).

Thus, we may have the following expression:

$$ZnS + NaCN \longrightarrow \underbrace{Na[Zn\,(CN)_4] + Na_2S}_{\text{Both are soluble}}$$

$$PbS + NaCN \longrightarrow \textbf{No Reaction}$$

***Step-2:* Roasting Step:** Following reactions come into play:

$$PbS + 3/2\,O_2 \longrightarrow PbO + SO_2^\uparrow$$

$$PbS + 2O_2 \longrightarrow PbSO_4\,[\textbf{Minor}]$$

$$PbO + SiO_2 \longrightarrow PbSiO_3\,[\textbf{Minor}]$$

Role of Lime: Because **lime (CaO)** happens to be more basic in nature *vis-a-vis* to **PbO**, the **CaO** does react in a preferential mode with **SiO₂**; and hence, its presence acts as **negative catalyst** towards the respective formation of these two compounds:

- **Lead Sulphate (PbSO₄)** and
- **Lead Silicate [PbSiO₃]**

NOTE	The above *two* compounds also go a long way to maintain the entire mass porons perceptively and help to complete the reaction satisfactorily.

***Step-3:* Smelting Step:** At the very outset '**coke**' is burnt carefully in a blast of fresh air to generate both CO and CO₂ profuselly:

$$C + O_2 \rightarrow CO_2^\uparrow$$

$$C + CO_2 \rightarrow 2CO \uparrow$$

Lead (Pb) Formation Reactions:

$$PbO + C \longrightarrow Pb(1) + CO \uparrow$$

$$PbO + C \longrightarrow Pb(1) + CO_2^\uparrow$$

$$PbSO_4 + 4C \longrightarrow PbS + 4CO \uparrow$$

$$\underset{\text{(Unreacted)}}{PbS} + 2PbO \longrightarrow 3Pb(1) + SO_2^\uparrow$$

$$PbS + \underset{\text{(If any)}}{PbSO_4} \longrightarrow 2Pb(1) + 2SO_2^\uparrow$$

Slag Formation

$$CaO + SiO_2 \longrightarrow \underset{\text{Calcium Silicate}}{CaSiO_4}$$

$$Fe_2O_3 + CO \longrightarrow 2FeO + CO_2^\uparrow$$

$$FeO + SiO_2 \longrightarrow \underset{\text{Iron Silicate}}{FeSiO_3}$$

Refining of Lead (Pb)—The **refining of Lead (Pb)** critically involves *three* **distinct steps**, namely:

- *Step-1*: **Softening of Lead (Pb)** i.e., the metal is melted in the critical presence of air. Thus, it helps to remove the seum of various oxides; and only **Ag** and **Au** left behind remains to be removed as for as possible.

- *Step-2*: **Desilverization** *i.e.*, exhaustive removal of **Silver (Ag)** either by **Parke's Process** or **Parkhinson's Process**.

- *Step-3*: **Bett's Electrolytic Process** *i.e.*, an **electrolyte** comprising: **PbSiF$_6$ + H$_2$SiF$_6$ + Gelatine***. Thus, at **Anode** the **Impure Lead (Pb)** gets collected and at **Cathode** the **Pure Lead (Pb)** is **obtained as a strip**.

Fig: 5.30 depicts the *three* **distinct steps** that are being engaged in a sequential episode for the refining of **Lead (Pb)** from **Crude Pb Molten**:

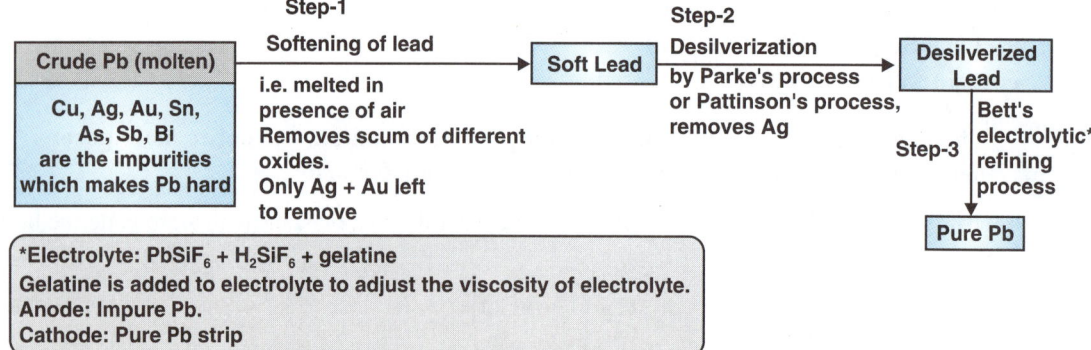

Fig. 5.30: The Elaborated Flow Chart for Refining of Lead from Crude Molten ore.

---

*   Gelatine–is added to the **Electrolyte** so as adjust the **Viscosity of Electrolyte** precisely.

2.2.14. *Extraction of Copper (Cu)*

Following are the *five* **common and important ores of Copper (Cu)**, such as:

(i) **Cu₂S [Copper Sulphide]: Copper Glance or Chalcocite,**

(ii) **CuFeS₂ [Copper Iron Sulphide]: Chalcopyrites or Copper Pyrite;**

(iii) **Cu₂O [Copper Oxide]: Ruby Copper or Cyprite;**

(iv) **Cu(OH)₂.CuCO₃: [Copper Hydroxide–Copper Carbonate]: Malachite; and**

(v) **Cu(OH)₂.2CuCO₃ [Copper Hydroxide-Copper Carbonate]: Azurite.** However, the relatively less important ores of copper (cu) are:

- **Malonite [CuO]** and • **Chrysocola [CuSiO₃.2H₂O].**

The most prevalent and major ore of **Copper (Cu) are** being employed is **Copper Pyrite [Cu₂S.FeS.FS₂].**

Fig: 5.31 shows the explicit and sequential steps organized flow chart for the critical extraction of **Copper (Cu)** from **Copper Pyrite.**

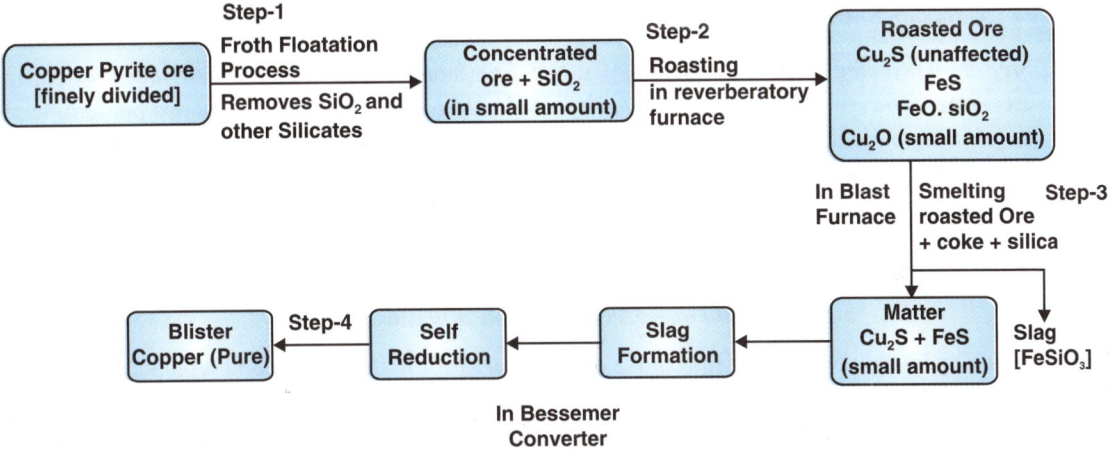

Fig. 5.31: The Elaborated Flow Chart for the Extraction of Copper (Cu) from Copper Pyrite.

The *three* individual steps involved in fig: 5.30 shall now be discussed briefly as under:

❏ **Step-1: Roasting Process**–In usual practice, **roasting** is being done with an excess of air at a **temperature slightly lower** than the so-called fusion-temperature of the ore chosen.

Thus, we may have the following expressions:

$$Cu_2S \cdot Fe_2S_3 + O_2 \longrightarrow Cu_2S + 2FeS + SO_2\uparrow$$

$$Cu_2 \cdot Fe_2S_3 + 4O_3 \longrightarrow Cu_2S + 2FeO + 3SO_2\uparrow$$

As we know that **Iron (Fe)** happens to be definitely *more* electropositive *vis-a-vis* to Copper (Cu)– hence its respective **sulphide (Fe₂S₃)** gets *oxidized preferentially*; and thereby **Cu₂S** remain virtually unaffected.

In case, any residual little quantum of **Cu₂O** gets formed, –it certainly reacts with the **untreated Iron Sulphide (FeS)** to **reproduce Cu₂S.** Hence, we may have the following reactions:

$$Cu_2S + 3/2O_2 \longrightarrow Cu_2O + SO_2\uparrow$$

$$Cu_2O + FeS \longrightarrow Cu_2S + Feo$$

❏ **Step-2: Smeting Process**–The preferred **fuel** in this particular instant is '**coke**'–that helps to maintain such temperature which essentially maintins the mixture in a Molten State.

Thus, we may have the following expressions:

$$FeS + 3/2O_2 \longrightarrow FeO + SO_2\uparrow$$

$$Cu_2O + FeS \longrightarrow Cu_2O + FeO$$

$$FeO + SiO_2 \longrightarrow FeSiO_3 \text{ [Iron Silicate as 'Slag']}$$

NOTE Amazingly, '*Slag*' being lighter in character floats critically as an '*invisible layer*' upon the top surface of matter (Cu_2S); and thus, could be removed via an altogether separate outlet.

❏ *Step-3: Bessemer Converter*–**Matte** [Cu_2S + FeS(small)] serves as the *raw material* for the **Bessemer Converter**. It is duly followed by air **blasting** at the very initial stage that predominantly helps to form the '**Slag**', and subsequently purified Sand (SiO_2) is incorporated from an external source appropriately.

Hence, the following reactions take place:

$$FeS + 3/2O_2 \longrightarrow FeO + SO_2\uparrow$$

$$SiO_2 + FeO \longrightarrow FeSiO_3 \text{ [Iron Silicate as 'Slag']}$$

Remarks: It may be observed that in the course of **Slag Formation**, the distinct and typical *Green Flame* appearing at the **mouth of the Bessemer Converter**–that eventually indicates the presence of **Iron (Fe)** in the state of **Ferrous Oxide (FeO)**. Hence, the ultimate disappearnace of the aforesaid **Green Flame** vividly implies that:

"the formation of 'Slag' is almost complete".

Now, the **air-blasting** is discontinued and the resulting '**Slag**' is removed finally.

Refinement of Blister Copper–The **Blister Copper (Cu)** actually comprises *impurity* ranging between 2 to 3% **(Chiefly Fe, As, and S)**.

Fig: 5.32 illustrates the various steps that are critically involved in the flow diagram pertaining to the refining of **Blister Copper (Cu)**.

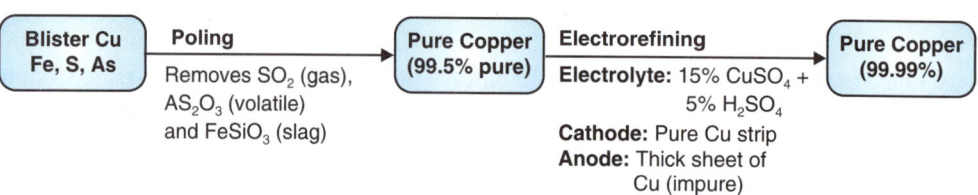

Figure: 5.32: Flow Diagram Showing the Refining of Blister Copper.

Remarks–In the very first **Poling Step**, the small quantum of **Cu₂O** soformed gets immediately reduced to metallic **Copper (Cu)** due to the effect of the so-called **Reducing Gases** generated from the so-called:

'charring of green wooden poles'.

Besdies, the spreading on the top surface of the '**molten mass**' with *powdered anthracite (Coke)* aids profusely in to produce a **Reducing Environment**.

However, in the *second* **Electrorefining step** such impurities as: Fe, Ni, and Zn do get easily dissolved in the solution; whereas, **Au, Ag,** and **Pt** get obviously deposited as '*anode mud*' just below the **Anode**.

2.2.15. *Extraction of Zinc (Zn)*

Following are the *six* most prevalent **ores of zinc (Zn)**, namely:

- **ZnS: Zinc Blende;** - **ZnO: Zincite** - **ZnO.Fe₂O₃: Franklinite**
- **ZnCO₃: Calamine;**
- **ZnSiO₃ : Willemite** and - **ZnSiO₃.ZnO.H₂O : Electric Calamine.**

However, **Zinc Blende**, also known as '**Black Jack**', is considered to be the *major ore of Zinc (Zn)*, which could be due to the usual intimate association of **Galena (PbS)**–that is invariably **Black in colouration**. Besides, **Calamine (ZnCO₃)** is also utilized to extract **Zinc (Zn)** by means of the so-called '**Carbon Reduction Process**'.

Fig: 5.33 depicts vividly the flow diagram for the '**Extraction**' and '**Refining**' of **Zinc (Zn)** from **Zinc Blende (ZnS)**.

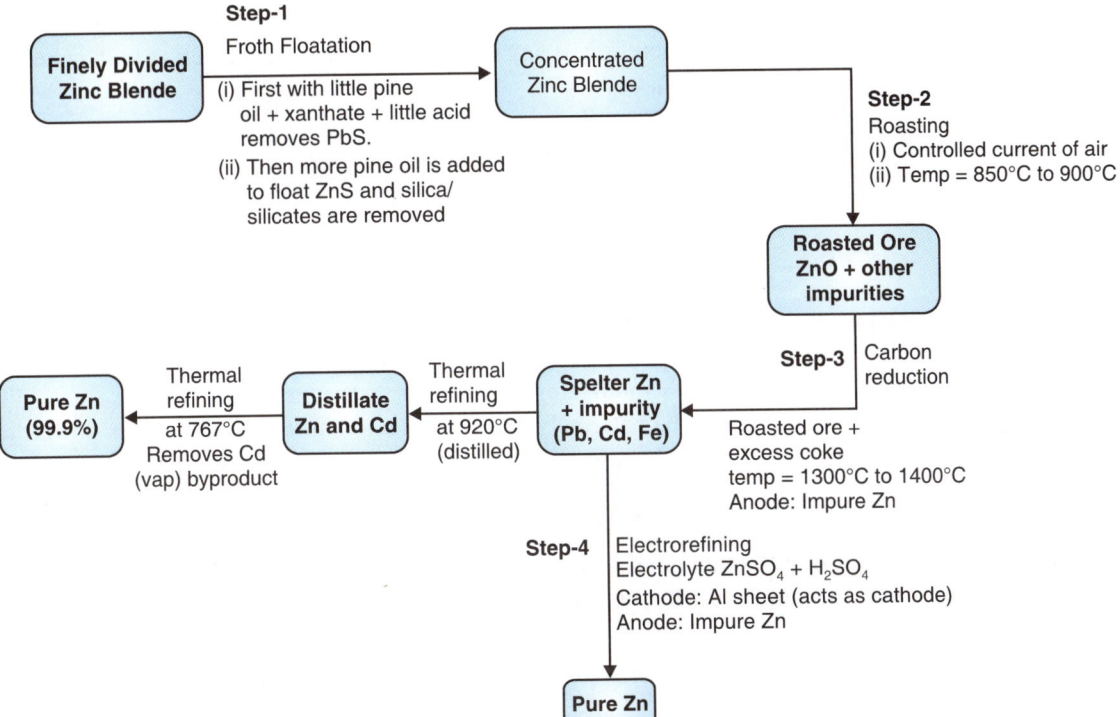

Fig. 5.33: The Explicit Representation of the Flow Chart for the Extraction of and Refining of Zinc (Zn) from its Ore Zinc Blende (ZnS).

There are **separate steps**–that essentially account for the *changes and reactions* being involved at different stages of **Zinc (Zn) extraction**, namely:

❑ *Step-1*: **Froth Floating**–It is performed actually in to steps in order to separate the following *two* chemical entities:

• **Lead Sulphide (PbS)** and   • **Zinc Sulphide (ZnS)**,

that are solely dependent upon their divergent **Floating Characteristic Features**. Amazingly, upon the addition of **Pine oil**,–the **Lead Sulphide (PbS)** floats first and is removed subsequently

❑ *Step-2*: **Roasting Process**–Importantly, in the course of Roasting Process–the prevailing temperature has get to be maintained above **850°C**; whereas,. the current of air should be controlled and monitored at a temperature **below 850°C** and in the presence of excess of air, zinc sulphide (ZnS) gets duly converted into **Zinc Sulphate (ZnSO₄)**–that eventually reverts to **Zinc Sulphide (ZnS)** in the course of ensuing carbon reduction of roasted ore.

Thus, we may have the following expressions:

$$ZnS + 3/2O_2 \xrightarrow{\;>850°\;} ZnO + SO_2\uparrow$$

$$ZnS + 2O_2 \xrightarrow{\;<850°C\;} ZnSO_4$$

$$ZnSO_4 + 4C \xrightarrow[\text{Carbon Reduction}]{\text{During}} ZnS + 4CO\uparrow$$

❑ *Step-3*: **Smelting Process**–In this **smelting process** usually an *excess of coke* is being employed in order to Stop the production of **carbon dioxide (CO₂)**,–rest the **Zn** so produced will be reverted to **Zinc oxide (ZnO)**. Therefore, if any **CO₂** by the subsequent reaction with coke.

Hence, we may have the following reactions:

$$ZnO + C \longrightarrow Zn + CO\uparrow$$

$$2ZnO + C \rightleftharpoons 2Zn + CO_2 \text{ (it is indeed a reversible reaction)}$$

$$Co_2 + C \longrightarrow 2CO\uparrow$$

Remarks: The temperature during **Smelting process** is invariably maintained **above 1300°C** (even though the BP of Zn is 920°C). However, in usual practice the prevailing temperature is always maintained much higher *vis-a-vis* to that actualy needed for **vapourizing zinc (Zn)** from the furnace. It is done so since the **reaction of CO₂ with coke** is found to be extremely endothermic in nature. Thus, it brings down the temperature to **less than 920°C**; and thereby affects directly the so-called **evaporation of Zinc (Zn)**.

NOTE   **Hence, the temperature is always maintained between 1300–1400°C.**

❑ **Step-4: Electrorefining Phenomenon**–In this particular instance, for the electrorefining of **Crude Zinc (Zn) metal**, **Aluminium-Sheet** is being employed as the **Cathode** in place of the usual **pure Zn-Strip**. It is so done since the electrolyte being employed is:

ZnSo₄ + H₂SO₄ (diluted); and hence in diluted H₂SO₄ the **Zinc (Zn) gets** dissolved easily; whereas, aluminium (Al) fails to do so:

Thus, we may have the following reactions:

$$Zn + H_2SO_4 \longrightarrow ZnSO_4 + H_2^\uparrow$$
(diluted)

$$Al + H_2SO_4 \longrightarrow \text{No Reaction}$$
(diluted)

$$2Al + H_2SO_4 \longrightarrow Al_2(SO_4)_3 + 3SO_2^\uparrow + H_2O$$
(concentrated)

Reactions taking place at the Electrode: $ZnSO_4 \longrightarrow Z_n^{2+} + SO_4^{2-}$

At Cathode: $Zn^{2+} + 2e^- \longrightarrow Zn$

At Anode: $^-OH \longrightarrow \dot{O}H + e$

$$4\dot{O}H \longrightarrow 2H_2O + O_2^\uparrow$$

Special Remarks–The crucial addition of **Sulphuric acid (H_2SO_4)** in the *electrolyte* along with **Zinc sulphate ($ZnSO_4$)** helps to enhance the **over voltage of H^+**. In this manner, it does help perceptively in the so-called meticulus deposition of **Zn^{2+} ion at the Cathode** exclusively–rest **hydrogen (H_2) shall be produced at the cathode**.

2.2.16. *Extraction of Iron (Fe)*

Following are the *five* most importnt **ores of Iron (Fe):**

  (i) **Haematite : Fe_2O_3;**
  (ii) **Magnetite : Fe_3O_4;**
  (iii) **Limonite [or Brown haematite]: $Fe_2O_3.3H_2O$;**
  (iv) **Siderite [or Spathic Iron Ore]: $FeCO_3$; and**
  (v) **Iron Pyrite: FeS_2**

However, the **major ore** that finds its wide use for the **extraction of Iron (Fe) is Haematite (Fe_2O_3);** whereas, **Iron Pyrite (FeS_2)** is never used commercaially since the **Iron (Fe)** duly recovered from this specific ore does comprise a lot of Sulphur (S) – that eventually renders it *more brittle* and **of no applicability status**.

Fig: 5.34 clearly depicts the elaborated **flow diagram** for the extraction of **Iron (Fe)** from *four* of its **recognized ores**.

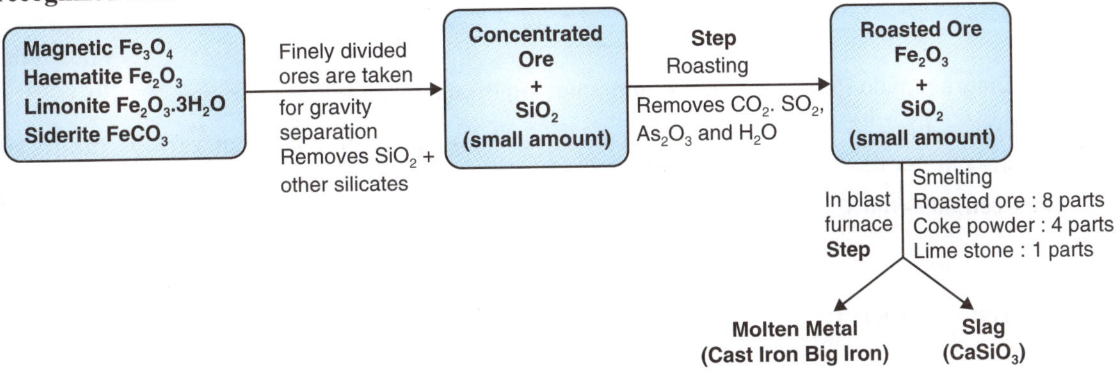

Fig. 5.34: The Descriptive Flow Chart for the Extraction of Iron **(Fe)** from: Magnetite, Haematite, Limonite, and Siderite.

There are *two* **major steps** that are duly involved in the reactions at different aforesaid steps of the **Iron Extraction phenomenon**, such as:

Step-1 **Roasting Process**–The **roasting process** occurs in the following *four* **sequential steps**:

$$Fe_3O_4 \longrightarrow FeO + Fe_2O_3$$

$$FeCO_3 \longrightarrow FeO + CO_2\uparrow$$

$$2FeO + \frac{1}{2}O_2 \longrightarrow Fe_2O_3$$

$$Fe_2O_3 \cdot 3H_2O \longrightarrow \underset{\text{Ferric Oxide}}{Fe_2O_3} + 3H_2O\uparrow$$

Thus, the *final product* **of roasting** is **Ferric Oxide [Fe$_2$O$_3$]**.

NOTE

1. **Even though the presence of the 'Sulphide Ore *viz.*, Iron Pyrite (FeS$_2$) yet *roasting process* is being adopted perceptively so that all FeO present may get converted into Ferric Oxide (Fe$_2$O$_3$).'**
2. **Since Fe$_2$O$_3$ fails to produce '*Slag*', hence it predominantly checks and prevents the so-called–'loss of FeO at *Slag* (FeSiO$_3$)'**

Step-2 **Smelting Process**–The different critical changes that do occur in the course of the **Smelting Process** in the *Blast Furnace* are illustrated explicitly in Fig: 5.35.

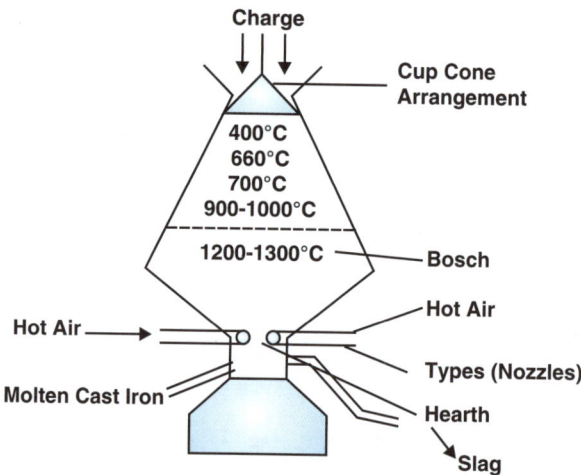

Fig. 5.35: Diagrammatic Representation of the Smelting Process in the Blast Furnace.

There are *three* different sets of chemical reactions that invariably take place at **various ranges of temperatures in the Blast Furnace**, such as:

❑ **Between 600–900°C**

$$\underset{\text{(Reduced Partially)}}{Fe_2O_3} + 3CO \longrightarrow 2Fe + +3CO_2\uparrow$$

❑ **Between 900–1000°C**

$$CaCO_3 \longrightarrow CaO + CO_2\uparrow$$

$$CO_2 + C \longrightarrow 2CO\uparrow$$

❑ **Between 1000–1300°C**

$$Fe_2O_3 + 3e \longrightarrow 2Fe + 3CO\uparrow$$

$$CaO + SiO_2 \longrightarrow CaSiO_3 \text{ (Slag)}$$

Remarks: However, at the critical temperature of **1500°C** (*viz.*, at the **Hearth**). At this point in time, the inherent *Coke Powder* that eventually **crossing the line of layers** fails to have a tendency of **burning any further. Thus**, it certainly reacts with MnO_2 and SiO_2 to yield **undescribed impurities** *e.g.*, Mn and Si – as shown under:

$$MnO_2 + 2e \longrightarrow Mn + 2CO\uparrow$$

$$SiO_2 + 2e \longrightarrow Si + 2CO\uparrow$$

NOTE	Thus, the *Cast iron* produced ultimately comprises such typical impurities as: C, S, P, Si, and Mn.

Purification of Iron (Fe)–The **purification of Iron (Fe)** is also referred to in the literature as the preparation of **Wrought Iron**.

It is, however, pertinent to state here that the so-called **Wrought Iron (Fe)** is invariably regarded to be:

"the *Purest Form of Iron (Fe)*–consisting of a product > 99.5% pure".

- **Carbon Content**–ranges **between 0.1 to 0.15%**; and
- **Other Impurities**–are found to be **< 0.3%**. (*e.g.*, **Mn, P, S, and Si**)

Fig: 5.36 shows intimately the various steps which are duly engged in the preparation of **Wrought-Iron (Fe)** in the so-called explicit *flow diagram*.

Fig. 5.36: Details of Flow Chart for the Preparation of Wrought Iron from Cast Iron.

The special effect and significance of the *Haematite Lining* lies in the specific and critical removal of:

"the inherent impurity plus production of Iron (Fe) at that site".

Following are the *seven* different reactions that occur the said process:

$$S + O_2 \longrightarrow SO_2\uparrow; C + O_2 \longrightarrow CO_2\uparrow$$

$$3S + 2Fe_2O_3 \longrightarrow 4Fe + 3SO_2\uparrow$$

$$3Si + 2Fe_2O_3 \longrightarrow 4Fe + 3SiO_2$$

$$3Mn + Fe_2O_3 \longrightarrow 2Fe + 3MnO$$

$$3C + 2Fe_2O_3 \longrightarrow 2Fe + 3CO\uparrow$$

$$4P + SO_2 \longrightarrow P_2O_5; Fe_2O_3 + P_2O_5 \longrightarrow 2FePO_4 \text{ [Slag]}$$

* **Puddling Process:** It involves the vigorous stirring of **Molten Iron** present in the *Reverberatory Furnace* by means of '*Rods*'– that are eventually being consumed in the said process.

Q.1. Discuss the following:

    (a)   **Characteristics of Ellingham Diagrams**

    (b)   **Applications of Ellingham Diagrams**

    (c)   **Limitations of Ellingham Diagrams**

Solution: **(a) Characteristics of Ellingham's Diagrams**–Following are the *five* most vital and important points that may be observed from the **Ellingham Diagrams** provided in Fig: 5.13.

1. Each and every 'plot' shows a perfect straight line.

2. Based on the fact that there is a perceptive decrease in the entropy (**ΔS is –ve**) critically during the **formation of oxide**, the ensuing value of $\Delta_f G$ either gets **enhanced** or becomes **less –ve** with the increase in temperature. Alternatively, the **curves** do have positive slopes. Nevertheless, whenever certain definite **changes in phase occurs** [*viz.*, **Solid → Liquid or Liquid → Gas**]– there is an sudden change in the so-called **positive slope of the curve**.

    Examples: Following are *two* **classical examples**, namely:

    • In the curve related to: **Zn → ZnO**, at the BP of Zn (~ 908°C)–there is an abrupt increase in the *+ve slope* of the observed (resultant) curve; and

    • One may also critically notice an abrupt change in the *+ ve slope* taking place in the **curve** related to: **Mg → MgO** at the *BP of Mg* (~ **1100°C**).

3. In the typical instance of the **oxidation reaction, such as: C(s) + O$_2$(g)→ CO$_2$(g)** for which **DS = 0** the slope of the **C → CO$_2$** curve is found to be 'zero'; and, therefore, this **curve is almost parallel to the temperature axis** perceptively.

4. Practically all the curves **slope upwards** because **D$_f$G^0 values** turns out to be **less negative** *vis-a-vis* **increase in temperature**.

5. In each curve there is a point below which the **Metal oxide remains fairly stable**; and above which the **oxide** is **unstable**.

    (b) Applications of Ellingham Diagrams-A critical examination of the **Ellingham Diagram** (*Fig: 5.13*) related to the ensuing conversion of the **metals** into their respective **Oxides** one may simply predict: which metal may be able to *reduce the oxide of the other metal(s)*.

    Thus, at a given temperature, a **metal (A)** may virtually reduce the oxide of **another metal (B)**, only when (**A → Oxide of A**) curve actually *lies below* (**B → Oxide of B**) curve in the **Ellingham Diagram**.

    (c) Limitations of Ellingham Diagram–There are *two* **distinct limitations of Ellingham Diagram**, such as:

    ❑ The **Ellingham Diagrams** simply suggest whether the reduction of the **metal oxide** by a given **RA** (*viz.*, other metal, **CO or CO**) is possible or not. However, these diagrams fail to explain about the rate at which the reduction reaction occurs.

    ❑ These **diagrams** are solely based upon the assumption that the **reactants** and **products of reaction** are both in an **equilibrium state**. However, it never always true since the **reactants and products of reaction** could be in **solid form**.

Q.2. By making use of the Ellingham Diagram how would you predict whether:

    (a)   **Fe can reduce Al$_2$O$_3$ to Al, and ZnO to Zn.**

    (b)   **CO and reduce ZnO to Zn.**

Solution: (a)

$$3Fe + Al_2O_3 \longrightarrow 3FeO + 2Al$$
$$\underset{RA}{}$$

$$Fe + ZnO \longrightarrow FeO + Zn$$
$$\underset{RA}{}$$

Amazingly, the **Ellingham Diagram (Fig: 5.13)** vividly depicts that since **Fe → FeO curve** very much lies above the **Al →Al$_2$O$_3$ curve**, Fe **cannot reduce Al$_2$O$_3$ to Al**. Likewise, the **Fe ® FeO curve** lies above the **Zn ® ZnO curve**. Hence, Fe is unable to **reduce ZnO to Zn**. Thus, Fe cannot reduce any of the aforesaid oxides.

(b) We have:

$$CO + ZnO \longrightarrow Zn + CO_2$$
$$\underset{RA}{}$$

since, the **CO → CO$_2$ curve** very much lies above **Zn → ZnO curve**; and hence, CO is unable to reduce **ZnO to Zn**.

Q.3. What would be the temperature at which:

(a) **Mg can reduce SiO$_2$ to Si**

(b) **Si can reduce MgO to Mg**

Solution:

(a) The specific **reduction of SiO$_2$ to Si by Mg** may be depicted clearly as under:

$$2Mg + SiO_2 \longrightarrow 2MgO + Si$$
$$\underset{RA}{}$$

In the above reaction **Mg** duly acts as a **RA** (reducing agent), since it helps to reduce **SiO$_2$ to Si**; and itself gets oxidized to **MgO**. The **Ellingham Diagram** (see *Fig: 5.13*) explicitly shows that at a temperature below **1966K**; because the **Mg → MgO curve** lies well below the corresponding **Si → SiO$_2$ curve**, and **Mg can reduce SiO$_2$ to Si** below **1966 K**.

Hence, we may have:

$$2Mg + SiO_2 \xrightarrow{<1966k} 2MgO + Si$$

(b) The reduction of **MgO to Mg by Si** may be depicted as under:

$$Si + 2MgO \longrightarrow SiO_2 + 2Mg$$

Thus, in the above reaction **Si** acts as a **RA**, because it reduces **MgO to Mg**, and gets itself oxidized to **SiO$_2$**. However, in Fig: 5.13 one may observe that a **temperature > 1966 K**, since **Si → SiO$_2$ curve** lies below the respective **Mg → MgO curve**, Si can reduce **MgO to Mg > 1966K**.

Hence, we may have the expression:

$$MgO \xrightarrow{>1966K} SiO_2 + Mg$$

Q.4. Write a brief note on the–'Thermodynamics of the metallurgy of Iron',

Solution: **Iron (Fe)** is obtained by the reduction of oxides of Fe (*viz.*, FeO, Fe$_2$O$_3$, Fe$_3$O$_4$). The reducing agent employed may be either **carbon or CO**.

(a) **Reduction of FeO to Fe by Carbon (C)**–In fact, the reduction of FeO by Fe by **carbon** may be depicted by the following reaction, wherein **C gets oxidised to CO**.

$$FeO + C \longrightarrow FeCO$$
$$\underset{RA}{}$$

From Fig: 5.13 it may be observed that the conversions: $C \rightarrow CO$ and $Fe ® FeO$ curves do cross each other at a temperature of **1073K**. Hence, above this very temperature, the ensuing $C \rightarrow CO$ **curve** lies below $Fe \rightarrow FeO$ **curve**. Besides, C is capable of reducing **FeO to Fe** at a **temperature > 1073K**.

Thus, we may curve the expression:

$$FeO + C \rightarrow Fe + CO$$

(b) Reduction of Fe_2O_3 (Haematite-ore of Iron) and Fe_3O_4 to Fe by CO

In this case, the reduction of Fe_2O_3 to Fe_3O_4 by CO may be shown by the following reactions clearly. In such reactions, CO is duly oxidised to CO_2; and hence, both **Iron oxides** are duly converted into FeO.

Thus, we may have the following expressions:

$$Fe_2O_3 + CO \longrightarrow 2FeO + CO_2$$

$$Fe_3O_4 + CO \longrightarrow 3FeO + CO_2$$

The **FeO** as obtained above gets duly reduced to Fe by CO.

$$FeO + CO \longrightarrow Fe + CO_2$$

It may be noticed in Fig: 5.13 that at a **temperature < 1073K**, since **CO ® CO_2** curve lies below **Fe→ FeO curve**, the respective oxides of Fe (FeO, Fe_2O_3, Fe_3O_4) may be reduced to **Fe** by CO.

Thus, we may have the following expression:

$$FeO + CO \xrightarrow{<1073K} Fe + CO_2$$

Q.5. The value of $\Delta_f G^{10}$ for the formation of Cr_2O_3 is found to be –540KJ.mole^{-1} and that of Al_2O_3 is –827KJ.mole^{-1}. Is the reduction of Cr_2O_3 is possible at all with Aluminium (Al)?

Solution: The reduction of **Cr_2O_3 (Chromium Oxide) with Al** may be vividly shown by the equation stated below:

$$Cr_2O_3(s) + 2Al(s) \longrightarrow 2Cr(s) + Al_2O_3(s)$$

The above reaction is quite feasible, if the values of **$\Delta_r G^o$** for the aforesaid reaction is negative.

Hence, let us calculate the value of **$\Delta_r G^o$** for the above reaction.

$$\Delta_r G^0 = \sum \Delta_f G^o \text{ (Products)} - \sum \Delta_f G^o \text{ (Reactants)}$$

$$= [\Delta_f G^0 (Al_2O_3) + 2 \times \Delta_f G^o(Cr)] - [\Delta_f G^0 (Cr_2O_2) + 2 \times \Delta_f G^o(Al)]$$

$$= [-827 + 2 \times 0] - [-540 + 2 \times 0]$$

$$= -827 + 540$$

$$= -287 \text{ KJ.mole}^{-1}$$

Since the value of $\Delta_r G^o$ is negative $(=-287$ KJ. mole$^{-1})$, the above reaction is absolutely feasible *i.e.,* the **reduction of Cr_2O_3 by Al is quite possible** perceptively.

PROBABLE QUESTIONS

1. Describe the following aspects in **Metallurgy:**
   (i) **Grinding of Ore**
   (ii) **Pulverizing of Ore**
   (iii) **Concentration of Ore**
2. Explain the following:
   (a) **Froth Floatation**
   (b) **Chemical Method of Separation**
   (c) **Various Practiced Reduction Processes**
3. Discuss the following:
   (i) **Magnesium Reduction Method**
   (ii) **Aluminium Reduction Method**
   (iii) **Auto-Reduction Method**
   (iv) **Electrolytic Reduction Method**
4. Write short notes on the following topics related to **Metallurgy**
   (a) **Amalgamation Phenomenon**
   (b) **Hydro-Metallurgy Process**
   (c) **Refining of Metal**
5. How would you explain the following:
   (i) **van Arkel De Boer Process**
   (ii) **Electrorefining Process**
6. Describe the **Thermodynamics of the Reduction Process**. Explain
7. Elaborate certain vital and important **characteristics Features of the Ellingham Diagram for Oxides** with the help of a neat and labelled diagram.
8. Explain the following in an explicit manner:
   (a) **Preparation of Alloys**
   (b) **Various Furnaces used in Metallurgy**
9. Attempt any *two* of the following:
   (i) **Extraction of Silver**
   (ii) **Extraction of Gold by Cyanide Method**
   (iii) **Extraction of Magnesium**
   (iv) **Extraction of Zinc**

UNIVERSITY QUESTIONS

1. (a) Explain briefly the principle involved in the following: (i) Electrolytic refining of metals (ii) van Arket Process (iii) Concentration of sulphide ores.

   (b) How does calcination differ from roasting? Illustrate using suitable examples.?   (Delhi 99)

2. Describe in short the methods for the extraction and isolation of metals from sulphide and oxide ores.
   (Kanpur 2000)

3. Write notes on: Roasting, Calcination, Smelting and Refining of metals.
   (Kumaon 2000, Delhi 2002, Lucknow 2002)

4. Discuss briefly cyanide process used for the extraction of metals from their ores.
   [Gauwhati (General) 2000]

5. The extraction of a metal from its ore is essentially a reduction process represented by:

$$M^n + ne^- \longrightarrow M$$

6. Comment on the above statement and discuss various types of reducing agents employed for the purpose.
   (Lucknow 2001)

7. (i) Mention the advantages of powder metallurgy.

    (ii) Briefly discuss any two methods used in the refining of metals.

    (iii) What is the role of flux in metallurgy?

    (iv) What is the reducing agent employed in the production of tungsten? Give reason.?
    (Bangalore 2001)

8. What different techniques are employed for the purification or refining of a metal?   (Delhi 2003)

9. Describe van Arket Process (Vapour phase process).   (Bangalore 2004)

10. Write in brief the general principles involved in the extraction of metals from their ores.
    (Purvanchal 2007)

11. Explain "Roasting" with examples.   (Meerut 2009)

SOLVOLYTIC REACTIONS
[HYDROLYSIS]

1. INTRODUCTION

Hydrolysis *i.e.*, the **Solvolytic Reactions** do refer to:

"**a chemical reaction that essentially involves the reaction of a molecule with water (H_2O); and hence, the process whereby the molecules are *cleared* into their constituents respectively by the addition of water.**"

Another school of thought advocates that a reaction wherein water (H_2O) serves as a *reactant* and brings about a definite change in the **pH**. In this conceptualized concept and analogy we have:

"**pH equals to-$\log_{\alpha H}$, where αH designates the activity profile of the H^+ ions by $C_H . \gamma \neq . \alpha H$ is the H^+ ion concentration and $\gamma \neq$ represents the mean activity coefficient of the H^+ ions; however, in dilute solutions $\neq = 1$**".

Alternative, the **hydrolysis phenomenon** may be regarded as:

"**the certain of an *hydrated ion* followed by the dissociations of the so-called aquoacid so formed**".

Hence, we may have the following expression:

$$Al^{3+} + 6H_2O \rightleftharpoons Al(H_2O)_6^{3+} \rightleftharpoons Al(H_2O)_5(OH)^{2+} + H^+$$

The Hydrolysis Reactions: In a broader perspective, it may be observed that the *more complicated nature* in this case is perhaps on account of the so-called '*polymerization*' of the '**aquated species**' in solutions. Besides almost all the '**metal alkyls**' as well as the '**metal hydrides**' do virtually undergo a **fast and rapid hydrolysis phenomenon**; whereas, by contrast it may be observed that:

"**both *alkyl* and *hydrogen derivatives* of the non-metals are fairly stable**".

Interestingly, the *halides* of most of the **non-metals** do have a tendency to undergo *hydrolysis phenomenon*; whereas, the ensuing **metal-halides** are indeed stable. In addition, the critical hydrolysis of the '**metal akyl**' is facilitated perceptively by:

"**the lowering of pH of the respective solutions; whereas, that of the so-called non-metal halides does proceed at a higher pH values**".

NOTE	Obviously, in a *hydrolysis reaction*, the H^+ ions preferentially go to the more electronegative atom.

Salient Features of Hydrolysis Reactions: These essentially include:

1. Hydrolysis reaction **may be expressed as the** Acid-Base Equilibrium reaction– **as given under:**

$$MY_n + {}_nHOH \rightleftharpoons M(OH)_n + nHY$$

2. Generally, the **hydrolysis phenomenon** is grossly favoured only when 'Y' is the **anion** of a relatively **weaker acid (HY)** and/or when **M** is a *cation* of a **weaker base, M(OH)**. In this way, the overall effect of the *anion* may be visualized by the so-called:

'hydrolytic sequential reactions of the Lithium Salts',

as expressed under explicitly:

$$Li_2C_2 + 2HOH \rightleftharpoons 2LiOH + C_2H_2$$

$$Li_2M_2 + 3HOH \rightleftharpoons 3LiOH + 2MR_2 [M = N, P, As, Sb$$

$$Li_2M + 2HOH \rightleftharpoons LiOH + LiMH [M = O, S, Se, Te]$$

$$LiX + HOH \rightleftharpoons Li(aq)^+ + X(aq)^- [X = F, Cl, Br, I]$$

3. In a particular instance, when the **cation actually combines** with the **hydroxyl ion [OH]** to result into the formation of an '**Insoluble Product**', hydrolysis gets facilitated easily:

$$Al_2Te_3 + 6HOH \longrightarrow + 2Al(OH)_3 + 3H_2Te$$

Thus, as and when the **pH of the ensuing solution** is lowered appreciately–even the so-called **anions of so called very strong acids** may be forced perceptively to accept a **Proton (H$^+$)** *viz.*, **HSO$_4^-$ formation from SO$_4^-$.**

Points to Ponder–It may, however, be observed and ascertained that as and when a substance specifically undergoes the so-called a **Nucleophilic Substitution Reaction** and also the **nucleophile** happens to be the **solvent itself,**–the overall ensuing reaction is usually termed as **solvolysis** [or **Hydrolysis** or **Solvolytic Reactions**].

Besides, in the *Solvolysis Reaction* whenever the **solvent** being employed is **water (H$_2$O), the** resulting reaction is known as **Hydrolysis**

Fig: 6.1 illustrates the **Schematic Representation of Hydrolysis.**

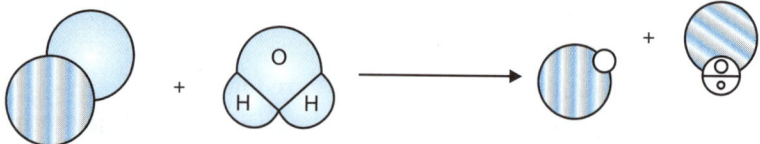

Fig. 6.1: The Schematic Representation of Hydrolysis.

Based upon the scientific revelations it may be observed that the hydrolysis (or **Solvolysis**) may eventually come into play *via* an array of **mechanisms,** namely:

(a) **Unimolecular Nucleophilic Substitution (SN^1),**

(b) **Bimolecular Nucleophilic Substitution (SN^2),**

(c) **Addition-Elimination Mechanism,**

(d) **Addition Mechanism,**

(e) **Redox Reaction**, and

(f) **Push-Pull Mechanism,**

which shall now be treated individually in the sections that follows:

2. HYDROLYSIS *via* SN^1 MECHANISM

The underlying mechanism is expatiated elegantly by means of the following *two* **classical examples,** such as:

(a) Hydrolysis of Nitrogen Fluoride (NF$_3$)–It has be duly proven that when the critical **hydrolysis of NF$_3$** is being performed at an *ordinary parameters*, practically **no reaction is observed by the SN^1– mechanism.**

It may be expressed as under:

$$NF_3 + H_2O \xrightarrow[\text{Parameters}]{\text{Under Normal}} \text{No Reaction}$$
Nitrogen Fluoride

The major reason for the **No Reaction** episode is due to the underlying fact that:

"the Nucleophile (H$_2$O) is not in a position to attack either at the *N-atom* or at the *F-atom* perhaps on account of the free access to the availability of the *energetically low-lying vacant orbitals*."

Nevertheless, **NF$_3$** gets duly hydrolyzed under *very drastic experimental parameters* to yield N$_2$O$_3$ [*i.e., $O \leftarrow N \equiv N = O$*] and **HF** *via* the **SN^1 Mechanism** as depicted in Fig: 6.2–in *three* **sequential steps:**

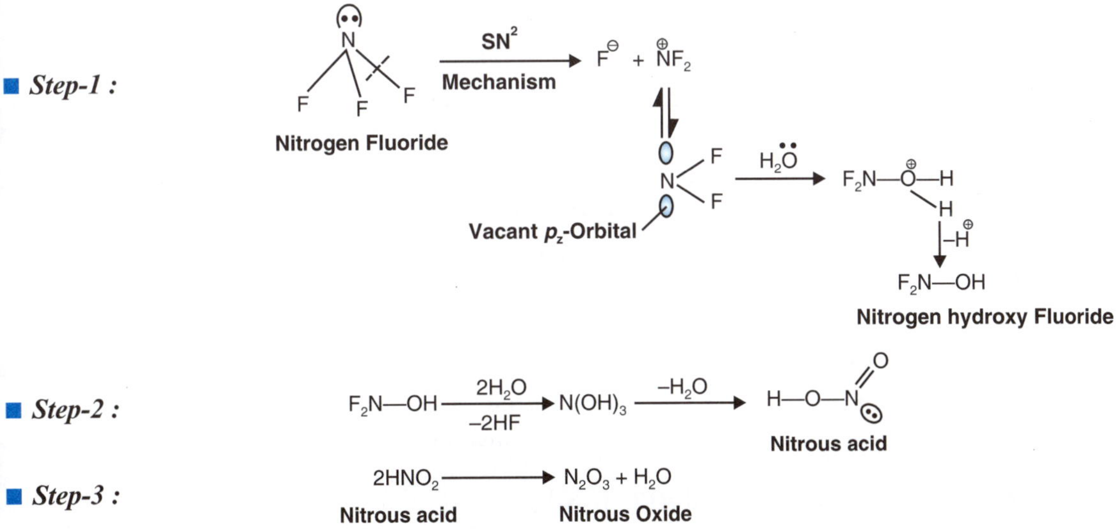

■ *Step-1 :*

■ *Step-2 :*

■ *Step-3 :*

$$2HNO_2 \longrightarrow N_2O_3 + H_2O$$
Nitrons acid   **Nitrogen Oxide**

Fig. 6.2: Hydrolysis of NF$_3$ (Nitrogen Fluoride) by the SN^1–Mechanism.

(b) **Hydrolysis of Carbon Tetrachloride [CCl₄]**–Obviously, **carbon tetrachloride (CCl₄)** is observed to be **absolutely inert** towards the **hydrolysis** phenomenon under ordinary parameters *via* the so-called **SN¹-Mechanism** based on the following *three* **specific reasons**, namely:

(i) **Non-availability** of **Vacant low-lying d-orbitals**.

(ii) **Steric Crowding phenomenon** caused by virtue of the critical presence of *form* **Cl-atoms** located strategically around the **C-atom**–that do not permit the *water (H₂O) molecule* to approach the so-called *antibonding orbitals* in the **C-Cl bonds** (*four* of them).

(iii) Amazingly, the **nucleophile** is not in a position to attack the ensuing **vacant 3d-orbital** that eventually has a tendency to repel the **incoming nucleophile** perceptively.

Nevertheless, **carbon tetrachloride (CCl₄)** definitely indulges in the critical **partial hydrolysis** particularly under the influence of drastic experimental parameters by making use of the **superheated steam** *via* **the SN¹ mechanism**, as shown below:

$$CCl_4 + H_2O \longrightarrow COCl_2 + 2HCl$$

| Carbon tetrachloride | Superheated steam | Carbonyl Chloride | |

Remarks: It may be observed that out of the **4-leaving moieties (Cl)**, only **2-moieties** do leave actually, whereas, the remaining **2-moieties** are indeed left upon the parent molecule.

Hence, the above expression of thoughts may stated as under:

Carbon tetrachloride   Carbonyl Chloride

3. HYDROLYSIS *via* SN² MECHANISM

Based on the scientific evidences and logical explanations we may broadly expatiate the mechanism of **hydrolysis** *via* **SN² mechanism** by the help of the following *six* **classical examples**, namely:

(i) **Hydrolysis of Silicon Tetrachloride [SiCl₄]**,

(ii) **Hydrolysis of Silicon Tetrafluoride [SiF₄] and Boron Trifluoride [BF₃]**,

(iii) **Hydrolysis of Boron Trichloride (BCl₃) and Berylium Chloride [BeCl₂]**,

(iv) **Hydrolysis of Sulphur Tetrafluoride [SF₄] and Sulphur Hexafluoride [SF₆]**,

(v) **Hydrolysis of Nitrogen Trichloride [NCl₃], Phosphorus Trichloride [PCl₃], Arsenic Trichloride [AsCl₃], Antimony Trichloride [SbCl₃], and Bismuth Trichloride [BiCl₃]**, and

(vi) **Hydrolysis of Interhalogen Chemical Entities (compounds)**,

which shall now be discussed separately in the sections that follows:

3.1. Hydrolysis of Silicon Tetrachloride [SiCl₄]

It has been proven and established beyond any reasonable doubt that:

"very much contrary to Carbon Tetrachloride (CCl₄), being actually insert to hydrolysis, SiCl₄ (Silicon Tetrachloride) promptly undergoes hydrolysis via the SN² mechanism."

Explanation: In the typical instance of **Silicon Tetrachloride [SiCl₄]**, the **Si atom** possesses a definite *vacant d orbital* so as to duly accommodate the so-called *incoming* Nucleophile, there by giving rise to a critical *transition state* on account of:

- *sp³d*-**hybridization**; and
- **trigonal bipyramidal geometry.**

The probable mechanism may be shown as under:

Silicon tetrachloride

Important Observations: Following are the *three* **vital and important observations** pertaining to the **hydrolysis of SiCl₄** by means of the **SN² mechanism**, as elaborated under:

1. In a situation when **R₁ R₂ R₃ Si Cl** undergoes hydrolysis, it is, in fact, intimately associated with an **inversion of configuration.**

2. However, when one compares the **critical rate of hydrolysis** for:
SiX₄ (where: **X = F, Cl, Br, I**), it almost maintains the some order comparable fairly to the corresponding **Lewis order Acid Strength Oder**–as given under:

$$\boxed{SiF_4 \rangle SiCl_4 \rangle SiBr_4 \rangle SiI_4}$$

NOTE	**Perhaps it could be by virtue of the effect of the so-called π-Back Bonding phenomenon**–that is found to be less dominant over the ensuing **Negative Inductive Effect** perceptively.

3. In the specific case of **Stannic Chloride [SnCl₄]** and **Stannous Chloride [SnCl₂]**–the observed *rate of hydrolysis* is as under:

$$\boxed{SnCl_4 \rangle SnCl_2}$$

which perhaps could be due to the higher level of inherent charges being possessed strategically upon the **former** (*SnCl₄*).

Nevertheless, for **SnCl₄ and Sn(CH₃)₄**–the observed **rate of hydrolysis** is as under:

$$\boxed{SnCl_4 \rangle Sn(CH_3)_4}$$

which is definitely due to the so-called **inherent positive inductive** effect of **methyl (CH₃) moiety** that critically reduces the **Lewis acidity** of the **stannic [Sn⁴⁺] ion**–that eventually leads to the ultimate reduction in:

"**observed rate of *Nucleophilic Attack* predominantly**".

3.2. Hydrolysis of Silicon Tetrafluoride [SiF₄] and Boron Trifluoride [BF₃]

Importantly, both the aforesaid molecules do exhibit explicitly *Partial Hydrolysis* perceptively, and the reactions involved may be shown as under:

$$SiF_4 + 4H_2O \longrightarrow Si(OH)_4 + 4HF$$
$$2SiF_4 + 4HF \longrightarrow 2H_2[SiF_4]$$

Summing up: $3SiF_4 + 4H_2O \longrightarrow Si(OH)_4 + 2H_2[SiF_6]$

Remarks: The above **complexation reaction** is observed to be so rapid and fast that unless and until all the **HF molecules** are duly consumed by **SiF$_4$**, the water (H_2O) molecule is unable to attack the immediate next molecule of SiF$_4$. Therefore, out of the *three* moles of SiF$_4$,–only *one mole of SiF$_4$* eventually gets hydrolyzed to yield the desired product, and, therefore, it is usually termed as the **Partial Hydrolysis**.

Likewise, we may also elaborate on the **hydrolysis of BF$_3$**, as given under:

$$BF_3 + 3H_2O \longrightarrow B(OH)_4 + 3HF$$
$$3BF_3 + 3HF \longrightarrow 3H^+[BF_4]$$

Summing up: $4BF_3 + 3H_2O \longrightarrow B(OH)_3 + 3H[BF_4]$

Thus, alternatively we may have:

$$4BF_3 + 12H_2O \longrightarrow 4B(OH)_3 + 12HF$$
$$12HF_3 + 3B(OH)_3 \longrightarrow 3H^+ 3BF_4 + 9H_2O$$

Summing up: $4BF_3 + 3H_2O \longrightarrow B(OH)_3 + 3[H + BF_4]$

Comment: Hence, once again we may take cognizance **Complexation Reaction** fact that *four* moles of BF$_3$ results the so-called **hydrolyzed product**; and hence, it may be vehemently called the **Partial Hydrolysis**.

3.3. Hydrolysis of Boron Trichloride [BCl$_3$] and Berylium Chloride [BeCl$_2$]

It is, however, important to state here that the *two* above mentioned chemical entities *viz.*, **BCl$_3$** and **BeCl$_2$** do virtually undergo complete hydrolysis–as depicted under:

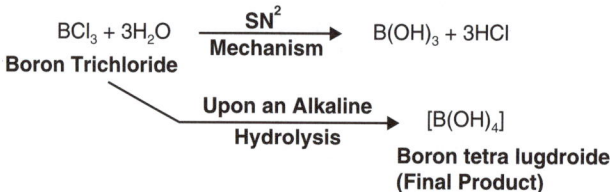

Remarks: The **rate of hydrolysis** for **BX$_3$ [X = F, Cl, Br, I]** is found to be in the following decreasing order:

$$\boxed{BF_3 < BCl_3 < BBr_3 < BI_3}$$

Likewise, for **Berylium Chloride [BeCl$_2$]** we may have the following sets of reactions:

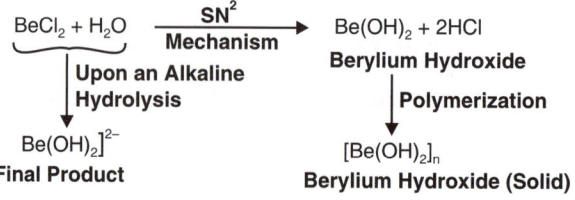

Comment: Hence, we may express the so-called **Final Product** in *two* **different configurations** as:

$$[Be(OH)_4]^{2-} \text{ and } [Be(OH)_2]_n$$

3.4. Hydrolysis of Sulphur Tetrafluoride [SF$_4$] and Sulphur Hexa Fluoride [SF$_6$]

It has been duly proven and ascertained that **Sulphur Tetrafluoride [SF$_4$]** critically undergoes hydrolysis rather rapidly so as to yield the products of reaction as: H_2SO_3 (**sulphurus acid**) and **HF** (**hydrofluoric acid**) *via.* the **SN^2 Mechanism** as shown in Fig: 6.3.

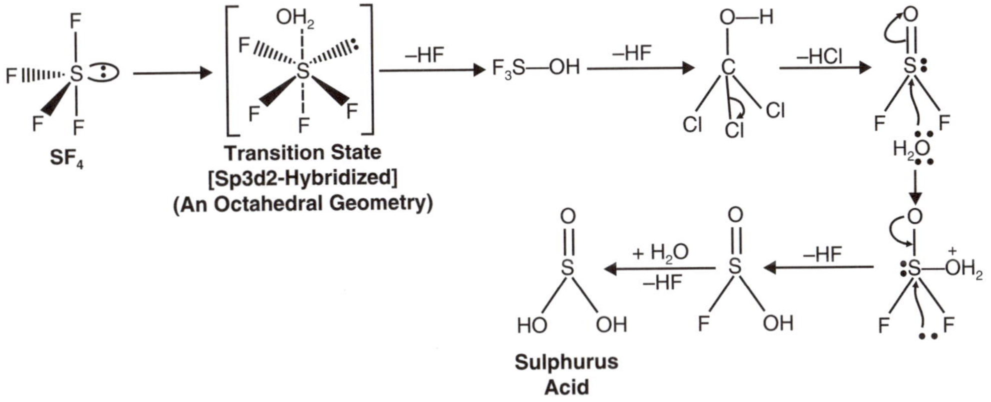

Fig. 6.3: The Hydrolysis of SF$_4$ by The SN^2-Mechanism.

Nevertheless, **SF$_6$** is found to be inert towards the **hydrolysis process** (*i.e.*, **unlike SF$_4$**), even though it certainly possesses the *altogether* **vacant *d* orbitals** that are duly available with the **central atom S.**

Explanation: Perhaps it may be duly explained based upon the so-called inherent and critical 'steric crowding' very much around the **S-atom.** Besides, this logical deliberation is strongly supported by the underlying fact that:

"**the prevailing order for the ensuing hydrolysis of *SF$_6$* and *TeF$_6$* is found to be: SF$_6$ < SeF$_6$ < TeF$_6$**".

Furthermore, we may also notice that as the prevailing size of the central atom enhances, the aforesaid *steric crowding* gets duly decreased; and hence, the surface of the so-called '**Central Atom**' is eventually rendered more exposed for the desired **Nucleophilic Attack** perceptively.

Points to Ponder: These essentially include:

1. Even though **SeF$_6$** is found to be absolutely inert very much akin to **SF$_6$**–it certainly undergoes **hydrolysis** to an extremely little extent; whereas, **TeF$_6$** gets hydrolyzed both promptly and completely. Thus, we may have the following expressions:

$$SeF6 + 4H_2O \longrightarrow H_2SeO_4 + 6HF$$

$$TeF6 + 6H_2O \longrightarrow Te(OH)6 + 6HF$$

2. Obviously, the actual characteristic of the products formed is found to be **different** perhaps due to the **steric crowding.** However, the so-called *intended product* obtained duly from SeF$_6$ **hydrolysis** is indeed **Se(OH)$_6$**; but in order to avoid the phenomenon of *steric crowding*–it ultimately gets condensed into **H$_2$SeO$_4$.**

3.5. Hydrolysis of Nitrogen Trichloride [NCl₃], Phosphorus Trichloride [PCl₃]₂ Arsenic Trichloride [AsCl₃], Antimony Trichloride [SbCl₃] and Bismuth Trichloride [BiCl₃]

The **hydrolysis** of these compounds *viz.*, NCl_3, PCl_3, $AsCl_3$, $SbCl_3$, and $BiCl_3$ have been discussed with logical explanations and reasonings in the sections that follows:

1. **Hydrolysis of Nitrogen Trichloride [NCl₃]**–The **hydrolysis** of NCl_3 gives rise to the production of **ammonia (NH₃)** and **hypochlorus acid (HOCl)**, as shown under in Fig: 6.4.

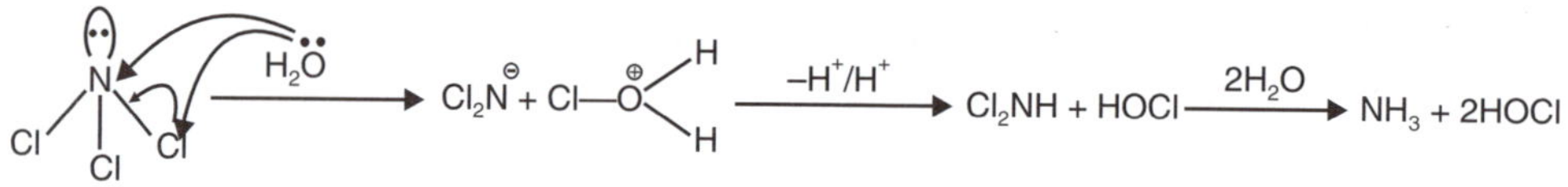

Fig. 6.4: The Hydrolysis of PCl₃ by SN²-Mechanism.

Remarks:

1. In this typical instance the **nucleophilic attack** upon the **N-atom** occur perhaps on account of the so-called:

"**non-availability of the *Vacant Orbital precisely*".**

2. Nevertheless, the crucial attack upon the **Cl-atom** is largely favoured since the prevailing **N-Cl bond** happens to be practically **non-polar in nature**; and thus, the attack occurs *via* the inherent **Electrometric Effect** perceptively.

(2) Hydrolysis of Phosphorus Trichloride (PCl₃)–The **hydrolysis of PCl₃** actually produces a mole each of **Phosphorous Acid (H₃PO₃)** and **Hydrochloric Acid (HCl)**,–as depicted in Fig: 6.5.

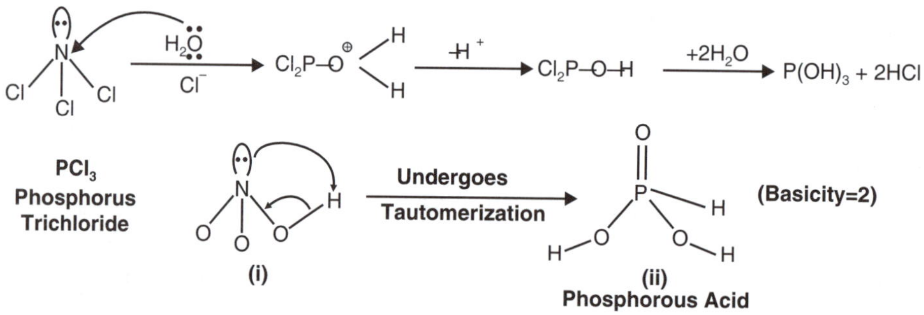

Fig. 6.5: Hydrolysis of PCl₃ by SN² -Mechanism.

Remarks: The most influential **driving force** for the aforesaid *Tautomerization Reaction* is solely by virtue of the appreciably **huge bond energy of II vis-a-vis to that of I.**

(3) Hydrolysis of Arsenic Trichloride [AsCl₃]–The critical **hydrolysis of AsCl₃** yields *two* **distinct products**, namely:

• **orthoarsenious acid [H₃AsO₃]** and
• **hydrochloric acid (HCl)**,

as illustrated in Fig: 6.6.

Remarks: In this particular case the **tautomerization** phenomenon fails to occur since the ($4dp - 2pp$) bond duly present in **As=O** is found to be not so effective and feasible *vis-a-vis* to the corresponding ($3dp–2pp$) bond duly present in the **P=O** of H_3PO_3 *i.e.*, **phosphorus acid.**

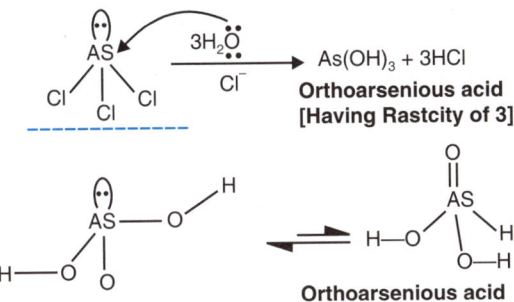

Fig. 6.6: The Hydrolysis of AsCl$_3$ by SN^2–Mechanism.

(4) **Hydrolysis of Antimony Trichloride [SbCl$_3$] and Bismuth Trichloride [BiCl$_3$]**–In general, both SbCl$_3$ and BiCl$_3$ upon hydrolysis yields a mole each of **HCl** and **MO$^+$Cl$^-$** (where: **M = Sb or Bi**) type of product; whereas, the so-called *anticipated product(s)* according to the SN^2-mechanism is:

$$M(OH)_3 + HCl$$

It is solely based on the fact that:

"**as one moves down the *Group* (in the Periodic Table) the observed electronegative character gets increased perceptively**",

and hence, there prevails an inherent tendency to form **Ionic Compound** also gets increased. Therefore, the resulting **MO$^+$Cl$^-$** so obtained predominatnly comprises **[M^{3+}O^{2-}]$^+$Cl$^-$** entity; and thus, ultimately we may have the following reaction:

$$MCl_3 + H_2O \rightleftharpoons \underset{\text{A White ppt}}{MO^+Cl\downarrow} + 2HCl \,[\text{M = Sb or Bi}]$$

3.6. Hydrolysis of Interhalogen Chemical Entities (Compounds)

As we know that the 'Interhalogens' do have the generalized formula **AX$_n$**; where, '**n**' is always on *odd number viz.*, 1, 3, 5, 7, and 9; and '**X**' designates a certainly **more electronegative halogen atom** (*viz.*, **F, Cl, I, Br**). Consequently, the entity '**A**' needs perceptively the **[d$^+$] charge**; and, therefore, evidently accepts the so-called **Nucleophilic Attack upon it,**–as shown under:

$$H_2\ddot{O} \longrightarrow A^{\delta+}\!\!-\!\!A^{\delta-} \longrightarrow \underset{\text{Transition State}}{[H_2O\text{-------}A\text{-------}X]} \xrightarrow{-HX} A\text{—O—H}$$

Likewise, we may also have the following expressions:

$$AX_3 + 2H_2O \longrightarrow HAO_2 + 3HX$$

$$AX_5 + 3H_2O \longrightarrow HAO_3 + 5HX$$

$$AX_7 + 4H_2O \longrightarrow HAO_4 + 7HX$$

Comment: In reality, these reaction products are produced invariably; and hence, occasionally certain *other* **side-products** are also obtained (but these are observed to be highly specific to the instance.)

4. HYDROLYSIS VIA ADDITION-ELIMINATION MECHANISM

The **hydrolysis *via* addition elimination mechanism** may be truly expatiated by considering the following *two* **classical examples**, namely:

- **Hydrolysis of Caro's Acid and Marshall's Acid**; and
- **Hydrolysis of Thionyl dichloride [SO_2Cl_2]**,

which shall now be treated separately as under:

4.1. Hydrolysis of Caro's Acid and Marshall's Acid

(a) The **addition-elimination mechanism for the Caro's Acid** is shown as under:

Therefore, we may have:

$$H_2SO_4 + H_2O \longrightarrow H_2SO_4 + H_2O_2$$

Likewise, for the Marshall;s Acid we may have the following reactions

Hence, we may have:

$$H_2SO_4 + 2H_2O \longrightarrow 2H_2SO_4 + H_2O_2$$

4.2. Hydrolysis of Thionyl Dichloride [SO_2Cl_2]

The exact and precise mechanism is shown as under in an explicit manner in *three* **sequential steps:**

Thionyl Dichloride

Explanation:

- ❏ *Step-1:* It simply accounts for the addition of the **nucleophile** since the π-**Bond** is found to be definitely far more flexible *vis-a-vis* the respective σ-Bond.
- ❏ *Step-2:* In this case, the so-called **leaving moiety (Cl⁻) gets duly eliminated.**
- ❏ *Step-3:* Here the phenomenon of **Deprotonation** comes into play *i.e.*, the removal of H^+.

| NOTE | Hence, the entire process is usually termed as **Addition-Elimination Mechanism.** |

Likewise, for **SOCl$_2$ and POCl$_3$** (*i.e.*, **Thionyl Chloride and Phosphorus Trichloride**);–as expressed under:

$$SOCl_2 + 2H_2O \longrightarrow H_2SO_4 + 2HCl$$

$$POCl_3 + 3H_2O \longrightarrow H_3PO_4 + 3HCl$$

5. HYDROLYSIS *VIA* ADDITION MECHANISM

The **hydrolysis** *via* the addition mechanism may be explained by considering the example of **sulphur trioxide (SO$_3$)**,–as given below:

It is, however, pertinent to state here that the so-called ensuing **hydrolysis of SO$_3$**,–only the **nucleophile** (*viz.*, **water**) duly incorporated to the respective **molecule (SO$_3$)** without any sort of *elimination* whatsoever of the **Leaving Moiety**. As a result, the entire phenomenon is known as–**Addition Mechanism** perceptively.

Likewise, for both **Sulphur Dioxide (SO$_2$)** and **Carbon Dioxide (CO$_2$)** we may have the following expressions:

$$SO_2 + H_2O \longrightarrow H_2SO_3 \text{ [Sulphurous Acid]}$$

$$CO_2 + H_2O \longrightarrow H_2CO_3 \text{ [Carbonic Acid] (\textit{Highly unstable})}$$

6. HYDROLYSIS *VIA* REDOX REACTION

The ensuing **hydrolysis** *via* **redox reaction** may be vividly explained by taking the classical examples related to the **hydrolysis of XeF$_3$ and XeF$_4$ [Xenon Trifluoride and Xenon Tetrafluoride]**, as shown under:

$$XeF_3 + H_2O \longrightarrow Xe + 2HF + \frac{1}{2}O_2$$

$$3XeF_4 + 6H_2O \longrightarrow 6H_2O + 2Xe + XeO_3 + 12HF + \frac{1}{2}O_2$$

Interestingly, **Xenon Hexafluoride (XeF$_6$)** undergoes the said hydrolysis process *via* the SN^2-mechanism (already explained earlier); and certainly *not via* the **Redox Mechanism**.

Thus, we may have the following reaction:

$$\underset{\substack{\textbf{Xenon} \\ \textbf{Hexafluoride}}}{XeF_6} + 3H_2O \longrightarrow \underset{\substack{\textbf{Xenon} \\ \textbf{Trizoxide}}}{XeO_3} + 6HF$$

7. HYDROLYSIS *VIA* THE PUSH-PULL MECHANISM

In a broader perspective, the aforesaid mechanism pertaining to the **hydrolysis** *via* **the Push-Pull Mechanism** may be expatiated by the following *three* typical examples, namely:

- **Hydrolysis of Silico Tetrahydride [SiH₄];**
- **Hydrolysis of Aluminium Hexamethyl [Al(CH₃)₆];** and
- **Hydrolysis of Carbides, Nitrides, and Phosphides,**

which shall now be treated individually in the sections that follows:

(a) Hydrolysis of Silico Tetrafluoride [SiH₄]–The **SiH₄ [Silanes]** are invariably hydrolyzed rather in a fast and rapid mode in the so-called critical presence of the *traces of alkali*. Hence, the precise and exact mechanism of the aforesaid reaction may be explained as under:

Therefore, we may have the following expression:

$$SiH_4 + H_2O \xrightarrow[\text{Traces only}]{(OH^-)} SiO_2 + nH_2O + 4H_2^\uparrow$$

$$Si_2H_6 + H_2O \xrightarrow[\text{Traces only}]{(OH^-)} 2SiO_2 + nH_2O + 6H_2^\uparrow$$

Nevertheless, **methane (CH₄)** fails to undergo hydrolysis:

$$CH_4 + H_2O \longrightarrow \text{No Reaction}$$

Remarks: The most probable and logical reason behind this being the so-called *difference in the polarities* of:

$$C^{\delta^-} - H^{\delta^+} \quad \text{and} \quad Si^{\delta^+} - H^{\delta^-} \text{ bonds;}$$

since the electronegativities of **C, Si,** and **H** are found to be **2.5, 1.8,** and **2.1 respectively.** Based on these facts one may arrive at the point that:

"**the C-atom present in Methane (CH₄) cannot be duly attacked by a Nucleophile**".

(b) Hydrolysis of Aluminium Hexamethyl [Al(CH₃)]–The reaction involved may be shown as under:

$$Al_2(CH_3)_6 + 6H_2O \xrightarrow{\text{Hydrolysis}} 2Al(OH)_3 + 6CH_4^\uparrow$$
$$\underset{\substack{\text{Aluminium}\\\text{Hexamethyl}}}{} \qquad\qquad \underset{\substack{\text{Aluminium}\\\text{Hydroxide}}}{}$$

(c) Hydrolysis of Carbides, Nitrides, and Phosphides–Following are one typical example each of Carbide, Nitride, and Phosphide:

- *Carbide:*

$$CaC_2 + 2H_2O \longrightarrow Ca(OH)_2 + C_2H_2^\uparrow$$
$$\underset{\substack{\text{Calcium}\\\text{Carbide}}}{} \qquad \underset{\substack{\text{Calcium}\\\text{Hydroxide}}}{} \underset{\text{Acetylene}}{}$$

- *Nitride:*

$$Mg_3N_2 + 6H_2O \longrightarrow Mg(OH)_2 + 2NH_3^\uparrow$$
$$\underset{\substack{\text{Magnesium}\\\text{Nitride}}}{} \qquad\qquad \underset{\substack{\text{Magnesium}\\\text{Hydroxide}}}{}$$

$$Ca_3P_2 + 6H_2O \longrightarrow 3Ca(OH)_2 + 2PH_3\uparrow$$

- **Phosphide:**   Calcium   Calcium   Phosphene
                    Carbide   Hydroxide

8. HYDROLYSIS *via* THE MIXED MECHANISM

Evidence from the scientific revelations and logical explanations we have observed critically and importantly that there are a good number of chemical entities (compounds) which essentially undergo the **hydrolysis phenomenon** by making use of:

"more than one Mechanisms or Mixed Mechanisms".

Following are *three* classical examples that amply demonstrate the **hydrolysis *via* the Mixed Mechanism**, such as:

- **Hydrolysis of P_4;**
- **Hydrolysis of P_4H_{10}; and**
- **Hydrolysis of PCl_5.**

(a) **Hydrolysis of P_4**–We may easily expatiate the so-called: "disproportionation of 'P_4' in an alkaline environment *via* the combination of *Addition Mechanism* plus *Addition-Elimination Mechanism* explicitly".
Hence, we may have the following expression:

[H₃PO₂
[Phosphorous Acid]

(b) **Hydrolysis of P_4H_{10}**–The critical hydrolysis of P_4H_{10} also takes place *via* the *two* mechanisms, namely;

- **Addition Mechanism**; and
- **Addition-Elimination Mechanism**,

as detailed under in Fig: 6.7.

Remarks:

1. In the **very first three steps (I, II, and III)** in the critical hydrolysis of P_4H_{10}, almost no elimination comes into play. Therefore, these **3-steps** essentially adopt the *Addition Mechanism* for the **nucleophile**.

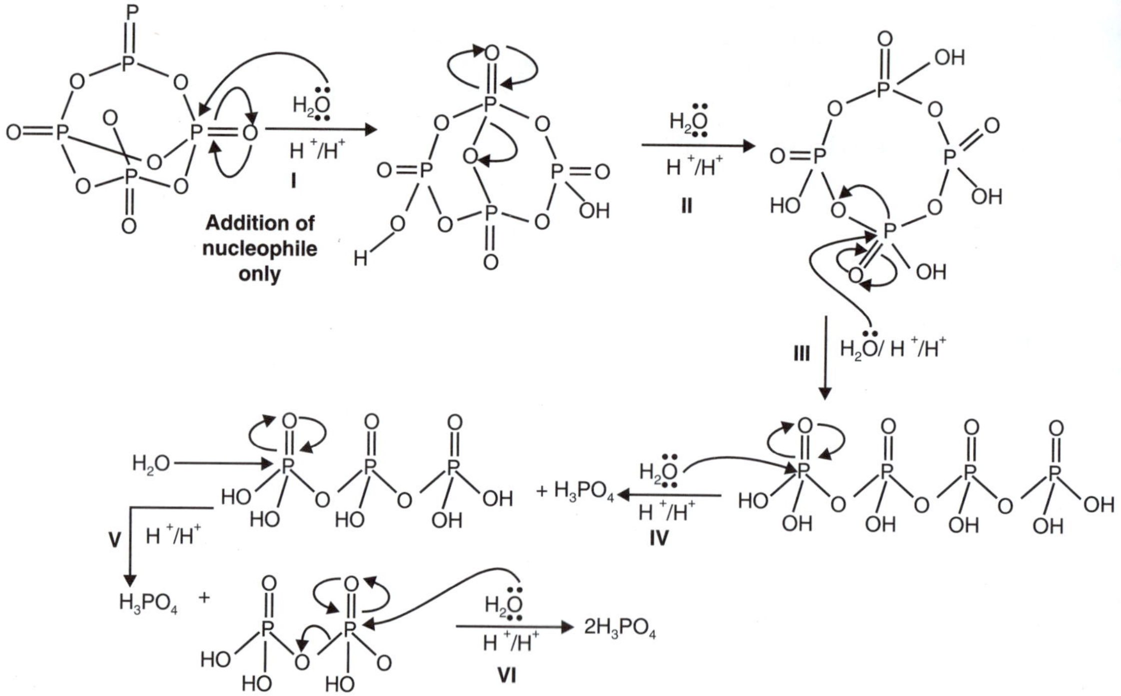

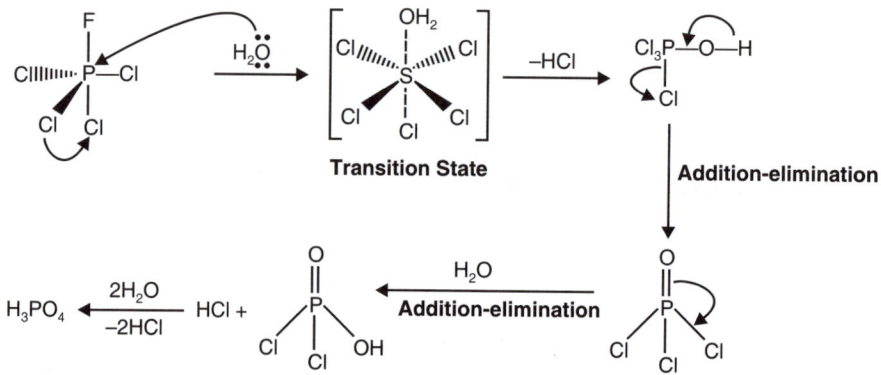

Fig. 6.7: The Hydrolysis of P_4H_{10} via Addition and Addition-Elimination Mechanism.

2. Nevertheless, in the **last *three* steps** (IV, V, and VI) only *'one PO$_4$ unit'* is virtually left behind that predominantly acts as the **leaving moiety**. Hence, the so-called **overall mechanism** is found to both *i.e.,*

- **Addition Mechanism** and
- **Addition-Elimination Mechanism**.

(c) Hydrolysis of Phosphorous Pentachloride (PCl$_5$)—Amazingly, the **hydrolysis of PCl$_5$** invariably proceeds *via* the **SN^2-mechanism** followed by the so-called **addition-elimination mechanism**, as depicted in Fig: 6.8.

Fig. 6.8: The Explicit Hydrolysis of Phosphorous Pentachloride (PCl$_5$) via SN^2–and Addition -Elimination Mechanisms.

SUGGESTED READING REFERENCES

Golub M *et al.*	: **The Chemistry of Pseudohalides**, Elsevier, Amsterdam, 1986.
Zollinger H	: **Diazo Chemistry, VCH**, New York, 1995.
Cotton FA *et al.*	: **Advanced Inorganic Chemistry**, 6th ed., Wiley India, New Delhi, 2004.
Sharpe AG	: **Inorganic Chemistry**, Pearson Edu. Ltd., New Delhi, 1992.

PROBABLE QUESTIONS

1. Give a brief account on the following:
   (a) **Hydrolysis Reactions**
   (b) **Salient Features of Hydrolysis Reactions**
   (c) **Schematic Representation of Hydrolysis**
2. Discuss the **Hydrolysis *via* SN^1-Mechanism**. Explain the hydrolysis of **Nitrogen Fluoride (NF$_3$)**.
3. How would you elaborate on the **Hydrolysis via SN^2-Mechanism**? Explain the **Hydrolysis of BCl$_3$ and BeCl$_2$**.
4. Explain in details the **Hydrolysis of Nitrogen Trichloride (NCl$_3$)**.
5. What is the actual difference between?
   (a) **Hydrolysis *via* Addition Mechanism**; and
   (b) **Hydrolysis *via* Addition-Elimination Mechanism**.
   Explain with suitable examples.
6. Give a comprehensive account on the **Hydrolysis *via* the Push-Pull Mechanism**. Support your answer with a suitable example.
7. How would you duly explain the **Hydrolysis of P$_4$ *via* the Mixed Mechanism**? Explain.
8. Attempt any **one** of the following:
   (i) **Hydrolysis of PCl$_5$**
   (ii) **Hydrolysis of P$_4$H$_{10}$**

INDEX